Microplastics in the Terrestrial Environment

This book aims to address details and research gaps in the impacts of microplastics in terrestrial ecosystems. It addresses the impact of microplastics on the soil environment and highlights and discusses their transport behavior, pollution level, and the combined effects of the microplastics with other pollutants on the soil ecology. Furthermore, it also highlights the effects of UV irradiations and mechanical abrasions from soil fauna and various agricultural practices.

FEATURES:

- Covers advances in plastic/micro-/nano-plastic pollution and possible pathways of pollution.
- Demonstrates the mitigation measures to minimize such pollution loads, with a special focus on the application of nanotechnology.
- Explores recycle and value-added products from waste plastic.
- Promotes development of alternate clean energy sources.
- Introduces appropriate alternatives and/or finding strategies to mitigate the existing microplastic crisis using suitable approaches.

This book is aimed at researchers and graduate students in environmental and chemical engineering, as well as remediation.

Microplastics in the Terrestrial Environment

Pathways and Remediation Strategies

Edited by Surajit Mondal, Papita Das, Arnab Mondal and Poushali Chakraborty

CRC Press is an imprint of the Taylor & Francis Group, an informa business

Designed cover image: www.shutterstock.com

First edition published 2025
by CRC Press
2385 NW Executive Center Drive, Suite 320, Boca Raton FL 33431

and by CRC Press
4 Park Square, Milton Park, Abingdon, Oxon, OX14 4RN

CRC Press is an imprint of Taylor & Francis Group, LLC

ISBN: 978-1-032-63410-4 (hbk)
ISBN: 978-1-032-68456-7 (pbk)
ISBN: 978-1-032-68457-4 (ebk)

DOI: 10.1201/9781032684574

Typeset in Times
by Apex CoVantage, LLC

Contents

Preface

Microplastics are tiny pieces of plastic that measure less than 5 mm and can be found in various environmental media, such as soil, water, air, and living organisms. They originate from the breakdown of larger plastic objects, the abrasion of synthetic fabrics and tires, the release of microbeads from cosmetic products, and the direct use of plastic granules and powders. Microplastics have raised global environmental concerns due to their widespread presence, persistence, and potential negative impacts on ecosystems and human health.

This book offers a comprehensive overview of the current state of knowledge on microplastics as emerging environmental pollutants, with a focus on their sources, distribution, fate, behavior, toxicity, remediation, and reuse strategies. It covers the latest advances in analytical methods, sampling techniques, and quality assurance for microplastic research. It also discusses the occurrence and transport of microplastics in different environmental compartments, such as freshwater, marine, terrestrial, and atmospheric systems. It assesses the ecological and health risks of microplastics, considering their interactions with other pollutants, such as organic compounds and heavy metals, and their effects on various biological levels, from microorganisms to humans. It reviews the existing and emerging technologies for the removal and degradation of microplastics from the environment, as well as the potential applications of microplastics as resources for energy, materials, and products. It also explores the social and economic aspects of microplastic pollution, such as public awareness, policy, regulation, and management.

This book is intended for researchers, students, professionals, and policymakers who are interested in the topic of microplastics and their impacts on the environment and human health. It provides a comprehensive and up-to-date reference for the current state of the art and future directions of microplastic research and management. It also highlights the challenges and opportunities for the development of sustainable and circular solutions for the plastic problem.

Editor Biographies

Dr. Surajit Mondal completed his masters in Energy Systems and PhD on the Renewable Energy domain in 2020. Currently he is teaching in University of Petroleum and Energy Studies to UG and PG students. He has published more than 32 international research and review articles as an author/co-author. He has published 42 patents and granted eight patents against his name. He has completed 2 DST (Govt. of India) funded projects in the field of Energy Systems/ Sustainability.

Professor Papita Das did her PhD in Chemical Engineering from Jadavpur University, India. She is Professor, Department of Chemical Engineering & Director, School of Advanced Studies in Industrial Pollution Control Engineering, Jadavpur University, Kolkata, India. She is known for her work in water treatment using different novel adsorbents materials. She is also working on biomass-based energy production and polymeric-nanocomposite synthesis and its degradation. She has published more than 175 International journal research articles and reviews and more than 50 book chapters in various SCI and Scopus indexed journals. She also Ranked in World Ranking among the top 2% Scientists (2020, 2021, and 2022), published by Stanford University, which represents the top 2% most cited scientists in various disciplines. She ranked 432 among 53,348 researchers in the field of Chemical Engineering (2022) based on career-long impact and 171 for a single year citations in 2021. She ranked 614 among 55,697 researchers in Chemical Engineering (2020) and 534 among 66,189 researchers in the field of Chemical Engineering (2021) based on career-long impact and ranked 217 for the single year 2020 discipline. She also guided ten PhD students (completed), two (submitted), and seven (ongoing). She was editor of two books published by Elsevier and Springer, and she is an editorial board member, Editor-in-Chief, and Associate Editor of various international journals. She has completed 17 projects funded by Indian Govt. agencies and industries.

Dr. Arnab Mondal earned his PhD from the Academy of Scientific and Innovative Research, conducting research at CSIR–National Physical Laboratory in New Delhi. Currently, he holds the position of Research Associate at Banaras Hindu University in Varanasi, India. Dr. Mondal has authored/co-authored over 15 international research and review articles. He pursued his master's degree at the Institute of Environment and Sustainable Development (IESD), Banaras Hindu University, Varanasi.

Ms. Poushali Chakraborty is currently pursuing her PhD in the Chemical Engineering Department of Jadavpur University. She has received her BSc degree from University of Calcutta in Zoology Hons in 2018 and her MSc degree in Biotechnology in 2020 from Jadavpur University. She has worked as Research

Assistant in Manovikas Kendra. She is currently working in the field of advanced technologies for water treatment, synthesis of biopolymers for drug delivery, packaging purposes and water treatment, and microbial bioremediation. Her other interests are synthesizing membranes, Aerogel, and advanced bioremediation techniques. She has presented her research papers at several international conferences. She has published book chapters and research papers in international journals.

Contributors

Banerjee, Soumitra
Centre for Incubation, Innovation, Research and Consultancy (CIIRC)
Department of Food Technology
Tataguni, Bangalore-560082, Karnataka, India

Bar, Anindita
Department of Life Science and Biotechnology
Jadavpur University
Kolkata, India

Barman, Nabasmita
Department of Chemical Engineering
Jadavpur University
Jadavpur, Kolkata-700032, India

Basak, Nilendu
Department of Microbiology
University of Kalyani
Kalyani, Nadia, West Bengal, India-741235

Belwal, Satyakirti
Department of Environmental Studies
The Maharaja Sayajirao University of Baroda
Vadodara-390002, Gujarat, India.

Bhattacharya, Sumanta
Research Scholar
MAKAUT, Haringhata, West Bengal, India

Billa, Shalini
Department of Biotechnology Engineering
Kolhapur Institute of Technology's College of Engineering (Autonomous)
Kolhapur, Maharashtra, India

Biswas, Anirban
Department of Environmental Science
Nabadwip Vidyasagar College
Nabadwip-741302, India

Biswas, Saroni
Centre for Sustainable Development and Research-700055
Kolkata, India

Bivin Ebenezer, S.
Department of Civil Engineering
National Institute of Technology
Tiruchirappalli, Tamilnadu-620015, India

Chakraborty, Poushali
Department of Chemical Engineering
School of Advanced Studies in Industrial Pollution Control and Engineering
Jadavpur University, Kolkata, India

Chatterjee, Ankita
Department of Biotechnology
School of Applied Sciences REVA University
Bengaluru, Karnataka-560064, India

Chowdhury, Atif Aziz
Department of Microbiology
University of Kalyani
Kalyani, Nadia, West Bengal-741235, India
Faculty of Agricultural
Environmental and Food Sciences
Free University of Bolzano
Piazza Università, 1, Bolzano-39100, Italy

Das, Gupta Parama
P.G Department of Microbiology
Bidhannagar College
EB-2, Sector 1, Salt Lake, Kolkata-700064, West Bengal, India

Das, Papita
Department of Chemical Engineering
Jadavpur University
Jadavpur, Kolkata-700032, India

Dhar, Bhowmick Gourav
Agricultural and Food Engineering Department
Indian Institute of Technology
Kharagpur-721302

Dutta, Ratna
Department of Chemical Engineering
Jadavpur University
Jadavpur, Kolkata-700032, India

Gandhimathi, R.
Department of Civil Engineering
National Institute of Technology
Tiruchirappalli, Tamilnadu-620015, India

Ganesan, Nirmala
Department of Bioengineering
School of Engineering Vels Institute of Science
Technology and Advanced Studies
Chennai, Tamil Nadu, India

Ghosh, Apurba Ratan
Professor and Head
Department of Environmental Science
The University of Burdwan
Burdwan, Golapbag, Purba Bardhaman, West Bengal-713104, India

Ghosh, Avijit
Department of Chemical Engineering
Heritage Institute of Technology
Kolkata, India

Goyal, Nishu
School of Health Sciences and Technology
University of Petroleum and Energy Studies
Dehradun, Uttarakhand-248001, India

Gupta, Parthapratim
Department of Chemical Engineering
National Institute of Technology (NIT) Durgapur
Mahatma Gandhi Avenue, A-Zone, Paschim Bardhhaman, West Bengal, India-713209

Hapani, Uma
Department of Environmental Studies
The Maharaja Sayajirao University of Baroda
Vadodara-390002, Gujarat, India

Harithra, V.
Department of Bioengineering
School of Engineering Vels Institute of Science
Technology and Advanced Studies
Chennai, Tamil Nadu, India

Hemavathi, A. B.
Department of Polymer Science and Technology
Sri Jayachamarajendra College of Engineering (SJCE)
Jagadguru Sri Shivarathreeshwara Science and Technology University
Mysore-570 006, Karnataka, India

Islam, Ekramul
Department of Microbiology
University of Kalyani
Kalyani, Nadia, West Bengal-741235, India

Jakka, Venkatalakshmi
Department of Chemistry
School of Applied Sciences and Humanities
Vignan's Foundation for Science Technology and Research (VFSTR) deemed to be university
Vadlamudi, Guntur, AP, India

Kagale, Sneha
Department of Biotechnology Engineering
Kolhapur Institute of Technology's College of Engineering (Autonomous)
Kolhapur, Maharashtra, India

Kandpal, Rahul
School of Health Sciences and Technology
University of Petroleum and Energy Studies
Dehradun, Uttarakhand-248001, India

Kumar, Vineet
Department of Microbiology,
School of Life Sciences
Central University of Rajasthan
NH-8, Bandarsindri, Kishangarh, Ajmer, Rajasthan, India

Kundu, Ishita
Department of Chemical Engineering
Heritage Institute of Technology
Kolkata, India

Mandal, Arghya
Research Scholar
Department of Environmental Science
The University of Burdwan
Burdwan, Golapbag, Purba Bardhaman, West Bengal-713104, India
SACT
Mankar College
Mankar, West Bengal-713144, India

Mandal, Tamal
Department of Chemical Engineering
National Institute of Technology (NIT) Durgapur
Mahatma Gandhi Avenue, A-Zone, Paschim Bardhhaman, West Bengal-713209, India

Mondal, Naba Kumar
Environmental Chemistry Laboratory
Department of Environmental Science
The University of Burdwan
West Bengal, India

Mondal, Niladri Sekhar
Department of Environmental Science
School of Sciences
Netaji Subhas Open University
DD-26, Sector-I, Salt Lake City, Kolkata—700 064.

Mukherjee, Arkajyoti
Agricultural and Food Engineering Department
Indian Institute of Technology
Kharagpur-721302

Nambiar, Anargha P.
Department of Biotechnology and Bioengineering
Institute of Advanced Research
Koba Institutional Area, Gandhinagar-382426, Gujarat, India

Paria, Kishalaya
Oriental Institute of Science and Technology
Vidyasagar University
Midnapore, West Bengal, India

Patra, Atanu
Department of Environmental Science
The University of Burdwan
Burdwan, Golapbag, Purba Bardhaman, West Bengal-713104, India

Mankar College
Mankar, West Bengal-713144, India

Patil, Anirudh Gururaj
Department of Food Technology
Annasaheb Dange College of Engineering & Technology
Ashta, Sangli-416301, Maharashtra, India

Patil, Gouri
Department of Polymer Science and Technology
Sri Jayachamarajendra College of Engineering (SJCE)
Jagadguru Sri Shivarathreeshwara Science and Technology University
Mysore-570006, Karnataka, India

Patil, Pallavi
Department of Biotechnology Engineering
Kolhapur Institute of Technology's College of Engineering (Autonomous)
Kolhapur, Maharashtra, India

Prince, Christina
Department of Bioengineering
School of Engineering
Vels Institute of Science
Technology and Advanced Studies
Chennai, Tamil Nadu, India

Raju, M.
Department of Civil Engineering
National Institute of Technology
Tiruchirappalli, Tamilnadu-620 015, India

Ranjan, Ved Prakash
CSIR–National Environmental Engineering Research Institute (CSIR-NEERI)
Nehru Marg, Nagpur-440020, India

Saha, Sudeshna
Department of Chemical Engineering
Jadavpur University
Kolkata, India

Sandil, Sirat
Institute of Aquatic Ecology
Centre for Ecological Research
National Laboratory for Water Science and Water Security
Institute of Aquatic Ecology
Centre for Ecological Research
Karolina út 29, 1113, Budapest, Hungary

Sarkhel, Rwiddhi
Department of Chemical Engineerin
National Institute of Technology (NIT) Durgapur
Mahatma Gandhi Avenue, A-Zone, Paschim Bardhhaman, West Bengal-713209, India

Sen, Kamalesh
Environmental Chemistry Laboratory
Department of Environmental Science
The University of Burdwan
West Bengal, India

Singh, Diksha
NIMS School of Chemical Engineering and Food Technology
NIMS University Rajasthan
Jaipur-303121, India

Singh, Sakshi
NIMS School of Chemical Engineering and Food Technology
NIMS University Rajasthan, Jaipur-303121, India

Sengupta, Shubhalakshmi
Department of Chemistry
School of Applied Sciences and Humanities
Vignan's Foundation for Science
Technology and Research (VFSTR)
deemed to be university
Vadlamudi, Guntur, AP, India.

Yadav, Snehal
Department of Food Technology
Annasaheb Dange College of Engineering & Technology
Ashta, Sangli-416 301, Maharashtra, India

Zameer, Farhan
Alva's Ayurveda Medical College & ATMA Research Centre
Vidyagiri, Moodubidire
Dakshina Kannada-574 227, Karnataka, India

Zaray, Gyula
Institute of Aquatic Ecology
Centre for Ecological Research
Karolina út 29, 1113, Budapest, Hungary
National Laboratory for Water Science and Water Security
Institute of Aquatic Ecology
Centre for Ecological Research
Institute of Chemistry
Eötvös Loránd University
Pázmány Péter sétány 1/A, H-1117,
Budapest, Hungary

1 Insights
Review of Electrochemical Sensors for Microplastics Monitoring and Their Remediation

Kamalesh Sen and Naba Kumar Mondal

1.1 INTRODUCTION

Owing to their pervasive presence in natural environments and their adverse impacts on both wildlife and human well-being, microplastics (MPs) have evolved into a prominent environmental challenge (Du et al., 2021; Yuan et al., 2022). Recognized as a relatively recent environmental pollutant, the imperative for effective monitoring and mitigation strategies for MPs has become paramount (Agbekpornu et al., 2023; Du et al., 2021; Duis and Coors, 2016). Conventional methods for cleanup and characterization have proven inadequate in addressing the challenges posed by these minuscule plastic particles (Kibria et al., 2023). Consequently, innovative approaches such as electrochemical sensors have garnered attention for their potential to provide practical solutions for both monitoring and remediating MPs (Hu et al., 2022; Reddy and Nair, 2022).

This review intends to provide a thorough overview of the latest developments in electrochemical sensors for monitoring microplastics (MPs) and their application in methods for mitigating MPs pollution (Baranwal et al., 2022; Piña et al., 2023; Reddy and Nair, 2022). Electrochemical techniques have demonstrated promise in various environmental monitoring applications due to their swift response, user-friendliness, portability, and cost-effectiveness (Baranwal et al., 2022). However, their application in the context of MPs detection and remediation is relatively uncharted territory, with only a limited number of studies conducted in recent years. The potential of electrochemical sensors as robust tools for detecting MPs in environmental samples is explored (Bayo et al., 2020; Reddy and Nair, 2022). These sensors utilize electrochemical principles to detect changes in electrical properties arising from interactions between microplastics (MPs) and the sensor's surface (Z. Chen et al., 2022; Grieshaber et al., 2008). Such interactions may encompass adsorption, binding, or changes in electrochemical behavior induced by the presence of MPs (Reddy and Nair, 2022). Notably, one of the key strengths of electrochemical sensors is their sensitivity, enabling the detection of even minute quantities of MPs within complex

DOI: 10.1201/9781032684574-1

environmental matrices (Piña et al., 2023). Their capacity to furnish real-time or nearly real-time data further enhances their utility in comprehending the dynamic nature of MPs pollution (Choran and Örmeci, 2023). Additionally, electrochemical sensors offer high selectivity, facilitating the differentiation of MPs from other interfering substances in the environment. This specificity is vital for accurate quantification and identification of MPs, which can vary in polymer type and shape (Z. Chen et al., 2022).

The amalgamation of electrochemical techniques with sensor design and development presents a unique opportunity to address the MPs conundrum (Ait-Touchente et al., 2020; Martic et al., 2022). This review delves into a range of electrochemical techniques employed, encompassing chronoamperometry measurements, impedimetric and amperometric sensors, electrochemical impedance spectroscopy (EIS), amperometry, voltammetry, serial faradaic ion concentration polarization (fICP), and several others (Z. Chen et al., 2022; Martic et al., 2022; Motalebizadeh et al., 2023; Raoof et al., 2006; Shimizu et al., 2017). By examining the strengths and limitations of these methods, this article aims to elucidate the potential of electrochemical sensors for precise and efficient MPs detection (Martic et al., 2022). In addition to monitoring, this review also explores the prospects of employing electrochemical strategies in the remediation of MPs (Baranwal et al., 2022). A unique approach involving the induction of cations (anodic dissolution) via metal electrodes to bind with MPs, leading to their aggregation into larger precipitates that can be readily removed, presents a promising avenue for mitigating MPs pollution (Baranwal et al., 2022).

The objective of this comprehensive review is to offer a detailed assessment of the current state of electrochemical sensors in the context of monitoring and remediating microplastics (MPs). Our discussion encompasses potential avenues for future research, opportunities for enhancement, and the challenges that must be surmounted to effectively harness these advanced approaches in tackling the worldwide MPs crisis. Ultimately, this assessment aspires to advance scientific and technological progress in this field, contributing to the promotion of a healthier and more sustainable environment for generations to come.

1.2 DIFFERENT MPs AND THEIR CHEMICAL NATURE

MPs encompass diminutive plastic particles, fibers, or fragments, spanning a size range from a few micrometers to several millimeters (Reddy and Nair, 2022). Their chemical composition exhibits variability contingent upon the specific plastic polymer they comprise and any accompanying additives or impurities. The prevalent MP constituents encompass polyethylene (PE), polypropylene (PP), polystyrene (PS), polyethylene terephthalate (PET), and polyvinyl chloride (PVC), among others. Fundamentally, MPs consist primarily of synthetic polymers, characterized by extended chains of repeating units (Esterhuizen and Kim, 2022; Mhiret Gela and Aragaw, 2022; O'Brien et al., 2023). These polymers originate from fossil fuel sources and are intentionally engineered for durability and resistance to degradation. Consequently, due to their innate non-biodegradable nature, MPs exhibit the capacity to endure within the environment for extended periods, possibly for decades or even centuries (Du et al., 2022; Kallenbach et al., 2022). Plastic polymers commonly integrate a spectrum of additives designed to augment their specific properties, including plasticizers, stabilizers, flame retardants, and pigments. These incorporated additives possess the potential to leach out from the plastic matrix, thereby potentially influencing the characteristics and conduct of MPs within the environment (Kallenbach et al., 2022; Ziani et al., 2023). Exposure to environmental factors, encompassing sunlight (UV radiation), elevated temperatures, and mechanical forces, instigates the process of weathering and degradation in MPs (Gunaalan et al., 2020). This transformative process leads to the disintegration of larger plastic items into MPs and further fragmentation into minute particles recognized as nanoplastics. Moreover, it is important to attribute that MPs may also encapsulate a medley of additives and impurities, including plasticizers, flame retardants, stabilizers, and pigments, contingent upon their intended purpose (Gunaalan et al., 2020; Smith et al., 2018; Yee et al., 2021). Consequently, as these MPs undergo disintegration or release into adjacent ecosystems, they have the potential to introduce additional chemical constituents into the environment (Joyce and Falkenberg, 2022). Additionally, as previously noted, the weathering and degradation of MPs can ultimately culminate in the generation of nanoplastics, which may exhibit distinctive chemical attributes and behaviors within the natural surroundings (Yee et al., 2021).

1.2.1 MPs' Relevance to the Environment

Since MPs are ubiquitous pollutants in terrestrial and aquatic settings, their relevance is based on their widespread dispersion and potential effects on a variety of environmental and health issues (Agbekpornu et al., 2023; Du et al., 2021; Duis and Coors, 2016). Their small size and buoyancy enable them to disperse widely, contaminating ecosystems worldwide (Du et al., 2021). They are found in oceans, rivers, lakes, soil, sediments, and even the atmosphere. MPs can have adverse effects on marine and terrestrial organisms (Abbasi and Turner, 2022; Duis and Coors, 2016). The consumption of MPs by marine animals can cause physical injury, digestive system blockages, and decreased feeding efficiency in aquatic environments (Kumar et al., 2022). MPs can also carry and transport harmful chemicals, potentially bioaccumulating in the food web (Abbasi and Turner, 2022; Joyce and Falkenberg, 2022). MPs have been exposed to food, food packaging, and even the air we breathe. Humans may unintentionally consume MPs through tainted food and water, though further research is needed to determine the full extent of any potential health impacts (Joyce and Falkenberg, 2022; P. Wu et al., 2022). MPs have a high surface-area-to-volume ratio, making them capable of adsorbing and accumulating persistent organic pollutants (POPs) and other toxic substances from the surrounding environment (Gunaalan et al., 2020; P. Wu et al., 2022). This adsorption can lead to the concentration of pollutants on the MP surface, potentially acting as carriers of contaminants. The impact of MPs on ecosystems and human health raises societal concerns about environmental conservation, sustainable waste management, and the need for better regulation of plastic use and disposal (Kumar et al., 2022; Yee et al., 2021). To reduce the influence of MPs on filter feeders and aquatic ecosystems, strategies encompass tackling plastic pollution at its origin, enhancing waste management practices, and overseeing and setting controls on MPs concentrations in water ecosystems (Joyce and Falkenberg, 2022; Walkinshaw et al., 2020). An improved comprehension of how filter feeders contribute to MPs contamination can provide valuable insights for shaping conservation and management approaches aimed at safeguarding these critical species and maintaining the well-being of aquatic environments (Leung et al., 2021; Walkinshaw et al., 2020).

The chemistry of MPs and their relevance in environmental systems are multifaceted subjects of global importance. It is essential to comprehend the chemical makeup and behavior of MPs in order to create efficient mitigation solutions for their detrimental effects on ecosystems and human health (P. Wu et al., 2022; Yee et al., 2021). Addressing the MP pollution issue requires collaborative efforts from researchers, policymakers, industries, and the public to reduce plastic waste generation, improve waste management practices, and develop sustainable alternatives to single-use plastics.

1.3 ELECTROCHEMICAL SENSORS FOR MPs

Electrochemical sensors have emerged as a promising and innovative tool for the precise and efficient detection of MPs in environmental samples (Z. Chen et al., 2022; Martic et al., 2022); hence, various sensor types are illustrated in Figure 1.1. These sensors exploit the principles of electrochemistry to measure changes in electrical properties resulting from the interaction between MPs and the sensing surface.

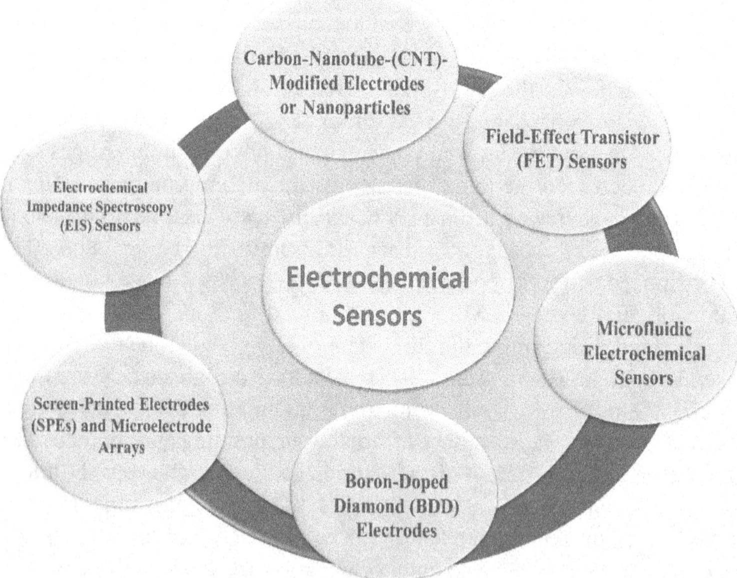

FIGURE 1.1 Electrochemical sensors applied on MPs detection.

Several electrochemical techniques have been applied for MPs detection, including electrochemical impedance spectroscopy (EIS), amperometry, voltammetry, and more. Here, we explore the key features and advantages of electrochemical sensors for MPs detection (Z. Chen et al., 2022; Lopes et al., 2022).

1.3.1 Different Types of Electrochemical Sensors Applied on MPs Detection and Advantages

High-sensitivity electrochemical sensors allow for the detection of even minute concentrations of MPs in intricate environmental matrices. For properly estimating MP contamination levels and comprehending their dispersion in varied ecosystems, this sensitivity is essential (Abbasi and Turner, 2022; Shimizu et al., 2017). The sensitivity of electrochemical sensors for MPs detection is substantially influenced by the electrode materials chosen and their surface changes. Tailoring the electrode surface to enhance MP adsorption and interaction is critical for increasing sensitivity. To increase the sensitivity of electrochemical sensors, a number of signal amplification techniques have been investigated, including enzyme- and nanomaterial-based methods (Baranwal et al., 2022; Grieshaber et al., 2008). These methods allow for the identification of MPs at lower quantities in intricate environmental matrices. The development of nano-electrodes and microelectrodes, which offer improved sensitivity due to their higher surface area-to-volume ratio, was made possible by developments in miniaturization and nanotechnology. Enhancing the proportion of signal to noise in electrochemical sensors requires lowering background noise and optimizing signal processing methods (González-González et al., 2023; X. Wang et al., 2022). This enables the detection of trace amounts of MPs amid background interference. Enhancing the selectivity of electrochemical sensors for MPs is crucial to avoid false positives caused by cross-reactivity with other environmental components. The development of sensor materials with specific affinity for MPs aids in achieving high selectivity (Avolio et al., 2022; Xu et al., 2023). Efficient sample preparation techniques that concentrate MPs from large volumes of environmental samples can significantly enhance sensitivity in electrochemical sensors. Synergy between sample preparation and sensor detection allows for more reliable and sensitive results. Integrating electrochemical sensors with real-time monitoring systems and automation improves data acquisition and analysis, enabling continuous and sensitive MPs detection (Adeniji and Stine, 2023; X. Wang et al., 2022). Enhancing sensitivity in electrochemical sensors for MPs detection is vital for accurately assessing MP pollution levels in the environment. Advances in electrode materials, surface modifications, signal amplification strategies, and miniaturization have contributed to improved sensitivity (Avolio et al., 2022; Rahn and Anand, 2021). However, challenges related to selectivity, cross-reactivity, and background noise remain to be addressed. Collaborative efforts between researchers, engineers, and environmental practitioners are essential to overcome these challenges and develop sensitive, reliable, and practical electrochemical sensors for MPs detection (González-González et al., 2023). Sensitive electrochemical sensors will be essential in determining the level of MP pollution and directing efficient mitigation efforts to protect our ecosystems and people's health as this technology continues to advance (Motalebizadeh et al., 2023). Due to their great sensitivity, low detection limits, potential for miniaturization, and portability, electrochemical sensors are useful tools for finding MPs (Yin et al., 2021). A variety of electrochemical sensors can be used to detect microplastics (MPs), each with unique benefits and working principles. Here are a few illustrations of electrochemical sensors with a reputation for being sensitive to microplastics.

1.3.1.1 Voltammetric Sensors

A voltage is applied to the working electrode while also measuring the resulting current in cyclic voltammetry (CV), a widely used electrochemical technique for detecting and analyzing a variety of analytes (Singh et al., 2023; Vidal et al., 2023). Researchers can use this method to learn more about the analyte's electrochemical behavior and, in the case of MPs detection, to establish the presence and concentration of MP particles. For MPs sensing, the working electrode material selection is essential. Glassy carbon, gold, platinum, and carbon-based materials like screen-printed electrodes are typical electrode materials (Kaya et al., 2020; Pandey and Chusuei, 2021; Vidal et al., 2023; Zhang et al., 2022). The selection is dependent on elements including the desired level of sensitivity and the microplastics' electrochemical characteristics (Hue et al., 2022; Zhang et al., 2022). To enhance sensitivity and selectivity to MPs, researchers can modify the surface of the working electrode in several ways: (1) The electrode surface can be functionalized with certain ligands that have a strong affinity for MPs, such as antibodies, aptamers, or molecularly imprinted polymers (MIPs). The electrochemical response of the electrode is altered when MPs bind to certain ligands. (2) Nanomaterials, such as nanoparticles, carbon nanotubes, or graphene, can be immobilized on the surface of an electrode to enhance surface area and offer a greater capacity for MP binding. (3) Changing the electrode's surface charge or employing charged polymers can aid in drawing in and capturing MP particles with opposing charges. Researchers are able to improve the sensitivity and selectivity of the CV-based MPs sensor by carefully choosing the modification technique and making the most of the experimental settings (Pandey and Chusuei, 2021). This makes it possible to identify MPs from other compounds that might be present in the sample and to detect MPs at low concentrations (Tarafdar et al., 2022; Zhang et al., 2022). In summary, cyclic voltammetry shows itself to be a flexible and highly sensitive approach for identifying MPs across a variety of environmental samples when combined with proper electrode surface improvements. Current research projects are focused on developing cutting-edge methods that improve sensor performance, ensuring even more accurate and reliable monitoring of MPs. Utilizing cyclic voltammetry, the particle-electrode impact approach was used to identify MPs in aquatic environments (L. Chen et al., 2022; Fu et al., 2020). Polyethylene and polypropylene can be detected in a phosphate buffer solution using cyclic voltammetry at 50 mV/s and a three-electrode setup with an Ag/AgCl reference electrode and gold wire counter electrode. Increased oxidation and reduction peaks were visible at different scan rates (100 – 450 mV/s). A CeO_2 NPs-modified glassy carbon electrode enhanced detection by interacting hydrophobically with MPs, enabling in-situ detection in aquatic media (Hue et al., 2022).

1.3.1.2 Amperometric Sensors

Electrochemical impedance spectroscopy (EIS) involves measuring the impedance of an electrochemical system across various frequencies. This technique is valuable for examining the interplay between MPs and the electrode's surface, offering crucial insights into the detection process (Martic et al., 2022; Tang et al., 2023; Vidal et al., 2023). EIS applies a sinusoidal AC voltage to the electrochemical cell or sensor. The impedance, represented as a complex number with both magnitude and phase components, is then measured across a range of frequencies. This allows for the assessment of the system's response at different timescales (Moazzenzade et al., 2023; Stradiotto et al., 2003; G. Wu et al., 2022). When MP particles are introduced into the electrochemical system, they can interact with the electrode's surface (H. Du et al., 2023; Farré, 2020). These interactions can involve physical adsorption, chemical binding, or other surface interactions. EIS measures impedance at different frequencies, ranging from very low frequencies (quasi-static) to high frequencies (dynamic) (Chu et al., 2022; G. Wu et al., 2022). By analyzing how impedance changes with frequency, researchers can gain insights into the nature of the MPs–electrode interaction. At high frequencies, the impedance measurement is influenced by processes that occur on a short timescale, such as charge transfer reactions at the electrode surface (H. Du et al., 2023). Changes in impedance at high frequencies may indicate alterations in the MPs' electronic properties or their effect on electron transfer processes (Hue et al., 2022; Martic et al., 2022; Vidal et al., 2023). Low-frequency impedance measurements are sensitive to slower processes, such as mass transport and diffusion of ions or molecules to and from the electrode surface. The presence of MPs can affect these processes, leading to changes in impedance (Chu et al., 2022). To improve the selectivity and sensitivity of MPs detection, surface modifications and sensor design optimization can be provided with EIS data. By understanding the impedance changes associated with MPs interactions, researchers can tailor sensors for improved performance (Bifano et al., 2023; G. Wu et al., 2022). Therefore, EIS is a valuable tool for studying the interplay between MPs and electrode surfaces in electrochemical sensors. It is a potential method for environmental monitoring and MPs detection applications because it provides detailed frequency-dependent information that can be used to further understand and improve the detection process (Chu et al., 2022; Vidal et al., 2023; G. Wu et al., 2022). Previously, various amperometric sensors were employed for the detection of MPs, as depicted in Table 1.1.

1.3.1.3 Potentiostatic Sensors

Potentiostatic sensors are a specialized type of electrochemical sensor employed to identify and quantify specific chemical compounds or analytes within a given sample (Motalebizadeh et al., 2023; Stradiotto et al., 2003). These sensors rely on potentiostatics, which involves careful measurement and control of the potential (voltage) differential between a working electrode and a reference electrode while also keeping an eye on the current flow that results from the working electrode's action (Miao et al., 2020; Shimizu et al., 2017). These sensors find widespread utility across various domains, including environmental monitoring, biomedical applications, and industrial processes (Farré, 2020). Integral to their operation is the use of chronoamperometry, an indispensable technique for gauging the current response

TABLE 1.1
Different Amperometric Sensors Applied in MPs Detection

Amperometric Sensors	Descriptions	References
Platinum working microelectrodes	Polystyrene MPs (320 for PS and 400 for EPS)	Vidal et al., 2023
Potential rapid sensor (micro-interdigitated electrode)	Polyethylene and polystyrene (20 μm and 150 μm)	Ching et al., 2023
Graphene electrode	Polystyrene	H. Du et al., 2023
Gold-plated electrode	Polyethylene (212–1000 μm)	Colson and Michel, 2021
EIS in combination with machine learning	Polypropylene (PP) and polyolefin (PO)	Bifano et al., 2023
Carbon nanohorn/rhodamine B modified electrode	Polystyrene MP	Zheng et al., 2023

arising from electrochemical reactions occurring at the working electrode (Martic et al., 2022; Puthongkham et al., 2021). By controlling the potential and tracking, the temporal evolution of the current, this integrated technique enables the sensitive and focused detection of specific analytes or chemical entities (Puthongkham et al., 2021). In chronoamperometry, the working electrode is subjected to a constant voltage, and the resulting current is monitored over a predetermined time period (Shimizu et al., 2017). This technique is commonly harnessed in manifold applications, including the detection of MPs (Durán et al., 2018). Typically, the working electrode is subject to modification or functionalization to heighten its sensitivity and specificity toward MPs (Cao et al., 2023; Durán et al., 2018). This alteration may involve the addition of certain recognition components or molecules, such as antibodies, aptamers, or molecularly imprinted polymers, which can distinguish and bind to MPs in a specific way. The working electrode is given a constant voltage that is selected to promote electrochemical interactions or reactions between the target MPs and the modified electrode surface (Martic et al., 2022). Concurrently, the resulting current is continuously scrutinized over the predetermined duration, often achieved by measuring the electron flow at the electrode–solution interface (Beduk et al., 2022; Cao et al., 2023; Miao et al., 2020). When MPs interact with the modified electrode surface, various phenomena come into play. MPs may adsorb or adhere to the surface, resulting in changes in current as the coverage of MPs on the surface evolves over time. In some cases, MPs may undergo electrochemical reactions at the electrode surface, leading to current alterations associated with their oxidation or reduction (Cao et al., 2023; Durán et al., 2018). Moreover, microplastics may physically block or hinder the access of other electroactive species to the electrode surface, thereby impacting the observed current (Puthongkham et al., 2021). In sum, when incorporated into potentiostatic sensors, chronoamperometry proves to be an invaluable electrochemical technique for the precise and discriminating identification of microplastics (MPs). This is accomplished by keeping a constant voltage at the working electrode while continuously monitoring current fluctuations over time. As a result, it makes it possible to distinguish between MPs based on their electrochemical characteristics and how they interact with the altered electrode surface (Martic et al., 2022).

1.3.1.4 Field-Effect Transistor (FET) sensors

FET-based sensors use a gate electrode that is sensitive to changes in the surrounding environment. They can be functionalized with receptors that specifically interact with MPs, causing changes in the transistor's conductivity (Rizzato et al., 2020; Sadighbayan et al., 2020; Tovar-Lopez, 2023). FET sensors are indeed a versatile class of sensors used for detecting a wide range of analytes, including MPs (Junhong Chen et al., 2022; Tovar-Lopez, 2023). These sensors work by applying an external electric field to a semiconductor channel, which modifies the conductivity of the channel (Guillaud et al., 1998; Sharma et al., 2020). When it comes to detecting MPs using FET sensors, the basic principle involves functionalizing the FET surface with a material that has an affinity for MPs (Elli et al., 2022; Martic et al., 2022; R. Sharma et al., 2021). The surface of the FET sensor is modified or coated with a material that can specifically bind to or interact with MPs (Martic et al., 2022; Tong et al., 2015; Yadav et al., 2021). This coating can be created to bind to MPs with a high affinity, allowing for selective binding. MPs will interact with the FET sensor's surface coating if they are present in the sample being analyzed. Depending on the composition of the coating, this contact may take the form of physical adsorption or chemical binding. The charge distribution or other characteristics of the semiconductor channel beneath the surface coating are altered by the binding or interaction of MPs with the coating (Le et al., 2017; Tamburri et al., 2022; Yadav et al., 2021). This change, in turn, affects the conductivity of the FET (Choi and Yoo, 2023; Martic et al., 2022; Rizzato et al., 2020). The conductivity changes are typically measured as an electrical signal, such as a change in current or voltage across the FET (Rizzato et al., 2020; Tovar-Lopez, 2023). The presence and quantity of MPs in the sample can then be associated with this signal. FET sensors' sensitivity and capacity to deliver real-time or nearly real-time monitoring make them advantageous for MP detection (Martic et al., 2022; Rizzato et al., 2020; Yadav et al., 2021). Additionally, FET sensors can be highly selective if the surface modification is tailored for MPs, reducing the likelihood of false positives. These sensors can be useful in various applications, including environmental monitoring, water quality assessment, and studies related to MP contamination in ecosystems (Elli et al., 2021; R. Sharma et al., 2021; Tovar-Lopez, 2023). Researchers and engineers continue to develop and improve FET-based sensors for specific applications, including the detection and quantification of MPs, contributing to our understanding of this global environmental issue.

1.3.1.5 Impedimetric Sensors

When MPs are adsorbed onto a quartz crystal's surface, quartz crystal microbalance (QCM) sensors track the change in resonance frequency of the crystal (Olaniyan et al., 2023; Pérez et al., 2021). Because the frequency shift is directly proportional to the mass of microplastics (MPs), it facilitates highly sensitive detection. When substances like MPs adhere to the surface of a quartz crystal, QCM sensors are employed to detect and analyze the resulting changes in mass (Olaniyan et al., 2023; Rastogi et al., 2021; Woo et al., 2022). QCM sensors operate on the principle of detecting changes in the resonance frequency of a quartz crystal caused by added mass (Huang et al., 2021; Lapointe et al., 2020). A QCM sensor consists of a thin quartz crystal, typically made of quartz or similar piezoelectric materials. When an alternating electrical voltage is applied across this crystal, it will resonate at a precise frequency because of the way it has been cut and polished (Rastogi et al., 2021; Xiao et al., 2020). The crystal vibrates at its natural resonant frequency. When MPs come into contact with the surface of the quartz crystal, they adhere to it due to various surface forces, notably van der Waals forces and electrostatic interactions (Wang et al., 2023). MPs contribute bulk to the crystal as they collect on its surface. The resonance frequency of the quartz crystal changes as a result of the additional mass from the adsorbed MPs. The mass of the adsorbed MPs has a direct impact on this shift (Shams et al., 2021; Wang et al., 2023). The greater the mass of MPs on the crystal, the larger the shift in resonance frequency. Electronic circuits in QCM sensors continuously track the crystal's resonance frequency (Olaniyan et al., 2023). Any change in frequency is detected and recorded. By calibrating the sensor, scientists can correlate the frequency shift with the mass of the adsorbed MPs. QCM sensors are extremely sensitive and can detect minute mass changes (Virkkala et al., 2023). This sensitivity makes them ideal for detecting low concentrations of MPs in various samples, including water, air, or biological fluids. They are particularly valuable for monitoring environmental samples where even small amounts of MPs can have significant impacts (Shams et al., 2021). Consequently, QCM sensors represent a reliable method for the highly sensitive detection of MPs. Their working theory is based on the observed changes in the resonance frequency of a quartz crystal, which occur due to the attachment of MPs to its surface (Huang et al., 2021; Shams et al., 2021). Because the frequency shift is inversely proportional to the mass of the adsorbed MPs, it allows for precise and sensitive quantification of MP contamination in various applications, including environmental monitoring and scientific research.

1.3.1.6 Electrochemical Biosensors

Biosensors can be designed with biological receptors such as antibodies or DNA aptamers that selectively bind to MPs (Naresh and Lee, 2021). When binding occurs, it can lead to changes in the electrochemical signal, which can be detected and quantified. Biosensors are analytical devices that incorporate biological receptors, such as antibodies or DNA aptamers, to selectively interact with specific target molecules, like MPs (Naresh and Lee, 2021; Pfeiffer and Mayer, 2016). These receptors have a high degree of specificity and affinity when identifying and attaching to their respective target molecules. When such binding events occur in a biosensor, they can trigger detectable changes in an electrochemical signal, which can then be quantified (Fredj et al., 2023). A biosensor is designed with a biological recognition element, which can be an antibody, DNA aptamer, enzyme, or other biomolecules depending on the target analyte. In the case of detecting MPs, receptors are selected or designed to specifically bind to certain characteristics or molecules on the surface of MPs (Das et al., 2014; Naresh and Lee, 2021). When the sample containing MPs is introduced to the biosensor, the biological recognition element selectively binds to the MPs present in the sample. This binding event is highly specific, ensuring that only MPs are captured by the biosensor (Fredj et al., 2023). The binding of MPs to the biological recognition element can lead to a signal transduction event. This can involve changes in the electrical conductivity, redox potential, or other electrochemical properties at the sensor's surface (Gongi et al., 2022). The sensitivity of the biosensor may occasionally be improved by incorporating signal amplification techniques (Campaña et al., 2019). For instance, enzyme-linked biosensors may use enzymatic reactions to amplify the electrochemical signal in response to MP binding (Choi and Yoon, 2023; Gongi et al., 2022). The biosensor's integrated transducer, which transforms the biochemical interaction into a quantifiable electrical signal, then detects variations in the electrochemical signal. The biosensor can reveal the presence and concentration of MPs in the sample by measuring changes in the electrochemical signal (Naresh and Lee, 2021). Typically, the degree of signal change is proportional to the number of MPs linked to the recognition element of the sensor (Gongi et al., 2022). The development of biosensors depends critically on the study of electron transfer pathways. These tools use the identification of particular biomolecules to reveal details about biological systems. In this setting, electron transfer is governed by the fundamental indirect electron transfer (IET) and direct electron transfer (DET) mechanisms (Fredj et al., 2023). In the same vein, a unique acetylcholine-detection biosensor has been created. The substrate of this biosensor is a polyaniline-polyvinylsulphonate electrochemically polymerized film on which the enzymes acetylcholinesterase and choline oxidase have become immobilized. The biosensor measures acetylcholine with high accuracy over a linear concentration range of 0.1–0.6 μM attributed to the enzymatic oxidation of hydrogen peroxide (H_2O_2) (Tunç et al., 2016). Nanostructure-based sensors have demonstrated remarkable capabilities for ultrasensitive detection (Dey et al., 2021; Huang et al., 2021). An electrochemical biosensor that uses graphene oxide nanowalls and achieves single-DNA resolution is one notable example. A graphene field-effect transistor biosensor has also been used to specifically identify custom-made micro- and nanostructured polymers (Tang et al., 2023). Biosensors designed to detect MPs using biological receptors offer several advantages, including high specificity and the potential for customization to target-specific types or sizes of MPs (Dey et al., 2021; Huang et al., 2021). They are valuable tools in environmental monitoring, research, and quality control, aiding in the assessment of MP contamination in various ecosystems and products.

1.3.1.7 Nanomaterial-Based Sensors

Electrochemical approaches can be used to locate and measure MPs in the environment using nanomaterial-based sensors (Dey et al., 2021; Tajik et al., 2021; Tavakoli et al., 2022). Materials such as graphene, carbon nanotubes, and various nanoparticles are preferred as the detection components due to their expansive surface area, excellent electrical conductivity, and chemical responsiveness (Junhong Chen et al., 2022; Gao et al., 2012; Khan et al., 2023; Lei and Ju, 2012; Rahman et al., 2021; Schiavi et al., 2023). These attributes render them well-suited for capturing and assessing MP particles. The nanomaterials can be customized or altered with specific molecules or coatings that possess a strong attraction to MPs (Dey et al., 2021; Gao et al., 2012). This customization enhances the sensor's selectivity, ensuring it primarily recognizes MPs amid other substances. To identify MPs in a sample, it is crucial to prepare the sample appropriately (Curulli, 2021; Tavakoli et al., 2022). This typically involves filtering and concentrating the sample to isolate MPs from other particles and debris. Once isolated, the MPs can be introduced to the sensor for analysis. The electrochemical cell of the sensor is where the isolated MPs interact with the customized nanomaterials. When MPs adhere to the nanomaterials, it induces changes in the sensor's electrical properties (Das et al., 2014; Gao et al., 2012). These alterations can be quantified as an electrical signal, which can then be correlated with the MPs' concentration in the sample (Curulli, 2021; Hong et al., 2023). Nanomaterial-based sensors exhibit remarkable sensitivity, enabling the detection of even minute concentrations of MPs. Moreover, the customization of nanomaterials guarantees selectivity, reducing the likelihood of false readings from other substances within the sample (Zhu et al., 2023). Certain nanomaterial-based sensors are designed for real-time monitoring, proving invaluable for ongoing environmental surveillance (Dey et al., 2021; Willner and Vikesland, 2018). They can be deployed in bodies of water, wastewater treatment facilities, or other relevant locations to continuously monitor MP contamination. Recent advancements in nanomaterial-based sensor technology have resulted in the creation of portable, miniaturized devices (Altug et al., 2022; Chauhan et al., 2021; Ren et al., 2022). These can be employed in the field for on-site MP detection, providing researchers and environmental agencies with user-friendly and convenient tools. In this context, nanomaterial-based sensors employing electrochemical methods hold significant promise as a solution for detecting MPs in environmental samples (Altug et al., 2022; Dey et al., 2021). Their high sensitivity, selectivity, and potential for real-time monitoring make them valuable assets in efforts to comprehend and mitigate the repercussions of MP pollution on ecosystems and human health (Altug et al., 2022; Willner and Vikesland, 2018).

1.3.1.8 Surface-Enhanced Raman Spectroscopy (SERS)

Though not strictly an electrochemical sensor, SERS can work in conjunction with electrochemical techniques to dramatically increase sensitivity in the detection and quantification of a variety of analytes, including MPs (Jaworska et al., 2016). This powerful analytical technique amalgamates conventional Raman spectroscopy with the signal enhancement capabilities of nanostructured surfaces (Ott et al., 2022). Functionalized nanoparticles, such as those composed of gold or silver, play a pivotal role in SERS applications (Jaworska et al., 2016; Ott et al., 2022). These nanoparticles undergo modification with specific ligands or coatings that have a selective affinity for MPs. This functionalization process entails affixing molecules or groups onto the nanoparticle surfaces that exhibit an attraction to MPs. Once functionalized, these nanoparticles can be introduced into a sample containing MPs (Dey, 2023; Mikac et al., 2023). They effectively capture the MPs from the sample matrix due to their specialized functional groups. Electrochemical methods can be strategically incorporated to further amplify the SERS signal (Mikac et al., 2023; Ott et al., 2022). For example, nanoparticles loaded with captured MPs can be immobilized on an electrode surface. By applying a voltage or current, the interaction between the captured MPs and the electrode is promoted, leading to amplified Raman signals through electrochemical amplification (Dey, 2023; Schiavi et al., 2023). Following electrochemical enhancement, the MPs, now firmly attached to the functionalized nanoparticles, can be subjected to SERS analysis. When a laser is directed at the electrode surface, it stimulates the Raman-active molecules on the nanoparticles, generating intensified Raman spectra (Dey, 2023; Jaworska et al., 2016). These spectra can then be gathered and scrutinized to discern and quantify the MPs based on their distinctive vibrational frequencies. The amalgamation of SERS with electrochemical techniques endows MP detection with several advantages (Terry et al., 2022). The synergy between SERS and electrochemistry significantly augments sensitivity, enabling the discernment of minute quantities of MPs within intricate environmental samples (Dey, 2023; Terry et al., 2022). Functionalized nanoparticles can be customized to selectively bind to specific types of MPs, thus facilitating precise detection and differentiation. The electrochemical dimension of this approach facilitates real-time monitoring and potentially continuous in-situ detection of MPs (Jaworska et al., 2016; Xu et al., 2020). The Raman spectra acquired through SERS afford quantitative insights into MP concentrations within the sample (Dey, 2023; Xu et al., 2020). This convergence of SERS and electrochemical techniques holds immense potential for combatting the escalating issue of MP pollution across diverse environmental settings, encompassing aquatic bodies and ecosystems (Terry et al., 2022; Xu et al., 2020). It provides a flexible and extremely sensitive method for identifying and quantifying MPs, with potential uses in environmental research and monitoring.

1.3.1.9 Microfluidic Electrochemical Sensors

Microfluidic electrochemical sensors offer a cutting-edge approach to MPs detection by seamlessly combining microfluidic devices with electrochemical sensors, resulting in lab-on-a-chip systems (Jaworska et al., 2016; Lokar et al., 2023). These integrated systems are designed for swift and automated MPs detection, offering exceptional advantages (Silva et al., 2023). By integrating microfluidic technology with electrochemical sensors, these lab-on-a-chip systems empower researchers and analysts to exercise meticulous authority over sample volumes and the precise conditions of reactions (Radhakrishnan et al., 2022; Rai et al., 2022). This level of control ensures accuracy and reproducibility in MPs detection assays. The microfluidic component allows for the efficient manipulation and transport of minute volumes of samples and reagents within microchannels, minimizing sample wastage and reducing analysis time (Ece et al., 2023; Lokar et al., 2023). This feature is especially advantageous when dealing with limited or precious samples. Furthermore, the integration of electrochemical sensors within the microfluidic setup enables real-time monitoring of electrochemical signals, offering rapid and quantitative insights into the presence and concentration of MPs (Alhalaili et al., 2022; Ece et al., 2023). Electrochemical detection is highly sensitive and selective, enhancing the reliability of MPs detection (Radhakrishnan et al., 2022; Rai et al., 2022; Shen et al., 2020). The lab-on-a-chip approach facilitates automation, streamlining the entire detection process, from sample introduction to data analysis. This automation not only accelerates the analysis but also reduces the potential for human error, making MPs detection more robust and dependable (Alhalaili et al., 2022; Silva et al., 2023). In summary, microfluidic electrochemical sensors represent an innovative and efficient solution for MPs detection (Alhalaili et al., 2022; Shen et al., 2020; Zhao et al., 2022). By combining microfluidic capabilities with electrochemical sensitivity, these lab-on-a-chip systems provide precise control, rapid analysis, and automation, making them a valuable tool for addressing the growing issue of MP pollution in various environmental settings (Rai et al., 2022).

1.3.1.10 Electrochemical Aptasensors for Sensing of MPs

In electrochemical sensors designed for precise detection of MPs, aptamers, which are artificial DNA or RNA sequences capable of exclusively binding to specific target molecules, prove to be invaluable (Das et al., 2014; Song et al., 2012). This binding event undergoes transduction into an electrochemical signal (Zheng et al., 2020). Electrochemical aptasensors, a captivating category of biosensors, leverage aptamers—artificial DNA or RNA sequences—to selectively detect and quantify target molecules such as MPs. In this context lies a more in-depth elucidation of their operation and the factors influencing their configuration. The initial step in crafting an electrochemical aptasensor entails the careful selection of an aptamer with a particular affinity for the target MPs or closely related compounds (Radi and Abd-Ellatief, 2021). Typically, these aptamers are developed through a process called systematic evolution of ligands by exponential enrichment (SELEX), wherein a diverse pool of DNA or RNA sequences undergoes iterative selection to attain enhanced affinity and specificity for the desired target (Radi and Abd-Ellatief, 2021; Dube et al., 2023). The choice of the electrochemical sensor hinges on several factors: (1) Electrode materials commonly encompass glassy carbon, gold, platinum, and carbon nanotubes. The selection depends on the electrochemical properties of the target MPs and the binding event facilitated by the aptamer. (2) Electrochemical aptasensors employ diverse transduction techniques, including amperometry, potentiometry, impedance spectroscopy, and cyclic voltammetry. The choice depends on the desired sensitivity and the specific electrochemical response characteristics requisite for the target MPs (Das et al., 2014; Radi and Abd-Ellatief, 2021; Zhao et al., 2022). Generally, Figure 1.2 provides a schematic representation (not to scale) of an electrochemical biosensor designed for detecting MPs, similarly to the observation diagram reflected in previous literature (Damiati and Schuster, 2020).

The composition and pH of the electrolyte solution wield influence over the sensor's performance and necessitate optimization to cater to the specific MPs under scrutiny (Hong et al., 2022). To ensure the selective binding of the target MPs, the selected electrode surface is modified with the aptamer through methods such as covalent attachment, adsorption, or other surface modification techniques (Ma et al., 2014; Saravanakumar et al., 2022). Upon the aptamer's interaction with the target MPs, a transformation in the electrochemical properties of the sensor materializes, manifesting as alterations in current, potential, or impedance (Radi, 2011; Song et al., 2012). This change in electrochemical characteristics undergoes conversion into an electrical signal, which could entail shifts in current, voltage, or impedance contingent on the chosen transduction method. Occasionally, signal amplification techniques are employed to increase the sensor's sensitivity and decrease its detection threshold. These techniques include using nanoparticles, enzymes, or redox mediators (Sohrabi et al., 2022). In sum, electrochemical aptasensors furnish a potent and selective means of discerning MPs and an array of other target molecules (Dube et al., 2023; Park et al., 2022). Researchers are continually advancing this technology, rendering it ever more versatile and applicable in diverse domains, including

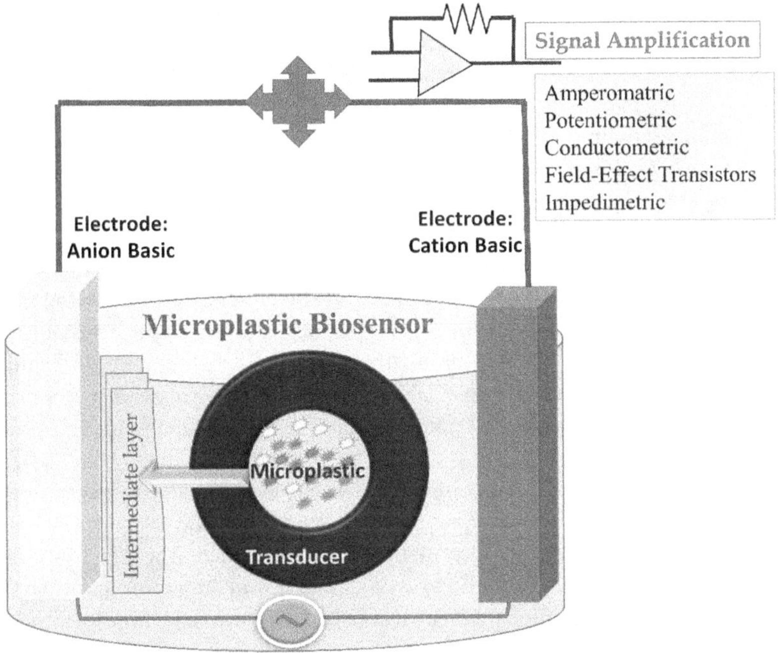

FIGURE 1.2 Unscaled schematic of an electrochemical biosensor for MPs (Damiati and Schuster, 2020).

environmental monitoring and healthcare. The selection of an electrochemical sensor hinges on factors like the specific MP particles slated for detection, the desired sensitivity, and the prevailing environmental conditions (Dube et al., 2023; Sohrabi et al., 2022). Consequently, researchers frequently customize sensor designs and functionalization strategies to optimize performance for their particular applications.

1.3.2 Selectivity of Electrochemical Sensors for MPs

Electrochemical sensors can be tailored to exhibit exceptional specificity for MPs, enabling them to differentiate between MPs and other interfering substances (Z. Chen et al., 2022; Zhu et al., 2015). This heightened selectivity ensures accurate recognition and measurement of MPs in environmental samples, emphasizing its pivotal role in bolstering the reliability and precision of MP detection through electrochemical sensors (Das et al., 2014). Advances in sensor material selection, surface functionalization, size and shape analysis, and the recognition of MP polymers have yielded promising outcomes in enhancing selectivity (Caldwell et al., 2022). Collaboration with spectroscopic techniques and multiparameter analysis further amplifies the sensor's capability to distinguish MPs from interfering substances (Flynn et al., 2023). Electrochemical sensors offer an appealing solution due to their real-time detection ability at low concentrations. However, enhancing selectivity is paramount to mitigate interference and achieve precise MPs detection. The selection of sensor material is of paramount importance in this endeavor (Caldwell et al., 2022).

In order to enhance selectivity, researchers have investigated a variety of materials, including molecularly imprinted polymers, metal oxides, and modified carbon-based nanomaterials (Flynn et al., 2023). Surface functionalization of sensor electrodes enhances selectivity by encouraging selective binding with MPs while minimizing interactions with interfering substances (Flynn et al., 2023; Naresh and Lee, 2021). Various strategies, including coating with specific ligands or functional groups, have been investigated for this purpose. Discriminating between MPs and non-plastic particles based on size and shape is fundamental for achieving selectivity (Zhu et al., 2015). Electrochemical sensors integrated with advanced data analysis techniques have demonstrated promising outcomes in accurately distinguishing MPs from other particulate matter (Caldwell et al., 2022). Recent advancements in harnessing molecularly imprinted polymers and aptamers have made it possible to recognize and selectively bind to different types of MP polymers, thus bolstering the sensor's selectivity for precise detection. The combination of electrochemical sensors with spectroscopic techniques, such as Raman spectroscopy or fluorescence imaging, offers complementary information, facilitating superior selectivity in MPs detection (Lapointe et al., 2020; Terry et al., 2022). Implementing multiparameter analysis, where electrochemical sensors concurrently measure multiple characteristics, augments selectivity by cross-verifying data from various sensor responses.

1.3.3 Rapid Response Time of Electrochemical Sensors for MPs Detection

Electrochemical sensors are ideal for real-time monitoring of MPs pollution due to their quick response times (Martic et al., 2022). Advancements in electrode material selection, micro-/nanostructured electrodes, and signal amplification strategies have significantly improved sensor response times (Dey, 2023; Radi, 2011). Integration with real-time data analysis and automation further enhances the potential of electrochemical sensors for rapid MPs detection in various environmental settings (Motalebizadeh et al., 2023). Continued research and technological advancements will pave the way for the widespread deployment of electrochemical sensors, supporting global efforts to combat MP pollution and preserve the health of our ecosystems. Electrochemical sensors provide rapid response times, enabling real-time or nearly real-time monitoring of MPs in the environment (Baranwal et al., 2022; Raykova et al., 2021). This fast response is particularly valuable for tracking MP pollution trends and evaluating the effectiveness of remediation efforts promptly. Electrode material plays a pivotal role in determining the response time of electrochemical sensors (Martic et al., 2022; Raykova et al., 2021; Simoska and Stevenson, 2019). Researchers have explored advanced materials, such as nanomaterials and conductive polymers, with high surface area and fast electron transfer kinetics, leading to faster response times (Y. Du et al., 2023; Raykova et al., 2021). Micro- and nanostructuring of electrodes facilitate enhanced surface interactions with MPs, enabling rapid analyte capture and reducing diffusion limitations, resulting in improved response times (Raykova et al., 2021). Miniaturization and microfabrication techniques have enabled the development of microelectrodes and microfluidic devices, reducing the diffusion path length and accelerating mass transport for faster detection (Alrifaiy et al., 2012;

Kuo and Chiu, 2011; Salva et al., 2021). Incorporating signal amplification strategies, such as nanoparticle labels or enzyme-based amplification, enhances signal intensity, enabling the detection of lower concentrations of MPs in shorter timeframes (Martic et al., 2022). Advancements in data processing and analysis techniques have facilitated real-time data interpretation, enabling instant feedback on MP concentrations during detection, supporting on-site decision making. Integration of electrochemical sensors into automated monitoring systems and sensor networks enables continuous and remote monitoring, providing rapid response time data on MP pollution dynamics (Christakis et al., 2023; Tovar-Lopez, 2023).

1.3.4 Portability and In-Field Testing, Cost-Effectiveness, and Easy Operation of Electrochemical Sensors

Many electrochemical sensors are compact and portable, making them suitable for on-site or in-field monitoring. This portability allows researchers and environmental practitioners to perform real-time measurements in remote or hard-to-reach locations. Electrochemical sensors are often cost-effective compared to other analytical methods, making them practical for large-scale monitoring projects (Z. Chen et al., 2022; Martic et al., 2022). This cost-effectiveness facilitates the deployment of sensor arrays, enhancing spatial coverage and data density. Electrochemical sensors are user-friendly and relatively simple to operate. They can be used by a wider spectrum of researchers and environmental experts because they require no substantial training (Joyce and Falkenberg, 2022; Radhakrishnan et al., 2022; Shams et al., 2021). Water, sediment, soil, and biota are just a few of the sample types that can be used with electrochemical sensors. This versatility allows for comprehensive monitoring of MPs in different environmental compartments. Electrochemical sensors can be integrated with the Internet of Things (IoT) and automated data collection systems, enabling real-time monitoring networks for MPs. Such networks provide continuous data streams, allowing for a better understanding of MP transport and dispersion patterns (Alhalaili et al., 2022; Ching et al., 2023; Leonard et al., 2022; Martic et al., 2022; Radhakrishnan et al., 2022).

Besides detection, electrochemical sensors can also be applied to MP remediation. Electrochemical methods, such as anodic dissolution of metal electrodes, can induce the production of cations that bond with MPs, causing them to congeal into larger precipitates for easier removal (Z. Chen et al., 2022; Oliveira et al., 2022). Therefore, electrochemical sensors offer numerous advantages for MPs detection, including sensitivity, selectivity, rapid response time, portability, cost-effectiveness, and compatibility with various sample types (Dey, 2023; Dey et al., 2021; Hue et al., 2022; Tavakoli et al., 2022). Electrochemical sensors are set to play a crucial role in tackling the worldwide issue of MP pollution and assisting successful mitigation efforts as research in this area advances.

1.4 ELECTROCHEMICAL REMEDIATION OF MPs

By successfully eliminating these persistent pollutants from different environmental matrices, electrochemical remediation is a viable strategy for reducing the environmental impact of MPs. This method uses electrochemistry to trigger processes that help MPs aggregate and be removed (Martic et al., 2022; Oliveira et al., 2022). A metal electrode generates cations (anodic dissolution) in the presence of an electric field. These cations then attach to MPs and cause them to solidify into bigger precipitates. Then these clumped MPs can be quickly removed via filtration or sedimentation (Kiendrebeogo et al., 2022; Kumar, 2023; Lu et al., 2023; Malinović et al., 2022; Patrício Silva, 2021). Next are some key aspects of electrochemical remediation of MPs.

1.4.1 Mechanistic Insights into Electrochemical Removal of MPs

MPs can be removed electrochemically by employing an electrochemical cell setup with particular electrodes, an electrolyte solution, and the introduction of an electric current or voltage (S. Wang et al., 2022). Electrochemical remediation relies on the formation of cations at the metal electrode surface, which react with MPs present in the surrounding environment. The electrochemical process induces the agglomeration of MPs, transforming them into larger particles that are more manageable for removal (S. Sharma et al., 2021; S. Wang et al., 2022). The core mechanism behind electrochemical removal of MPs is electrochemical agglomeration (Girish et al., 2023). By applying an electric field to metal electrodes immersed in the contaminated sample, cations (anodic dissolution) are generated (Kiendrebeogo et al., 2021; Wala and Simka, 2021). These cations interact with MPs, leading to their agglomeration into larger particles that can be easily separated and removed. The effectiveness and selectivity of electrochemical agglomeration are strongly influenced by the electrode material selected (Sadia et al., 2022). Different metal electrodes exhibit varying catalytic properties, affecting the kinetics of cation generation and the subsequent interactions with MPs. Understanding the impact of electrode materials is crucial for optimizing the removal process (Tang et al., 2022; Zhang et al., 2021). Electrochemical remediation offers efficient agglomeration of MPs, leading to increased particle size and improved separation from the environmental matrix (V. K. Sharma et al., 2021). The technique is particularly useful for dealing with MPs that are challenging to remove using conventional cleanup methods. Electrochemical remediation can be performed in-situ, directly at the polluted site, which reduces the need for extensive sample transportation and processing (Lu et al., 2023; Zhang et al., 2021). This capability makes it a practical option for remediating MPs in hard-to-reach or remote areas.

Metal electrodes, often made of materials like aluminum or iron, are used. When an electric current is applied, these electrodes dissolve into metal ions (Alkhadra et al., 2022; Wala and Simka, 2021). These metal ions play a critical role in destabilizing the MPs' surface charges. The metal ions generated during electrocoagulation reduce the electrostatic repulsion between MP particles (Kiendrebeogo et al., 2021). MP particles typically carry a negative charge due to their surface functional groups, and the metal ions neutralize this charge. Reduced electrostatic repulsion allows MP particles to come into closer proximity. As a result, MP particles start to aggregate or agglomerate due to attractive forces such as van der Waals forces and hydrophobic interactions (Agboola and Benson, 2021; Wang et al., 2020). Agglomerates of MPs may further join together through flocculation (Agboola and Benson, 2021; Hu and

Xu, 2020). Flocculation is a process in which these agglomerates form larger and heavier flocs, making it easier for them to settle out of the water. Overall, the electrochemical process reduces the mobility of MPs in the water (Jiang et al., 2021; Kiran et al., 2022). Larger agglomerates and flocs are less likely to remain suspended in the water and are more easily separated. After the electrochemical treatment, the settled flocs, which now contain the MPs, can be separated from the water (Dey et al., 2021; Mao et al., 2023). Separation methods such as sedimentation or filtration can be employed for this purpose. In some cases, particularly with conductive or electroactive MPs, electro-oxidation may occur. Electro-oxidation involves the direct oxidation of MPs at the electrode surface, breaking them down into smaller, less harmful compounds (Agboola and Benson, 2021; Kiendrebeogo et al., 2021). It is crucial to remember that the kind and size of MPs, the water chemistry, the materials used for the electrodes, and the particular circumstances of the electrochemical setup can all affect how effectively MPs are removed electrochemically (Ammar et al., 2023; Dey et al., 2021). Optimizing these parameters is crucial for achieving effective MP removal. Researchers continue to explore and refine the mechanisms involved in electrochemical MP removal to improve its efficiency and applicability in addressing MP pollution in aquatic environments (Hu and Xu, 2020; Lu et al., 2023).

1.4.2 Efficiency and Reusability for Electrochemical Remediation of MPs

Efficiency and reusability are vital considerations in tackling the urgent issue of MPs pollution through electrochemical remediation. Electrochemical methods hold promise for removing MPs from water systems, with efficiency reflecting how effectively they achieve this (Alkhadra et al., 2022; Zhu et al., 2015). Key factors contributing to efficiency include electrode material choice, voltage, current density, and electrolyte composition. Reusability is equally crucial, entailing the ability to regenerate and reuse electrodes without performance degradation. Maintaining electrode lifespan and managing removed MPs responsibly are integral. Cost-effective solutions are essential for reusability (H. Du et al., 2023; Lu et al., 2023). Ongoing research and development aim to enhance efficiency and reusability, prioritizing economically viable and environmentally sustainable MPs remediation in water systems (Dey et al., 2021; Pandey et al., 2023). Electrochemical remediation is selective for MPs, minimizing interference with non-plastic materials present in the environment. The technology is more cost-effective and sustainable because the metal electrodes used in the procedure may be recycled. Electrochemical remediation has the potential for large-scale applications due to its scalability and adaptability to different environmental settings (Alkhadra et al., 2022; Boinpally et al., 2023). It can be employed in bodies of water, soil, sediments, and other contaminated areas. Selectivity and reusability are key advantages of electrochemical remediation methods when it comes to addressing MP pollution (Miranda Zoppas et al., 2023; Osman et al., 2023). Electrochemical remediation techniques, such as electrocoagulation, leverage the surface charge properties of MPs. MPs typically carry negative surface charges due to functional groups on their surfaces. Metal electrodes in the electrochemical cell generate metal ions (e.g., aluminum or iron ions) that neutralize these negative charges, specifically targeting MPs (Boinpally et al., 2023; Rai et al., 2021). This charge-based selectivity allows the method to differentiate between MPs and other non-plastic materials in the environment. Electrochemical processes can also exhibit size-based selectivity. Larger MP particles tend to settle more quickly as they agglomerate and form flocs (Alkhadra et al., 2022). Smaller MPs may require longer treatment times for effective removal. This size-based selectivity can be advantageous in addressing different MP-sized fractions. The selectivity of electrochemical remediation can be further tuned by adjusting operating parameters, such as the type of electrode material, current density, and treatment time (Hu and Xu, 2020). This enables customization to target particular MP sizes and types. The capacity to regenerate and reuse the metal electrodes is one of the key benefits of electrochemical remediation (Alkhadra et al., 2022; Boinpally et al., 2023). MP removal is made easier by the dissolution of metal electrodes into metal ions during the electrochemical process. The electrodes can be cleaned and regenerated for usage later after the treatment, and the metal ions can be collected. This reusability reduces the operational costs associated with electrode replacement. The reusability of electrodes not only reduces costs but also contributes to the sustainability of the remediation method by minimizing waste generation and resource consumption (Ahmad et al., 2023; Boinpally et al., 2023). It aligns with the principles of environmentally friendly and resource-efficient technologies. The initial investment in electrode materials and equipment can be significant, but the long-term cost savings through electrode regeneration and reuse can make electrochemical remediation economically attractive, especially for continuous or large-scale applications (Lu et al., 2023; Pandey et al., 2023). Thus the selectivity of electrochemical remediation methods, based on surface charge and size, allows for the targeted removal of MPs while minimizing interference with non-plastic materials (Boinpally et al., 2023; Junliang Chen et al., 2022). The reusability of metal electrodes enhances the sustainability and cost-effectiveness of the method, making it a practical and environmentally friendly approach to addressing MP pollution in various environmental settings.

1.5 THE FUTURE: ENVIRONMENTAL IMPLICATIONS AND CHALLENGES

The future of electrochemical MPs sensors and their remediation techniques holds great promise while also presenting notable challenges within the context of environmental impact. The development of advanced electrochemical MPs sensors is poised to revolutionize the precision and reach of MP contamination monitoring in natural ecosystems, encompassing oceans, rivers, and lakes. This heightened monitoring capability will yield invaluable insights for gauging the scope and origins of MP pollution. As the environmental ramifications of MPs become increasingly evident, there is a likelihood of heightened regulatory and policy backing for innovative technologies like electrochemical sensors, aimed at assessing and mitigating MP pollution. This support has the potential to catalyze intensified research endeavors and commercialization initiatives. Furthermore, mounting public awareness regarding the adverse effects of MP pollution on ecosystems and human well-being will fuel demand for effective monitoring and remediation solutions. Electrochemical sensors can actively contribute to citizen science initiatives, further elevating awareness and community engagement in this crucial issue. However, it is important to acknowledge that developing electrochemical MPs sensors possessing exceptional sensitivity and selectivity across a wide spectrum of MP characteristics and environmental conditions represents a formidable challenge. The reliability and consistency of sensor performance in complex real-world samples are paramount. Simultaneously, the creation of compact, portable, and field-deployable electrochemical sensors suitable for in-situ monitoring remains an ongoing challenge. These sensors must exhibit robustness to withstand diverse environmental conditions while maintaining precision. Standardization of methodologies for MP detection and quantification via electrochemical sensors is imperative to ensure data comparability across various studies and regions. Uniform protocols can facilitate data interpretation and underpin policy development. Environmental samples are often rife with diverse interfering substances, including organic matter and inorganic particles. Developing sensors capable of discerning MPs from these interferences is indispensable to ensure accuracy in assessments.

While electrochemical methods exhibit potential for MP removal, optimizing their efficiency and scalability for large-scale remediation continues to be a complex task. Balancing treatment effectiveness with considerations of energy consumption and cost-effectiveness remains pivotal. Understanding the fate and environmental impact of residual MPs generated during the remediation process is a critical aspect. This entails assessing whether any breakdown products pose novel environmental concerns. Addressing the multifaceted issue of MP pollution necessitates seamless collaboration among scientists, engineers, policymakers, and environmental organizations. Bridging the gap between research findings and the implementation of effective policies is essential. Moreover, ensuring that electrochemical sensor technologies and remediation methods are economically viable for widespread adoption presents its set of challenges. Cost-effective solutions are essential to incentivize their utilization. Finally, the sustainability of the supply chain for materials integral to electrochemical sensors, including electrode materials, is paramount for their long-term viability. The electrochemical MPs sensors and their remediation techniques have the potential to play a pivotal role in mitigating MP pollution. Nonetheless, surmounting technical obstacles, achieving standardization, and harmonizing environmental considerations with practicality will be pivotal for their successful integration into environmental monitoring and mitigation initiatives. Collaborative efforts involving researchers, policymakers, and industry stakeholders will be instrumental in shaping the trajectory of these technologies.

1.6 CONCLUSIONS

As a result, electrochemical sensors have become a potent and cutting-edge tool for the accurate and successful detection of MPs in environmental samples. These sensors offer high sensitivity, enabling the detection of even trace amounts of MPs in complex matrices. The sensitivity of electrochemical sensors has substantially increased thanks to developments in electrode materials, surface modifications, signal amplification techniques, and miniaturization. While electrochemical sensors offer remarkable sensitivity, challenges related to selectivity, cross-reactivity, and background noise remain to be addressed. Collaborative efforts among researchers, engineers, and environmental practitioners are essential to overcome these challenges and develop sensitive, reliable, and practical electrochemical sensors for MPs detection. Sensitive electrochemical sensors will be essential in determining the level of MPs pollution and directing efficient mitigation efforts to protect our ecosystems and people's health as this technology continues to advance. These sensors, whether using voltammetry, amperometry, impedance spectroscopy, potentiostatic methods, FET sensors, quartz crystal microbalance, biosensors, nanomaterials, SERS, microfluidics, or aptamers, offer diverse approaches to address the global issue of MP pollution. Their high sensitivity, low detection limits, and potential for real-time monitoring make them valuable assets in efforts to combat this environmental challenge.

Electrochemical remediation offers a promising solution to combat MP pollution. It utilizes electrochemistry to efficiently remove MPs from various environments. This method's mechanistic insights, selectivity, reusability, and adaptability make it a valuable tool in addressing this pervasive issue. Electrochemical remediation's ability to agglomerate MPs, its selective nature, and the potential for electrode regeneration contribute to its sustainability and cost-effectiveness. As research continues, this innovative approach may become a cornerstone in the fight against MP contamination, providing an efficient and environmentally friendly means of safeguarding our ecosystems.

ACKNOWLEDGMENTS

I would like to express my gratitude to the distinguished faculty members of The University of Burdwan's Department of Environmental Science for their invaluable support of this endeavor.

CREDIT AUTHORSHIP CONTRIBUTION STATEMENT

Dr. Kamalesh Sen: Conceptualization, methodology, writing original draft. Dr. Naba Kumar Mondal: Conceptualization, review methodology, and writing original draft, Supervision, review, and revise.

REFERENCES

Abbasi, S., Turner, A., 2022. Sources, concentrations, distributions, fluxes and fate of microplastics in a hypersaline lake: Maharloo, south-west Iran. *Science of The Total Environment* 823, 153721. https://doi.org/10.1016/j.scitotenv.2022.153721

Adeniji, T.M., Stine, K.J., 2023. Nanostructure modified electrodes for electrochemical detection of contaminants of emerging concern. *Coatings* 13, 381. https://doi.org/10.3390/coatings13020381

Agbekpornu, P., Kevudo, I., Agbekpornu, P., Kevudo, I., 2023. *The Risks of Microplastic Pollution in the Aquatic Ecosystem.* IntechOpen. https://doi.org/10.5772/intechopen.108717

Agboola, O.D., Benson, N.U., 2021. Physisorption and chemisorption mechanisms influencing micro (nano) plastics-organic chemical contaminants interactions: A review. *Frontiers in Environmental Science* 9.

Ahmad, A., Priyadarshini, M., Yadav, S., Ghangrekar, M.M., Surampalli, R.Y., 2023. Evaluation of waste medicine wrappers as an efficacious low-cost novel electrode material in electrocoagulation for the remediation of Coomassie brilliant blue from wastewater. *Journal of Environmental Chemical Engineering* 11, 110484. https://doi.org/10.1016/j.jece.2023.110484

Ait-Touchente, Z., Falah, S., Scavetta, E., Chehimi, M.M., Touzani, R., Tonelli, D., Taleb, A., 2020. Different electrochemical sensor designs based on diazonium salts and gold nanoparticles for Pico molar detection of metals. *Molecules* 25, 3903. https://doi.org/10.3390/molecules25173903

Alhalaili, B., Popescu, I.N., Rusanescu, C.O., Vidu, R., 2022. Microfluidic devices and microfluidics-integrated electrochemical and optical (bio)sensors for pollution analysis: A review. *Sustainability* 14, 12844. https://doi.org/10.3390/su141912844

Alkhadra, M.A., Su, X., Suss, M.E., Tian, H., Guyes, E.N., Shocron, A.N., Conforti, K.M., de Souza, J.P., Kim, N., Tedesco, M., Khoiruddin, K., Wenten, I.G., Santiago, J.G., Hatton, T.A., Bazant, M.Z., 2022. Electrochemical methods for water purification, ion separations, and energy conversion. *Chemical Reviews* 122, 13547–13635. https://doi.org/10.1021/acs.chemrev.1c00396

Alrifaiy, A., Lindahl, O.A., Ramser, K., 2012. Polymer-based microfluidic devices for pharmacy, biology and tissue engineering. *Polymers* 4, 1349–1398. https://doi.org/10.3390/polym4031349

Altug, H., Oh, S.-H., Maier, S.A., Homola, J., 2022. Advances and applications of nanophotonic biosensors. *Nature Nanotechnology* 17, 5–16. https://doi.org/10.1038/s41565-021-01045-5

Ammar, M., Yousef, E., Mahmoud, M.A., Ashraf, S., Baltrusaitis, J., 2023. A comprehensive review of the developments in electrocoagulation for the removal of contaminants from wastewater. *Separations* 10, 337. https://doi.org/10.3390/separations10060337

Avolio, R., Grozdanov, A., Avella, M., Barton, J., Cocca, M., De Falco, F., Dimitrov, A.T., Errico, M.E., Fanjul-Bolado, P., Gentile, G., Paunovic, P., Ribotti, A., Magni, P., 2022. Review of pH sensing materials from macro- to nano-scale: Recent developments and examples of seawater applications. *Critical Reviews in Environmental Science and Technology* 52, 979–1021. https://doi.org/10.1080/10643389.2020.1843312

Baranwal, J., Barsc, B., Gatto, G., Broncova, G., Kumar, A., 2022. Electrochemical sensors and their applications: A review. *Chemosensors* 10, 363. https://doi.org/10.3390/chemosensors10090363

Bayo, J., Olmos, S., López-Castellanos, J., 2020. Microplastics in an urban wastewater treatment plant: The influence of physicochemical parameters and environmental factors. *Chemosphere* 238, 124593. https://doi.org/10.1016/j.chemosphere.2019.124593

Beduk, T., Gomes, M., De Oliveira Filho, J.I., Shetty, S.S., Khushaim, W., Garcia-Ramirez, R., Durmus, C., Ait Lahcen, A., Salama, K.N., 2022. A portable molecularly imprinted sensor for on-site and wireless environmental bisphenol a monitoring. *Frontiers in Chemistry* 10.

Bifano, L., Meiler, V., Peter, R., Fischerauer, G., 2023. Detection of microplastics in water using electrical impedance spectroscopy and support vector machines. *tm—Technisches Messen* 90, 374–387. https://doi.org/10.1515/teme-2022-0095

Boinpally, S., Kolla, A., Kainthola, J., Kodali, R., Vemuri, J., 2023. A state-of-the-art review of the electrocoagulation technology for wastewater treatment. *Water Cycle* 4, 26–36. https://doi.org/10.1016/j.watcyc.2023.01.001

Caldwell, J., Taladriz-Blanco, P., Lehner, R., Lubskyy, A., Ortuso, R.D., Rothen-Rutishauser, B., Petri-Fink, A., 2022. The micro-, submicron-, and nanoplastic hunt: A review of detection methods for plastic particles. *Chemosphere* 293, 133514. https://doi.org/10.1016/j.chemosphere.2022.133514

Campaña, A.L., Florez, S.L., Noguera, M.J., Fuentes, O.P., Ruiz Puentes, P., Cruz, J.C., Osma, J.F., 2019. Enzyme-based electrochemical biosensors for microfluidic platforms to detect pharmaceutical residues in wastewater. *Biosensors (Basel)* 9, 41. https://doi.org/10.3390/bios9010041

Cao, H., Pavitt, A.S., Hudson, J.M., Tratnyek, P.G., Xu, W., 2023. Electron exchange capacity of pyrogenic dissolved organic matter (pyDOM): Complementarity of square-wave voltammetry in DMSO and mediated chronoamperometry in water. *Environmental Science: Processes & Impacts* 25, 767–780. https://doi.org/10.1039/D3EM00009E

Chauhan, N., Saxena, K., Tikadar, M., Jain, U., 2021. Recent advances in the design of biosensors based on novel nanomaterials: An insight. *Nanotechnology and Precision Engineering (NPE)* 4, 045003. https://doi.org/10.1063/10.0006524

Chen, J., Pu, H., Hersam, M.C., Westerhoff, P., 2022. Molecular engineering of 2D nanomaterial field-effect transistor sensors: Fundamentals and translation across the innovation spectrum. *Advanced Materials* 34, 2106975. https://doi.org/10.1002/adma.202106975

Chen, J., Wu, J., Sherrell, P.C., Chen, J., Wang, H., Zhang, W., Yang, J., 2022. How to build a microplastics-free environment: Strategies for microplastics degradation and plastics recycling. *Advanced Science (Weinh)* 9, 2103764. https://doi.org/10.1002/advs.202103764

Chen, L., Wang, D., Sun, T., Fan, T., Wu, S., Fang, G., Yang, M., Zhou, D., 2022. Quantification of the redox properties of microplastics and their effect on arsenite oxidation. *Fundamental Research*. https://doi.org/10.1016/j.fmre.2022.03.015

Chen, Z., Wei, W., Liu, X., Ni, B.-J., 2022. Emerging electrochemical techniques for identifying and removing micro/nanoplastics in urban waters. *Water Research* 221, 118846. https://doi.org/10.1016/j.watres.2022.118846

Ching, C.T.S., Lee, P.-Y., Hieu, N.V., Chou, H.-H., Yao, F.Y.-D., Cheng, S.-Y., Lin, Y.-K., Phan, T.L., 2023. Real-time, economical identification of microplastics using impedance-based interdigital array microelectrodes and k-nearest neighbor model. *Biotechnology and Bioprocess Engineering* 28, 459–466. https://doi.org/10.1007/s12257-022-0262-y

Choi, H.K., Yoon, J., 2023. Enzymatic electrochemical/fluorescent nanobiosensor for detection of small chemicals. *Biosensors* 13, 492. https://doi.org/10.3390/bios13040492

Choi, J., Yoo, H., 2023. Combination of polymer gate dielectric and two-dimensional semiconductor for emerging field-effect transistors. *Polymers* 15, 1395. https://doi.org/10.3390/polym15061395

Choran, N., Örmeci, B., 2023. Micro-flow imaging for in-situ and real-time enumeration and identification of microplastics in water. *Frontiers in Water* 5.

Christakis, I., Tsakiridis, O., Kandris, D., Stavrakas, I., 2023. Air pollution monitoring via wireless sensor networks: The investigation and correction of the aging behavior of electrochemical gaseous pollutant sensors. *Electronics* 12, 1842. https://doi.org/10.3390/electronics12081842

Chu, N., Cai, J., Li, Z., Gao, Y., Liang, Q., Hao, W., Liu, P., Jiang, Y., Zeng, R.J., 2022. Indicators of water biotoxicity obtained from turn-off microbial electrochemical sensors. *Chemosphere* 286, 131725. https://doi.org/10.1016/j.chemosphere.2021.131725

Colson, B.C., Michel, A.P.M., 2021. Flow-through quantification of microplastics using impedance spectroscopy. *ACS Sensors* 6, 238–244. https://doi.org/10.1021/acssensors.0c02223

Curulli, A., 2021. Electrochemical biosensors in food safety: Challenges and perspectives. *Molecules* 26, 2940. https://doi.org/10.3390/molecules26102940

Damiati, S., Schuster, B., 2020. Electrochemical biosensors based on S-layer proteins. *Sensors* 20, 1721. https://doi.org/10.3390/s20061721

Das, R., Sharma, M.K., Rao, V.K., Bhattacharya, B.K., Garg, I., Venkatesh, V., Upadhyay, S., 2014. An electrochemical genosensor for Salmonella typhi on gold nanoparticles-mercaptosilane modified screen printed electrode. *Journal of Biotechnology* 188, 9–16. https://doi.org/10.1016/j.jbiotec.2014.08.002

Dey, T.K., 2023. Microplastic pollutant detection by surface enhanced Raman spectroscopy (SERS): A mini-review. *Nanotechnology for Environmental Engineering* 8, 41–48. https://doi.org/10.1007/s41204-022-00223-7

Dey, T.K., Uddin, M.E., Jamal, M., 2021. Detection and removal of microplastics in wastewater: Evolution and impact. *Environmental Science and Pollution Research* 28, 16925–16947. https://doi.org/10.1007/s11356-021-12943-5

Du, H., Chen, G., Wang, J., 2023. Highly selective electrochemical impedance spectroscopy-based graphene electrode for rapid detection of microplastics. *Science of the Total Environment* 862, 160873. https://doi.org/10.1016/j.scitotenv.2022.160873

Du, S., Zhu, R., Cai, Y., Xu, N., Yap, P.-S., Zhang, Yunhai, He, Y., Zhang, Y., 2021. Environmental fate and impacts of microplastics in aquatic ecosystems: A review. *RSC Advances* 11, 15762–15784. https://doi.org/10.1039/D1RA00880C

Du, Y., Liu, X., Dong, X., Yin, Z., 2022. A review on marine plastisphere: Biodiversity, formation, and role in degradation. *Computational and Structural Biotechnology Journal* 20, 975–988. https://doi.org/10.1016/j.csbj.2022.02.008

Du, Y., Yu, D.-G., Yi, T., 2023. Electrospun nanofibers as chemosensors for detecting environmental pollutants: A review. *Chemosensors* 11, 208. https://doi.org/10.3390/chemosensors11040208

Dube, S., Satish, S., Rawtani, D., 2023. Aptasensors in environmental forensics: Tracking the silent killers. *WIREs Forensic Science* 5, e1482. https://doi.org/10.1002/wfs2.1482

Duis, K., Coors, A., 2016. Microplastics in the aquatic and terrestrial environment: Sources (with a specific focus on personal care products), fate and effects. *Environmental Sciences Europe* 28, 2. https://doi.org/10.1186/s12302-015-0069-y

Durán, F.E., de Araújo, D.M., do Nascimento Brito, C., Santos, E.V., Ganiyu, S.O., Martínez-Huitle, C.A., 2018. Electrochemical technology for the treatment of real washing machine effluent at pre-pilot plant scale by using active and non-active anodes. *Journal of Electroanalytical Chemistry* 818, 216–222. https://doi.org/10.1016/j.jelechem.2018.04.029

Ece, E., Hacıosmanoğlu, N., Inci, F., 2023. Microfluidics as a Ray of hope for microplastic pollution. *Biosensors* 13, 332. https://doi.org/10.3390/bios13030332

Elli, G., Ciocca, M., Lugli, P., Petti, L., 2021. Field-effect-transistor based biosensors: A review of their use in environmental monitoring applications, in: *2021 IEEE International Workshop on Metrology for Agriculture and Forestry (MetroAgriFor). Presented at the 2021 IEEE International Workshop on Metrology for Agriculture and Forestry (MetroAgriFor)*, pp. 102–107. https://doi.org/10.1109/MetroAgriFor52389.2021.9628685

Elli, G., Hamed, S., Petrelli, M., Ibba, P., Ciocca, M., Lugli, P., Petti, L., 2022. Field-effect transistor-based biosensors for environmental and agricultural monitoring. *Sensors* 22, 4178. https://doi.org/10.3390/s22114178

Esterhuizen, M., Kim, Y.J., 2022. Effects of polypropylene, polyvinyl chloride, polyethylene terephthalate, polyurethane, high-density polyethylene, and polystyrene microplastic on Nelumbo nucifera (Lotus) in water and sediment. *Environmental Science and Pollution Research* 29, 17580–17590. https://doi.org/10.1007/s11356-021-17033-0

Farré, M., 2020. Remote and in situ devices for the assessment of marine contaminants of emerging concern and plastic debris detection. *Current Opinion in Environmental Science & Health, Environmental Chemistry: Innovative Approaches and Instrumentation in Environmental Chemistry* 18, 79–94. https://doi.org/10.1016/j.coesh.2020.10.002

Flynn, C.D., Chang, D., Mahmud, A., Yousefi, H., Das, J., Riordan, K.T., Sargent, E.H., Kelley, S.O., 2023. Biomolecular sensors for advanced physiological monitoring. *Nature Reviews Bioengineering* 1, 560–575. https://doi.org/10.1038/s44222-023-00067-z

Fredj, Z., Singh, B., Bahri, M., Qin, P., Sawan, M., 2023. Enzymatic electrochemical biosensors for neurotransmitters detection: Recent achievements and trends. *Chemosensors* 11, 388. https://doi.org/10.3390/chemosensors11070388

Fu, W., Min, J., Jiang, W., Li, Y., Zhang, W., 2020. Separation, characterization and identification of microplastics and nanoplastics in the environment. *Science of the Total Environment* 721, 137561. https://doi.org/10.1016/j.scitotenv.2020.137561

Gao, C., Guo, Z., Liu, J.-H., Huang, X.-J., 2012. The new age of carbon nanotubes: An updated review of functionalized carbon nanotubes in electrochemical sensors. *Nanoscale* 4, 1948–1963. https://doi.org/10.1039/C2NR11757F

Girish, N., Parashar, N., Hait, S., 2023. Coagulative removal of microplastics from aqueous matrices: Recent progresses and future perspectives. *Science of the Total Environment* 899, 165723. https://doi.org/10.1016/j.scitotenv.2023.165723

Gongi, W., Touzi, H., Sadly, I., Ben Ouada, H., Tamarin, O., Ben Ouada, H., 2022. A novel impedimetric sensor based on cyanobacterial extracellular polymeric substances for microplastics detection. *Journal of Polymers and the Environment* 30, 4738–4748. https://doi.org/10.1007/s10924-022-02555-6

González-González, R.B., Flores-Contreras, E.A., González-González, E., Torres Castillo, N.E., Parra-Saldívar, R., Iqbal, H.M.N., 2023. Biosensor constructs for the monitoring of persistent emerging pollutants in environmental matrices. *Industrial & Engineering Chemistry Research* 62, 4503–4520. https://doi.org/10.1021/acs.iecr.2c00421

Grieshaber, D., MacKenzie, R., Vörös, J., Reimhult, E., 2008. Electrochemical biosensors—sensor principles and architectures. *Sensors* 8, 1400–1458. https://doi.org/10.3390/s80314000

Guillaud, G., Simon, J., Germain, J.P., 1998. Metallophthalocyanines: Gas sensors, resistors and field effect transistors in memory of Christine Maleysson. *Coordination Chemistry Reviews* 178–180, 1433–1484. https://doi.org/10.1016/S0010-8545(98)00177-5

Gunaalan, K., Fabbri, E., Capolupo, M., 2020. The hidden threat of plastic leachates: A critical review on their impacts on aquatic organisms. *Water Research* 184, 116170. https://doi.org/10.1016/j.watres.2020.116170

Hong, J., Lee, B., Park, C., Kim, Y., 2022. A colorimetric detection of polystyrene nanoplastics with gold nanoparticles in the aqueous phase. *Science of the Total Environment* 850, 158058. https://doi.org/10.1016/j.scitotenv.2022.158058

Hong, J., Su, M., Zhao, K., Zhou, Y., Wang, J., Zhou, S.-F., Lin, X., 2023. A minireview for recent development of nanomaterial-based detection of antibiotics. *Biosensors* 13, 327. https://doi.org/10.3390/bios13030327

Hu, E., Sun, C., Yang, F., Wang, Y., Hu, L., Wang, L., Li, M., Gao, L., 2022. Microplastics in 48 wastewater treatment plants reveal regional differences in physical characteristics and shape-dependent removal in the transition zone between North and South China. *Science of the Total Environment* 834, 155320. https://doi.org/10.1016/j.scitotenv.2022.155320

Hu, H., Xu, K., 2020. Chapter 8—Physicochemical technologies for HRPs and risk control, in: Ren, H., Zhang, X. (Eds.), *High-Risk Pollutants in Wastewater*. Elsevier, pp. 169–207. https://doi.org/10.1016/B978-0-12-816448-8.00008-3

Huang, C.-J., Narasimha, G.V., Chen, Y.-C., Chen, J.-K., Dong, G.-C., 2021. Measurement of low concentration of micro-plastics by detection of bioaffinity-induced particle retention using surface plasmon resonance biosensors. *Biosensors* 11, 219. https://doi.org/10.3390/bios11070219

Hue, N.T.H., Imran, M., Choi, Y.-H., Akhtar, M.S., Kwak, D.-H., 2022. Microplastics contaminant detection in aquatic environment by hydrophobic cerium oxide nanoparticles. https://doi.org/10.2139/ssrn.4267720

Jaworska, A., Fornasaro, S., Sergo, V., Bonifacio, A., 2016. Potential of surface enhanced raman spectroscopy (SERS) in therapeutic drug monitoring (TDM). A critical review. *Biosensors* 6, 47. https://doi.org/10.3390/bios6030047

Jiang, Y., Zhao, H., Liang, J., Yue, L., Li, T., Luo, Y., Liu, Q., Lu, S., Asiri, A.M., Gong, Z., Sun, X., 2021. Anodic oxidation for the degradation of organic pollutants: Anode materials, operating conditions and mechanisms. A mini review. *Electrochemistry Communications* 123, 106912. https://doi.org/10.1016/j.elecom.2020.106912

Joyce, P.W.S., Falkenberg, L.J., 2022. Microplastics, both non-biodegradable and biodegradable, do not affect the whole organism functioning of a marine mussel. *Science of the Total Environment* 839, 156204. https://doi.org/10.1016/j.scitotenv.2022.156204

Kallenbach, E.M.F., Rødland, E.S., Buenaventura, N.T., Hurley, R., 2022. Microplastics in terrestrial and freshwater environments, in: Bank, M.S. (Ed.), *Microplastic in the Environment: Pattern and Process, Environmental Contamination Remediation and Management*. Springer International Publishing, pp. 87–130. https://doi.org/10.1007/978-3-030-78627-4_4

Kaya, S.I., Cetinkaya, A., Bakirhan, N.K., Ozkan, S.A., 2020. Trends in sensitive electrochemical sensors for endocrine disruptive compounds. *Trends in Environmental Analytical Chemistry* 28, e00106. https://doi.org/10.1016/j.teac.2020.e00106

Khan, A., Ezati, P., Kim, J.-T., Rhim, J.-W., 2023. Biocompatible carbon quantum dots for intelligent sensing in food safety applications: Opportunities and sustainability. *Materials Today Sustainability* 21, 100306. https://doi.org/10.1016/j.mtsust.2022.100306

Kibria, M.G., Masuk, N.I., Safayet, R., Nguyen, H.Q., Mourshed, M., 2023. Plastic waste: Challenges and opportunities to mitigate pollution and effective management. *International Journal of Environmental Research* 17, 20. https://doi.org/10.1007/s41742-023-00507-z

Kiendrebeogo, M., Karimi Estahbanati, M.R., Khosravanipour Mostafazadeh, A., Drogui, P., Tyagi, R.D., 2021. Treatment of microplastics in water by anodic oxidation: A case study for polystyrene. *Environmental Pollution* 269, 116168. https://doi.org/10.1016/j.envpol.2020.116168

Kiendrebeogo, M., Karimi Estahbanati, M.R., Ouarda, Y., Drogui, P., Tyagi, R.D., 2022. Electrochemical degradation of nanoplastics in water: Analysis of the role of reactive oxygen species. *Science of the Total Environment* 808, 151897. https://doi.org/10.1016/j.scitotenv.2021.151897

Kiran, B.R., Kopperi, H., Venkata Mohan, S., 2022. Micro/nano-plastics occurrence, identification, risk analysis and mitigation: Challenges and perspectives. *Reviews in Environmental Science and Biotechnology* 21, 169–203. https://doi.org/10.1007/s11157-021-09609-6

Kumar, R., 2023. Metal oxides-based nano/microstructures for photodegradation of microplastics. *Advanced Sustainable Systems* 7, 2300033. https://doi.org/10.1002/adsu.202300033

Kumar, R., Manna, C., Padha, S., Verma, A., Sharma, P., Dhar, A., Ghosh, A., Bhattacharya, P., 2022. Micro(nano)plastics pollution and human health: How plastics can induce carcinogenesis to humans? *Chemosphere* 298, 134267. https://doi.org/10.1016/j.chemosphere.2022.134267

Kuo, J.S., Chiu, D.T., 2011. Controlling mass transport in microfluidic devices. *Annual Review of Analytical Chemistry (Palo Alto California)* 4, 275–296. https://doi.org/10.1146/annurev-anchem-061010-113926

Lapointe, M., Farner, J.M., Hernandez, L.M., Tufenkji, N., 2020. Understanding and improving microplastic removal during water treatment: Impact of coagulation and flocculation. *Environmental Science & Technology* 54, 8719–8727. https://doi.org/10.1021/acs.est.0c00712

Le, T.-H., Kim, Y., Yoon, H., 2017. Electrical and electrochemical properties of conducting polymers. *Polymers* 9, 150. https://doi.org/10.3390/polym9040150

Lei, J., Ju, H., 2012. Signal amplification using functional nanomaterials for biosensing. *Chemical Society Reviews* 41, 2122–2134. https://doi.org/10.1039/C1CS15274B

Leonard, J., Koydemir, H.C., Koutnik, V.S., Tseng, D., Ozcan, A., Mohanty, S.K., 2022. Smartphone-enabled rapid quantification of microplastics. *Journal of Hazardous Materials Letters* 3, 100052. https://doi.org/10.1016/j.hazl.2022.100052

Leung, M.M.-L., Ho, Y.-W., Maboloc, E.A., Lee, C.-H., Wang, Y., Hu, M., Cheung, S.-G., Fang, J.K.-H., 2021. Determination of microplastics in the edible green-lipped mussel Perna viridis using an automated mapping technique of Raman microspectroscopy. *Journal of Hazardous Materials* 420, 126541. https://doi.org/10.1016/j.jhazmat.2021.126541

Lokar, N., Pečar, B., Možek, M., Vrtačnik, D., 2023. Microfluidic electrochemical glucose biosensor with in situ enzyme immobilization. *Biosensors* 13, 364. https://doi.org/10.3390/bios13030364

Lopes, L.C., Santos, A., Bueno, P.R., 2022. An outlook on electrochemical approaches for molecular diagnostics assays and discussions on the limitations of miniaturized technologies for point-of-care devices. *Sensors and Actuators Reports* 4, 100087. https://doi.org/10.1016/j.snr.2022.100087

Lu, Y., Li, M.-C., Lee, J., Liu, C., Mei, C., 2023. Microplastic remediation technologies in water and wastewater treatment processes: Current status and future perspectives. *Science of the Total Environment* 868, 161618. https://doi.org/10.1016/j.scitotenv.2023.161618

Ma, X., Jiang, Y., Jia, F., Yu, Y., Chen, J., Wang, Z., 2014. An aptamer-based electrochemical biosensor for the detection of Salmonella. *Journal of Microbiological Methods* 98, 94–98. https://doi.org/10.1016/j.mimet.2014.01.003

Malinović, B.N., Markelj, J., Žgajnar Gotvajn, A., Kralj Cigić, I., Prosen, H., 2022. Electrochemical treatment of wastewater to remove contaminants from the production and disposal of plastics: A review. *Environmental Research Letters* 20, 3765–3787. https://doi.org/10.1007/s10311-022-01497-8

Mao, Y., Zhao, Y., Cotterill, S., 2023. Examining current and future applications of electrocoagulation in wastewater treatment. *Water* 15, 1455. https://doi.org/10.3390/w15081455

Martic, S., Tabobondung, M., Gao, S., Lewis, T., 2022. Emerging electrochemical tools for microplastics remediation and sensing. *Frontiers in Sensors* 3.

Mhiret Gela, S., Aragaw, T.A., 2022. Abundance and characterization of microplastics in main urban ditches across the Bahir Dar City, Ethiopia. *Frontiers in Environmental Science* 10.

Miao, F., Liu, Y., Gao, M., Yu, X., Xiao, P., Wang, M., Wang, S., Wang, X., 2020. Degradation of polyvinyl chloride microplastics via an electro-Fenton-like system with a TiO2/graphite cathode. *Journal of Hazardous Materials* 399, 123023. https://doi.org/10.1016/j.jhazmat.2020.123023

Mikac, L., Rigó, I., Himics, L., Tolić, A., Ivanda, M., Veres, M., 2023. Surface-enhanced Raman spectroscopy for the detection of microplastics. *Applied Surface Science* 608, 155239. https://doi.org/10.1016/j.apsusc.2022.155239

Miranda Zoppas, F., Sacco, N., Soffietti, J., Devard, A., Akhter, F., Marchesini, F.A., 2023. Catalytic approaches for the removal of microplastics from water: Recent advances and future opportunities. *Chemical Engineering Journal Advances* 16, 100529. https://doi.org/10.1016/j.ceja.2023.100529

Moazzenzade, T., Huskens, J., Lemay, S.G., 2023. Utilizing the Oxygen reduction reaction in particle impact electrochemistry: A step toward mediator-free digital electrochemical sensors. *ACS Omega* 8, 31265–31270. https://doi.org/10.1021/acsomega.3c03576

Motalebizadeh, A., Fardindoost, S., Jungwirth, J., Tasnim, N., Hoorfar, M., 2023. Microplastic in situ detection based on a portable triboelectric microfluidic sensor. *Analytical Methods*. https://doi.org/10.1039/D3AY01286G

Naresh, V., Lee, N., 2021. A review on biosensors and recent development of nanostructured materials-enabled biosensors. *Sensors* 21, 1109. https://doi.org/10.3390/s21041109

O'Brien, S., Rauert, C., Ribeiro, F., Okoffo, E.D., Burrows, S.D., O'Brien, J.W., Wang, X., Wright, S.L., Thomas, K.V., 2023. There's something in the air: A review of sources, prevalence and behaviour of microplastics in the atmosphere. *Science of the Total Environment* 874, 162193. https://doi.org/10.1016/j.scitotenv.2023.162193

Olaniyan, P.O., Nadim, M.-M., Subir, M., 2023. Detection and binding interactions of pharmaceutical contaminants using quartz crystal microbalance—role of adsorbate structure and surface functional group on adsorption. *Chemosphere* 311, 137075. https://doi.org/10.1016/j.chemosphere.2022.137075

Oliveira, T.M.B.F., Ribeiro, F.W.P., Morais, S., de Lima-Neto, P., Correia, A.N., 2022. Removal and sensing of emerging pollutants released from (micro)plastic degradation: Strategies based on boron-doped diamond electrodes. *Current Opinion in Electrochemistry* 31, 100866. https://doi.org/10.1016/j.coelec.2021.100866

Osman, A.I., Hosny, M., Eltaweil, A.S., Omar, S., Elgarahy, A.M., Farghali, M., Yap, P.-S., Wu, Y.-S., Nagandran, S., Batumalaie, K., Gopinath, S.C.B., John, O.D., Sekar, M., Saikia, T., Karunanithi, P., Hatta, M.H.M., Akinyede, K.A., 2023. Microplastic sources, formation, toxicity and remediation: A review. *Environmental Chemistry Letters* 21, 2129–2169. https://doi.org/10.1007/s10311-023-01593-3

Ott, C.E., Perez-Estebanez, M., Hernandez, S., Kelly, K., Dalzell, K.A., Arcos-Martinez, M.J., Heras, A., Colina, A., Arroyo, L.E., 2022. Forensic identification of fentanyl and its analogs by electrochemical-surface enhanced Raman spectroscopy (EC-SERS) for the screening of seized drugs of abuse. *Frontiers in Analytical Science* 2.

Pandey, P., Dhiman, M., Kansal, A., Subudhi, S.P., 2023. Plastic waste management for sustainable environment: Techniques and approaches. *Waste Disposal and Sustainable Energy* 5, 205–222. https://doi.org/10.1007/s42768-023-00134-6

Pandey, R.R., Chusuei, C.C., 2021. Carbon nanotubes, graphene, and carbon dots as electrochemical biosensing composites. *Molecules* 26, 6674. https://doi.org/10.3390/molecules26216674

Park, J., Kim, M., Kim, W., Jo, S., Kim, W., Kim, C., Park, H., Lee, W., Park, J., 2022. Ultrasensitive detection of 25-hydroxy vitamin D3 in real saliva using sandwich-type electrochemical aptasensor. *Sensors and Actuators B: Chemical* 355, 131239. https://doi.org/10.1016/j.snb.2021.131239

Patrício Silva, A.L., 2021. New frontiers in remediation of (micro)plastics. *Current Opinion in Green and Sustainable Chemistry* 28, 100443. https://doi.org/10.1016/j.cogsc.2020.100443

Pérez, R.L., Ayala, C.E., Park, J.-Y., Choi, J.-W., Warner, I.M., 2021. Coating-based quartz crystal microbalance detection methods of environmentally relevant volatile organic compounds. *Chemosensors* 9, 153. https://doi.org/10.3390/chemosensors9070153

Pfeiffer, F., Mayer, G., 2016. Selection and biosensor application of aptamers for small molecules. *Frontiers in Chemistry* 4, 188250. https://doi.org/10.3389/fchem.2016.00025

Piña, S., Sepúlveda, P., Garcia-Garcia, A., Moreno-Bárcenas, A., Toledo-Neira, C., Salazar-González, R., 2023. Fast simultaneous electrochemical detection of bisphenol-A and bisphenol-S in urban wastewater using a graphene oxide-iron nanoparticles hybrid sensor. *Electrochimica Acta*, 143164. https://doi.org/10.1016/j.electacta.2023.143164

Puthongkham, P., Wirojsaengthong, S., Suea-Ngam, A., 2021. Machine learning and chemometrics for electrochemical sensors: Moving forward to the future of analytical chemistry. *Analyst* 146, 6351–6364. https://doi.org/10.1039/D1AN01148K

Radhakrishnan, S., Mathew, M., Sekhar Rout, C., 2022. Microfluidic sensors based on two-dimensional materials for chemical and biological assessments. *Materials Advances* 3, 1874–1904. https://doi.org/10.1039/D1MA00929J

Radi, A.-E., 2011. Electrochemical aptamer-based biosensors: Recent advances and perspectives. *International Journal of Electrochemistry* 2011, e863196. https://doi.org/10.4061/2011/863196

Radi, A.-E., Abd-Ellatief, M.R., 2021. Electrochemical Aptasensors: Current status and future perspectives. *Diagnostics* 11, 104. https://doi.org/10.3390/diagnostics11010104

Rahman, A., Kang, S., Wang, W., Garg, A., Maile-Moskowitz, A., Vikesland, P.J., 2021. Nanobiotechnology enabled approaches for wastewater based epidemiology. *TrAC Trends in Analytical Chemistry* 143, 116400. https://doi.org/10.1016/j.trac.2021.116400

Rahn, K.L., Anand, R.K., 2021. Recent advancements in bipolar electrochemical methods of analysis. *Analytical Chemistry* 93, 103–123. https://doi.org/10.1021/acs.analchem.0c04524

Rai, P.K., Islam, M., Gupta, A., 2022. Microfluidic devices for the detection of contamination in water samples: A review. *Sensors and Actuators A: Physical* 347, 113926. https://doi.org/10.1016/j.sna.2022.113926

Rai, V., Liu, D., Xia, D., Jayaraman, Y., Gabriel, J.-C.P., 2021. Electrochemical approaches for the recovery of metals from electronic waste: A critical review. *Recycling* 6, 53. https://doi.org/10.3390/recycling6030053

Raoof, J.-B., Ojani, R., Ramine, M., 2006. Electrocatalytic oxidation and voltammetric determination of L-Cysteic acid at the surface of p-Bromanil modified carbon paste electrode. *Electroanalysis* 18, 1722–1726. https://doi.org/10.1002/elan.200603580

Rastogi, R., Beggiato, M., Dogbe Foli, E.A., Vincent, R., Dupont-Gillain, C., Adam, P.-M., Krishnamoorthy, S., 2021. Quantifying analyte surface densities and their distribution with respect to electromagnetic hot spots in plasmon-enhanced spectroscopic biosensors. *The Journal of Physical Chemistry C* 125, 9866–9874. https://doi.org/10.1021/acs.jpcc.1c00793

Raykova, M.R., Corrigan, D.K., Holdsworth, M., Henriquez, F.L., Ward, A.C., 2021. Emerging electrochemical sensors for real-time detection of tetracyclines in milk. *Biosensors* 11, 232. https://doi.org/10.3390/bios11070232

Reddy, A.S., Nair, A.T., 2022. The fate of microplastics in wastewater treatment plants: An overview of source and remediation technologies. *Environmental Technology & Innovation* 28, 102815. https://doi.org/10.1016/j.eti.2022.102815

Ren, J., Yin, X., Hu, H., Wang, S., Tian, Y., Chen, Y., Li, Y., Wang, J., Zhang, D., 2022. A multi-scenario dip-stick immunoassay of 17β-estradiol based on multifunctional and non-composite nanoparticles with colorimetric-nanozyme-magnetic properties. *Sensors and Actuators B: Chemical* 367, 132150. https://doi.org/10.1016/j.snb.2022.132150

Rizzato, S., Leo, A., Monteduro, A.G., Chiriacò, M.S., Primiceri, E., Sirsi, F., Milone, A., Maruccio, G., 2020. Advances in the development of innovative sensor platforms for field analysis. *Micromachines* 11, 491. https://doi.org/10.3390/mi11050491

Sadia, M., Mahmood, A., Ibrahim, M., Irshad, M.K., Quddusi, A.H.A., Bokhari, A., Mubashir, M., Chuah, L.F., Show, P.L., 2022. Microplastics pollution from wastewater treatment plants: A critical review on challenges, detection, sustainable removal techniques and circular economy. *Environmental Technology & Innovation* 28, 102946. https://doi.org/10.1016/j.eti.2022.102946

Sadighbayan, D., Hasanzadeh, M., Ghafar-Zadeh, E., 2020. Biosensing based on field-effect transistors (FET): Recent progress and challenges. *TrAC Trends in Analytical Chemistry* 133, 116067. https://doi.org/10.1016/j.trac.2020.116067

Salva, M.L., Rocca, M., Niemeyer, C.M., Delamarche, E., 2021. Methods for immobilizing receptors in microfluidic devices: A review. *Micro and Nano Engineering* 11, 100085. https://doi.org/10.1016/j.mne.2021.100085

Saravanakumar, K., SivaSantosh, S., Sathiyaseelan, A., Naveen, K.V., Afaan Ahamed, M.A., Zhang, X., Priya, V.V., MubarakAli, D., Wang, M.-H., 2022. Unraveling the hazardous impact of diverse contaminants in the marine environment: Detection and remedial approach through nanomaterials and nano-biosensors. *Journal of Hazardous Materials* 433, 128720. https://doi.org/10.1016/j.jhazmat.2022.128720

Schiavi, S., Parmigiani, M., Galinetto, P., Albini, B., Taglietti, A., Dacarro, G., 2023. Plasmonic nanomaterials for micro- and nanoplastics detection. *Applied Sciences* 13, 9291. https://doi.org/10.3390/app13169291

Shams, M., Alam, I., Chowdhury, I., 2021. Interactions of nanoscale plastics with natural organic matter and silica surfaces using a quartz crystal microbalance. *Water Research* 197, 117066. https://doi.org/10.1016/j.watres.2021.117066

Sharma, R., Verma, N., Lugani, Y., Kumar, S., Asadnia, M., 2021. Chapter 1—Conventional and advanced techniques of wastewater monitoring and treatment, in: Inamuddin, R.B., Asiri, A.M. (Eds.), *Green Sustainable Process for Chemical and Environmental Engineering and Science*. Elsevier, pp. 1–48. https://doi.org/10.1016/B978-0-12-821883-9.00009-6

Sharma, S., Basu, S., Shetti, N.P., Nadagouda, M.N., Aminabhavi, T.M., 2021. Microplastics in the environment: Occurrence, perils, and eradication. *Chemical Engineering Journal* 408, 127317. https://doi.org/10.1016/j.cej.2020.127317

Sharma, S., Shekhar, S., Gautam, S., Sharma, B., Kumar, A., Jain, P., 2020. Chapter 13—Carbon-based nanomaterials as novel nanosensors, in: Pal, K., Gomes, F. (Eds.), *Nanofabrication for Smart Nanosensor Applications, Micro and Nano Technologies*. Elsevier, pp. 323–347. https://doi.org/10.1016/B978-0-12-820702-4.00014-3

Sharma, V.K., Ma, X., Guo, B., Zhang, K., 2021. Environmental factors-mediated behavior of microplastics and nanoplastics in water: A review. *Chemosphere* 271, 129597. https://doi.org/10.1016/j.chemosphere.2021.129597

Shen, L.-L., Zhang, G.-R., Etzold, B.J.M., 2020. Paper-based microfluidics for electrochemical applications. *ChemElectroChem* 7, 10–30. https://doi.org/10.1002/celc.201901495

Shimizu, K., Sokolov, S.V., Kätelhön, E., Holter, J., Young, N.P., Compton, R.G., 2017. In situ detection of microplastics: Single microparticle-electrode impacts. *Electroanalysis* 29, 2200–2207. https://doi.org/10.1002/elan.201700213

Silva, L.R.G., Stefano, J.S., Crapnell, R.D., Banks, C.E., Janegitz, B.C., 2023. Additive manufactured microfluidic device for electrochemical detection of carbendazim in honey samples. *Talanta Open* 7, 100213. https://doi.org/10.1016/j.talo.2023.100213

Simoska, O., Stevenson, K.J., 2019. Electrochemical sensors for rapid diagnosis of pathogens in real time. *Analyst* 144, 6461–6478. https://doi.org/10.1039/C9AN01747J

Singh, S., Naithani, A., Kandari, K., Roy, S., Sain, S., Roy, S.S., Wadhwa, S., Tauseef, S.M., Mathur, A., 2023. Oxygenated graphitic carbon nitride based electrochemical sensor for dibenzofuran detection. *Diamond and Related Materials* 139, 110276. https://doi.org/10.1016/j.diamond.2023.110276

Smith, M., Love, D.C., Rochman, C.M., Neff, R.A., 2018. Microplastics in seafood and the implications for human health. *Current Environmental Health Reports* 5, 375–386. https://doi.org/10.1007/s40572-018-0206-z

Sohrabi, H., Arbabzadeh, O., Khaaki, P., Khataee, A., Majidi, M.R., Orooji, Y., 2022. Patulin and Trichothecene: Characteristics, occurrence, toxic effects and detection capabilities via clinical, analytical and nanostructured electrochemical sensing/biosensing assays in foodstuffs. *Critical Reviews in Food Science and Nutrition* 62, 5540–5568. https://doi.org/10.1080/10408398.2021.1887077

Song, K.-M., Lee, S., Ban, C., 2012. Aptamers and their biological applications. *Sensors* 12, 612–631. https://doi.org/10.3390/s120100612

Stradiotto, N.R., Yamanaka, H., Zanoni, M.V.B., 2003. Electrochemical sensors: A powerful tool in analytical chemistry. *Journal of the Brazilian Chemical Society* 14, 159–173. https://doi.org/10.1590/S0103-50532003000200003

Tajik, S., Beitollahi, H., Nejad, F.G., Dourandish, Z., Khalilzadeh, M.A., Jang, H.W., Venditti, R.A., Varma, R.S., Shokouhimehr, M., 2021. Recent developments in polymer nanocomposite-based electrochemical sensors for detecting environmental pollutants. *Industrial & Engineering Chemistry Research* 60, 1112–1136. https://doi.org/10.1021/acs.iecr.0c04952

Tamburri, M.N., Soon, Z.Y., Scianni, C., Øpstad, C.L., Oxtoby, N.S., Doran, S., Drake, L.A., 2022. Understanding the potential release of microplastics from coatings used on commercial ships. *Frontiers in Marine Science* 9.

Tang, W., Li, H., Fei, L., Wei, B., Zhou, T., Zhang, H., 2022. The removal of microplastics from water by coagulation: A comprehensive review. *Science of the Total Environment* 851, 158224. https://doi.org/10.1016/j.scitotenv.2022.158224

Tang, Y., Hardy, T.J., Yoon, J.-Y., 2023. Receptor-based detection of microplastics and nanoplastics: Current and future. *Biosensors and Bioelectronics* 234, 115361. https://doi.org/10.1016/j.bios.2023.115361

Tarafdar, A., Choi, S.-H., Kwon, J.-H., 2022. Differential staining lowers the false positive detection in a novel volumetric measurement technique of microplastics. *Journal of Hazardous Materials* 432, 128755. https://doi.org/10.1016/j.jhazmat.2022.128755

Tavakoli, H., Mohammadi, S., Li, X., Fu, G., Li, X.J., 2022. Microfluidic platforms integrated with nano-sensors for point-of-care bioanalysis. *TrAC Trends in Analytical Chemistry* 157, 116806. https://doi.org/10.1016/j.trac.2022.116806

Terry, L.R., Sanders, S., Potoff, R.H., Kruel, J.W., Jain, M., Guo, H., 2022. Applications of surface-enhanced Raman spectroscopy in environmental detection. *Analytical Science Advances* 3, 113–145. https://doi.org/10.1002/ansa.202200003

Tong, X., Ashalley, E., Lin, F., Li, H., Wang, Z.M., 2015. Advances in MoS2-based field effect transistors (FETs). *Nano-Micro Letters* 7, 203–218. https://doi.org/10.1007/s40820-015-0034-8

Tovar-Lopez, F.J., 2023. Recent progress in micro- and nanotechnology-enabled sensors for biomedical and environmental challenges. *Sensors* 23, 5406. https://doi.org/10.3390/s23125406

Tunç, A.T., Aynacı Koyuncu, E., Arslan, F., 2016. Development of an acetylcholinesterase-choline oxidase based biosensor for acetylcholine determination. *Artificial Cells, Nanomedicine, and Biotechnology* 44, 1659–1664. https://doi.org/10.3109/21691401.2015.1080167

Vidal, J.C., Midón, J., Vidal, A.B., Ciomaga, D., Laborda, F., 2023. Detection, quantification, and characterization of polystyrene microplastics and adsorbed bisphenol A contaminant using electroanalytical techniques. *Microchim Acta* 190, 203. https://doi.org/10.1007/s00604-023-05780-5

Virkkala, T., Kosourov, S., Rissanen, V., Siitonen, V., Arola, S., Allahverdiyeva, Y., Tammelin, T., 2023. Bioinspired mechanically stable all-polysaccharide based scaffold for photosynthetic production. *Journal of Materials Chemistry B*. https://doi.org/10.1039/D3TB00919J

Wala, M., Simka, W., 2021. Effect of anode material on electrochemical oxidation of low molecular weight alcohols—a review. *Molecules* 26, 2144. https://doi.org/10.3390/molecules26082144

Walkinshaw, C., Lindeque, P.K., Thompson, R., Tolhurst, T., Cole, M., 2020. Microplastics and seafood: Lower trophic organisms at highest risk of contamination. *Ecotoxicology and Environmental Safety* 190, 110066. https://doi.org/10.1016/j.ecoenv.2019.110066

Wang, F., Zhang, M., Sha, W., Wang, Y., Hao, H., Dou, Y., Li, Y., 2020. Sorption behavior and mechanisms of organic contaminants to nano and microplastics. *Molecules* 25, 1827. https://doi.org/10.3390/molecules25081827

Wang, J., Waltmann, C., Harms, C., Hu, S., Hegarty, J., Shindel, B., Wang, Q., Dravid, V., Shull, K.R., Torkelson, J.M., Olvera de la Cruz, M., 2023. Tailoring interactions of random copolymer polyelectrolyte complexes to remove nanoplastic contaminants from water. *Langmuir* 39, 7514–7523. https://doi.org/10.1021/acs.langmuir.3c01028

Wang, S., Xu, M., Jin, B., Wünsch, U.J., Su, Y., Zhang, Y., 2022. Electrochemical and microbiological response of exoelectrogenic biofilm to polyethylene microplastics in water. *Water Research* 211, 118046. https://doi.org/10.1016/j.watres.2022.118046

Wang, X., Lu, D., Liu, Y., Wang, W., Ren, R., Li, M., Liu, D., Liu, Y., Liu, Y., Pang, G., 2022. Electrochemical signal amplification strategies and their use in olfactory and taste evaluation. *Biosensors* 12, 566. https://doi.org/10.3390/bios12080566

Willner, M.R., Vikesland, P.J., 2018. Nanomaterial enabled sensors for environmental contaminants. *Journal of Nanobiotechnology* 16, 95. https://doi.org/10.1186/s12951-018-0419-1

Woo, H., Hyun Kang, S., Kwon, Y., Choi, Y., Kim, J., Ha, D.-H., Tanaka, M., Okochi, M., Su Kim, J., Koo Kim, H., Choi, J., 2022. Sensitive and specific capture of polystyrene and polypropylene microplastics using engineered peptide biosensors. *RSC Advances* 12, 7680–7688. https://doi.org/10.1039/D1RA08701K

Wu, G., Zheng, H., Xing, Y., Wang, C., Yuan, X., Zhu, X., 2022. A sensitive electrochemical sensor for environmental toxicity monitoring based on tungsten disulfide nanosheets/hydroxylated carbon nanotubes nanocomposite. *Chemosphere* 286, 131602. https://doi.org/10.1016/j.chemosphere.2021.131602

Wu, P., Lin, S., Cao, G., Wu, J., Jin, H., Wang, C., Wong, M.H., Yang, Z., Cai, Z., 2022. Absorption, distribution, metabolism, excretion and toxicity of microplastics in the human body and health implications. *Journal of Hazardous Materials* 437, 129361. https://doi.org/10.1016/j.jhazmat.2022.129361

Xiao, L., Zheng, Z., Irgum, K., Andersson, P.L., 2020. Studies of emission processes of polymer additives into water using quartz crystal microbalance—a case study on organophosphate esters. *Environmental Science & Technology* 54, 4876–4885. https://doi.org/10.1021/acs.est.9b07607

Xu, G., Cheng, H., Jones, R., Feng, Y., Gong, K., Li, K., Fang, X., Tahir, M.A., Valev, V.K., Zhang, L., 2020. Surface-enhanced Raman spectroscopy facilitates the detection of microplastics <1 μm in the environment. *Environmental Science & Technology* 54, 15594–15603. https://doi.org/10.1021/acs.est.0c02317

Xu, J., Li, Y., Liu, Y., Wu, X., Huang, K., 2023. Ultra-sensitive self-powered sensor based on the hollow Rubik's cube substrate and multiple signal amplification strategy for real-time heavy metal detection. *Sensors and Actuators B: Chemical* 394, 134377. https://doi.org/10.1016/j.snb.2023.134377

Yadav, N., Garg, V.K., Chhillar, A.K., Rana, J.S., 2021. Detection and remediation of pollutants to maintain ecosustainability employing nanotechnology: A review. *Chemosphere* 280, 130792. https://doi.org/10.1016/j.chemosphere.2021.130792

Yee, M.S.-L., Hii, L.-W., Looi, C.K., Lim, W.-M., Wong, S.-F., Kok, Y.-Y., Tan, B.-K., Wong, C.-Y., Leong, C.-O., 2021. Impact of microplastics and nanoplastics on human health. *Nanomaterials (Basel)* 11, 496. https://doi.org/10.3390/nano11020496

Yin, R., Ge, H., Chen, H., Du, J., Sun, Z., Tan, H., Wang, S., 2021. Sensitive and rapid detection of trace microplastics concentrated through Au-nanoparticle-decorated sponge on the basis of surface-enhanced Raman spectroscopy. *Environmental Advances* 5, 100096. https://doi.org/10.1016/j.envadv.2021.100096

Yuan, Z., Nag, R., Cummins, E., 2022. Human health concerns regarding microplastics in the aquatic environment—from marine to food systems. *Science of the Total Environment* 823, 153730. https://doi.org/10.1016/j.scitotenv.2022.153730

Zhang, J., Li, H., Li, Y., Li, S., Xu, Y., Li, H., 2022. Boron-doped carbon nanoparticles for identification and tracing of microplastics in "turn-on" fluorescence mode. *Chemical Engineering Journal* 435, 135075. https://doi.org/10.1016/j.cej.2022.135075

Zhang, Y., Jiang, H., Bian, K., Wang, H., Wang, C., 2021. A critical review of control and removal strategies for microplastics from aquatic environments. *Journal of Environmental Chemical Engineering* 9, 105463. https://doi.org/10.1016/j.jece.2021.105463

Zhao, K., Wei, Y., Dong, J., Zhao, P., Wang, Y., Pan, X., Wang, J., 2022. Separation and characterization of microplastic and nanoplastic particles in marine environment. *Environmental Pollution* 297, 118773. https://doi.org/10.1016/j.envpol.2021.118773

Zheng, H., Sun, H., Zhang, Z., Qian, Y., Zhu, X., Qu, J., 2023. A sensitive biosensor based on carbon nanohorn/rhodamine B for toxicity detection of polystyrene microplastics and typical pollutants. *Microchemical Journal* 193, 109036. https://doi.org/10.1016/j.microc.2023.109036

Zheng, X., Gao, S., Wu, J., Hu, X., 2020. Recent advances in aptamer-based biosensors for detection of pseudomonas aeruginosa. *Frontiers in Microbiology* 11.

Zhu, C., Yang, G., Li, H., Du, D., Lin, Y., 2015. Electrochemical sensors and biosensors based on nanomaterials and nanostructures. *Analytical Chemistry* 87, 230–249. https://doi.org/10.1021/ac5039863

Zhu, H., Huang, X., Deng, Y., Chen, H., Fan, M., Gong, Z., 2023. Applications of nanomaterial-based chemiluminescence sensors in environmental analysis. *TrAC Trends in Analytical Chemistry* 158, 116879. https://doi.org/10.1016/j.trac.2022.116879

Ziani, K., Ioniță-Mîndrican, C.-B., Mititelu, M., Neacşu, S.M., Negrei, C., Moroşan, E., Drăgănescu, D., Preda, O.-T., 2023. Microplastics: A real global threat for environment and food safety: A state of the art review. *Nutrients* 15, 617. https://doi.org/10.3390/nu15030617

2 Microplastics Degradation Technologies and Remediation Techniques from Aquatic Systems

M. Raju, R. Gandhimathi, and S. Bivin Ebenezer

2.1 INTRODUCTION

Plastic pollution has emerged to be the most significant problem of the 21st century. Plastic waste is a major contributor to municipal solid waste. Annually around 21 million tons of plastic are introduced into municipal waste in the European Union alone; 25% of that is dumped in landfills. Also, million tons of mismanaged plastics are generated and discarded every year. All these plastics get disintegrated due to the presence of sunlight and form smaller sized plastic known as microplastics (Calero et al., 2021). The plastic particles ranging from 1 μm to 5 mm are known to be microplastics (MPs) (Samandra et al., 2022). MPs plastics have similar characteristics to plastics; they are non-degradable, synthetic in composition, and insoluble in water. They are classified into two types based on the source: primary MPs and secondary MPs. Primary MPs are purposefully added to items such as personal care items and cleaning products like detergents (Raju et al., 2023). Secondary MPs are those disintegrated from the larger plastics due to physical and chemical factors (Figure 2.1). Approximately 2,900 to 7,400 metric tons of primary MPs are released into the European Union's (EU's) environment each year, with a significant portion originating from personal care products. In addition, an estimated 1.5–2 million metric tons of secondary microplastics are deposited within the EU, with tire wear being the major source of this environmental pollution (Calero et al., 2021). Based on the forms, the MPs are classified as fibers, fragments, and beads, and, based on the polymer type, they are classified into polyethylene (PE), polypropylene (PE), polystyrene (PS), polyethylene terephthalate (PET), polyvinyl chloride (PVC), polyamide (PA), etc. (Priya et al., 2022).

The high quantity of plastic waste significantly affects the ecosystem and human health. The MPs penetrate throughout the human system, and they induce impairment in human and living species ranging from minimal biological disturbance to severe unfavorable consequences leading to mortality (Bhuyan, 2022). It also causes oxidative stress, inflammation, and toxicity, which may lead to increased translocation of

DOI: 10.1201/9781032684574-2

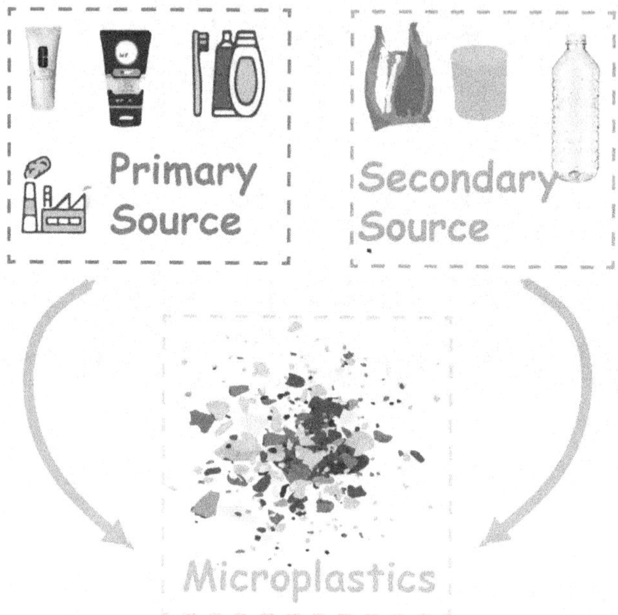

FIGURE 2.1 Source of microplastics.

particles and neoplasia. MPs are also capable of disrupting the immune system of human beings (Prata, da Costa, et al., 2020).

MPs also serve as a vector in transporting organic and other pollutants to the aquatic ecosystem. MPs gets transported to the aquatic medium through various sources such as rivers and wastewater treatment plants (WWTP). In the aquatic environment, MPs are ingested by lower trophic-level organisms and bioaccumulated to higher trophic-level organisms, thereby quickly entering the food chain (Priya et al., 2022). Hence, it becomes essential to study the fate of MPs in the aquatic environment. There are limited research articles discussing the degradation of MPs using various technologies. This chapter discusses the various degradation technologies used to degrade MPs, their merits and shortcomings, and the control measures that can be adopted to overcome MPs pollution.

2.2 MPs DEGRADATION MECHANISMS

Plastic degradation occurs mainly by biotic and abiotic processes. Biotic degradation occurs by the living organisms present in the environment, and abiotic degradation is due to physical conditions such as temperature, wind, light, etc. (Gewert et al., 2015). In abiotic degradation, the chemical reaction occurs on the surface of the plastic material. The MPs that are smaller in size and have high surface-to-volume ratios tend to degrade faster than the larger plastic particles (Tofa et al., 2019a). During the initial stage of the degradation, the surface of the plastic gets bent, and color changes

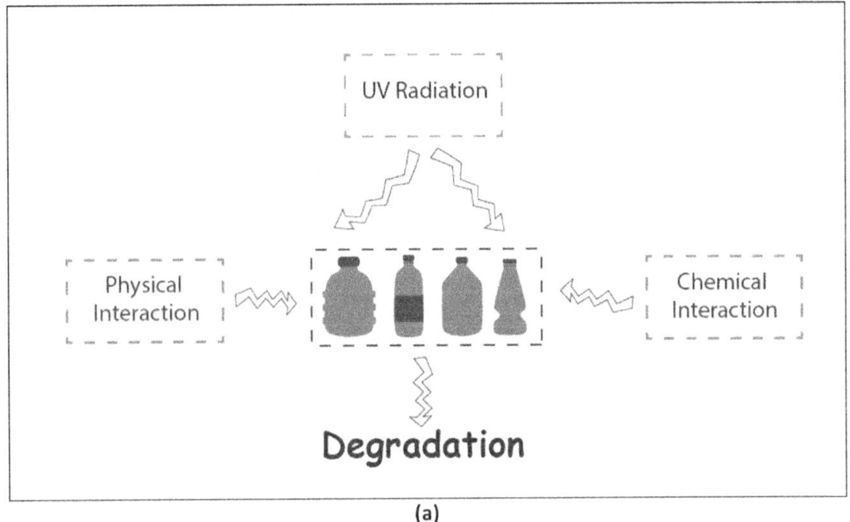

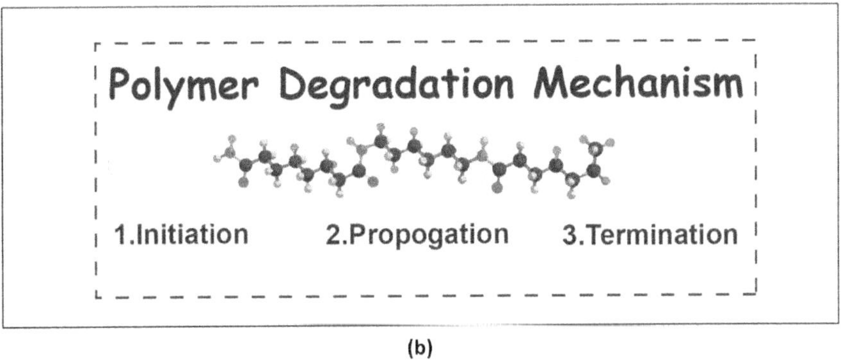

FIGURE 2.2 (a) UV, physical, and chemical interaction with plastics. (b) Three steps involved in polymer degradation.

has been observed, which will further lead to cracking on the surface of plastic. The sunlight of wavelength 300 nm or more will accelerate the rate of degradation of plastics (Woo et al., 2008). The physical and chemical interaction of plastic is also responsible for plastic degradation (Figure 2.2[a]). In nature, the degradation of polymer has three steps: initiation, propagation, and termination (Figure 2.2 [b]).

- In the initiation step, the main polymer is broken down due to the presence of light. When the light falls on the surface of the plastics, the polymer chain absorbs the light and photoinitiation happens, which promotes the chain breakage reaction, leading to the formation of oligomers and initial radicles (Feldman, 2002). Mostly, polyethylene and polypropylene do not have a chromophoric group and are most likely unaffected by photoinitiation.

However, structural abnormalities and the presence of small impurities on the surface of the plastic will initiate the photo reaction to lesser extent (Yousif et al., 2023). Sunlight provides the light energy needed for the photoinitiation.

- In the propagation step, radicals react with the oxygen and form peroxy radicals, leading to a complex reaction involving radicals.
- Termination occurs when radicals react with one another and form inert products.

2.3 TECHNOLOGIES FOR DEGRADING MPs

Cutting-edge technology, such as like AOPs (advanced oxidation process), MBR (membrane bioreactor), biological degradation, supercritical water degradation (Figure 2.3), is discussed in this chapter. These technologies can be used to degrade MPs in the aquatic environment.

2.3.1 Advanced Oxidation Processes

Advanced oxidation processes are widely adopted in wastewater treatment plants. Hydroxyl radicals $(HO\bullet)$ and sulfate radical $(HO\bullet)$ are the two major free radicals responsible for the degradation of organic pollutants in AOPs. $HO\bullet$ is a highly reactive, short-lived, and non-selective oxidant with an oxidation potential of $E^0 = 2.730$ V. They attack organic pollutants by electron transfer mechanisms or by abstraction of hydrogen atoms and hydroxylation (Bule Možar et al., 2023). The $HO\bullet$ are produced mainly using hydrogen peroxide, ozonation, and Fenton's reagent and photocatalysis. The sulfate radicals $(SO_4\bullet\text{-})$ are produced by activating persulfate or

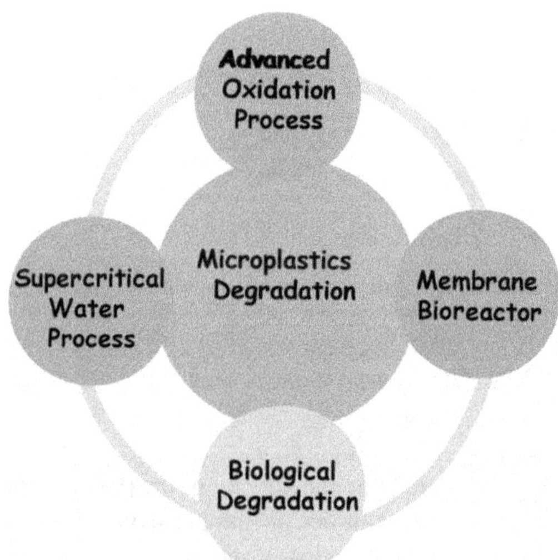

FIGURE 2.3 Various technologies for MPs degradation.

peroxymonosulfate. The lifetime of SO_4 •- is high when compared to HO•. The physical activation of persulfate produces HO• along with SO_4 •-, but its rate is significantly less (Bule Možar et al., 2023).

MPs are made up of long-chain polymers that are very strong and difficult to breakdown through chemical and biological methods (Shen, Song, et al., 2022). During the AOP process, under the action of free radicals, the long polymeric molecules are decomposed into small organic compounds, forming H_2O and CO_2 (Zhou et al., 2021). Some intermediates that are formed during the AOP process can be used as raw material for industrial production (Shen, Song, et al., 2022). Various types of AOPs, such as the Fenton process (Huang et al., 2017; Kida et al., 2019), electro-Fenton (Miao et al., 2020), photocatalysis (Lee et al., 2020; Tofa et al., 2019a), and UV-based AOPs (Ouyang et al., 2022; Zhang et al., 2023), are tested in the laboratory to degrade MPs.

The degradation rate of MPs using AOPs varies depending on the type of process involved. The electro-oxidation process using electrodes such as boron-doped diamond, mixed metal oxide, iridium oxide degraded around 58%of polystyrene MPs in just 1 hour (Kiendrebeogo et al., 2021). The photo-Fenton process degraded polypropylene and polyvinyl MPs up to 95% in 1 week of treatment (Piazza et al., 2022). Peroxymonosulfate (PMS) degraded TiO_2 of polyethylene MPs in 48 hours. In this study, PMS was activated using vacuum ultraviolet rays (Ling et al., 2023). The electro-Fenton process using TiO_2 graphite as a cathode degraded 58% of PVC plastics (Miao et al., 2020). AOPs have also been tested for degradation of plasticizers such as bisphenol A (BPA), which is added during the production of plastics as a stabilizing agent; PMS could degrade about 99.1% of BPA in 60 minutes contact time (John et al., 2023). The mechanism of degradation is not the same for all types of MPs, and the toxicity of byproducts formed as an intermediate needs much attention (Bule Možar et al., 2023). The coexistence of different kinds of MPs is a significant problem during the degradation process of MPs in an aqueous environment (Shen et al., 2022).

2.3.2 Photocatalytic Degradation

Photocatalytic degradation is a type of advanced oxidation process (AOP) (Schwarze et al., 2023). This method of degradation involves the generation of reactive oxygen species (ROS), which plays an important role in the degradation of microplastics. Superoxide ions (O_2 •-) and hydroxyl radicals (HO•) participate in the photocatalytic degradation, leading to the breakage of polymer chains and production of intermediates (Ge et al., 2022). Other methods, such as thermochemical and biochemical approaches, demand higher energy consumption and the use of harsh chemicals at substantially elevated temperatures and pressures. In contrast, photocatalysis is considered a more environmentally friendly and cost-effective option due to its appropriate energy requirements, moderate reaction conditions, and its utilization of sunlight or UV light as an energy source (Li et al., 2023). Photocatalysis can be classified into two groups: homogeneous and heterogeneous photocatalysis. Homogeneous photocatalysts exist in same phase as that of reactants, while heterogeneous catalysts exist in a different phase from that of reactants. Heterogeneous photocatalysis are known for their high reaction rate, short process time, and high accessibility but are limited by their high cost, cumbersome extraction, and complicated recycling procedures. Heterogeneous photocatalysis is preferred over homogeneous photocatalysis for practical applications due to their higher stability and ease of recovering (Hamd et al., 2022).

The mechanism of photocatalysis can be defined as:

$$\text{Semiconductor} + \text{hv} \rightarrow \text{h}^+ + \text{e}^- \quad (2.1)$$

$$\text{h}^+ + \text{H}_2\text{O} \rightarrow \text{HO}\bullet \quad (2.2)$$

$$e^- + O_2 \rightarrow \text{O}_2\bullet^- \quad (2.3)$$

$$\text{Pollutant} + \text{ROS} \rightarrow \text{CO}_2 + \text{H}_2\text{O} \ldots \text{(Llorente-García et al., 2020)} \quad (2.4)$$

When plastics are exposed to visible or ultraviolet radiation, they absorb photons and enter higher excited states, which can result in chain scission, branching, cross-linking, and oxidation reactions. However, this degradation process can extend over a span of up to a century (Li et al., 2023). In order to increase the rate of reaction, catalysts have been added, and in the presence of catalyst, their degradation occurs in four important steps: (1) light absorption for the generation of the electron hole pair, (2) separation of excited charges, (3) transfer of electrons and holes to the surface of photocatalysts, and (4) utilization of charges on the surface of redox reactions (Zhu & Wang, 2017).

Tofa et al. (2019b) studied the degradation of fragmented, low-density polyethylene (LDPE) in the presence of visible light using zinc oxide nanorods as the catalyst. As a result, a 30% increase in carbonyl index of residues and an increase in brittleness with a large number of wrinkles, cracks, and cavities on the surface were obtained when exposed for about 175 hours. This study further yielded that the degradation is directly proportional to the catalyst surface. The degradation led to the formation of low-weight compounds such as hydroperoxides, peroxides, carbonyl, and unsaturated groups, which causes brittleness along with wrinkles, cracks, and cavities. In the study conducted by Khairudin et al. (2022), a 64% degradation of polystyrene microplastic was obtained when it was illuminated in visible light at an intensity of 3.6 W/cm^2 for 120 hours using a flake-like BiOI-Fe_3O_4 microswimmer as a catalyst.

Though various photocatalysts are developed for the degradation of MPs, the following challenges are to be taken care of. Photocatalytic degradation produces intermediate products that may be toxic in nature depending on the type of photocatalyst used. These photocatalysts are non-selective and cannot attack the reactive sites of microplastics such as functional groups. Additionally, more effective methods should be developed for purifying the degraded products (Zhu & Wang, 2017). The degradation mechanism is shown in Figure 2.4.

2.3.3 Membrane Bioreactor

A membrane bioreactor (MBR) couples the biological process with membrane technology to remove pollutants. It has wide application in municipal and industrial wastewater treatment plants. First, degradation of organic matter occurs in the

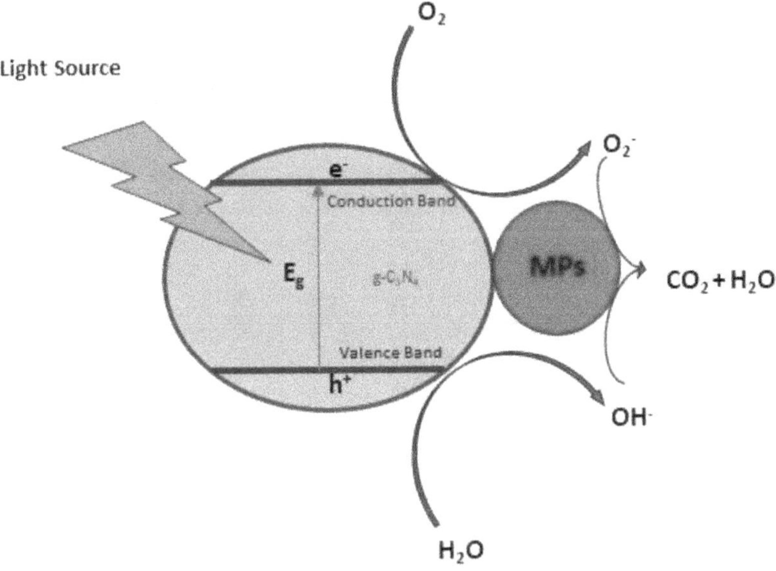

FIGURE 2.4 Photocatalysis mechanisms for MPs degradation.

bioreactor tank; then the separation happens in the membrane module (Al-Asheh et al., 2021; Raju et al., 2023). Sun et al. (2021) have investigated the performance of MBR for the removal of MPs present in landfill leachate. The results show that MBR has the capability to separate the MPs load from leachate; the initial load of MPs was 235.4 ± 17.1 item/L, which was reduced to 3.8 ± 0.5 item/L after MBR treatment. The performance of MBR is better in removing MPs when compared to the reverse osmosis (RO) system (Sun et al., 2021). The MPs removal by two different technologies, such as MBR and rapid sand filter (RSF) have been compared by Bayo et al. (2020). The results show that the MPs removal efficiency of MBR is 79.01%, which is higher than that of RSF (75.49%). According to this study, the removal percentage by advanced technologies (MBR) is not much higher when compared to the conventional method (RSF), and microfibers present in the leachate easily escaped through both MBR and RSF systems (Bayo et al., 2020). Cai et al. (2021) have also reported that fibers that are < 200 µm could easily pass through the membrane systems. They used an integrated membrane system (IMS) to remove the MPs present in the wastewater. The IMS consists of a pretreatment unit followed by MBR and RO systems. In this study, the removal efficiency of 93.2% was achieved after MBR treatment, which was increased to 98% after RO treatment.

Though MBR is capable of removing MPs efficiently, its major drawback is membrane fouling, which increases the cost of treatment. Li et al. (2020) conducted the laboratory study using polyvinylchloride (10 particles/L) to evaluate the performance of MBR. The performance of MBR was inhibited after the addition of PVC MPs, and the presence of MPs leads to higher membrane fouling. The membrane fouling is mainly due to tiny MPs entering the pores of the membrane, which increases membrane resistance. In addition, MPs are capable of altering the performance of MBR.

Wang et al. (2022) investigated the pollutant removal and membrane fouling capacity due to the presence of synthetic MPs. The results reveal that the pollutant removal capacity does not change due to the MPs. However, MPs mitigated the membrane fouling by continuously scouring the membrane surface of the MBR. In contrast, the same study reported that the presence of microbeads increased the secretion of a soluble microbial product and extracellular polymeric substance, which finally leads to an increase in membrane fouling.

2.3.4 Biological Degradation

Bacteria, fungi, algae, biofilms, and bacteria consortia were tested in the laboratory scale for the degradation of microplastics (Tiwari et al., 2020; Yang et al., 2014). The biological degradation of MPs includes various steps such as the breakdown of larger polymers into smaller ones, conversion from polymer to oligomers, dimers, and monomers, and finally mineralization with the help of microbes. On complete mineralization, carbon dioxide and several other intermediate compounds are formed. The intermediate compounds are used to grow microbes (Anand et al., 2023). Some of the microbes can grow in extremely stressful conditions where they use polymers as an energy source. In such cases, it is vital to study the particular bacteria in an isolated condition. Many scientists worldwide have tried studying the biodegradation potential of MPs using various microorganisms. Polypropylene degradation was studied by Auta et al. (2017) in a Bushnell Haas medium with the bacteria isolated from mangrove sediments. The degradation was found to be 6.4% and 4% using *Rhodococcus species* strain 36 and *Bacillus species strain* 27 in 40 days. The degradation of polypropylene was confirmed using FTIR and SEM analysis. Yang et al. (2014) reported that the degradation efficiencies of polyethylene MPs were 10.70 ± 0.2% and 6.10 ± 0.3% using *Bacillus species* YP1 and *Enterobacter asburiae* YT1 in 60 days. Another study by (Park & Kim, 2019) used *Bacillus* sp. and *Paenibacillus* sp. for degrading polyethylene MPs, and the results show that it could degrade around 14.7% after 60 days. These bacteria were isolated from municipal landfill sites, and the SEM analysis of the samples clearly showed microbes on the surface of MPs, which indicates that the microbes utilized MPs as a carbon source. The mean diameter of the MPs also decreased by 70% after 60 days (Park & Kim, 2019). Both bacterial consortiums and pure culture are used in the biodegradation process of MPs. However, by using pure culture, the degradation pathway can be identified clearly. Moreover, pure culture is easy to monitor the environmental factors affecting the biodegradation process (Anand et al., 2023).

Though many studies have been carried out on microbial degradation of MPs, the mechanism by which microbe's uptake MPs is still unclear. Particularly, the pathway involved in the uptake of monomers by microbes after the polymer becomes fragmented is complex to understand. For efficient biodegradation of MPs, several key factors are required, like potential organisms, suitable pathway, and environmental conditions such as pH, moisture content, and salinity. The surface structure of the MPs also influences the biodegradation of MPs (Anand et al., 2023). For studying the degradation of MPs, many techniques such as weight loss, CO_2 release due to degradation of polymers, and loss of additives on the MPs have been adopted. Among these methods, mass loss measurement is the most widely used approach for monitoring the efficiency of biodegradation. Nevertheless, it requires careful execution due to the inherently slow nature of biological processes (Raddadi & Fava, 2019). Additionally, to confirm the degradation of MPs, advanced techniques such as SEM, FTIR, XRD, and gas chromatography are employed.

2.3.5 Nanotechnology-Based Degradation

Nanotechnology-based solutions for water and wastewater treatment has been practiced for several years (Baruah et al., 2019). Specific surface area, high functionality, and high reactivity are the major advantages of using this technology (Chaturvedi et al., 2020). Nanomaterials have been incorporated into several water treatment methods to enhance removal efficiency. Majorly, nanomaterials are being incorporated with treatment processes like adsorption, membrane technology, and photocatalysis to eliminate MPs from the environment (Ouda et al., 2023). Recently, nanomagnetic and non-magnetic adsorbents have been adopted to remove MPs. Magnetic carbon nanotubes (CNT) were used as an adsorbent to remove MPs from wastewater and achieved 100% removal efficiency due to high adsorbance rate of the catalyst. CNT have large surface area and are hydrophobic in nature, which makes them suitable materials to adsorb organic particles like MPs (Tang et al., 2021). Modified biochar with Mg and Zn has been used as an adsorbent material, and the results show that metals that are positively charged attract negatively charged MPs (Wang et al., 2021). Graphene-oxide-coated sponge has been tested to adsorb MPs, and, because of hydrogen bonds and electrostatic interaction, the removal rate was very high (Sun et al., 2020). Iron-modified biochar adsorbed 206.46 mg/g of nanoplastics of diameter 30nm (Singh et al., 2021).

Recently, many studies have been done by incorporating nanomaterials into membrane to remove MPs. Gnanasekaran et al. (2021) incorporated metal organic frameworks (MOFs) with polysulfone membrane to form a composite mixed matrix membrane (MMMs). Metal-organic frameworks consist of a metal center linked by organic ligands, and their unique properties, such as high surface area, pore functionality, extended porosity, make them suitable for pollutant removal. These MOFs-incorporated MMMs had high electrostatic repulsion and high stability, which gave high MPs removal efficiency (Gnanasekaran et al., 2021). The electro-spun nano membrane is widely used in removing particulate matter such as PM 2.5 or PM 10, and it can potentially remove MPs from aqueous bodies. Pervez et al. (2022) used electro-spun membrane to remove micro- and nanoplastics from wastewater, and the removal efficiency was found to be 90%. These electro-spun membranes had a positive charge, high flux, and less fouling compared to conventional membranes.

Nanomaterials are also used in the biological process to enhance MPs degradation. Although bioremediation can achieve the degradation of MPs, it is an extremely slow process. Hence the combination of nanomaterials and biotechnology offers a new approach for the degradation of MPs. The biomaterial-based nanofiltration technique removed MPs as small as 50 nm; the biomaterial was based on cellulose fabric incorporated with polysaccharide nanocrystals. The coating of biomaterials also improved the tensile strength, hydrophilicity, elasticity, and surface charge of the fabric (Jalvo et al., 2021).

2.3.6 Supercritical Water Oxidation Processes

Supercritical water oxidation (SCW) is a combination of the physical and thermal processes, where water is used above its critical temperature, and the pressure will reach around 374.5°C and 220 bars, respectively. SCW has a viscosity similar to gas and a density similar to water; due to this property, it has a better dissolution and mass transfer property. This SCW can degrade almost all organic compounds and provides conditions to degrade plastics (Zhao et al., 2022). The energy required for SCW is much less when compared to pyrolysis experimentation (Zaker et al., 2019). Bai et al. (2020) performed the supercritical oxidation reaction to degrade the polycarbonate MPs and found that strong free radicals are formed during the reaction, destroying the structure of polycarbonates. The degradation of polystyrene using SCW under CO_2 conditions was studied, and the results show that around 47.6%was converted to carbon at 700°C (Zhao et al., 2022). Chand et al. (2022) conducted an experiment on MPs degradation in sewage sludge using hydrothermal liquefaction under supercritical conditions. It was found that about 76% reduction in the MPs number and a 97% reduction in mass could be achieved. A similar kind of study using three different MPs in sewage sludge was carried out by Li et al. (2022) and found that the main degradation pathway was depolymerization and leaching of additives. Supercritical CO_2 with NaOH and HCl were used to degrade polystyrene (PS) MPs, and researchers found that the degradation efficiency depends on reaction time, temperature, and NaOH/HCl concentration. In this work, better degradation efficiency was achieved in the presence of an acid-base solution (Liu et al., 2023). Hydrothermal degradation of PP and PE plastics was carried out using supercritical water, and, as a result, oil and gas were produced at 425 and 450°C. The components present in the oil were alkenes, alkanes, aromatic hydrocarbons, alcohols, and cycloalkanes. The oil and gas yields were maximum for PP, and only 8–9% of the PP plastics were undegraded, whereas more than 70% of the (PE were undegraded at the end of the reaction time (Čolnik et al., 2022).

2.4 MPs CONTROL MEASURES

The following control measures can be taken to avoid or eliminate MPs pollution in the environment.

2.4.1 Reducing the Plastic Production

The four Rs principles to reduce, reuse, recycle, and recover will aid in controlling plastic pollution (Figure 2.5). Reducing the production of plastics significantly limits the secondary MPs pollution in the environment. Recycling and reusing the existing plastic will reduce the need for new plastics. There are several benefits of using plastics as recycled material for different materials (Mourshed et al., 2017):

1. By using plastic material in a sustainable way, the size of the landfill is reduced. Today the landfill is expanding very fast, especially in cities. On the other hand, landfills are acting as a source of MPs. So this constructive

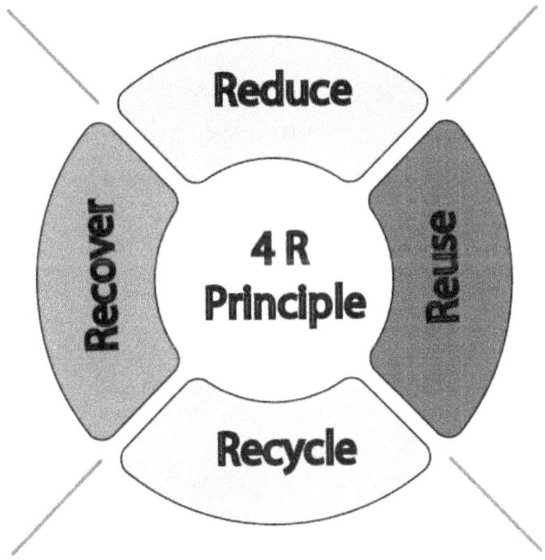

FIGURE 2.5 4R principle of plastic waste management.

way of using plastic will reduce the plastic accumulated in landfills and MPs pollution in the environment.

2. The demand for energy is increasing day by day. So using plastic waste materials in the cement, steel, and glass industries will partially fulfill the energy requirement.
3. It also helps to reduce the release of pollutants to the water and other parts of the environment.
4. The reusing option reduces the need to pay for the scarce material or resource. For example, the need to purchase fossil fuel from other countries can be reduced, which will strengthen the country's economy.

While the first 3 Rs are well-known, the last R stands for recover or rethink. Rethink means people have to think about why they are buying plastic products and what is the use of it. Recover implies recovering the valuable things from the waste plastic materials and producing energy out of it: for example, methane gas production from the waste material (Mourshed et al., 2017).

2.4.2 Ban on Single-Use Plastics

Single-use plastics are made of polyethylene for films, carry bags, food packaging, etc. and can be used only once. Single-use plastics contains hazardous chemicals and have an adverse impact on human health and environment (Clayton et al., 2021). The 'European Strategy for Plastics in a Circular Economy' proposed a single-use plastic directive. The aim of this directive is to reuse or recycle packing plastics in a cost-effective manner by 2030 (Matthews et al., 2021). Table 2.1 shows the list of countries that have implemented various regulations on plastics.

TABLE 2.1
Countries and Their Regulations on Plastic Usage

S. No.	Country	Description	Reference
1.	Albania	Banned lightweight plastics since 2016	Emerging Europe, 2018
2.	Bahamas	Ban on single-use plastic bags since 2020	Turbine, 2020
3.	Belize	Ban on single-use plastics since 2019	Clayton et al., 2021
4.	Canada	Ban on single-use plastics <50 micron since 2018	Xanthos & Walker, 2017
5.	Denmark	Tax for plastics packing bags since 1993	Jacobsen et al., 2003
6.	India	Single-use plastic ban from 2022	Nøklebye et al., 2023
7.	Jersey	Plastic bags ban from July 2022; charge on reusable bags	BBC, 2022
8.	Japan	Implemented charge for using plastics	JFS, 2015
9.	Maldives	Single-use plastic ban since 2022	Ministry of Environment, Climate Change and Technology, 2022
10.	New Zealand	Ban on single-use plastic since 2019	Newshub, 2018
11.	Republic of Korea	Single-use plastic ban since 2018	UPI, 2018

2.4.3 Policy-Level Changes

The US government draft national strategy has framed three goals to prevent plastic pollution. Goal A involves reduction of plastic pollution, Goal B proper plastic management, and Goal C preventing plastic pollution in the waterways (USEPA, 2023). The Indian government banned the use of single-use plastics from July 1, 2022 (Nøklebye et al., 2023). There are also policies related to bans on microbeads. The Microbeads Free Act 2015 was proposed by the US government to control the use of microbeads in personal care products (US Microbead Free Water Act, 2015). Likewise, many policies have been proposed, and few of them have been implemented across the globe. Existing plastic management strategies contain flaws that enable microplastics to be released into the environment. Simply reinforcing the existing policies will not resolve this issue. New policies should be developed and enforced to control plastic and MPs pollution (Hettiarachchi & Meegoda, 2023). Policies related to plastic can be categorized into four types:

1. **Product Design:** Policies for eco-friendly production, certification, and labeling of plastic materials fall under this category.
2. **Consumption:** This includes bans on certain kinds of plastics, policies related to reduction in consumption.
3. **Disposal and End Life:** Policies related to transboundary movement between countries, collection targets, etc.
4. **Circular Economy:** Policies related to circular approaches such as expended producer responsibility and deposit refund system.

The ban first started in single-use plastic and is slowly getting extended to other products. Some policymakers feel that policy-level changes in product bans have limited success but that suitable alternatives could not be found (Knoblauch & Mederake, 2021).

2.4.4 Behavioral Change of People

While there is a lot of data on plastic recycling and reuse, yet very little is known about human and plastic interactions. Plastic still follows a linear economy, and changing it to a circular economy runs into many difficulties; one such difficulty is people's behavior. The sociodemographic factors such as gender, education, income, marital status, and age influence antiplastic behavior and plastic policies implementation among the people. The view toward plastic, whether it is considered a convenience for modern life or as a polluting factor due to its environmental concern, depends mostly on people's perceptions (Macdonald et al., 2023). Most individuals do not consider the impact of disposal despite their awareness and education (Heidbreder et al., 2019). Recently, COVID 19 raised some alarm regarding public health among the public. However, during the pandemic, many restaurants encouraged the use of plastics, which has changed the perception of plastic as a healthy product rather than an environmental pollutant, thus making people use plastics more frequently (Prata et al., 2020).

Motivation, capability, and opportunity is needed to improve the behavior that reduces the plastic waste generation and overall awareness among the public. Information about the environmental effects and material incentives also promote behavior that reduces plastic waste consumption (Allison et al., 2022).

2.4.5 Finding Alternatives for Plastics

Renewable-biomass-based plastic material known as bioplastic made up of agricultural waste, plants, and algae have been used as an alternative for polymer-based plastics (Otaigbe et al., 1999; Poirier et al., 1992). The usage of bioplastics reduces the usage of fossil fuels and lowers the carbon footprint. Most bioplastics are also biodegradable and reduce toxicity (Atiwesh et al., 2021). Though they have several advantages, they face difficulties such as cost of production, resource availability, variation in biodegradability. Instead of using plastic materials, paper can be used for packing. Glass and stainless steel can be used to carry things from restaurants and hotels. This can be possible only by changing the behavior of society.

2.5 CONCLUSIONS

This chapter has provided an overview of innovative technologies that hold the potential to combat MPs pollution, and the final section of the chapter discusses control measures that can be taken to avoid MPs in the environment. Degrading microplastics in aquatic media is crucial for preserving the health and sustainability of aquatic ecosystems, protecting human health, maintaining the aesthetic value of natural water bodies, and mitigating the environmental and ecological impacts associated with plastic pollution. Advanced technologies like AOPs, MBR, biological degradation, nanotechnology, and supercritical water oxidation are the possible ways to degrade MPs in aquatic environment.

- AOPs have the ability to produce radicals and oxidants, which can effectively break down MPs into less harmful components. Combined with UV radiation or ozone, AOPs have demonstrated the potential for MPs degradation.
- Membrane bioreactors represent an advancement in eliminating MP pollution. These systems combine biological processes with membrane filtration resulting in the removal of MPs from wastewater. The available study reveals that the synergistic combination of degradation and filtration can achieve better outcomes.
- Photocatalysis, utilizing catalyst for degradation of MPs, offers a friendly solution. Using photocatalysts like titanium dioxide, zinc oxide can reduce MP pollution across environmental contexts.
- Nanomaterials incorporated in various technologies, such as AOPs or photocatalysis, have displayed effectiveness in MPs removal due to their large surface area and reactivity. The integration of nanotechnology with degradation processes presents an innovative solution for addressing the issue of microplastics.
- Furthermore, supercritical water oxidation has emerged as an exciting approach. Supercritical water can efficiently degrade MPs by operating at critical temperatures and pressures, offering an energy-efficient solution for removing the contaminants.

The choice of the most suitable technology for removing microplastics (MPs) from aquatic environments depends on several factors, including the specific characteristics of the environment, the type of water body, the quantity of MPs present, and the presence of other pollutants. To truly address the issue, further research, interdisciplinary collaborations, and continued innovation are necessary to upgrade these technologies to a large-scale application and safeguard future generations from MPs pollution.

REFERENCES

Al-Asheh, S., Bagheri, M., & Aidan, A. (2021). Membrane bioreactor for wastewater treatment: A review. In *Case Studies in Chemical and Environmental Engineering* (Vol. 4). Elsevier Ltd. https://doi.org/10.1016/j.cscee.2021.100109

Allison, A. L., Baird, H. M., Lorencatto, F., Webb, T. L., & Michie, S. (2022). Reducing plastic waste: A meta-analysis of influences on behaviour and interventions. *Journal of Cleaner Production*, *380*, 134860. https://doi.org/10.1016/J.JCLEPRO.2022.134860

Anand, U., Dey, S., Bontempi, E., Ducoli, S., Vethaak, A. D., Dey, A., & Federici, S. (2023). Biotechnological methods to remove microplastics: A review. *Environmental Chemistry Letters*, *21*(3), 1787–1810. https://doi.org/10.1007/s10311-022-01552-4

Atiwesh, G., Mikhael, A., Parrish, C. C., Banoub, J., & Le, T. A. T. (2021). Environmental impact of bioplastic use: A review. *Heliyon*, *7*(9), e07918. https://doi.org/10.1016/j.heliyon.2021.e07918

Auta, H. S., Emenike, C. U., & Fauziah, S. H. (2017). Screening of Bacillus strains isolated from mangrove ecosystems in Peninsular Malaysia for microplastic degradation. *Environmental Pollution*. https://doi.org/10.1016/j.envpol.2017.09.043

Bai, B., Liu, Y., Meng, X., Liu, C., Zhang, H., Zhang, W., & Jin, H. (2020). Experimental investigation on gasification characteristics of polycarbonate (PC) microplastics in supercritical water. *Journal of the Energy Institute*, *93*(2), 624–633. https://doi.org/10.1016/J.JOEI.2019.06.003

Baruah, A., Chaudhary, V., Malik, R., & Tomer, V. K. (2019). Nanotechnology based solutions for wastewater treatment. *Nanotechnology in Water and Wastewater Treatment: Theory and Applications*, 337–368. https://doi.org/10.1016/B978-0-12-813902-8.00017-4

Bayo, J., López-Castellanos, J., & Olmos, S. (2020). Membrane bioreactor and rapid sand filtration for the removal of microplastics in an urban wastewater treatment plant. *Marine Pollution Bulletin*, *156*, 111211. https://doi.org/10.1016/J.MARPOLBUL.2020.111211

BBC. (2022). Single-use plastic carrier bag ban in Jersey begins—BBC News. Single-use plastic carrier bag ban in Jersey begins. *BBC News*. www.bbc.com/news/world-europe-jersey-62238573

Bhuyan, M. S. (2022). Effects of microplastics on fish and in human health. *Frontiers in Environmental Science*, *10*, 827289. https://doi.org/10.3389/FENVS.2022.827289/BIBTEX

Bule Možar, K., Miloloža, M., Martinjak, V., Cvetnić, M., Kušić, H., Bolanča, T., Kučić Grgić, D., & Ukić, Š. (2023). Potential of advanced oxidation as pretreatment for microplastics biodegradation. In *Separations* (Vol. 10, Issue 2). MDPI. https://doi.org/10.3390/separations10020132

Cai, Y., Wu, J., Lu, J., Wang, J., & Zhang, C. (2021). Fate of microplastics in a coastal wastewater treatment plant: Microfibers could partially break through the integrated membrane system. *Frontiers of Environmental Science & Engineering*. https://doi.org/10.1007/s11783-021-1517-0

Calero, M., Godoy, V., Quesada, L., & Martín-Lara, M. Á. (2021). Green strategies for microplastics reduction. *Current Opinion in Green and Sustainable Chemistry*, *28*, 100442. https://doi.org/10.1016/J.COGSC.2020.100442

Chand, R., Kohansal, K., Toor, S., Pedersen, T. H., & Vollertsen, J. (2022). Microplastics degradation through hydrothermal liquefaction of wastewater treatment sludge. *Journal of Cleaner Production*, *335*(January), 130383. https://doi.org/10.1016/j.jclepro.2022.130383

Chaturvedi, V. K., Kushwaha, A., Maurya, S., Tabassum, N., Chaurasia, H., & Singh, M. P. (2020). Wastewater treatment through nanotechnology: Role and prospects. *Restoration of Wetland Ecosystem: A Trajectory Towards a Sustainable Environment*, 227–247. https://doi.org/10.1007/978-981-13-7665-8_14

Clayton, C. A., Walker, T. R., Bezerra, J. C., & Adam, I. (2021). Policy responses to reduce single-use plastic marine pollution in the Caribbean. *Marine Pollution Bulletin*, *162*, 111833. https://doi.org/10.1016/J.MARPOLBUL.2020.111833

Čolnik, M., Kotnik, P., Knez, Ž., & Škerget, M. (2022). Chemical recycling of polyolefins waste materials using supercritical water. *Polymers*, *14*(20), 4415. https://doi.org/10.3390/POLYM14204415/S1

Emerging Europe. (2018). *Albania Bans Lightweight Plastic Bags*. https://emerging-europe.com/news/albania-bans-lightweight-plastic-bags/

Feldman, D. (2002). Polymer weathering: Photo-oxidation. *Journal of Polymers and the Environment*, *10*(4), 163–173. https://doi.org/10.1023/A:1021148205366/METRICS

Ge, J., Zhang, Z., Ouyang, Z., Shang, M., Liu, P., Li, H., & Guo, X. (2022). Photocatalytic degradation of (micro)plastics using TiO2-based and other catalysts: Properties, influencing factor, and mechanism. *Environmental Research*, *209*. https://doi.org/10.1016/J.ENVRES.2022.112729

Gewert, B., Plassmann, M. M., & Macleod, M. (2015). Pathways for degradation of plastic polymers floating in the marine environment. *Environmental Science: Processes & Impacts*, *17*(9), 1513–1521. https://doi.org/10.1039/C5EM00207A

Gnanasekaran, G., Arthanareeswaran, G., & Mok, Y. S. (2021). A high-flux metal-organic framework membrane (PSF/MIL-100 (Fe)) for the removal of microplastics adsorbing dye contaminants from textile wastewater. *Separation and Purification Technology*, *277*, 119655. https://doi.org/10.1016/J.SEPPUR.2021.119655

Hamd, W., Daher, E. A., Tofa, T. S., & Dutta, J. (2022). Recent advances in photocatalytic removal of microplastics: Mechanisms, kinetic degradation, and reactor design. *Frontiers in Marine Science*, *9*, 885614. https://doi.org/10.3389/FMARS.2022.885614/BIBTEX

Heidbreder, L. M., Bablok, I., Drews, S., & Menzel, C. (2019). Tackling the plastic problem: A review on perceptions, behaviors, and interventions. *The Science of the Total Environment*, *668*, 1077–1093. https://doi.org/10.1016/J.SCITOTENV.2019.02.437

Hettiarachchi, H., & Meegoda, J. N. (2023). Microplastic pollution prevention: The need for robust policy interventions to close the loopholes in current waste management practices. *International Journal of Environmental Research and Public Health*, *20*(14), 6434. https://doi.org/10.3390/IJERPH20146434

Huang, W., Luo, M., Wei, C., Wang, Y., Hanna, K., & Mailhot, G. (2017). Enhanced heterogeneous photo-Fenton process modified by magnetite and EDDS: BPA degradation. *Environmental Science and Pollution Research International*, *24*(11), 10421–10429. https://doi.org/10.1007/S11356-017-8728-8

Jacobsen, H. K., Birr-Pedersen, K., & Wier, M. (2003). Distributional implications of environmental taxation in Denmark. *Fiscal Studies*, *24*(4), 477–499. https://doi.org/10.1111/J.1475-5890.2003.TB00092.X

Jalvo, B., Aguilar-Sanchez, A., Ruiz-Caldas, M. X., & Mathew, A. P. (2021). Water filtration membranes based on non-woven cellulose fabrics: Effect of nanopolysaccharide coatings on selective particle rejection, antifouling, and antibacterial properties. *Nanomaterials*, *11*(7), 1752. https://doi.org/10.3390/NANO11071752/S1

JFS. (2015). *Current Status of Plastic Bag Reduction Efforts in Japan | Japan for Sustainability. Japan for Sustainability*. www.japanfs.org/en/news/archives/news_id035126.html

John, K. I., Omorogie, M. O., Bayode, A. A., Adeleye, A. T., & Helmreich, B. (2023). Environmental microplastics and their additives—a critical review on advanced oxidative techniques for their removal. *Chemical Papers*, *77*(2), 657–676. https://doi.org/10.1007/s11696-022-02505-5

Khairudin, K., Abu Bakar, N. F., & Osman, M. S. (2022). Magnetically recyclable flake-like BiOI-Fe3O4 micro swimmers for fast and efficient degradation of microplastics. *Journal of Environmental Chemical Engineering*, *10*(5), 108275. https://doi.org/10.1016/j.jece.2022.108275

Kida, M., Ziembowicz, S., & Koszelnik, P. (2019). Impact of a modified fenton process on the degradation of a component leached from microplastics in bottom sediments. *Catalysts*, *9*(11). https://doi.org/10.3390/catal9110932

Kiendrebeogo, M., Karimi Estahbanati, M. R., Khosravanipour Mostafazadeh, A., Drogui, P., & Tyagi, R. D. (2021). Treatment of microplastics in water by anodic oxidation: A case study for polystyrene. *Environmental Pollution*, *269*, 116168. https://doi.org/10.1016/j.envpol.2020.116168

Knoblauch, D., & Mederake, L. (2021). Government policies combatting plastic pollution. *Current Opinion in Toxicology*, *28*, 87–96. https://doi.org/10.1016/J.COTOX.2021.10.003

Lee, J. M., Busquets, R., Choi, I. C., Lee, S. H., Kim, J. K., & Campos, L. C. (2020). Photocatalytic degradation of Polyamide 66; Evaluating the feasibility of photocatalysis as a microfibre-targeting technology. *Water*, *12*(12), 3551. https://doi.org/10.3390/W12123551

Li, L., Liu, D., Song, K., & Zhou, Y. (2020). Performance evaluation of MBR in treating microplastics polyvinylchloride contaminated polluted surface water. *Marine Pollution Bulletin, 150*. https://doi.org/10.1016/j.marpolbul.2019.110724

Li, W., Zhao, W., Zhu, H., Li, Z. J., & Wang, W. (2023). State of the art in the photochemical degradation of (micro)plastics: From fundamental principles to catalysts and applications. *Journal of Materials Chemistry A, 11*(6), 2503–2527. https://doi.org/10.1039/D2TA09523H

Li, X., Wang, X., Chen, L., Huang, X., Pan, F., Liu, L., Dong, B., Liu, H., Li, H., Dai, X., & Hur, J. (2022). Changes in physicochemical and leachate characteristics of microplastics during hydrothermal treatment of sewage sludge. *Water Research, 222*, 118876. https://doi.org/10.1016/J.WATRES.2022.118876

Ling, C., Li, C., Liang, A., & Wang, W. (2023). Efficient degradation of polyethylene microplastics with VUV/UV/PMS: The critical role of VUV and mechanism. *Separation and Purification Technology, 316*(March), 123812. https://doi.org/10.1016/j.seppur.2023.123812

Liu, Y., Shi, J., Mao, L., Lu, B., Kang, X., & Jin, H. (2023). Base- or acid-assisted polystyrene plastic degradation in supercritical CO2. *Waste Disposal and Sustainable Energy, 5*(2), 165–175. https://doi.org/10.1007/S42768-023-00139-1/FIGURES/7

Llorente-García, B. E., Hernández-López, J. M., Zaldívar-Cadena, A. A., Siligardi, C., & Cedillo-González, E. I. (2020). First insights into photocatalytic degradation of HDPE and LDPE microplastics by a mesoporous N-TiO2 coating: Effect of size and shape of microplastics. *Coatings, 10*(7). https://doi.org/10.3390/coatings10070658

Macdonald, A., Allen, D., Williams, L., Flowers, P., & Walker, T. R. (2023). People, plastic, and behaviour change-a comment on drivers of plastic pollution, barriers to change and targeted behaviour change interventions. *Environmental Science: Advances*. https://doi.org/10.1039/d2va00248e

Matthews, C., Moran, F., & Jaiswal, A. K. (2021). A review on European Union's strategy for plastics in a circular economy and its impact on food safety. *Journal of Cleaner Production, 283*, 125263. https://doi.org/10.1016/J.JCLEPRO.2020.125263

Miao, F., Liu, Y., Gao, M., Yu, X., Xiao, P., Wang, M., Wang, S., & Wang, X. (2020). Degradation of polyvinyl chloride microplastics via an electro-Fenton-like system with a TiO2/graphite cathode. *Journal of Hazardous Materials, 399*. https://doi.org/10.1016/j.jhazmat.2020.123023

Ministry of Environment, Climate Change and Technology, M. (2022). *The Maldives Bans Production and Sales of Single-Use Plastics Effective from 1st June 2022*. Ministry of Environment, Climate Change and Technology. www.environment.gov.mv/v2/en/news/15184

Mourshed, M., Masud, M. H., Rashid, F., & Joardder, M. U. H. (2017). Towards the effective plastic waste management in Bangladesh: A review. *Environmental Science and Pollution Research, 24*(35), 27021–27046. https://doi.org/10.1007/S11356-017-0429-9/FIGURES/25

Newshub. (2018). *Single-Use Plastic Bags Banned from July 1, Government Confirms*. www.newshub.co.nz/home/politics/2018/12/single-use-plastic-bags-banned-from-july-1-government-confirms.html

Nøklebye, E., Adam, H. N., Roy-Basu, A., Bharat, G. K., & Steindal, E. H. (2023). Plastic bans in India—addressing the socio-economic and environmental complexities. *Environmental Science & Policy, 139*, 219–227. https://doi.org/10.1016/J.ENVSCI.2022.11.005

Otaigbe, J. U., Goel, H., Babcock, T., & Jane, J. (1999). Processability and properties of biodegradable plastics made from agricultural biopolymers. *Journal of Elastomers & Plastics, 31*(1), 56–71. https://doi.org/10.1177/009524439903100104

Ouda, M., Banat, F., Hasan, S. W., & Karanikolos, G. N. (2023). Recent advances on nanotechnology-driven strategies for remediation of microplastics and nanoplastics from aqueous environments. *Journal of Water Process Engineering*, *52*(January), 103543. https://doi.org/10.1016/j.jwpe.2023.103543

Ouyang, Z., Li, S., Zhao, M., Wangmu, Q., Ding, R., Xiao, C., & Guo, X. (2022). The aging behavior of polyvinyl chloride microplastics promoted by UV-activated persulfate process. *Journal of Hazardous Materials*, *424*(PB), 127461. https://doi.org/10.1016/j.jhazmat.2021.127461

Park, S. Y., & Kim, C. G. (2019). Biodegradation of micro-polyethylene particles by bacterial colonization of a mixed microbial consortium isolated from a landfill site. *Chemosphere*, *222*, 527–533. https://doi.org/10.1016/j.chemosphere.2019.01.159

Pervez, M. N., Mishu, M. R., Talukder, M. E., Stylios, G. K., Buonerba, A., Hasan, S. W., Cai, Y., Zhao, Y., Figoli, A., Zarra, T., Belgiorno, V., & Naddeo, V. (2022). Electrospun nanofiber membranes for the control of micro/nanoplastics in the environment. *Water Emerging Contaminants & Nanoplastics*, *1*(2), 10. https://doi.org/10.20517/WECN.2022.05

Piazza, V., Uheida, A., Gambardella, C., Garaventa, F., Faimali, M., & Dutta, J. (2022). Ecosafety screening of photo-Fenton process for the degradation of microplastics in water. *Frontiers in Marine Science*, *8*(February), 1–15. https://doi.org/10.3389/fmars.2021.791431

Poirier, Y., Dennis, D. E., Klomparens, K., & Somerville, C. (1992). Polyhydroxybutyrate, a biodegradable thermoplastic, produced in transgenic plants. *Science (New York, NY)*, *256*(5056), 520–523. https://doi.org/10.1126/SCIENCE.256.5056.520

Prata, J. C., da Costa, J. P., Lopes, I., Duarte, A. C., & Rocha-Santos, T. (2020). Environmental exposure to microplastics: An overview on possible human health effects. *Science of the Total Environment*, *702*, 134455. https://doi.org/10.1016/J.SCITOTENV.2019.134455

Prata, J. C., Silva, A. L. P., Walker, T. R., Duarte, A. C., & Rocha-Santos, T. (2020). COVID-19 pandemic repercussions on the use and management of plastics. *Environmental Science and Technology*, *54*(13), 7760–7765. https://doi.org/10.1021/acs.est.0c02178

Priya, K. L., Renjith, K. R., Joseph, C. J., Indu, M. S., Srinivas, R., & Haddout, S. (2022). Fate, transport and degradation pathway of microplastics in aquatic environment—a critical review. *Regional Studies in Marine Science*, *56*, 102647. https://doi.org/10.1016/J.RSMA.2022.102647

Raddadi, N., & Fava, F. (2019). Biodegradation of oil-based plastics in the environment: Existing knowledge and needs of research and innovation. *Science of the Total Environment*, *679*, 148–158. https://doi.org/10.1016/J.SCITOTENV.2019.04.419

Raju, M., Gandhimathi, R., & Nidheesh, P. V. (2023). The cause, fate and effect of microplastics in freshwater ecosystem: Ways to overcome the challenge. *Journal of Water Process Engineering*, *55*, 104199. https://doi.org/10.1016/j.jwpe.2023.104199

Samandra, S., Johnston, J. M., Jaeger, J. E., Symons, B., Xie, S., Currell, M., Ellis, A. V., & Clarke, B. O. (2022). Microplastic contamination of an unconfined groundwater aquifer in Victoria, Australia. *Science of the Total Environment*, *802*. https://doi.org/10.1016/j.scitotenv.2021.149727

Schwarze, M., Borchardt, S., Frisch, M. L., Collis, J., Walter, C., Menezes, P. W., Strasser, P., Driess, M., & Tasbihi, M. (2023). Degradation of Phenol via an advanced oxidation process (AOP) with immobilized commercial titanium dioxide (TiO2) photocatalysts. *Nanomaterials*, *13*(7), 1249. https://doi.org/10.3390/NANO13071249/S1

Shen, M., Song, B., Zhou, C., Hu, T., Zeng, G., & Zhang, Y. (2022). Advanced oxidation processes for the elimination of microplastics from aqueous systems: Assessment of efficiency, perspectives and limitations. In *Science of the Total Environment* (Vol. 842). Elsevier B.V. https://doi.org/10.1016/j.scitotenv.2022.156723

Shen, M., Zhang, Y., Almatrafi, E., Hu, T., Zhou, C., Song, B., Zeng, Z., & Zeng, G. (2022). Efficient removal of microplastics from wastewater by an electrocoagulation process. *Chemical Engineering Journal*, *428*. https://doi.org/10.1016/j.cej.2021.131161

Singh, N., Khandelwal, N., Ganie, Z. A., Tiwari, E., & Darbha, G. K. (2021). Eco-friendly magnetic biochar: An effective trap for nanoplastics of varying surface functionality and size in the aqueous environment. *Chemical Engineering Journal, 418*. https://doi.org/10.1016/J.CEJ.2021.129405

Sun, C., Wang, Z., Chen, L., & Li, F. (2020). Fabrication of robust and compressive chitin and graphene oxide sponges for removal of microplastics with different functional groups. *Chemical Engineering Journal, 393*. https://doi.org/10.1016/J.CEJ.2020.124796

Sun, J., Zhu, Z. R., Li, W. H., Yan, X., Wang, L. K., Zhang, L., Jin, J., Dai, X., & Ni, B. J. (2021). Revisiting microplastics in landfill leachate: Unnoticed tiny microplastics and their fate in treatment works. *Water Research, 190*. https://doi.org/10.1016/j.watres.2020.116784

Tang, Y., Zhang, S., Su, Y., Wu, D., Zhao, Y., & Xie, B. (2021). Removal of microplastics from aqueous solutions by magnetic carbon nanotubes. *Chemical Engineering Journal, 406*, 126804. https://doi.org/10.1016/J.CEJ.2020.126804

Tiwari, N., Santhiya, D., & Sharma, J. G. (2020). Microbial remediation of micro-nano plastics: Current knowledge and future trends. *Environmental Pollution, 265*, 115044. https://doi.org/10.1016/j.envpol.2020.115044

Tofa, T. S., Kunjali, K. L., Paul, S., & Dutta, J. (2019a). Visible light photocatalytic degradation of microplastic residues with zinc oxide nanorods. *Environmental Chemistry Letters, 17*(3), 1341–1346. https://doi.org/10.1007/S10311-019-00859-Z/TABLES/1

Tofa, T. S., Kunjali, K. L., Paul, S., & Dutta, J. (2019b). Visible light photocatalytic degradation of microplastic residues with zinc oxide nanorods. *Environmental Chemistry Letters, 17*(3), 1341–1346. https://doi.org/10.1007/S10311-019-00859-Z/TABLES/1

Turbine, T. (2020). Customer complaints as single use plastic ban comes into effect. *The Tribune*. www.tribune242.com/news/2020/jan/03/customer-complaints-single-use-plastic-ban-comes-e/

UPI. (2018). South Korea bans disposable coffee cups, plastic bags. *UPI.com*. www.upi.com/Top_News/World-News/2018/08/02/South-Korea-bans-disposable-coffee-cups-plastic-bags/6841533181415/

US Congress. Microbead-Free Waters Act of 2015, Public Law H.R.1321 - Microbead-Free Waters Act of 2015. (2015). https://www.congress.gov/bill/114thcongress/house-bill/1321/text?q=%7B%22search%22:[%22microbead+free+waters+act%22]%7D&r=1

USEPA. (2023). *Draft National Strategy to Prevent Plastic Pollution: Part of a Series on Building a Circular Economy for All: Executive Summary*. United States Environmental Protection Agency. www.epa.gov/circulareconomy/draft-national-strategy-prevent-plastic-pollution

Wang, J., Sun, C., Huang, Q. X., Chi, Y., & Yan, J. H. (2021). Adsorption and thermal degradation of microplastics from aqueous solutions by Mg/Zn modified magnetic biochars. *Journal of Hazardous Materials, 419*. https://doi.org/10.1016/J.JHAZMAT.2021.126486

Wang, Q. Y., Li, Y. L., Liu, Y. Y., Zhou, Z., Hu, W. J., Lin, L. F., & Wu, Z. C. (2022). Effects of microplastics accumulation on performance of membrane bioreactor for wastewater treatment. *Chemosphere, 287*, 131968. https://doi.org/10.1016/J.CHEMOSPHERE.2021.131968

Woo, R. S. C., Zhu, H., Leung, C. K. Y., & Kim, J. K. (2008). Environmental degradation of epoxy-organoclay nanocomposites due to UV exposure: Part II residual mechanical properties. *Composites Science and Technology, 68*(9), 2149–2155. https://doi.org/10.1016/J.COMPSCITECH.2008.03.020

Xanthos, D., & Walker, T. R. (2017). International policies to reduce plastic marine pollution from single-use plastics (plastic bags and microbeads): A review. *Marine Pollution Bulletin, 118*(1–2), 17–26. https://doi.org/10.1016/J.MARPOLBUL.2017.02.048

Yang, J., Yang, Y., Wu, W. M., Zhao, J., & Jiang, L. (2014). Evidence of polyethylene biodegradation by bacterial strains from the guts of plastic-eating waxworms. *Environmental Science and Technology, 48*(23), 13776–13784. https://doi.org/10.1021/es504038a

Yousif, E., Ahmed, D., Zainulabdeen, K., & Jawad, A. (2023). Photo-physical and morphological study of polymers: A review. *Physical Chemistry Research*, *11*(2), 409–424. https://doi.org/10.22036/PCR.2022.342751.2105

Zaker, A., Chen, Z., Wang, X., & Zhang, Q. (2019). Microwave-assisted pyrolysis of sewage sludge: A review. *Fuel Processing Technology*, *187*, 84–104. https://doi.org/10.1016/J.FUPROC.2018.12.011

Zhang, X., Peng, M., Zhang, Q., Ma, X., Song, J., Sun, M., Du, E., & Xu, X. (2023). UV-photoaging behavior of polystyrene microplastics enhanced by thermally-activated persulfate. *Journal of Environmental Chemical Engineering*, *11*(5). https://doi.org/10.1016/j.jece.2023.110508

Zhao, S., Wang, C., Bai, B., Jin, H., & Wei, W. (2022). Study on the polystyrene plastic degradation in supercritical water/CO2 mixed environment and carbon fixation of polystyrene plastic in CO2 environment. *Journal of Hazardous Materials*, *421*(July 2021), 126763. https://doi.org/10.1016/j.jhazmat.2021.126763

Zhou, D., Chen, J., Wu, J., Yang, J., & Wang, H. (2021). Biodegradation and catalytic-chemical degradation strategies to mitigate microplastic pollution. In *Sustainable Materials and Technologies* (Vol. 28). Elsevier B.V. https://doi.org/10.1016/j.susmat.2021.e00251

Zhu, S., & Wang, D. (2017). Photocatalysis: Basic principles, diverse forms of implementations and emerging scientific opportunities. *Advanced Energy Materials*, *7*(23). https://doi.org/10.1002/AENM.201700841

3 Efficacy of Bacterial Consortium on Microplastic Mineralization at Municipal Dumping Grounds

Anindita Bar, Kishalaya Paria, and Sudeshna Saha

3.1 INTRODUCTION

Primitive men have learned to use metals after using stones for daily needs. But nowadays plastic is being used for not only low cost and durability but also for its excellent thermal stability and light weight. Unknowingly, people are creating danger to themselves by using plastic. Today we use a lot of plastic, which converts into microplastics during natural and forceful weatherization. Microplastic not only disturbs the global ecosystems, it can percolate through our cellular membrane. That's why it has been reported that microplastics are found in the infant brain via the mother's breast milk. On the other hand, several aquatic animals and plants also carry a significant amount of microplastics. As a result, various animals including humans have been suffering from several life-threatening diseases, even cancer.

Urban areas are more affected by plastic pollution than rural areas. Large amounts of plastic waste are found mainly in municipal areas, which eventually end up in waste dumping sites and in rivers or seawater. In the modern view, plastic waste is being used as an alternative source of geotextile for new building construction and in road building. We are hard put overcome microplastics pollution in our environment. The rapid increasing rate of plastic production contributes to huge amounts of plastic debris. This plastic debris is highly resistant to environmental degradation processes and can adsorb toxic heavy metals and other pollutants. Moreover, plastics or microplastics are directly or indirectly responsible for several morbidity and mortality of aquatic as well as terrestrial animals.

In the physicochemical weatherization of plastics, they are grouped into several categories. Based on the size of the polymers, plastics can be grouped into macro- (>25 mm), meso- (<25–5 mm), micro- (5 mm to 0.1 μm), and nano- (<0.1 μm) plastics

DOI: 10.1201/9781032684574-3

(Crawford and Quinn, 2016). Out of them, micro- and nanoplastics are now a global concern due to the smaller size and more invasive nature. Due to rapid unplanned urbanization and excessive population outburst, developing countries produce a huge amount of municipal solid waste. According to a report published by the Ministry of Housing and Urban Affairs, Govt. of India, 2023 solid waste generation in India was 152,245 MT/D. The presence of microplastics in this huge amount of municipal solid waste is well documented (Shi et al., 2022; Petrović et al., 2022). During natural calamities, several anthropogenic factors (e.g., runoff of water, wind) can influence the entering of microplastic into surrounding aquatic and terrestrial ecosystems. Despite contaminating the subsurface water, groundwater also get contaminated by microplastic leaching (Singha and Bhagwat, 2022). Humans ingest biomagnified microplastics from both terrestrial and aquatic food chains and also directly through inhalation.

In this situation, either we should stop using plastic materials in our daily lives, or new technology should be developed for the degradation of plastic waste. For the reduction of plastic waste, several countries have taken some steps such as increasing the use of jute and biodegradable polymer for several purposes. On the other hand, to mitigate the microplastic pollution, microbial biodegradation is now another point of interest because it is the only process that can mineralize microplastics completely and produce carbon dioxide, methane, water. Various species of bacteria, fungus, and algae have been identified as having the ability to colonize on microplastic surfaces and degrade them (Auta et al., 2018; Lin et al., 2022). Generally, biodegradation is a slow process; hence, it increases the persistence time of microplastics in the environment. To increase the biodegradation and decrease the persistence time, biostimulation and bioaugmentation are effective practices. In general, bioaugmentation is mainly involved in using pure and/or consortia polymer-degrading cultures and the addition of genetically engineered microorganisms to increase biodegradation activities (Kalogerakis et al., 2015). Researchers have isolated indigenous bacteria from mangrove rhizosphere and enhanced them in the laboratory in order to reintroduce them to the original environment for better degradation performance (Auta et al., 2022).

Municipal solid waste (MSW) contains a vast variety (structure, size, color, shapes, surface morphology) of microplastics; hence a large variety of microorganisms must be involved in the in-situ biodegradation process. Although microplastic biodegradation in aquatic and soil environments is well studied and reported, fewer studies are available on solid waste microplastic degradation (Canopoli et al., 2018). This chapter mainly emphasize combining laboratory methods and in-situ remediation of microplastic with the involvement of microbial consortia at municipal solid waste dumping sites, as presented by Figure 3.1.

3.2 TYPES OF MICROPLASTIC-ASSOCIATED CHEMICALS

Up to 92.4% of plastic waste types are able to convert to microplastics (Santana et al., 2016) that are primarily composed by polyethylene, polypropylene, polystyrene, polyethylene terephthalate, polyvinyl chloride, nylons, and polyamide (Carr

FIGURE 3.1 General concept of microplastic degradation through bacterial consortium.

et al., 2016). The main plastic polymers are generally harmless, but for commercial purpose the polymers are compounded with:

1. Additives or monomeric ingredients, such as antioxidants (irganox, Chimassorb, Eastman OABS, etc.), colorants (carbon black, cobalt blue, polymethyl methacrylate), plasticizer and softeners, lubricants and flow promoters, impact modifier, fire retardants (antimony oxides, ammonium polyphosphate) (Al-Malaika et al., 2017)
2. Chemicals absorbed from the surroundings, including heavy metals, like lead, mercury, antimony, chromium, etc., organophosphates, organochlorides, and detergents.

3.3 PRODUCTION OF MICROPLASTICS IN THE GLOBAL SCENARIO

Microplastics (MPs) are potential contaminants in all environmental compartments, including atmospheric, terrestrial, and aquatic environments (Klein et al., 2018; Blair Espinoza, 2019). It was estimated that around 368 million tons (Mt) of plastic waste was generated globally as of 2019 (Walker and Fequet, 2023). Right now, more or less 9% may be recyclable, while 12%was incinerated, and the remaining amount was used for land-fill purposes or discarded directly (Wang et al., 2021). The cumulative global production of plastics is forecast to reach 2600 Mt by 2050 (Liu et al., 2022).

3.4 SOURCES AND TYPES OF MP IN MUNICIPAL SOLID WASTE

Generally, microplastics vary in size, shape, types, colors, and sources. There are three main sources adding microplastics to MSW dumping sites:

1. **Landfill Refuses:** Landfills are the major source as well as the major sink of microplastics. Almost $21-42\%$ of global plastic production ends in landfill (He et al., 2019). Large plastics of landfill break down into small pieces through several physical, chemical, and biochemical reactions, thus producing secondary microplastics. one study shows that microplastic abundance in landfill is in the range of 20,000–91,000 items/kg, which is much higher than the amount present in agricultural soils (Golwala et al., 2021). The high concentration of microplastics in solid waste contaminate the water bodies, sediments, and aquatic animals (e.g., mussels) near landfills (Carr et al., 2016). Landfills primarily contain weatherized microplastics with a small amount of newly entered plastics. This weathered and aged plastic debris contributes secondary microplastics to the landfill.
2. **Sludge and Biosolids:** Industrial sludge and biosolids are another source of microplastic to MSW. Wastewater treatment plants (WWTp), the textile industries, and the plastic industries are the main contributors of microplastics to dumping sites. Among all these sources, WWT plants are mainly important because they treat wastewaters from various kinds of industry with a huge amount of MP, which remain almost undegraded in WWTp (De Falco et al., 2018). Treatment processes like membrane bioreactor or reverse osmosis can remove $99.4-99.9\%$ and 90.4% of microplastics, respectively (Lares et al., 2018; Ziajahromi et al., 2017). These removal efficiencies represent the presence of microplastics in discharged sludge and biosolids. Hence a large amount of microplastics can enter the environment through WWTp effluents, depending on waste handling practices. The abundance of microplastics was found to be greater in primary sludge $(8600-21{,}200 \text{ items/kg})$ compared to secondary sludge $(1500-7300 \text{ items/kg})$, which indicate a greater removal of microplastics during primary sedimentation (Gies et al., 2018).
3. **Food Waste:** Abundant use of plastic materials in the food industries for production, processing, and packaging can contribute a huge amount of microplastics in the municipal solid waste dumping sites. This microplastics contamination from pre- and post-consumer food waste has varying degrees of toxicity depending on population density, human engagement, and waste management practices (Porterfield et al., 2023). All human food and food products derived from aquatic or terrestrial food chains are the major source of microplastic. These microplastics reached to the food waste via wasted food (Rist et al., 2018; Wright and Kelly, 2017). Almost all consumable

food items are contaminated by microplastics. Microplastic concentration in marine fish was found to be 0–15 items/fish, in mussels 2.7 – 3.7 items/mussel, in oysters 1.4 – 7 items/oyster (Golwala et al., 2021). Sea salt contains 550 – 681 items of MP/kg (Jin et al., 2021). Conti et al. noted that fresh fruits contain more microplastic then vegetables (Conti et al., 2020). Moreover, microplastic particles were also detected in drinking water, beverages, sugar, and honey (Jin et al., 2021). Food wastes collected from grocery stores contain about 300,000 pieces of microplastics per kilogram (Golwala et al., 2021).

3.5 MIGRATION OF MICROPLASTICS FROM DUMPING SITES

Microplastics, once deposited in the dumping sites, can be biomagnified to the natural ecosystem as depicted in Figure 3.2.

1. **By Vertical and Horizontal Movement:** Soil is the primary component that comes in direct contact with dumpsite microplastics. Soil characteristics (soil macropores, aggregation, cracking, agronomic practices, biotic community) influence the horizontal and vertical migration of microplastic (Guo et al., 2020). With increasing exposure time, the migration increases, and the load of adsorbed materials, like other pollutants, toxic heavy metals, antibiotics, increases. These toxin- and pollutant-loaded microplastics can affect and alter the soil texture and soil profile, the micro- and macrobiota

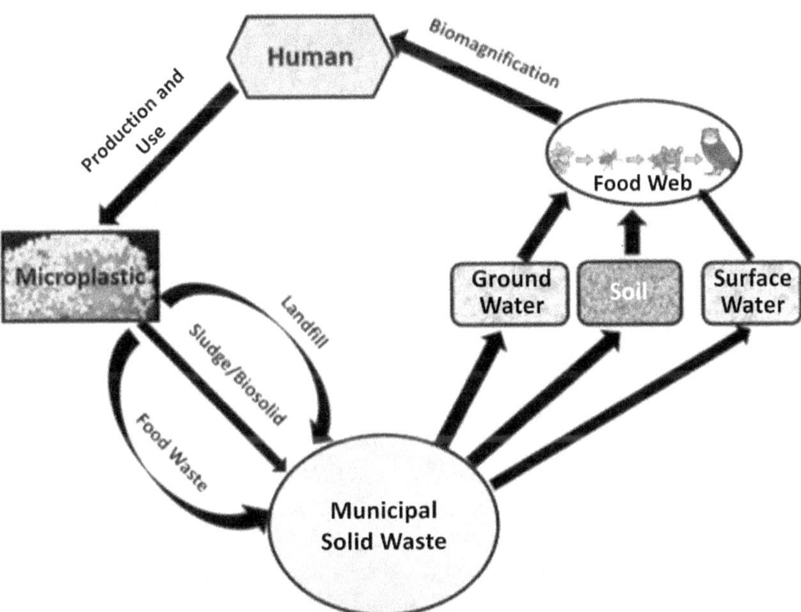

FIGURE 3.2 Source and migration of microplastic to and from municipal solid waste.

of rhizosphere soil, and the aquatic and terrestrial food chains (de Souza Machado et al., 2018; Li et al., 2022; Kaur et al., 2022; Groh et al., 2019; Deng et al., 2017). Microplastic incorporation into soil clumps and soil aggregation can change soil porosity, thus altering the water content of the soil (Wang et al., 2020). Desiccation cracking was also observed on the soil surface due to the disturbance of the water-holding capacity of soil (Wan et al., 2019). This can also lead to the migration of pollutants into deep soil layers along the cracks (Kumari et al., 2022). Being a carbon-rich polymer, MPs have a considerable effect on soil bulk density (de Souza Machado et al., 2018). They also have a significant effect on soil-inhabitant microbial enzymes like urease, catalase activities, fluorescein diacetate hydrolase (FDAse), and phenol oxidase (Huang et al., 2019). High concentrations of PP MPs (28% weight/weight) have been reported to increase significantly the enzymatic activity and dissolve organic matter (DOM) and pools of organic C, N, and P (Liu et al., 2017). Moreover, MP-induced soil parameter changes may lead to microhabitat loss and the extinction of indigenous microorganisms (Guo et al., 2020). The agricultural lands are susceptible to contamination via runoff.

2. **By Runoff and Wind:** Runoff and wind can transport microplastics from dumpsites to surface water. The high microplastic concentration in urban runoff is considered a potential threat for surface water (Kilponen, 2016). Microplastic concentrations in runoff from landfills need more investigations. Wind is another component, which can carry the plastic particles and then mix them with water and soil (Rezaei et al., 2016). Due to their light weight and low density, microplastics can easily float on water and get mixed with the water by wave action (Qiu et al., 2020). Wind can carry these lightweight plastics and microplastics from dumpsites to nearby agricultural lands or open land.
3. **By Leachate:** Leachate from dumpsites and landfills has been identified as a major source for surface water as well as groundwater contamination. Microplastics <1 μm are more capable of moving through the soil columns during leaching and contaminating the ground water (Mackevica and Hartmann, 2018). The vertical movement of microplastics from dumpsites can contaminate groundwater aquifer systems. The most common type of microplastic found in groundwater are polyethylene (PE) and polyethylene terephthalate (PET) (Khant and Kim, 2022). The microplastic abundance in the groundwater sample around the landfill site are in the range of 3–23 items/kg (Bharath K et al., 2023). Though, by using an impervious liner in a landfill, leaching can be controlled to a greater extent but eventually will increase the cost of handling MSW.
4. **By Terrestrial Organisms:** Microbes and terrestrial organisms such as earthworms (*Pheretima* sp.), mole rats (*Heterocephalus* sp.), field mice (*Bandicota* sp.), various scavenging birds related to the dumpsite can act as carriers of microplastics. They intake microplastics through the food chain and become a potential reservoir of it for their predators (Rillig et al., 2017). By attaching to the outer surface of the organisms, microplastics get carried

freely to another place. When the organisms forage in the plastic-loaded sites, microplastics get attached to their body fur, feather, or outer body surface and get transported to other places.

5. **By Using the Compost Prepared from Dumpsite Solids:** Studies show that almost 90% of MSW disposed of in a non-scientific manner impart negative effects on human health and the environment (Meena et al., 2023). To address these problems, the 3R approach (reduce, reuse, recycle) of an integrated waste management system was used to emphasize composting and recycling. Composting is a viable solution to deal with the organic part $(40-60\%)$ of the solid waste. Studies demonstrated the presence of toxic heavy metals in microplastics in commercially prepared compost from MSW. The concentration of microplastic in this type of compost amounted to 5–20 items/g of dry weight (Edo et al., 2022). Zhang et al. observed that after an annual application of 30 tons/ha of sludge-based fertilizer, the microplastic abundance was 545.9 items/kg (Zhang et al., 2020). Depending on the source and preparation method, the amount can vary. Hence there is a plausible opportunity for the microplastics to get transported and contaminate the agricultural land through the compost.

3.6 BIODEGRADATION OF MICROPLASTICS AND BACTERIA INVOLVED IN IT

Biodegradation is the process where microplastics undergo depolymerization and mineralization by microorganisms. Numerous microbes are mediated microplastic biodegradation pathways, not all of which are yet understood well (Yuan et al., 2020) and are under extensive research. Generally, in a natural environment, the biodegradation process involves four major steps:

1. **Colonization of Microbes on Microplastic Surface and Biofilm Formation:** Fungus hyphae get attached to the hydrophobic surface of plastics and create cracks on the surface by producing hydrophobins (Sánchez, 2020). Most of the bacteria need this crack for attachment to plastic surfaces (Sander, 2019). By colonizing on plastic surfaces, microbes form a thin layer known as biofilm. This biofilm, also known as plastisphere, is composed of orderly or complex arrangements of microorganisms that are highly dependent on the composition and types of additives present in the microplastics. The biofilm formed on the microplastics alter the structure by changing the carbonyl indices and double bonds present in it (Tu et al., 2020).
2. **Depolymerization of the Long Polymer by Extracellular Enzymes to Produce Oligomers, Dimers, Monomers:** During colonization, microbes can secrete some extracellular enzymes, like depolymerases, hydrolases, esterases, lipases, laccase, to break down the polymer backbone (Miri et al., 2022). These enzymes can convert side chains to functional groups and increase the hydrophilicity for further microbial attachment (Taniguchi et al., 2019). They also enhance side chain cleavage, resulting cracks on plastic surfaces (Miri et al., 2022).

3. **Assimilation of the Smaller Fragments by Microbial Cells:** After depolymerization, the metabolic intermediates get transported into the microbial cytoplasm for further metabolism (Restrepo-Florez et al., 2014). Intracellular enzymes like mono-oxygenase, hydrolases, and dioxygenases catalyze the assimilation process (Yuan et al., 2020).
4. **Complete Mineralization of Microplastics by Intracellular Enzymes:** Microbes use the polymer as a carbon source for the TCA cycle and produce byproducts like water, carbon-dioxide, and methane. The assimilated MPs with carbonyl or hydroxyl functional groups get metabolized by intracellular enzymes to produce completely oxidized metabolized-like water, CO_2 in aerobic conditions, and CH4 in anaerobic conditions (Yuan et al., 2020).

After reaching the environment microplastic undergo several physical and chemical breakdown processes like UV irradiation, photothermal oxidation, wind flow. These abiotic processes are comparatively faster than biodegradation, but they can only produce smaller pieces of plastics. Nevertheless, these processes facilitate effective biodegradation (Zhang et al., 2021). In solid waste dumping sites, microplastics get a brief exposure to an aerobic environment and then a prolonged period of anaerobic conditions, along with mechanical stress, varying pH, and temperature (Kjeldsen et al., 2002). The degradation procedure is also enhanced with the longer residential time of the microplastics (Canopoli et al., 2020) at the MSW dumping site.

For industrial-level microplastic treatment and separation, various techniques like microalgal adsorption, adaptable membrane filtration, photocatalytic degradation, electrocoagulation are available. Nevertheless, biodegradation is a relatively slower process, is the only complete degradation pathway, and is cost-effective.

3.7 BACTERIAL CONSORTIA INVOLVED IN BIODEGRADATION

Different microbial species coexist symbiotically, use the same resource, and are considered a microbial consortia. The in-situ biodegradation process of microplastic is associated with the involvement of a several microorganisms. In contrast to the in-vitro condition, where a single species of bacteria is being recruited to degrade the microplastic, bacterial consortia show higher efficiency of degradation. The vast diversity of microplastic in terms of source, structure, color, forms, additives, and the presence of other copollutants make the solid waste a harbor for microbial consortia. Microbial species and diversity play an important role in biodegradation (Canopoli et al., 2018). Several studies have demonstrated the involvement of biodegrading microorganisms in landfill and sludge dumping sites and are presented in Table 3.1. Park and Kim observed the association of *Bacillus* sp. and *Paenibacillus* sp. can degrade polyethylene effectively with a 22.8% reduction of PE particle diameter and 14.7% weight loss. A consortium of the fungus *Aspergillus clavatus* and the bacteria *Lysinibacillus xylanilyticus*, isolated from various municipal landfill sites of India and Iran, has proved to degrade LDPE (Gajendiran et al., 2016; Esmaeili et al., 2013). Muenmee et al. (2015) observed a good growth of types I and II methanotrophs associated with microplastic biodegradation in a semiaerobic landfill site. The anaerobic sludge consortium of *Cloacamonales* sp. and *Thermotogales* sp. were

TABLE 3.1
Microbial Consortia Involved in Biodegradation in Landfill and Solid Waste

Source of Microplastic	Microplastic Types	Microorganisms	References
Municipal landfill refuse India, Iran	LDPE	*Aspergillus clavatus* and *Lysinibacillus xylanilyticus*	Gajendiran et al., 2016; Esmaeili et al., 2013
Municipal landfill, Korea	PE	*Bacillus velezensis*, *Bacillus pseudomycoides*, *Paenibacillus alvei*, *Paenibacillus motobuensis*	Park and Kim, 2019
Simulated open landfill, Thailand	HDPE, LDPE, PP, PS	Type I/II methanotrophs (*Methylobacter* sp./*Methylocella* sp.)	Muenmee et al., 2015
Anaerobic sludge, USA	poly(3HB-co-3HHx) thermoplastics	*Cloacamonales* sp. And *Thermotogales* sp.	Wang et al., 2018
Municipal solid waste and soil bed, India	HDPE, LDPE	*Bacillus cereus*, *Bacillus pumilus*, and *Arthrobacter* sp.	Satlewal et al., 2008

associated with poly(3HB-co-3HHx) thermoplastics degradation (Wang et al., 2018). Satlewal et al. isolated a microbial consortium of *Bacillus cereus*, *Bacillus pumilus*, and *Arthrobacter* sp. from the municipal solid waste and soil bed of Pantnagar, India, which can effectively degrade HDPE and LDPE.

3.8 NEW MICROBIAL CONSORTIUM: INTRODUCTION TO SOLID WASTE DISPOSAL SITE

Researchers observed and documented the prevalence of polyethylene, polystyrene, polypropylene, and polyethylene terephthalate in landfill and municipal solid waste dumping sites of India (Kabir et al., 2023; Sekar and Sundaram, 2023). The in-situ degradation of these predominant microplastics need a group of symbiotic microorganisms with intercellular communication by secreted enzymes and byproducts (Jaiswal et al., 2019). This symbiotic characteristic of microbial consortia is important for plastic mineralization, which is performed by various enzymes and their synthetic pathways. That synergistic activity usually does not exist in a single strain, but microbial consortium easily performed it by division of labor. In general, fungi show a faster rate of degradation than bacteria, probably due to their complex enzymatic activities. Hence an effective consortium for biodegradation generally consists of both fungus and bacteria (Zhu et al., 2023). The microbial consortium is also intentionally optimized and tailored as per the experimental need (Navarro-Díaz et al., 2016). However, in solid waste and landfill sites, biodegradation of plastics is executed by bacterial consortia due to the presence of a variety of microplastics. Interestingly, all examples of practical application of microplastic-degrading microbial consortia were isolated from landfill refuses, sludge, and solid waste (Yuan et al., 2020). As biodegradation is a comparatively slower process, an organized and tailored microbial consortia can be effective. Careful selection of indigenous microorganisms depending on the prevalent MP types and the physicochemical parameters of the dumping site can create an effective biome for faster degradation.

Studies from various Indian landfill sites with LDPE prevalence show the presence of microorganisms like *Aspergillus clavatus*, *Lysinibacillus xylanilyticus*, *Bacillus cereus*, *Bacillus pumilus*, and *Arthrobacter* (Gajendiran et al., 2016; Esmaeili et al., 2013; Satlewal et al., 2008). Designing a microbial consortium with these organisms and introducing it to a solid waste disposal site can enhance the biodegradation rate of LDPE of that site. Optimum pH, temperature, humidity of the selected organisms and the same parameters of the desired site is important for organizing the biota. Auta et al. isolated eight different bacteria from mangrove soil and tested their degradation capability of PET and PS. They further investigated the in-situ bioremediation of mangrove soil that was artificially contaminated with PET and PS microplastics using indigenous microbial consortium (Auta et al., 2022).

3.9 CHALLENGES TO OVERCOME

Lots of difficulties have to be overcome for the successful introduction of the designed microbial consortia and enhancement of the biodegradation rate.

- Selection of effective and correct microbes for designing the consortia is a tough job and needs immense research to understand the characteristics of microbes and microplastics. Particularly, designing a consortium for solid waste dumping grounds is challenging because they possess a vast variety of microplastics. These sites are generally rich in indigenous microbes with biodegradation capability.
- The introduction of a single or group of exogenous microorganisms to a site and their interaction to the indigenous microbe species are a complex procedure. It can be symbiotic, competitive, or harmful for the native species. A harmful association can damage the preexisting microbial biome by producing degrading enzymes and by altering the pH, humidity, and altering other parameters. Hence the introduced microbes should be symbiotic with the indigenous microbes to enhance the degradation rate.
- Preexisting pollutants, antibiotics, and heavy metals can negatively affect the biodegradation capability of the new microbes. These agents have a complex interaction pattern with themselves and with the microplastics.

3.10 CONCLUSION

Increasing production of plastic and poor waste management are the major causes of plastic pollution. Some 55% of used plastics culminate in solid waste dumping sites and landfills. Therefore, the municipal solid waste disposal sites turn into reservoirs as well as the main source of microplastic pollution for both the aquatic and terrestrial ecosystems. MSW finds wide application as compost in agricultural fields, which is usually obtained by commercial composting of organic solid waste. Microplastic-rich MSW results in high microplastic loading in the compost, which is a significant source of plastic in soil, surface water, and groundwater. Bioremediation is the most promising solution of this problem as only microbes can completely mineralize the plastics by utilizing them as a nutrient source. Ex-situ biodegradation experiments at the laboratories can build an understanding of the mechanisms, pathways, and molecular basis of enzymatic reactions, but in-situ degradation of microplastics at the reservoirs like the sea, mangrove sediment, and urban solid waste dumping grounds is a complex phenomenon due to various interconnecting factors. The multifaceted nature of these sites, with respect to the biotic components like a large variety of macro- and microorganisms and the abiotic components like mechanical stress, various pH levels, temperature, anaerobic conditions, make in-situ biodegradation challenging for scientists. The slow biodegradation rate of plastic is another challenge that needs to be overcome for effective microplastic degradation. Bioaugmentation by introducing indigenous and exogenous plastic degrading microbes can solve the problem to a great extent. Creation of an effective microbial consortium by selecting the right microbes and introducing them to the plastic dumping sites can enhance the degradation rate. The large-scale application of this process needs more intense research work and is an ongoing process.

REFERENCES

Al-Malaika, S., Axtell, F., Rothon, R., and Gilbert, M. 2017. Additives for plastics. In *Brydson's plastics materials* (pp. 127–168). Butterworth-Heinemann.

Auta, H.S., Abioye, O.P., Aransiola, S.A., Bala, J.D., Chukwuemeka, V.I., Hassan, A., Aziz, A., and Fauziah, S.H. 2022. Enhanced microbial degradation of PET and PS microplastics under natural conditions in mangrove environment. *Journal of Environmental Management*, 304, 114273.

Auta, H.S., Emenike, C.U., Jayanthi, B., and Fauziah, S.H. 2018. Growth kinetics and biodeterioration of polypropylene microplastics by Bacillus sp. and Rhodococcus sp. isolated from mangrove sediment. *Marine Pollution Bulletin*, 127, 15–21.

Bharath, K.M., Muthulakshmi, A.L., and Natesan, U. 2023. Microplastic contamination around the landfills: Distribution, characterization and threats: A review. *Current Opinion in Environmental Science & Health*, 31, 100422.

Blair Espinoza, R.M. 2019. Microplastics in wastewater treatment systems and receiving waters (Doctoral dissertation, University of Glasgow).

Canopoli, L., Beatriz, F., Frederic, C., and Stuart, T.W. 2018. Physico-chemical properties of excavated plastic from landfill mining and current recycling routes. *Waste Management*, 76, 55–67.

Canopoli, L., Coulon, F., and Wagland, S.T. 2020. Degradation of excavated polyethylene andpolypropylene waste from landfill. *Science of the Total Environment*, 698, 134125.

Carr, S.A., Liu, J., and Tesoro, A.G. 2016. Transport and fate of microplastic particles in wastewater treatment plants. Water Research, 91, 174–182.

Conti, G.O., Ferrante, M., Banni, M., Favara, C., Nicolosi, I., Cristaldi, A., . . . Zuccarello, P. 2020. Micro-and nano-plastics in edible fruit and vegetables. The first diet risks assessment for the general population. *Environmental Research*, 187, 109677.

Crawford, C.B., and Quinn, B. 2016. *Microplastic pollutants*. Elsevier Limited.

de Souza Machado, A.A., Kloas, W., Zarfl, C., Hempel, S., and Rillig, M.C. 2018. Microplastics as an emerging threat to terrestrial ecosystems. *Global Change Biology*, 24(4), 1405–1416.

Deng, Y., Zhang, Y., Lemos, B., and Ren, H. 2017. Tissue accumulation of microplastics in mice and biomarker responses suggest widespread health risks of exposure. *Scientific Reports*, 7(1), 46687.

Edo, C., Fernández-Piñas, F., and Rosal, R. 2022. Microplastics identification and quantification in the composted organic fraction of municipal solid waste. *Science of the Total Environment*, *813*, 151902.

Esmaeili, A., Pourbabaee, A.A., Alikhani, H.A., Shabani, F., and Esmaeili, E. 2013. Biodegradation of low-density polyethylene (LDPE) by mixed culture of Lysinibacillusxylanilyticus and Aspergillus niger in soil. *PLoS One*, 8(9), e71720.

Falco, D.F., Gullo, M.P., Gentile, G., Pace, E.D., Cocca, M., Gelabert, L., Agnésa, M.B., Rovira, A., Escuder, R., Villalba, R., Mossotti, R., Montarsolo, A., Gavignano, S., Tonin, C., and Avella, M. 2018. Evaluation of microplastic release caused by textile washing processes of synthetic fabrics. *Environmental Pollution*, 236, 916–925.

Gajendiran, A., Krishnamoorthy, S., and Abraham, J. 2016. Microbial degradation of low-density polyethylene (LDPE) by *Aspergillus clavatus* strain JASK1 isolated from landfill soil. *3 Biotech*, 6, 1–6.

Gies, E.A., LeNoble, J.L., Noël, M., Etemadifar, A., Bishay, F., Hall, E.R., and Ross, P.S. 2018. Retention of microplastics in a major secondary wastewater treatment plant in Vancouver, Canada. *Marine Pollution Bulletin*, 133, 553–561.

Golwala, H., Zhang, X., Iskander, S.M., and Smith, A.L. 2021. Solid waste: An overlooked source of microplastics to the environment. *Science of the Total Environment*, 769, 144581.

Groh, K.J., Backhaus, T., Carney-Almroth, B., Geueke, B., Inostroza, P.A., Lennquist, A., . . . Muncke, J. 2019. Overview of known plastic packaging-associated chemicals and their hazards. *Science of the Total Environment*, 651, 3253–3268.

Guo, J.J., Huang, X.P., Xiang, L., Wang, Y. Z., Li, Y.W., Li, H., . . . Wong, M.H. 2020. Source, migration and toxicology of microplastics in soil. *Environment International*, 137, 105263.

He, P., Chen, L., Shao, L., Zhang, H., and Lü, F. 2019. Municipal solid waste (MSW) landfill: A source of microplastics?—evidence of microplastics in landfill leachate. *Water Research*, 159, 38–45. https://pib.gov.in/PressReleaselframePage.aspx

Huang, Y., Zhao, Y., Wang, J., Zhang, M., Jia, W., and Qin, X. 2019. LDPE microplastic films alter microbial community composition and enzymatic activities in soil. *Environmental Pollution*, 254, 112983.

Jaiswal, S., Sharma, B., and Shukla, P. 2019. Integrated approaches in microbial degradation of plastics. *Environmental Technology & Innovation*, 100567. https://doi.org/10.1016/j.eti.2019.100567.

Jin, M., Wang, X., Ren, T., Wang, J., and Shan, J. 2021. Microplastics contamination in food and beverages: Direct exposure to humans. *Journal of Food Science*, 86(7), 2816–2837.

Kabir, M.S., Wang, H., Luster-Teasley, S., Zhang, L., and Zhao, R. 2023. Microplastics in landfill leachate: Sources, detection, occurrence, and removal. *Environmental Science and Ecotechnology*, 16, 100256. https://doi.org/10.1016/j.ese.2023.100256

Kalogerakis, N., Arff, J., Banat, I.M., Broch, O.J., Daffonchio, D., Edvardsen, T., . . . Fava, F. 2015. The role of environmental biotechnology in exploring, exploiting, monitoring, preserving, protecting and decontaminating the marine environment. *New Biotechnology*, 32(1), 157–167.

Kaur, K., Reddy, S., Barathe, P., Oak, U., Shriram, V., Kharat, S.S., . . . Kumar, V. 2022. Microplastic-associated pathogens and antimicrobial resistance in environment. *Chemosphere*, 291, 133005.

Khant, N.A., and Kim, H. 2022. Review of current issues and management strategies of microplastics in groundwater environments. *Water*, 14(7), 1020.

Kilponen, J. 2016. *Microplastics and harmful substances in urban runoffs and landfill leachates: Possible emission sources to marine environment*; Faculty of Technology, Environmental Technology, Lahti University of Applied Sciences: Lahti, Finland.

Kjeldsen, P., Barlaz, A.M., Rooker, A.P., Baun, A., Ledin, A., and Christensen, T H. 2002. Present and long-term composition of MSW landfill leachate: A review. *Critical Reviews in Environmental Science and Technology*, 32(4), 297–336.

Klein, S., Dimzon, I.K., Eubeler, J., and Knepper, T.P. (2018). Analysis, occurrence, and degradation of microplastics in the aqueous environment. In: Wagner, M., Lambert, S. (Eds.), *Freshwater microplastics: Emerging environmental contaminants?* Springer International Publishing, Cham, pp. 51–67.

Kumari, A., Rajput, V.D., Mandzhieva, S.S., Rajput, S., Minkina, T., Kaur, R., Glinushkin, A.P. 2022. Microplastic pollution: An emerging threat to terrestrial plants and insights into its remediation strategies. *Plants*, 11(3), 340.

Lares, M., Ncibi, M.C., Sillanpää, M., and Sillanpää, M. 2018. Occurrence, identification and removal of microplastic particles and fibers in conventional activated sludge process and advanced MBR technology. *Water Research*, 133, 236–246.

Li, J., Yu, S., Yu, Y., and Xu, M. 2022. Effects of microplastics on higher plants: A review. *Bulletin of Environmental Contamination and Toxicology*, 109(2), 241–265.

Lin, L., Li, H., Hong, H., Yuan, B., Sun, X., He, L., . . . Yan, C. 2022. Enhanced heavy metal adsorption on microplastics by incorporating flame retardant hexabromocyclododecanes: Mechanisms and potential migration risks. *Water Research*, 225, 119144.

Liu, H., Yang, X., Liu, G., Liang, C., Xue, S., Chen, H., Geissen, V. 2017. Response of soil dissolved organic matter to microplastic addition in Chinese loess soil. *Chemosphere*, 185, 907–917.

Liu, X., Lei, T., Boré, A., Lou, Z., Abdouraman, B., and Ma, W. 2022. Evolution of global plastic waste trade flows from 2000 to 2020 and its predicted trade sinks in 2030. *Journal of Cleaner Production*, 376, 134373.

Mackevica, A., and Hartmann, N.B. 2018. *Mikroplast i grundvand-En vurdering af potentialet for forekomst af mikroplast i dansk grundvand*. Danmarks Tekniske Universitet (DTU). https://mst.dk/media/148257/bilag-3-notatmikroplast-i-grundvand.pdf

Meena, M.D., Dotaniya, M.L., Meena, B.L., Rai, P.K., Antil, R.S., Meena, H.S., . . . Meena, R.B. 2023. Municipal solid waste: Opportunities, challenges and management policies in India: A review. *Waste Management Bulletin*, 1(1), 4–18.

Miri, S., Saini, R., Davoodi, S.M., Pulicharla, R., Brar, S.K., and Magdouli, S. 2022. Biodegradation of microplastics: Better late than never. *Chemosphere*, 286, 131670.

Muenmee, S., Chiemchaisri, W., and Chiemchaisri, C. 2015. Microbial consortium involving biological methane oxidation in relation to the biodegradation of waste plastics in a solid waste disposal open dump site. *International Biodeterioration & Biodegradation*, 172–181.

Navarro-Díaz, M., Valdez-Vazquez, I., and Escalante, A.E. 2016. Ecological perspectives of hydrogen fermentation by microbial consortia: What we have learned and the way forward. *International Journal of Hydrogen Energy*, 41, 17297–17308.

Park, S.Y., and Kim, C.G. 2019. Biodegradation of micro-polyethylene particles by bacterial colonization of a mixed microbial consortium isolated from a landfill site. *Chemosphere*, 222, 527–533.

Petrović, M., Mihajlović, I., Tubić, A., and Novaković, M. 2022. Microplastic in municipal solid waste landfills. *Current Opinion in Environmental Science & Health*, 100428.

Porterfield, K.K., Hobson, S.A., Neher, D.A., Niles, M.T., and Roy, E.D. 2023. Microplastics in composts, digestates, and food wastes: A review. *Journal of Environmental Quality*, 52(2), 225–240.

Qiu, R., Song, Y., Zhang, X., Xie, B., & He, D. 2020. Microplastics in urban environments: Sources, pathways, and distribution. *Microplastics in Terrestrial Environments: Emerging Contaminants and Major Challenges*, 41–61.

Restrepo-Flórez, J.M., Bassi, A., and Thompson, M.R. 2014. Microbial degradation and deterioration of polyethylene–A review. *International Biodeterioration & Biodegradation*, 88, 83–90.

Rezaei, M., Riksen, M.J., Sirjani, E., Sameni, A., and Geissen, V. 2019. Wind erosion as a driver for transport of light density microplastics. *Science of the Total Environment*, 669, 273–281.

Rillig, M.C., Ingraffia, R., and de Souza Machado, A.A. 2017. Microplastic incorporation into soil in agroecosystems. *Frontiers in Plant Science*, 8, 1805.

Rist, S., Almroth, B.C., Hartmann, N.B., and Karlsson, T.M. 2018. A critical perspective on early communications concerning human health aspects of microplastics. *Science of the Total Environment*, 626, 720–726.

Sánchez, C. 2020. Fungal potential for the degradation of petroleum-based polymers: An overview of macro-and microplastics biodegradation. *Biotechnology Advances*, 40, 107501.

Sander, M., Kohler, H.-P.E., and McNeill, K. 2019. Assessing the environmental transformation of nanoplastic through 13C-labelled polymers. *Nature Nanotechnology*, 14, 301–303.

Santana, M.F.M., Ascer, L.G., Custódio, M.R., Moreira, F.T., and Turra, A. 2016. Microplastic contamination in natural mussel beds from a Brazilian urbanized coastal region: Rapid evaluation through bioassessment. *Marine Pollution Bulletin*, 106(1–2), 183–189.

Satlewal, A., Soni, R., Zaidi, M., Shouche, Y., and Goel, R. 2008. Comparative biodegradation of HDPE and LDPE using an indigenously developed microbial consortium. *Journal of Microbiology and Biotechnology*, 18(3), 477–482.

Sekar, V., and Sundaram, B. 2023. Preliminary evidence of microplastics in landfill leachate, Hyderabad, India. *Process Safety and Environmental Protection*, 175, 369–376.

Shi, X., Chen, Z., Wu, L., Wei, W., and Ni, B.J. 2022. Microplastics in municipal solid waste landfills: Detection, formation and potential environmental risks. *Current Opinion in Environmental Science & Health*, 100433.

Singh, S., and Bhagwat, A. (2022). Microplastics: A potential threat to groundwater resources. *Groundwater for Sustainable Development*, *19*, 100852.

Taniguchi, I., Yoshida, S., Hiraga, K., Miyamoto, K., Kimura, Y., and Oda, K. 2019. Biodegradation of PET: Current status and application aspects. *ACS Catalysis*, 9(5), 4089–4105.

Tu, C., Chen, T., Zhou, Q., Liu, Y., Wei, J., Waniek, J.J., and Luo, Y. 2020. Biofilm formation and its influences on the properties of microplastics as affected by exposure time and depth in the seawater. *Science of the Total Environment*, 734, 139237.

Walker, T. R., and Fequet, L. 2023. Current trends of unsustainable plastic production and micro (nano) plastic pollution. *TrAC Trends in Analytical Chemistry*, 116984.

Wan, Y., Wu, C., Xue, Q., and Hui, X. 2019. Effects of plastic contamination on water evaporation and desiccation cracking in soil. *Science of the Total Environment*, 654, 576–582.

Wang, C., Liu, Y., Chen, W.Q., Zhu, B., Qu, S., and Xu, M. 2021. Critical review of global plastics stock and flow data. *Journal of Industrial Ecology*, 25(5), 1300–1317.

Wang, S., Lydon, K.A., White, E.M., Grubbs, J.B., Lipp, E.K., Locklin, J., and Jambeck, J.R. 2018. Biodegradation of poly(3-hydroxybutyrate- co-3-hydroxyhexanoate)plastic under anaerobic sludge and aerobic seawater conditions: Gas evolution and microbial diversity. *Environmental Science & Technology,* 52(10), 5700–5709.

Wang, T., Wang, L., Chen, Q., Kalogerakis, N., Ji, R., and Ma, Y. 2020. Interactions between microplastics and organic pollutants: Effects on toxicity, bioaccumulation, degradation, and transport. *Science of the Total Environment*, 748, 142427.

Wright, S.L., and Kelly, F.J. 2017. Plastic and human health: A micro issue? *Environmental Science & Technology*, 51(12), 6634–6647.

Yuan, J., Ma, J., Sun, Y., Zhou, T., Zhao, Y., and Yu, F. 2020. Microbial degradation and other environmental aspects of microplastics/plastics. *Science of the Total Environment*, 715, Article 136968.

Ziajahromi, S., Neale, P.A., Rintoul, L., and Leusch, F.D. 2017. Wastewater treatment plants as a pathway for microplastics: Development of a new approach to sample wastewater-based microplastics. *Water Research*, 112, 93–99.

Zhang, L., Xie, Y., Liu, J., Zhong, S., Qian, Y., and Gao, P. 2020. An overlooked entry pathway of microplastics into agricultural soils from application of sludge-based fertilizers. *Environmental Science & Technology*, 54(7), 4248–4255.

Zhang, S., Wang, J., Yan, P., Hao, X., Xu, B., Wang, W., and Aurangzeib, M. 2021. Non-biodegradable microplastics in soils: A brief review and challenge. *Journal of Hazardous Materials*, *409*, 124525.

Zhu, J., Dong, G., Feng, F., Ye, J., Liao, C.H., Wu, C.H., and Chen, S.C. 2023. Microplastics in the soil environment: Focusing on the sources, its transformation and change in morphology. *Science of the Total Environment*, *896*, 165291. https://doi.org/10.1016/j.scitotenv.2023.165291

4 Microplastics in Atmospheric Pathways, Depositions, and Remediation Techniques

Nabasmita Barman, Ratna Dutta, and Papita Das

4.1 INTRODUCTION

Microplastics (MPs) are tiny pieces of plastic with regular or irregular shapes and sizes ranging from 1 µm to 5 mm (NOAA, 2023); that is about the size of a grain of rice or smaller. Microplastics comprise mainly carbon and hydrogen atoms bounded together in a polymeric chain and also contain other chemicals like phthalates, polybrominated diphenyl ethers, tetrabromobisphenol, and many more to regulate and improve its quality. These small synthetic plastic particles can be present in different environments, including the aquatic environment (oceans, rivers, lakes), terrestrial environment (soil), and even the atmosphere. They are produced from the fragmentation of larger plastic items, such as plastic bottles, plastic carry bags, and packaging materials (Fadare *et al.*, 2020). Additionally, they also generated through natural processes like weathering and UV radiation, as well as human activities like manufacturing chemicals and disposal of wastes. On the basis of their origin, MPs can be present in the environment in two forms as primary MPs and secondary MPs:

4.1.1 Primary MPs

These are the small plastic pieces that are intentionally produced at a very tiny particle size for various industrial, commercial, and consumer purposes. Unlike secondary MPs, which manufactured from the disintegration of larger plastic products, primary MPs are designed and produced as micro-sized materials for specific applications.

4.1.1.1 Types and Sources

- **Microbeads:** These are tiny, spherical plastic particles commonly present in individual care products and an individual's cosmetic products (Duis and Coors, 2016), like exfoliating scrubs, toothpaste, and shower gels, to

DOI: 10.1201/9781032684574-4

provide texture and abrasiveness. They are typically made of polyethylene or other synthetic polymers.

- **Pellets or Nurdles:** These are preproduction plastic pellets used as raw materials for the manufacturing of plastic products. They can accidentally escape into the environment during transport, handling, or the production process.
- **Microfibers:** Tiny synthetic fibers are shed from textiles made of materials like polyester, nylon, and acrylic. These fibers can be released during the washing of clothing and textiles.
- **Powders and Granules:** Small plastic particles, often used in industrial processes like sandblasting, as well as in various products like cleaning agents and paints.

4.1.1.2 Environmental Impact

Primary MPs pose significant environmental concerns because they are designed to be persistent and durable. They do not readily break down in the environment and can persist for long periods. These small particles can be ingested by aquatic organisms, entering the food chain and potentially affecting the health of marine life. Microbeads and microfibers, in particular, are of concern because they are prevalent in aquatic ecosystems, posing risks to aquatic organisms and potentially making their way into human diets.

4.1.1.3 Regulations and Bans

Many countries and regions have recognized the environmental threat posed by primary MPs and have taken steps to regulate or ban their use. For instance, the United States passed the Microbead-Free Waters Act in 2015, which prohibited the manufacture and sale of personal care products containing plastic microbeads (www.fda.gov/cosmetics/cosmetics-laws-regulations/microbead-free-waters-act-faqs). The European Union has also adopted measures to restrict the intentional use of MPs in certain products, such as cosmetics and detergents.

4.1.1.4 Alternatives to Reduce Formation of Primary MPs

In response to concerns about primary MPs, some industries have sought alternatives. For example, natural exfoliants like ground nutshells, salt, or sugar can replace microbeads in personal care products. Manufacturers are exploring ways to reduce microfiber shedding from textiles by modifying production techniques or developing fabrics with fewer microfibers. Improved waste management, recycling, and reduction in the use of single-use plastics can also help mitigate the release of primary MPs into the environment.

4.1.2 Secondary MPs

Secondary MPs are tiny plastic particles that result from the breakdown of larger plastic items through various physical, chemical, or biological processes. These larger plastic items can be anything from bottles and bags to fishing nets and industrial waste.

4.1.2.1 Sources

- **Mechanical Breakdown:** This occurs when larger plastic items like bottles, packaging materials, and fishing gear are exposed to environmental forces such as sunlight, wind, waves, and abrasion. Over time, these forces can cause the plastic to break into smaller and smaller pieces.
- **Chemical Degradation:** Exposure to sunlight, heat, and environmental chemicals can cause chemical reactions that weaken and break down plastic polymers. This process is known as photodegradation.
- **Biological Processes:** Some marine organisms, such as bacteria and algae, can produce enzymes that break down plastic, leading to the formation of smaller plastic fragments.

4.1.2.2 Environmental Impact

- **Marine Environment:** Secondary MPs are a major concern in the world's oceans and water bodies. Marine animals often mistake these tiny plastic particles for food, leading to ingestion. This ingestion can harm marine life and potentially enter the food chain, posing risks to human health.
- **Terrestrial Environment:** Secondary MPs can also accumulate in terrestrial environments through various mechanisms, including the breakdown of plastic waste in landfills and the spread of plastic particles through the atmosphere.

4.1.2.3 Transport and Persistence

Secondary MPs can be transported over long distances by wind, water currents, and wildlife, contributing to their widespread distribution. Due to their small size and resistance to biodegradation, they can persist in the environment for extended periods.

4.1.2.4 Mitigation and Management

Efforts to address the issue of secondary MPs include reducing plastic waste generation, improving waste management practices, and promoting recycling and the use of biodegradable plastics. Additionally, research is ongoing to develop technologies for the efficient removal and monitoring of MPs in natural environments.

Secondary MPs are a pervasive environmental issue resulting from the breakdown of larger plastic materials. Their small size, widespread distribution, and potential ecological and human health risks underscore the importance of addressing this problem through a combination of regulatory measures, waste reduction, and scientific research to better understand their impact on ecosystems and human well-being.

4.1.3 MPs Cycle

The cycle of airborne MPs in the environment is complex and can have significant impacts on ecosystems and human health (refer to Figure 4.1).

- **Direct Atmospheric Emissions:** MPs can be released directly into the atmosphere through various human activities, such as industrial processes,

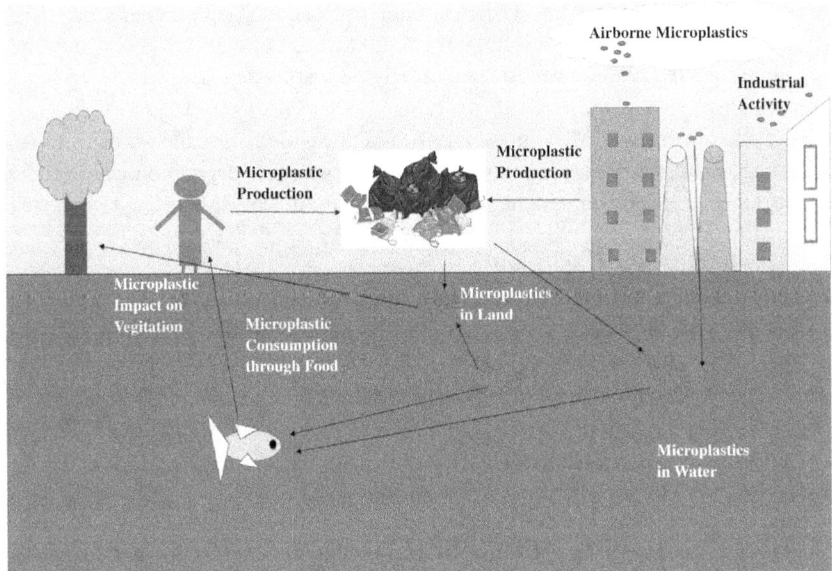

FIGURE 4.1 Microplastics cycle.

road traffic, and construction. These sources often produce larger plastic debris that can break down into smaller particles due to weathering and mechanical forces.

- **Weathering and Degradation:** Larger plastic items, such as bottles and bags, break down into smaller particles due to weathering, UV radiation, and mechanical wear and tear. These smaller plastic fragments can become airborne.
- **Transport in the Atmosphere:** Once in the atmosphere, airborne MPs can be transported over long distances by wind (Munyaneza *et al.*, 2022). They can remain suspended in the air for extended periods, depending on their size and weight. Atmospheric conditions, such as temperature, humidity, and air currents, influence the movement and distribution of these particles (Alfonso *et al.*, 2021).
- **Deposition:** MPs eventually settle out of the air and deposit onto the Earth's surface. Deposition can occur through dry deposition (the settling out of airborne MPs onto surfaces without the assistance of moisture) (Szewc *et al.*, 2021), wet deposition (carried to the ground by precipitation, such as rain and snow) (Sun *et al.*, 2022), gravity settling (heavier MPs settle faster and may accumulate on the ground or water bodies).
- **Ecological Impact:** MPs can enter terrestrial and aquatic ecosystems, potentially harming wildlife. In aquatic environments, they can be ingested by aquatic organisms, entering the food web and posing health risks to both aquatic life and humans. On land, MPs can accumulate in soil, affecting soil health and potentially impacting plants and terrestrial animals. Humans

can be exposed to airborne MPs through inhalation (Karbalaei *et al.*, 2018). Although the health effects are not yet fully understood, there is concern that inhaling MPs could have adverse effects on respiratory health.

The cycle of airborne MPs in the environment involves the release of tiny plastic particles into the air, their transport over long distances, deposition onto land and water surfaces, ecological impacts, and potential human exposure in a cycling process.

4.1.3.1 Effect of MPs Pollution

MPs pose several environmental and health concerns. MPs are pervasive in the environment and can be ingested by a wide range of organisms, including fish, birds, and even microscopic marine organisms. This ingestion can lead to physical harm, blockages, and potential transfer of toxic chemicals up the food chain. MPs can disrupt ecosystems by altering nutrient cycling and contaminating habitats. They can also transport harmful pathogens and invasive species. Following two ecosystems affected by MPs contamination is worth mentioning:

- **Terrestrial Ecosystem Contamination:** Airborne MPs can settle on land and water surfaces, contaminating ecosystems. This contamination can disrupt natural processes and potentially harm wildlife.
- **Aquatic Ecosystems Contamination:** When airborne MPs are deposited in water bodies, they can have detrimental effects on aquatic life, including ingestion by aquatic organisms and the introduction of MPs into aquatic food webs.

There is growing concern about the potential health risks associated with MPs too. While research is ongoing, there is evidence that MPs can enter the human body through various sources, including food, water, and air. The long-term health effects of this exposure are as follows:

- **Respiratory Health:** When people inhale (Karbalaei *et al.*, 2018) airborne MPs, these particles can reach the respiratory system, including the lungs. This may lead to inflammation and irritation of the airways, similar to the effects of inhaling other particulate matter. Respiratory problems, such as coughing and asthma-like symptoms, could result from exposure to high levels of airborne MPs.
- **Cardiovascular Health:** There is some concern that the inflammation triggered by exposure to airborne MPs could also have systemic effects on cardiovascular health (Prata, 2018). Inflammation is associated with various heart and vascular conditions.
- **Toxic Chemical Exposure:** MPs can act as carriers for toxic chemicals. They can absorb and concentrate pollutants from the environment, such as pesticides and heavy metals. When these MPs are inhaled, there is the potential for the release of these chemicals in the body, which could have adverse health effects.

- **Immune System Effects:** Prolonged exposure to airborne MPs could potentially affect the immune system. The body may react to the presence of foreign MP particles by triggering immune responses, although the extent and nature of these responses are still being studied.

MPs can accumulate in soil, potentially affecting plant growth and agricultural productivity. Their presence may also lead to the transfer of MPs to crops, posing a risk to food safety.

Efforts to mitigate the impact of MPs include reducing plastic waste through recycling and responsible disposal, banning or phasing out certain plastic products, and developing technologies to capture and remove MPs from the environment. Additionally, there is a need for continued research to better understand its direct and indirect threat and its potential consequences for ecosystems, the environment, and human health. Therefore, the objective of the article is to present a detailed focus on MPs atmospheric corridors, its pattern of deposition, and its remediation outline to make strategic plan to control MPs pollution.

4.2 SOURCES

MPs, which are ubiquitously present in the environment, have become one of the emerging major environmental contaminants worldwide in recent times. Though they are normally associated with marine pollution, their existence in the atmosphere is similarly quite alarming. MPs can be generated through various processes like weathering, wind erosion, etc. These airborne MPs are produced from numerous sources in different environmental conditions. These sources can be divided into two types: active source and passive source. Additionally, their dispersion through the air gives rise to potential risks to ecosystems and human health both. This study examines the potential sources of airborne MPs. It also highlights the gravity of the addressed environmental issues. The two types of sources of airborne MPs are discussed next.

4.2.1 Active Sources of Airborne MPs

Active sources of MPs are those source from which MPs are released into the environment directly. The main active sources of MPs are as follows:

- **Vehicular Emissions:** Vehicular emission is predominantly one of the potential active sources of airborne MPs originating from various forms as follows:

  Tire Wear: When the vehicle travels over the road, the abrasion between the road surface and tire of the car results in tear and tire wire (Kole *et al.*, 2017). Additionally, this process releases tiny particles of rubber that contain MPs (also referred to as 'tire wear particles' or 'tire dust') into the air. Tire dust generally remains as suspended matter in the atmosphere, which then easily carried long distances by winds.

Road Surface Degradation: Plastic-based materials are used in the maintenance and construction of road surfaces. The materials can be broken down over time to release MPs in the atmosphere.

Brake Dust: The disc brake present in the breaking system of any modern vehicle can produce MPs containing dust. Moreover, the plastic containing pads of the brake also contribute to the production of airborne MPs.

Vehicle Body Degradation: The parts of the vehicle body, after degrading over time in the harsh conditions of the environment, lead to the generation of millions of MPs into the air. After released into the air, the small MPs which can travel over long distances through the air.

- **Industrial Activities:** Another prominent source of MPs is industrial emission. For example:

  Plastic Manufacturing: During the production of plastic products, which includes the processing of the raw materials and the extrusion or molding of the plastic products, the release of MP particles into the atmosphere occurs. Plastic manufacturing industries, especially those involving production of single-use plastic product such as drinking bottles, plastic bags, straws, cups, are recognized as potential source of MPs (Fadare *et al.,* 2020).

  Plastic Processing: Plastic manufacturing equipment and instruments can indeed contribute to the generation of airborne MPs. Many plastic manufacturing processes involve mechanical actions such as cutting, grinding, and sanding plastic materials. These actions can create dust and particles that are released into the air, potentially including MPs. For example, in plastic fabrication shops or recycling facilities, cutting or grinding plastic materials can generate airborne MP particles. Moreover, in plastic processing, plastic pellets are heated and shaped into various products that are generally well contained within machines, but there may be instances of plastic dust or particles escaping into the surrounding environment, especially if the equipment is not properly maintained or if there are design flaws. Additionally, over time, the machinery and equipment used in plastic manufacturing degrade and release MPs to the atmosphere.

  Waste Management and Recycling: Inappropriate handling of plastic waste at various industrial facilities like landfilling and recycling accelerates the scattering of the MPs into the air.

- **Atmospheric Weathering:** Generally, after exposure to the outdoor environment from the industrial settings, plastics can undergo weathering because of temperature fluctuation, wind. and sunlight. This process gives rise to the breakdown of larger plastic particles into smaller particles and their release into the air. Once airborne, these MPs can be transported over long distances by wind currents, potentially leading to widespread distribution in the atmosphere. They can settle back to the ground or be deposited into water bodies.
- **Agricultural Practices:** One of the emerging source of MPs and a matter of concern is agricultural practices (Tian *et al.,* 2022). Here are some ways in which agricultural practices can contribute to the generation and dissemination of airborne MPs:

Plastic Mulch: In conventional agriculture, plastic mulches are mostly used for enhancing crop production. They are used for controlling weeds, improving the soil temperature, and retaining moisture. Due to long time exposures to sunlight and various mechanical stresses, the degradation of plastic mulch starts, which results in the release of MP particles into the air.

Irrigation and Rainwater Runoff: Plastic films, beads, or materials used in agricultural practices can become disintegrates due to weathering or physical stress. These fragmented pieces are carried out by rainwater runoff and wind, becoming airborne MPs.

Plastic-Based Pesticide and Fertilizer Packaging: Various agricultural inputs, like many fertilizer, pesticides come in plastic packaging. Littering of these plastic containers results in the breakdown and release of MPs in the environment. Moreover, MP-containing additives are sprayed into the agricultural field, which potentially act as a source of airborne MPs.

Plastic Residue from Agricultural Machinery: The machinery, equipment used in agricultural can undergo wear and tear during regular use, which can cause MP particles to disperse into the air.

- **Construction and Demolition Activities:** Another active source is construction and demolition activities (Prasittisopin *et al.,* 2023). If this waste is not properly managed, it could potentially contribute to the overall MP pollution in the environment. For example, if plastic debris from construction materials (such as plastic bags, packaging, paints, insulation, sealants, or fragments of plastic products) ends up in the soil or near water bodies, it may eventually break down into smaller particles and become airborne under certain conditions. Additionally, the use of plastic-based construction materials like PVC pipes or plastic composites could also be a source of MPs. Furthermore, if these materials deteriorate over time or are not adequately managed, it can produce airborne MPs.
- **Personal Care Products:** Everyday use products containing microbeads, MP glitter, and other MP-containing ingredients like water bottles, cosmetic product bottles (Duis and Coors, 2016), face masks (mainly used during COVID-19) (Fadare and Okoffo, 2020) become one of the major active sources for the MPs in the atmosphere.

  Exfoliating Scrubs: Several facial and body scrubs are produced by using plastic microbeads to exfoliate the skin. These microbeads contribute to the MP load in the environment.

  Toothpaste: Many toothpaste formulations include MP particles to aid in teeth polishing. If these are digested accidentally while brushing, they contribute to MP pollution.

  Shower Gels and Body Washes: Most shower gels and body washes contain MP beads or glitter for aesthetic purposes. These particles can also produce MPs in the air.

  Cosmetics: Some makeup products, such as eyeshadows, nail polishes, and lipsticks, contain MP glitter to add shimmer and shine.

Due to growing environmental concerns and increased awareness about MP pollution, many countries and regions have taken steps to ban or restrict the use of MPs in personal care products. For example, in the United States, the Microbead-Free Waters Act of 2015 (www.fda.gov/cosmetics/cosmetics-laws-regulations/microbead-free-waters-act-faqs) prohibits the manufacture and sale of rinse-off cosmetics containing plastic microbeads. Several other countries have also enacted similar legislation to address this issue.

Manufacturers have responded by developing MP-free alternatives, such as using natural exfoliants like sugar, salt, or crushed fruit seeds in place of plastic microbeads.

4.2.2 Passive Sources of Airborne MPs

Passive sources of MPs are those sources from which MPs are generated indirectly. The major sources are as follows:

- **Atmospheric Deposition from Oceans and Water Bodies:** Approximately 10–20% of total MPs are found in oceans and water bodies (Osman *et al.*, 2023). Water bodies like oceans, rivers, and lakes act as reservoirs of MPs. Plastic waste enters these waterbodies through improper waste disposal, stormwater runoff, and accidental spillage. In the case of oceans, especially in coastal areas, strong winds and sea spray can generate aerosols that contain water droplets mixed with MP particles from the water surface. These aerosols can travel through the atmosphere and potentially be transported over long distances before eventually settling back to the Earth's surface. The process of the airborne transport of MPs through sea spray and aerosols is known as atmospheric deposition.
- **Land Runoff and Sewage Treatment:** Urban runoff is one of the most significant sources of MPs in air. MPs can enter into the air in two ways:

  Urban Runoff: Rainwater or melted snow in urban areas can pick up MPs from various surfaces such as roads, pavements, and plastic waste. This runoff eventually reaches stormwater drains, rivers, and coastal areas. During heavy rainfall events, the force of the water can resuspend the deposited MPs from surfaces, leading to their release into the air. These airborne MPs can then be transported over long distances before settling back to the ground or being washed into water bodies (Wang *et al.*, 2022a).

  Sewage Treatment Plants: Wastewater treatment plants are designed to treat sewage and remove solid particles, but they are not specifically equipped to capture MPs. As a result, MPs present in domestic and industrial wastewater can pass through the treatment process and be discharged into receiving waters (Ruffell *et al.*, 2021). From there, they may eventually become airborne through processes such as wave action, water agitation, or microbial degradation that release MPs into the air. The steps included in this process are as follows:

  Aerosolization During Treatment: During the wastewater treatment process, agitation, aeration, and other mechanical processes can lead to the release of MPs into the air in the form of tiny particles or aerosols. These MPs can then be carried by air currents to surrounding areas.

Sewage Sludge Management: Wastewater treatment plants produce sewage sludge, which is a byproduct that is often treated and disposed of in various ways. When sewage sludge is dried, processed, or transported, it can release MP particles into the air.

Stormwater Runoff: Wastewater treatment plants may not be able to completely remove all MPs from the wastewater. During heavy rain events, stormwater runoff can carry these MPs from the treatment plant and discharge them into nearby water bodies or directly into the air. The frequent presence of MPs in agricultural soil and stormwater in urban areas can be transported through rainwater Moreover, these small particles move through the air to different regions in the world.

- **Synthetic Fiber Shedding**: Clothing made from synthetic fibers, like nylon and polyester, can discharge tiny MP fibers during drying, washing, and normal wear (Dris *et al.*, 2016; Liu *et al.*, 2019a). These fibers can enter the air through indoor and outdoor ventilation systems and become airborne pollutants.

  Synthetic Fibers: When synthetic fibers like polyesters, nylon, acrylic, which are commonly used in the production of textiles, polypropylene is being damaged can shed small MP fibers.

  Mechanical Shredding: In various industrial processes, textiles made from synthetic fibers can be mechanically shredded to recycle or repurpose them. Shredding is done to break down the materials into smaller pieces for further processing. However, during this shredding process, MP fibers can be released into the air.

  Fabric Degradation: Even in everyday use, synthetic textiles can degrade over time due to factors such as friction, washing, and exposure to UV light. As a result, MP fibers break off from the fabric and can be released into the surrounding environment.

  Indoor Air Pollution: Synthetic fibers used in carpets, upholstery, and clothing can contribute to indoor air pollution. The MP fibers from these materials can become airborne through activities like walking, vacuuming, or simply moving within an indoor space.

Once released into the air, these MP fibers can be transported over long distances through atmospheric processes such as wind currents (Munyaneza *et al.*, 2022). This can lead to widespread distribution and potential deposition in various environments, including marine and terrestrial ecosystems.

- **Fragmentation of Larger Plastic Debris:** The process of fragmentation occurs when large plastic items, such as bottles, bags, or fishing nets, are exposed to various environmental stressors, such as sunlight, waves, and mechanical abrasion. Several factors can contribute to the release of airborne MPs through fragmentation:

  Weathering: Exposure to UV radiation and other weathering processes can weaken the structure of plastic debris, making it more susceptible to breaking apart into smaller particles.

Wave Action: Plastic debris floating in oceans can be subjected to continuous wave action, leading to mechanical stress and breaking the items into smaller fragments.

Wind: Winds can pick up and transport lightweight MP particles, especially from coastal areas and open water bodies.

Human Activities: Human activities, such as vehicle traffic on roads, can lead to the breakdown of plastic waste, which can then become airborne due to vehicular movement or wind.

- **Landfill:** Landfill is a significant MP source (Afrin *et al.*, 2020) as the fragmentation of larger pieces of plastic take place by physical, chemical, and biological processes over a long period. When plastic waste is disposed of in landfills, it undergoes a process of degradation due to exposure to environmental factors such as sunlight, heat, and moisture. Over time, larger plastic items can break down into smaller and smaller particles, eventually leading to the generation of MPs. Wind and other weather conditions can transport these MPs from the landfill sites into the surrounding environment, where they become airborne and can be carried over long distances. Additionally, incineration in sanitary landfill also become one of the suitable sources of airborne MPs. In this process, MPs are released with the incinerated gaseous particles.
- **Tourism:** Tourism is an important passive source of MP pollution. The main factors include commercial fishing (disposed fishing nets near sea shore,) seaside tourism (canned fish, water bottles, seafood eatery plates) (Karbalaei *et al.*, 2018), offshore industries, marine vessels. MPs present in oceans and other water bodies can be transported into the atmosphere through processes like wave action, wind-driven resuspension, and sea spray. These MPs can then be carried over long distances by atmospheric currents. Potential sources of airborne MPs in tourism settings might include:

Beaches: Coastal tourism areas often experience high concentrations of plastic waste on beaches. As these plastics break down over time, they can release MP particles into the air, especially during windy conditions or human activities like beach cleaning.

Plastic Waste Mismanagement: Improperly managed plastic waste from tourists and local residents in popular tourist destinations can end up being fragmented into MPs and transported into the air by wind or other mechanical processes.

Recreational Activities: Some tourism-related activities, such as water sports, can result in the release of MPs directly into the environment. For example, the degradation of plastic equipment, like paddleboards or kayaks, can contribute to the release of MPs into the water and eventually into the air.

Infrastructure and Facilities: Tourism infrastructure, including hotels, resorts, and recreational facilities, often generate plastic waste from single-use items like straws, cups, and food containers. If not properly managed, this waste can degrade and release MPs.

The sources of MPs are elaborately presented in a flowchart (Figure 4.2).

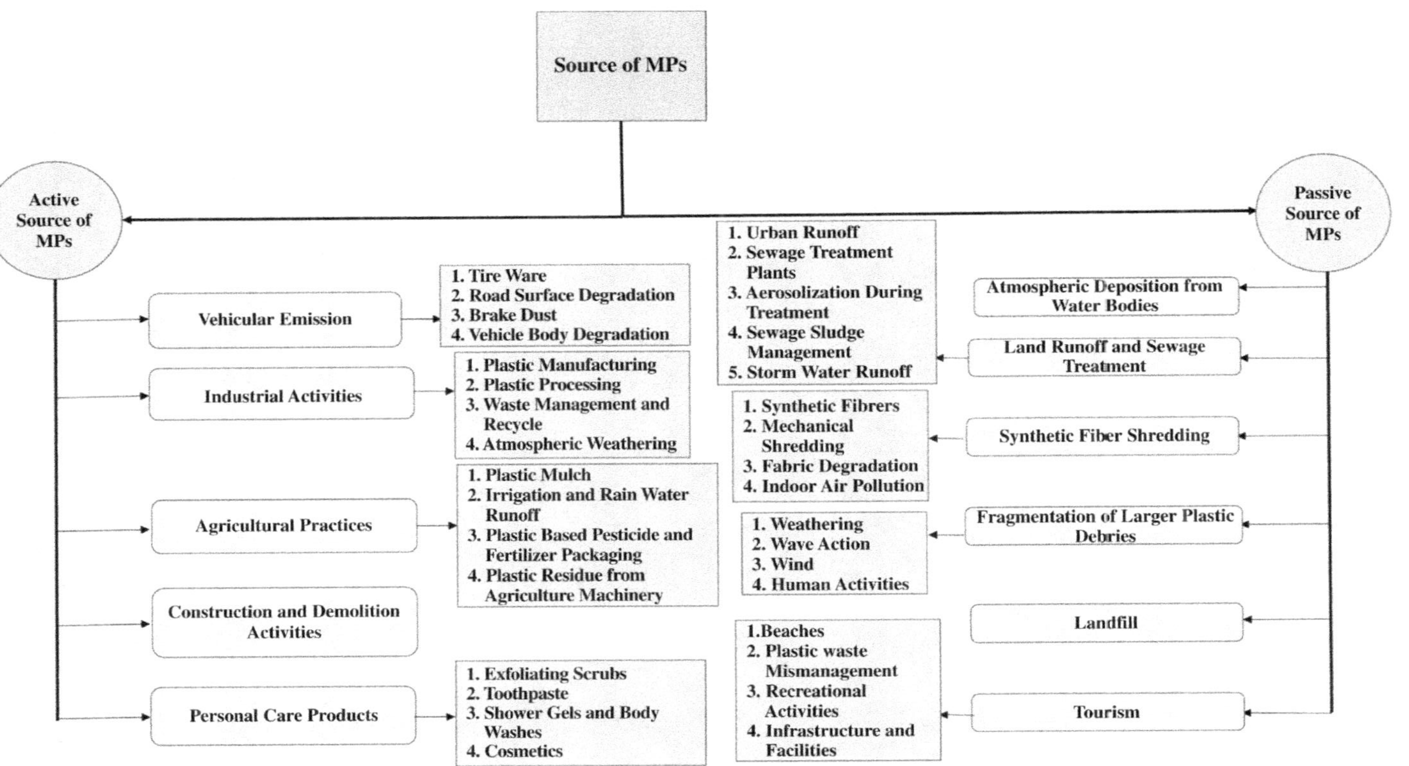

FIGURE 4.2 Flowchart of sources of microplastics.

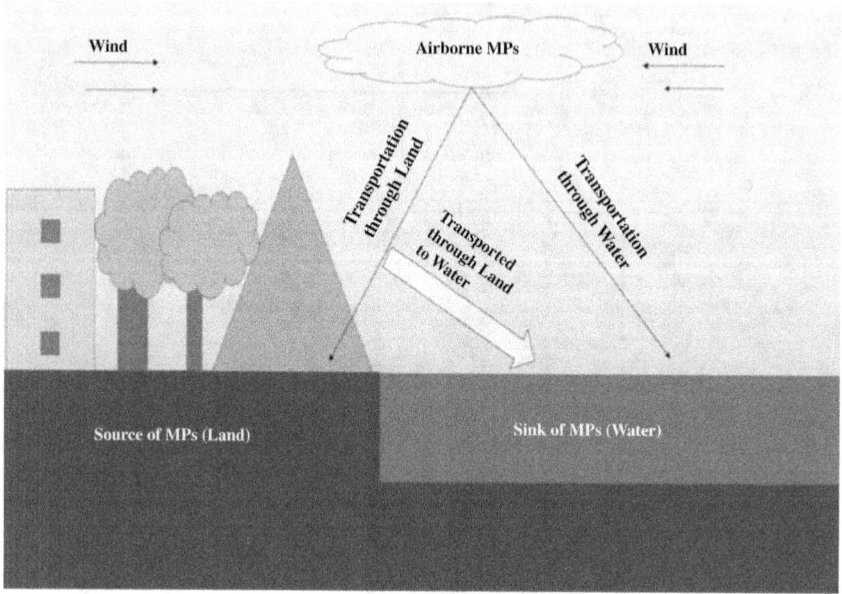

FIGURE 4.3 Transportation of microplastics.

Addressing the issue of airborne MPs requires a comprehensive approach involving regulatory measures, sustainable waste management practices, and public awareness campaigns. By understanding and mitigating these sources, we can minimize the impact of airborne MPs on both the environment and human health.

4.3 TRANSPORT OF MICROPLASTICS

MPs eventually find their way into different ecosystems after being released from a variety of sources. The dissemination of MPs is thought to be aided by air currents, which can contaminate both land and aquatic systems (Figure 4.3). Although the mechanisms by which MPs are transported through the air are not completely understood, it is thought that atmospheric transport is an important method for moving MPs over great distances (Munyaneza *et al.*, 2022). Several mechanisms that encourage the passage of such particles from the atmosphere to different water bodies are involved in the transportation of airborne MPs through aquatic ecosystems. MPs can be carried through the atmosphere over long or short distances. This transportation is influenced by dynamic circulation processes like air currents moving both upward and downward, convection drift, turbulence, and the direction and speed of the wind. Mobility, length, size and shape, and density play major roles (Qian *et al.*, 2017; Horton and Dixon, 2018; Bullard *et al.*, 2021). Small particles of MPs can be transported long distances in comparison to larger particles of MPs. This transportation can be regional or global (Dris *et al.*, 2016; Munyaneza *et al.*, 2022). MP transportation by air is seen as an important route. This mode of transport is influenced by air movement directions and is not as restricted by environmental boundaries as terrestrial and aquatic pathways. However, as there hasn't been much research on the transit of MPs in the atmosphere, more research is required. Weather factors had an impact on how atmospheric MPs were distributed. This suggests that the movement and distribution of MPs in the air may be influenced by elements such as downward and upward drafts, wind patterns, precipitation, and other atmospheric conditions (Ding *et al.*, 2022).

The elements involved in the movement of MPs both on land and in the sea are interrelated and not stand-alone activities. The interaction between air movement, deposition, and emissions from the sea hints that these processes are not distinct from one another. Instead, they may occur simultaneously and have an effect on the substance's concentrations in both the marine environment and the atmosphere (Ferrero *et al.*, 2022). The final atmospheric and marine concentrations of the substance are the result of the combined effects of all these processes working together, rather than being completely separate phenomena. Generally, the flux of transport of suspended atmospheric microplastics (SAMP) in a horizontal direction is calculated (Liu *et al.*, 2019b) as follows:

$$\text{SAMPs flux} = \int_0^{h_1} n_\delta w_\theta d_Z$$

where $n_\delta w_\theta d_Z$ are SAMP abundance (number/cubic meter of air, wind speed in θ direction, and approximate sampling height (generally 40 m), respectively. During mobilization by strong winds, SAMPs may encounter the so-called grasshopper effect like other persistent organic pollutants. SAMPs settlement may be influenced by a humidity difference between daytime and nighttime. This terrestrial transport may be toward seaside (in water bodies), and reverse movement is also reported by some authors, resulting in deposition in various natural sinks like soil, fresh water, and seas.

4.3.1 Transportation Through Water

Airborne MPs come into the contact with precipitation like snow or rain. Then they are carried along to various water bodies like rivers, lakes, seas, and oceans. Additionally, they settle onto the land or water surface through this process. MPs can be carried by atmospheric circulations, potentially over long distances. This indicates that MPs can travel through the air and be deposited in different regions, including remote areas like glaciers. According to the modeling of the retrograde air mass trajectory and the chemical makeup of the MPs, the MPs in the Demula Glacier were mostly transported by long-range atmospheric processes. This implies that the glacier temporarily absorbs MPs from the atmosphere (Wang *et al.*, 2022b). It is well-known that MPs enter the ecosystem through freshwater and terrestrial channels. They often end up in the ocean, where they are considered a major sink for these particles. Rivers are identified as significant contributors, carrying MPs from land to sea.

Wind can transport airborne MPs over long distances before they settle back to the Earth's surface. Once deposited in water bodies, these particles can be further transported by water currents. Although more MPs might theoretically be transported downstream by higher horizontal wind speeds, there is a roughly inverse relationship between MP abundance and wind speed. This suggests that elements other than horizontal wind speed, including vertical transfer from higher elevations, could be influencing the occurrence of the detected particles. Between MP abundance and wind speed, there is mostly a positive link. This indicates that certain wind directions might be linked to higher MP levels. Similarly, a positive relationship is observed between MP abundance and barometric pressure. This implies that, in addition to horizontal mobility, vertical transport may also affect the distribution of MP (Horton and Dixon, 2018; Liu *et al.*, 2019b). After deposition, MPs on land can be carried into water bodies by surface runoff during rainfall or snowmelt events. Rainwater can wash MPs from streets, sidewalks, and other surfaces into storm drains, ultimately leading to rivers and seas. MPs that have already been deposited in water bodies can be resuspended by turbulence, wave action, or other hydrodynamic processes. This resuspension can cause them to be transported to different locations within the water column or to other water bodies. When consuming waterborne suspended particles, some aquatic organisms, including filter-feeding creatures like mussels and zooplankton, may inadvertently or intentionally consume airborne MPs. As a result, MPs can be transported through the food web. MPs are transported to deeper marine settings in large part by marine creatures. They primarily do this via changing the MPs' density and adsorption (attachment) characteristics. For instance, MP surface characteristics and density might be altered as a result of biofouling (the buildup of organisms on surfaces). Different types of marine organisms contribute to the movement of MPs. Macroalgae can trap MPs through chemical or electrostatic interactions. Swimmer organisms ingest MPs and later excrete them in fecal matter. The transmission of MPs from top sediments to deeper layers may be aided by benthic animals that live on the ocean floor (Li *et al.*, 2023). Freshwater ecosystems and soils serve as both entry points for MPs and sinks for them, holding onto a sizable portion of the MP pollution they take in. Due to the proximity and volume of plastic inputs, some locations on land and in freshwater habitats may acquire MPs at higher concentrations than in the ocean. MPs in rivers are subject to transport mechanisms similar to those of other sediment particles. The rate of river flow determines the energy available for particle transport. Rivers are likely to have a limited supply of MPs, which means that they can transport all plastics that are given to them. MPs can settle out in rivers' slower-moving parts, and sediment buildup may result in their burial. The behavior of MPs particles during sediment movement is influenced by elements like their density, shape, and size. Dense particles can sink, while buoyant particles can be retained underwater due to their irregular shapes. Particle behavior is complex and depends on various factors, making modeling of MPs transport challenging. Sediment transport and deposition in rivers vary over time and space due to factors like local turbulence, flood events, and changes in river channel morphology. Remobilization of previously deposited MPs may result from these variations. MPs can disperse widely once they enter the ocean due to its strong winds, currents, and wide geographic coverage. Biofouling (the attachment of marine creatures), egestion in fecal pellets, and integration into sinking debris are a few examples of the processes that cause vertical motion. Deep sea, Southern Ocean, and Arctic ice cores are just a few of the ocean places where MPs have been discovered. It is significant to emphasize that research on the transportation of airborne MPs via water is currently underway, and our knowledge of the underlying mechanisms is still developing. MPs can have a variety of negative environmental effects when present in aquatic habitats, including potential harm to marine life and ecosystems. Researchers and decision makers are attempting to resolve this problem by taking steps to lessen the release of MPs into the environment and thereby lessen their effects on aquatic ecosystems.

4.3.2 Transport over Land

The main routes by which MPs move throughout the continental environment and accumulate include airborne fallout, groundwater migration, runoff from land catchments, and water movement through rivers and effluents. Size, shape, polymer composition, and density of MPs interaction with weather conditions, such as soil structure, precipitation, wind, humidity, water temperature, and salinity, affect how they move through the environment (Alfonso *et al.*, 2021). According to an air mass trajectory study, MPs can travel substantial distances through the atmosphere, with the documented highest distance reaching 95 km (Allen *et al.*, 2019). MPs can use atmospheric transport to travel to distant and sparsely populated areas. This suggests that MPs have an impact even in places remote from significant sources of plastic pollution, underscoring the significance of comprehending this mechanism of pollution. Larger plastic objects or materials like synthetic fabrics or rubber that contain plastic can produce MPs through physical weathering. In dry and semiarid areas, dust storms can significantly contribute to the transportation and redistribution of MPs. For MPs, these storms provide a route to various locations (Abbasi *et al.*, 2022). Weather factors have an impact on how atmospheric MPs are distributed. This suggests that the movement and distribution of MPs in the air may be influenced by variables such as wind patterns, precipitation, and other atmospheric conditions. Certain source areas like Vietnam, the Philippines, and Malaysia contribute atmospheric MPs to southeastern China through the East Asian summer monsoon. These particles can be effectively transported back onto the continent by the monsoon. The precise route by which MPs are transported from the source regions to China was determined with the assistance of the wind field and backward trajectory investigations. This shows that some airborne MPs are transferred inland and some do not stay over the ocean (Wang *et al.*, 2021). MPs can be released into the atmosphere from land surfaces and may do so as a result of a number of natural and human-caused events, including wave action, wind erosion, and human activity. The researchers proposed that atmospheric (airborne) and oceanic (waterborne) vectors both contribute to the transmission of MPs in the Antarctic region. MPs fibers were likely delivered to the Antarctic from southern South America by air masses or winds, according to an analysis of air mass back trajectories. Additionally, the work highlights the potential involvement of ocean currents in MPs dispersion by suggesting that fibers may also be transported into the Antarctic via subterranean waters (Cunningham *et al.*, 2022). MPs can be found in agricultural soils when sewage sludge is utilized as fertilizer. However, a sizable amount of MPs might also be kept in these soils for a long time, which could result in concentrated areas of buildup. MPs may also be released into the air as a result of specific farming techniques and the use of fertilizers made of plastic. Agriculture soils are significant locations where MPs can gather and spread. Different land management techniques may have an impact on how MPs are stored and distributed in these soils (Lwanga *et al.*, 2023). The addition of organic compost has increased the diversity and complexity of MPs in the soil, whereas long-term plastic mulching (at least 17 years) contributed most to MPs accumulation (Wang *et al.*, 2023). Researchers are looking into several methods for quantifying airborne MPs and analyzing their possible impacts on the environment and human health. To devise successful tactics to reduce their effects, it is essential to comprehend how MPs are transported through the air and how they may be deposited back onto the ground or into aquatic bodies.

4.4 DEPOSITION OF AIRBORNE MPs

The process through which minute plastic particles, or MPs, are carried through the atmosphere and finally land on terrestrial surfaces like soil, vegetation, and water bodies is known as airborne MPs deposition (Figure 4.4). When these bodies are close to urban or industrial regions where plastic pollution is a problem, airborne MPs can fall directly onto land or sea surfaces. Different geographic regions have significantly varying MP abundances in atmospheric deposition, principally due to differences in sources and transport dynamics. These variations are attributed to anthropogenic activities, population densities, industrialization levels, and characteristics of land use and land cover. When it rains, MPs that have been dumped on surfaces like roads, sidewalks, and buildings can be washed into stormwater systems and ultimately end up in rivers, lakes, and oceans. While the remainder may infiltrate urban rivers by runoff, these atmospheric MPs immediately deposit into them. Rainfall events were observed to promote the deposition of specific types of MPs. Rain episodes increased the likelihood of the deposition of smaller and more

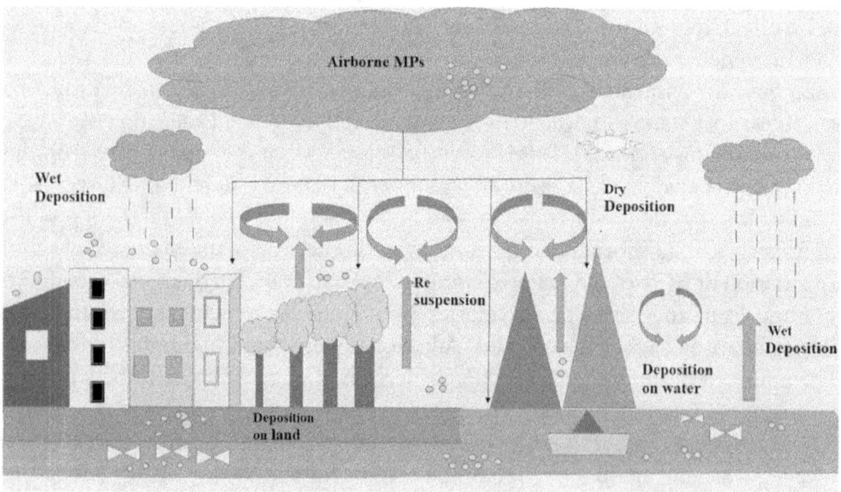

FIGURE 4.4 Deposition of microplastics.

fibrous MPs (Sun *et al.*, 2022). The use of plastic mulch in fields is one agricultural practice that might emit MPs into the atmosphere. The wind can then carry these particles and deposit them in surrounding bodies of water. The wind has the ability to transport MPs over great distances. Additionally, weather patterns, climatic conditions, and wind patterns also influence airborne MP concentrations (Liao *et al.*, 2021; Parashar and Hait, 2023) These particles can be lifted from urban areas, agricultural fields, and water bodies through processes like turbulence, convection, and advection. Once in the atmosphere, they are capable of traveling over great distances before returning to earth. The trajectory of air masses, as well as considerations related to MP transport and settling, indicate that the emission sources of MPs are likely regional (Allen *et al.*, 2019). As an example, most of the MP deposited in the Gdynia area were traced back to local sources within a distance of less than 100 km (Szewc *et al.*, 2021). As air slows down near the ground, MPs carried by the wind lose their momentum and settle due to gravity. This is more likely to occur in areas with obstacles that hinder wind flow, such as urban canyons and areas with dense vegetation. Vegetation can play a significant role in intercepting airborne MPs. Plant leaves and stems can act as surfaces where MPs can settle, especially during rain or dew events. Over time, these particles can become incorporated into the soil through processes like leaf litter decomposition. The lipophilic nature of MP shows a great affinity to suspend in the waxy leaf surface (Bi *et al.*, 2020). The multiple pathways of MP deposition, including through fall, stem flow, and litter fall flux, including dry and wet deposition, is very crucial. The canopy effect of forests could also influence MP deposition (Klein and Fischer, 2019). The deposition of airborne MPs is significantly influenced by the concentration of particulate matter. The atmospheric particulate matter concentrations, especially the concentration of $PM_{2.5}$ (fine particulate matter), were closely linked with the deposition of MPs. This suggests that the concentration of $PM_{2.5}$ could potentially be used as an indicator to estimate MP deposition (Sun *et al.*, 2022). According to the following formula (Parashar and Hait, 2023), the deposition rate is determined:

$$Df = \frac{NMPs}{S * T}$$

where Df represents the deposition flux of MPs in each sample (MPs/m²/day), $NMPs$ is the number of MPs in each sample, S is the top sectional area of the sample collector (m²), and T is the sampling period.

4.4.1 Deposition in Land

Photo degradation or mechanical degradation eventually leads to the MPs settling down onto the earth's surface. As the atmospheric currents weaken or change direction, the MP particles begin to settle out of the air and onto the Earth's surface. This settling can occur through dry or wet deposition. Dry deposition happens when MPs simply settle onto surfaces due to gravitational forces. This can occur on soil, vegetation, human-made structures, etc. Wet deposition occurs when MPs are brought down by precipitation, such as rain or snow. The particles are captured by raindrops or snowflakes and carried to the ground as the precipitation falls. Over time, the deposited MPs can accumulate on land surfaces. They can become embedded in soils, potentially affecting soil health and fertility by altering soil properties, impacting nutrient cycling, and potentially harming soil-dwelling organisms. MPs can be ingested by plants and animals, leading to potential health effects and ecosystem disruptions. The potential sources of atmospheric MPs include vehicular emissions, construction activities, and waste mismanagement (Parashar and Hait, 2023). Airborne MP distribution and abundance appear to be influenced by wind patterns. A study focused on the earliest evidence of MP deposition in a highly populated city like urban London, a European metropolis (Wright *et al.*, 2020), showed that the level is higher than those of prior studies conducted in Paris (Dris *et al.*, 2016), Dongguan (Cai *et al.*, 2017), and Hamburg (Klein and Fischer, 2019). The MP deposition rate in Jakarta, Indonesia, is extremely low, averaging only 15 particles per square meter each day. The height at which samples are obtained, the characteristics of the sampling site, and the techniques employed to filter the samples are some of the variables that could account for this variation (Purwiyanto *et al.*, 2022). The amount of MPs that are deposited is influenced by variables like population density, forest cover, infrastructure, and dust emissions. The correlation between wind speed and storm events helps to evaluate the MP presence in the environment, but no significant links were found between precipitation amount and MP deposition (Klein and Fischer, 2019; Wright *et al.*, 2020). In contrast, another study (Allen *et al.*, 2019) in a nonurban area like a remote pristine mountain catchment in the French Pyrenees proved the link with precipitation. This study also suggested that MP pollution is not limited to urban areas but also depends on the nonurban area. But in urban areas, MP deposition is higher than in rural/peri-urban areas (Liao *et al.*, 2021; Parashar and Hait, 2023). Additionally, arid and windy environments facilitate the transport of particles through the atmosphere. Areas situated downwind from known sources, such as large cities, tend to receive higher atmospheric MP deposition compared to upwind regions (Liao *et al.*, 2021). The MPs can exhibit surface degradation and can accumulate metal contaminants (Parashar and Hait, 2023).

4.4.2 Deposition in Water

The process by which minute plastic particles suspended in the air are deposited into water bodies like rivers, lakes, oceans, and even groundwater is referred to as airborne MP deposition in water. Due to the possible effects that MPs may have on the environment and human health, their presence in water bodies is a source of rising worry. MPs can be consumed by aquatic species, making their way into the food chain and perhaps making their way to people through the ingestion of seafood. They can also adsorb toxic chemicals, making them carriers of pollutants. The accumulation of MPs in aquatic ecosystems can disrupt marine life, impact water quality, and have broader ecological consequences.

A model was applied to atmospheric MPs in the ocean to trace the likely origins of the airborne MPs, and it indicated that most of them originated from both land-based sources and the adjacent oceanic atmosphere. The MPs in the northwestern Pacific Ocean's atmosphere were predominantly produced from sources on land and the nearby maritime environment, according to research that examined the types of plastic polymers included in the MPs and air mass trajectories (Ding *et al.*, 2022). Land-based or terrestrial deposition is substantially higher than ocean-based deposition (Szewc *et al.*, 2021). Furthermore, compared to those discovered along the shore, atmospheric MPs were found to be more prevalent in pelagic (open ocean) regions, and they also tended to be smaller in size. This implies that smaller MPs in the atmosphere may be able to travel over the open ocean over long distances (Ding *et al.*, 2022). More MPs are introduced through treated wastewater than are deposited in the urban environment. A prominent source of MPs in urban waters, particularly small fibrous MPse, is wet atmospheric deposition (Sun *et al.*, 2022). The city rivers might act as a secondary sink of airborne MPs. This means that MPs might be entering the atmosphere from water bodies like rivers, possibly due to various processes like erosion or runoff (Huang *et al.*, 2021).

4.4.3 Wet and Dry Deposition

MPs can be deposited onto land surfaces through both wet and dry deposition. Wet deposition occurs when MPs are captured by raindrops or snowflakes and brought down to the ground during precipitation events. Dry deposition, on the other hand, involves the settling of MPs directly from the air onto surfaces due to gravitational forces. The rainy season has a significantly higher deposition rate compared to the dry season, indicating that rain may play a role in transporting MPs to the ground and water. This increase is positively correlated with both rainfall and wind speed. High relative humidity and faster wind speeds during dry times intensify MPs deposition (Szewc *et al.*, 2021). This implies that local weather influences the distribution and deposition of MPs (Dris *et al.*, 2016; Allen *et al.*, 2019; Purwiyanto *et al.*, 2022; Sun *et al.*, 2022). But in addition to the weather, their origins also had an impact on the distribution of atmospheric MP (Ding *et al.*, 2022).

4.4.4 Indoor and Outdoor Deposition

The process by which minute plastic particles are carried to and settle in both indoor and outdoor settings is referred to as indoor and outdoor deposition of airborne MPs. Indoor MPs can originate from various sources, including indoor use of plastic products, such as synthetic textiles, furniture, and consumer goods. Additionally, they can become airborne indoors through activities like dusting, vacuuming, or even just walking on synthetic carpets or using plastic-based products. Heating, ventilation, and air conditioning (HVAC) systems can also distribute outdoor MPs to indoors. Once suspended in the indoor air, MPs can settle on indoor surfaces like floors, furniture, and countertops. Additionally, they may become stuck in air filters, which will eventually accumulate. People can be exposed to indoor MPs via breathing in contaminated air, eating dust or other surface-borne particles, or coming into direct touch with contaminated things through their skin. The degree of exposure is affected by things like ventilation and interior air quality. Outdoor MPs, on the other hand, are generally produced by breaking down larger plastic products like plastic bottles, bags, and packaging. Additionally, they may come from numerous industrial and agricultural operations, synthetic garment fibers, and tire wear particles. Wind and weather patterns help MPs move through the atmosphere. They have the ability to travel great distances before settling down. The soil, water (such as rivers and oceans), plants, and even buildings and structures can all become surfaces where outside MPs can settle. Additionally, they are present in aquatic habitats and sediments. According to studies on the presence of MPs in indoor air (Liao *et al.*, 2021; Jenner *et al.*, 2021; Perera *et al.*, 2022; Yao *et al.*, 2022), the indoor MP concentration is higher than the outside concentration. The migration of MPs throughout the indoor environment was accelerated by the turbulence brought on by air conditioner circulation. It stresses how important it is for things like the amount of textile material present in indoor environments and airflow patterns to play a part in how MPs are distributed there. Closed windows make it difficult for indoor MPs to leave, which causes them to amass and suspend in small areas over time. Increased human activity adds more MP sources and speeds up their resuscitation (Zhang *et al.*, 2020). The types of MPs discovered indoors were connected to sources in the interior environment and human behaviors. In contrast, metropolitan and inland locales in high-density areas had the highest outdoor MP abundance close to an industrial zone. It was discovered that the quantity and dispersion of airborne MPs were correlated with human activity, industrialization, and population density. In other words, a higher number of people and greater industrial activity tended to be associated with higher airborne MP concentrations. Additionally, it is hypothesized that the concentration of airborne MPs in less developed nations is somewhat lower than what has been found in more developed nations (Perera *et al.*, 2022). Airborne MPs can exhibit concentration change in a variety of indoor spatial situations, including workplaces, hallways, classrooms, and single-family homes. It describes the kinds and numbers of MPs present in various areas. For fibers, single-family homes have the highest deposition rates; for film-like polymers, schools have the highest deposition rates. This indicates that residential dwellings are a major source of MP fibers, but schools have greater rates of film-like plastic deposition, probably as a result of the frequent usage of materials like polyethylene (PE). MPs have comparable textures but varying sizes in both indoor and outdoor settings. This implies that, with time, these particles may degrade from being larger MPs to smaller MPs (Yao *et al.*, 2022). Studies have shown that people may inhale MPs, making interior settings like homes a substantial source of MP exposure for people (Jenner *et al.*, 2021; Perera *et al.*, 2022).

Deposition of MPs by airborne means requires a multifaceted strategy. To address this new environmental issue, there is an increasing need for monitoring and regulation of airborne MPs, including their properties and deposition rates. In order to control the release of MPs into the environment, this would include the creation of standardized measurement techniques and rules.

4.5 REMEDIATION

An emerging environmental concern, remediation of airborne MPs entails the elimination or reduction of minute plastic particles dispersed in the atmosphere. Remediating airborne MPs is a complex and multifaceted task that requires a combination of preventive measures, cleanup technologies, and policy interventions. Here are some general approaches for the remediation of MPs:

- **Prevention:** Preventing the production of airborne MPs in the first place is the best strategy for dealing with them. This can be achieved through various means, such as reducing plastic production and consumption, promoting sustainable product design, and minimizing plastic waste through recycling and responsible disposal.
- **Source Control:** Targeting the sources of airborne MPs is crucial. For example, implementing filtration systems in wastewater treatment plants can help capture MPs released from washing synthetic textiles. Similarly, covering landfills to prevent the wind from dispersing plastic particles can reduce their release into the atmosphere.
- **Air Filtration:** Installation of air filtration systems in industrial facilities, indoor spaces, and even vehicles can help to capture airborne MPs. HEPA (high-efficiency particulate air) filters are effective at capturing airborne particles, including MPs.
- **Natural Remediation:** An emerging area of research is using natural mechanisms to remove MPs. Some microbes and plants have demonstrated the capacity to degrade or absorb MPs from the environment. Environmentally friendly repair techniques might be provided by utilizing these biological systems.
- **Innovative Technologies:** Researchers are developing innovative technologies specifically designed to capture airborne MPs. Some of these technologies include electrostatic precipitators, cyclone separators, and passive air samplers. These systems can be deployed in areas with high MP concentrations.
- **Monitoring and Assessment:** Regular monitoring and assessment of air quality are essential to track the presence and concentration of airborne MPs. This data can inform remediation efforts and help evaluate their effectiveness.
- **Regulation and Policy:** By enacting tougher controls on plastic manufacture, disposal, and pollution, governments and regulatory agencies can significantly contribute to the solution of the problem of airborne MPs. Bans on single-use plastics, extended producer responsibility (EPR) programs, and pollution control regulations are examples of possible policy interventions.
- **Public Awareness and Education:** Raising awareness among the public about the impact of MPs and the importance of responsible plastic use and disposal can contribute to reducing their generation and dispersal. The cleanup of airborne MPs is a complicated and developing topic and is difficult to remember. Additionally, a holistic approach that combines prevention, mitigation, and policy measures is crucial for addressing this environmental challenge effectively.

An attempt to reduce MP contaminants from the air is not as much a feasible technique as reductions in water and soil. Also the fate of airborne MP ends up in aquatic ecosystem or in land. Therefore, by treating the MPs in water and soil, we can reduce the MPs in the atmosphere. Nevertheless, some approaches, like source reduction and industrial air filtration, help to directly remove MPs from the air, but ultimately it is mainly reduced by treating water and soil. There are various procedures (refer

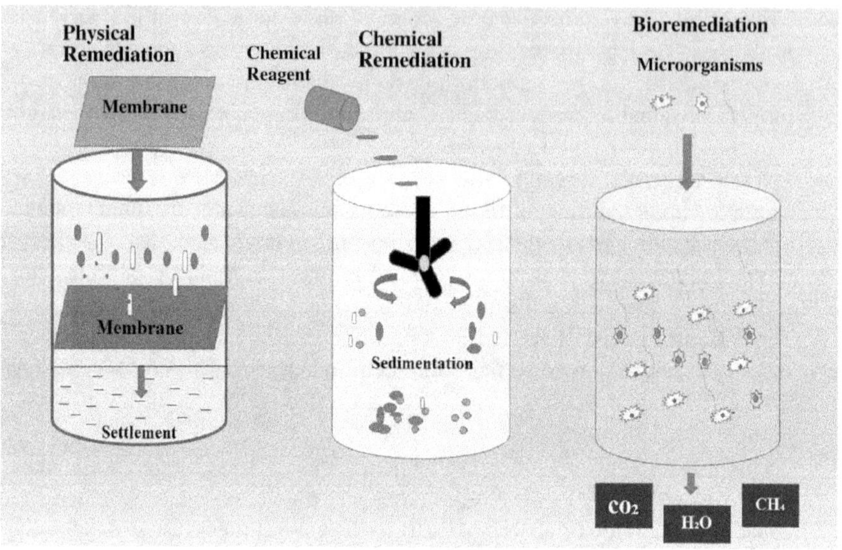

FIGURE 4.5 Remediation techniques of microplastics.

to Figure 4.5) to remove MPs from the environment. The three main procedures are discussed next.

4.5.1 Physical Remediation

Airborne MPs are microscopic plastic particles suspended in the atmosphere that have the potential to harm both people and the environment. Physical methods (refer to Table 4.1) of remediation involve the use of various techniques to physically capture or separate MPs from the air. Physical treatment methods like air filtration, cyclone separator, electrostatic precipitator, membrane technology, adsorption are very popular. HVAC (heating, ventilation, and air conditioning) systems and air purifiers frequently use high-efficiency particulate air (HEPA) filters. According to Gorji *et al.* (2017), they can collect particles as tiny as 0.3 m, which include numerous MPs. Another physical method employs an instrument like an electrostatic precipitator that uses electrostatic charges to attract and arrest airborne particles, including MPs (Enders *et al.*, 2020). They are often used in industrial settings. Cyclone separators and gravity settlers are another two approachable effective methods for reducing MPs. Cyclone separators are those devices that use centrifugal force to separate particles (Air Pollution Control Fact Sheets 1 by USEPA [https://nepis.epa.gov/Exe/ZyPDF.cgi?Dockey=P100C75Q.PDF],) They are effective at capturing larger MPs and can be integrated into industrial processes to prevent MP emissions. The gravity settling method is used in some cases, which allows particles to settle out of the air due to gravity, and it can be an effective remediation method (Air Pollution Control Fact Sheet 2 by USEPA, https://www3.epa.gov/ttncatc1/dir1/fsetling.pdf). This is typically more suitable for larger MPs that are heavy enough to settle quickly. Also

TABLE 4.1
Different Methods of Removal of Airborne MPs

Methods	Separation Techniques	Types	Source	Efficiency	Literature Reference
Physical method	Adsorption	Polystyrene	Wastewater treatment plant	95%	Wang *et al.*, 2020
		All MPs	Water treatment plant	100%	Siipola *et al.*, 2020
		Polystyrene	Wastewater	89.8%—neat polystyrene 72.4%—carboxylate-modified polystyrene 88.9%—amine modified polystyrene	Sun *et al.*, 2020
		Polystyrene	Water	90%	Misra *et al.*, 2020
		Polystyrene	Water	94%	Zhang *et al.*, 2021
	Membrane technology	All MPs	Municipal wastewater treatment plant	94%	Talvitie *et al.*, 2017
		All MPs	—	~95.5%	Chen *et al.*, 2020
		All MPs	Treated Wastewater Plant	79%–89%	Liu *et al.*, 2020
	Sedimentation	Mostly polyamide	Municipal wastewater treatment plant	64.4%	Liu *et al.*, 2019c
		All MPs	Wastewater treatment plant	Primary treatment—72% Secondary treatment—88% Tertiary treatment—94%	Iyare *et al.*, 2020
	Filtration	Polyamide nylon Polystyrene	—	94%	Pizzichetti *et al.*,2021
		Polyethylene Polyamide	Wastewater treatment plant	>96%	Shen *et al.*, 2021

(Continued)

TABLE 4.1 (*Continued*)
Different Methods of Removal of Airborne MPs

Methods	Separation Techniques	Types	Source	Efficiency	Literature Reference
Chemical Method	Agglomeration and flocculation	Polyethylene	Drinking water treatment plant	Below 40%	Ma *et al.*, 2019
		Polystyrene	wastewater treatment plant	99.4%	Rajala *et al.*, 2020
		Polyethylene	Drinking water treatment plant	92.7%	Shahi *et al.*, 2020
		Polystyrene Polyethylene	Water treatment plant	—	Zhou *et al.*, 2021
		Polystyrene	—	90%	Park *et al.*, 2021
		Polyethylene	Graywater	70.56%–96.10%	Esfandiari and Mowla, 2021
		Polyethylene Terephthalate	Printing and dyeing wastewater	93%	Duan *et al.*, 2022
	Photocatalysis	Low density Polyethylene	—	30%	Tofa *et al.*, 2019
		High density Polyethylene	—	—	Ariza-Tarazona *et al.*, 2019
	Chemical degradation	All type	Wastewater treatment plant	89.9%	Hidayaturrahman and Lee, 2019
	Nanomaterial	Polyethylene	—	—	Tofa *et al.*, 2019
		Polyethylene	Aqueous medium	35.66–50.46%	Uogintė *et al.*,2023
Biological method	Bacteria	Polyethylene	Aqueous medium	8%	Orr *et al.*, 2004
		Polyethylene	—	9.26%	Vimala and Mathew, 2016
		Polyethylene	Domestic sewage water	—	Sarmah and Rout, 2019
	Fungi	Polyurethane	—	55%	Oprea and Doroftei, 2011
		Polyethylene pellets	—	43%	Paco *et al.*, 2017
	Algae	—	Red Sea	—	Martin *et al.*, 2019
		Polystyrene	Wastewater treatment plant	—	Cunha *et al.*, 2020
		Polyethylene	Maldives	2.3 ± 0.7%	Corona *et al.*, *2020*
	Biodegradable polymers	—	—	—	Mitrano and Wohlleben, 2020

passive collectors are devices designed to capture airborne MPs without the need for external energy sources. These can include specially designed surfaces or materials that attract and retain MPs, which can then be collected periodically. But the favorable physical methods are adsorption, membrane technology, filtration, and sedimentation. The adsorption method involves using adsorbent materials like modified biochar, nanoparticles, and nanocomposites to capture and trap MP particles (Wang *et al.*,2020; Siipola *et al.*,2020; Sun *et al.*, 2020; Misra *et al.*,2020; Zhang *et al.*, 2021). MPs are physically removed from a variety of sources, including wastewater treatment plants, using filtration filters, such as magnetic biochar and wood chip biochar embedded in sand filters (Talvitie *et al.*, 2017; Chen *et al.*, 2020; Liu *et al.*, 2020). Composite filters and membranes, such as one of the improved cationic surfactant filter media made of ammonium silicate, are highly effective MPs removers (greater than 96%) from wastewater treatment plants, whereas cost-effective membranes produced from cellulose acetate and polycarbonate remove approximately 94% of MPs (Pizzichetti *et al.*, 2021; Shen *et al.*, 2021). The primary sedimentation phase in wastewater treatment facilities is when different high-density MPs can be effectively removed from wastewater via sedimentation (Liu *et al.*, 2019c; Iyare *et al.*, 2020). Along with these desirable techniques, some obscured techniques like natural processes such as rain and snow can also help to remove MPs from the air by causing them to settle onto surfaces or get washed out of the atmosphere. However, this process is not very efficient and relies on environmental conditions. Improving ventilation systems in indoor spaces can help reduce the concentration of airborne MPs. This may involve better air exchange rates, the use of air curtains, or other techniques to prevent the entry of outdoor pollutants. In some cases, it may be possible to encapsulate MPs using materials that trap and immobilize the particles, preventing them from becoming airborne. This approach is more focused on preventing MPs from becoming airborne in the first place. In sensitive environments such as laboratories or clean rooms, specialized air filtration and ventilation systems are used to minimize the presence of airborne contaminants, including MPs.

It is crucial to keep in mind that the efficacy of these physical remediation techniques can change depending on the elements nature, size, and shape of MPs, the environment in which they are present, and the severity of the issue. To effectively remediate airborne MPs pollution, a mix of physical and additional remediation techniques, such as chemical or biological treatments, may be required.

4.5.2 Chemical Methods

Chemical methods for the remediation of airborne MPs involve using various chemical substances or processes to capture, break down, or otherwise mitigate the presence of MPs particles in the air. Chemical methods offer potential solutions to address this issue, but they come with their own challenges and limitations. Some key chemical methods (refer to Table 4.1) for the remediation of airborne MPs are sedimentation, coagulation, and flocculation, where chemicals called coagulants and flocculants can be added to the air to encourage MP particles to clump together form larger aggregates and exchange of ligands, making them easier to remove. Then, by a sedimentation process, agglomerates are broken down and the particles are separated

(Zhou *et al.*, 2021). This method is frequently employed in the treatment of wastewater (Park *et al.*, 2021; Chellasamy *et al.*, 2022), and it might be modified for the remediation of air. The most conventionally used coagulant is alum, which mainly consists of aluminum and iron (Ma *et al.*, 2019; Rajala *et al.*, 2020; Esfandiari and Mowla, 2021). Alum-coated sand is also used to remove MPs by the agglomeration and flocculation method (Shahi *et al.*, 2020; Duan *et al.*, 2022). Chemical substances known as adsorbents can be used to capture MPs from the air. The removal of MPs from wastewater has been successfully accomplished using a combination of adsorption, flocculation, and filtration (Sturm *et al.*, 2020). Activated carbon, zeolites, and certain polymers are examples of adsorbents that can attract and bind MP particles. The air is passed through a filter or bed of the adsorbent material, and MPs eventually adhere to the surface. These captured MPs can later be removed and disposed of properly. Chemical degradation by some chemicals, such as ozone $\left(O_3\right)$, hydroxyl radicals $\left(OH^{\bullet}\right)$, and certain enzymes, can be used to divide MPs into more manageable, less harmful substances. Certain chemicals can enhance the photodegradation process. Photocatalytic degradation is a specific advanced oxidation method that involves activating by absorbing light, a semiconducting substance (photocatalyst) produces free radicals like hydroxyl and superoxide ions. The MPs are oxidized and weakened by these radicals (Cao *et al.*, 2022). This includes using ozone, ultraviolet radiation with illumination, catalysts, heat, and hydrogen peroxide to initiate the breakdown of MPs (Kim *et al.*, 2022). For example, ozone treatment can be used to degrade MPs by breaking the polymer chains and reducing them to smaller molecules. However, this method may produce potentially harmful byproducts. Titanium oxide and its modified forms like N-TiO_2 coating (Ariza-Tarazona *et al.*, 2019) and graphite-doped TiO_2 (Uogintė *et al.*, 2023) are commonly used as catalysts in this process with suitable degradation efficiency. Additionally, low-density MPs can be broken down by zinc-based nanomaterials (Tofa *et al.*, 2019; Cao *et al.*, 2022; Chellasamy *et al.*, 2022). Another strategy entails employing chemicals to change the MPs' characteristics and make them easier to remove. For instance, surface treatments with chemicals like surfactants can be used to increase the hydrophilicity of MPs, making them easier to capture using water-based methods. Electrostatic precipitators can also be used to remove charged MP particles from the air. These devices use high-voltage electric fields to charge the particles, causing them to be attracted to collector plates with the opposite charge. However, this method may not be practical for large-scale remediation and may require specific conditions.

Various variables, including particle size, plastic type, and environmental conditions, can affect how effectively MPs are captured or degraded. Additionally, some chemical methods may generate secondary pollutants or waste that need to be managed carefully. The choice of a remediation method should consider the specific circumstances, MP composition, resource availability, and environmental effects. In many instances, a mix of physical, chemical, and biological techniques may be the best course of action to handle the intricate problem of airborne MPs contamination.

4.5.3 Bioremediation

Bioremediation is a process that uses biological organisms to break down or remove pollutants from the environment. When it comes to the bioremediation of airborne

MPs, the goal is to utilize microorganisms or other biological agents to degrade or capture these tiny plastic particles suspended in the air. Some approaches for bioremediation (refer to Table 4.1.) of airborne MPs are through microorganisms like bacteria and fungi that have the natural ability to break down various organic compounds, including plastics. It can be aerobic degradation where CO_2 and water are produced or anaerobic digestion where, along with CO_2 and water, methane and hydrogen sulfide are also produced (Chandra *et al.*, 2020). Bacteria like *Bacillus subtilis* (Vimala and Mathew, 2016), *Rhodococcus rubber* (Orr *et al.*, 2004), cynobacterial species (*Phormidium lucidum, Oscillatoria subbrevis*) (Sarmah and Rout, 2019), fungus like *Chaetomium globosum* (Oprea and Doroftei, 2011) and *Zalerion maritimum* (Paço *et al.*, 2017) are used, but these organisms show very low efficiency except *Zalarion meritimum*, which shows 43% efficiency of removing MPs. Another effective technique is biofilm that are communities of microorganisms that grow on surfaces. They can be engineered or selected to colonize on specific materials, including surfaces exposed to airborne MPs. These genetic engineering techniques can be used to enhance the plastic-degrading abilities of certain microbes. Scientists have identified enzymes like PETase that can degrade PET (polyethylene terephthalate) plastics and can be introduced into bacteria to improve plastic degradation. Biofilms can potentially trap and degrade MPs as they settle on surfaces, such as buildings or urban infrastructure. For instance, *Pseudomonus aeruginosa* have trapped MPs by a sticky exopolymorphic substance and then released them during resource recovery (Liu *et al.*, 2021). These methods are suitable for all the MPs (Badola *et al.*, 2022.) Phytoplankton and algae are aquatic microorganisms, but they can also be utilized in bioremediation efforts. For example, macroalgae and marine microalgae can physically entangle and adsorb MPs. Algae can be grown in controlled environments and then released into the atmosphere in specific areas to capture MPs. Algae can produce sticky substances that can adhere to MPs, making it easier to collect and remove them from the air. Microalgae is one of the significant biological tools for removing MPs (Cunha *et al.*, 2020). Corals are another useful living organism utilized to remove MPs as they are the primary ocean sink for MPs (Martin et al., 2019). For example, mushroom coral is significantly effective to remove bio-fouled MPs (Corona *et al.*, 2020). Certain plants, such as mosses and lichens, have been shown to capture airborne particles effectively. Researchers are investigating the potential for these natural air-filtering systems to capture MPs. Green walls and green roofs, where plants are grown on building surfaces, could serve as passive bioremediation systems that capture airborne MPs as part of their filtration process. Another approach to mitigating airborne MPs is to replace traditional plastics with biodegradable polymers that are less persistent in the environment (Mitrano and Wohlleben, 2020). These polymers can be designed to break down more readily when exposed to environmental conditions.

Advanced sensors and sampling techniques are used to track MP concentrations in the air. Additionally, the feasibility and effectiveness of these methods may need to be evaluated further through field trials and long-term studies. Combining bioremediation with other pollution control measures, such as reducing plastic use and improving waste management, is also essential for addressing the broader issue of MP pollution.

Apart from these three methods another preferable method is recycling plastics. Socioeconomic advancement and inefficient management or incorrect disposal of plastic waste causes the production of plastics at higher rates that contributes to global plastic pollution (Brown *et al.*, 2023). Governments have primarily responded to concerns about plastic pollution by implementing bans and prohibitions on plastics. This reflects the prevailing approach of many governments worldwide. An alternative perspective, arguing that there doesn't have to be an 'all or nothing' approach to deal with MPs use. In other words, instead of completely banning MPs, regulations can serve as a catalyst for change and encourage the development of better practices (Mitrano and Wohlleben, 2020). Therefore, the reduce-reuse-and-recycle principle becomes essential for managing plastic waste. These principles promote a more sustainable approach to deal with plastics.

The omnipresence of MPs and their environmental impact, emphasizing the need for careful assessment, responsible waste management, and innovation in addressing the challenges posed by plastic waste in our environment are one of the greatest concerns. Therefore, future study is most essential for all organism to live sustainably in nature.

4.6 CONCLUSION

In conclusion, the presence of airborne MPs is a concerning and complex environmental issue that demands our attention and immediate action. This emerging field of research has revealed that MP particles have become ubiquitous in our atmosphere. The implications of airborne MPs are multifaceted and far-reaching. First and foremost, there are potential health concerns for both humans and wildlife. These particles have the ability to enter the respiratory system deeply when inhaled, which could result in a variety of health problems, such as inflammation, respiratory disorders, and cardiovascular ailments. Additionally, MPs may carry toxic chemicals and pollutants, further exacerbating their potential harm. Beyond health concerns, airborne MPs contribute to the growing plastic pollution crisis. As these particles settle back to Earth, they can contaminate terrestrial and aquatic ecosystems, posing risks to flora and fauna. Marine environments are particularly susceptible, as airborne MPs can wash into oceans and add to the already massive problem of plastic pollution in our seas. Addressing the issue of airborne MPs requires a multifaceted approach. To further understand the sources, distribution, and potential hazards related to these particles, extensive research must first be conducted. Steps including strengthening waste management procedures, promoting sustainable alternatives, and limiting the manufacture of plastic are crucial. Furthermore, individual actions can make a difference. Reducing personal plastic consumption, properly disposing of plastics, and supporting initiatives that seek to clean up plastic waste are small steps that can collectively have a significant impact. A complex and worrying component of the wider global problem of plastic pollution is represented by airborne MPs. Their existence in the atmosphere emphasizes the urgent need for coordinated actions at the local, national, and international levels to reduce the hazards to the environment and public health posed by these microscopic plastic particles. Only a combination of research, regulation, and responsible consumer choices can help to address this pressing environmental challenge and work toward a more sustainable and plastic-free future.

REFERENCES

Abbasi, S., Rezaei, M., Ahmadi, F. and Turner, A., 2022. Atmospheric transport of microplastics during a dust storm. *Chemosphere*, 292, p. 133456. https://doi.org/10.1016/j.chemosphere.2021.133456

Afrin, S., Uddin, M.K. and Rahman, M.M., 2020. Microplastics contamination in the soil from Urban Landfill site, Dhaka, Bangladesh. *Heliyon*, 6(11). https://doi.org/10.1016/j.heliyon.2020.e05572

Air Pollution Control Technology Fact Sheet 1 (Cyclons). *USEPA*. https://nepis.epa.gov/Exe/ZyPDF.cgi?Dockey=P100C75Q.PDF

Air Pollution Control Technology Fact Sheet 2 (Settling Chambers). *USEPA*. https://www3.epa.gov/ttncatc1/dir1/fsetling.pdf

Alfonso, M.B., Arias, A.H., Ronda, A.C. and Piccolo, M.C., 2021. Continental microplastics: Presence, features, and environmental transport pathways. *Science of the Total Environment*, 799, p. 149447. https://doi.org/10.1016/j.scitotenv.2021.149447

Allen, S., Allen, D., Phoenix, V.R., Le Roux, G., Durántez Jiménez, P., Simonneau, A., Binet, S. and Galop, D., 2019. Atmospheric transport and deposition of microplastics in a remote mountain catchment. *Nature Geoscience*, 12(5), pp. 339–344. https://doi.org/10.1038/s41561-019-0335-5

Ariza-Tarazona, M.C., Villarreal-Chiu, J.F., Barbieri, V., Siligardi, C. and Cedillo-González, E.I., 2019. New strategy for microplastic degradation: Green photocatalysis using a protein-based porous N-TiO_2 semiconductor. *Ceramics International*, 45(7), pp. 9618–9624. https://doi.org/10.1016/j.ceramint.2018.10.208

Badola, N., Bahuguna, A., Sasson, Y. and Chauhan, J.S., 2022. Microplastics removal strategies: A step toward finding the solution. *Frontiers of Environmental Science & Engineering*, 16, pp. 1–18. https://doi.org/10.1007/s11783-021-1441-3

Bi, M., He, Q. and Chen, Y., 2020. What roles are terrestrial plants playing in global microplastic cycling? *Environmental Science & Technology*, 54(9), pp. 5325–5327. https://dx.doi.org/10.1021/acs.est.0c01009

Brown, E., MacDonald, A., Allen, S. and Allen, D., 2023. The potential for a plastic recycling facility to release microplastic pollution and possible filtration remediation effectiveness. *Journal of Hazardous Materials Advances*, 10, p. 100309. https://doi.org/10.1016/j.hazadv.2023.100309

Bullard, J.E., Ockelford, A., O'Brien, P. and Neuman, C.M., 2021. Preferential transport of microplastics by wind. *Atmospheric Environment*, 245, p. 118038. https://doi.org/10.1016/j.atmosenv.2020.118038

Cai, L., Wang, J., Peng, J., Tan, Z., Zhan, Z., Tan, X. and Chen, Q., 2017. Characteristic of microplastics in the atmospheric fallout from Dongguan city, China: Preliminary research and first evidence. *Environmental Science and Pollution Research*, 24, pp. 24928–24935. https://doi.org/10.1007/s11356-017-0116-x

Cao, B., Wan, S., Wang, Y., Guo, H., Ou, M. and Zhong, Q., 2022. Highly-efficient visible-light-driven photocatalytic H_2 evolution integrated with microplastic degradation over MXene/ZnxCd1-xS photocatalyst. *Journal of Colloid and Interface Science*, 605, pp. 311–319. https://doi.org/10.1016/j.jcis.2021.07.113

Chandra, P. and Singh, D.P., 2020. Microplastic degradation by bacteria in aquatic ecosystem. In *Microorganisms for sustainable environment and health* (pp. 431–467). Elsevier. https://doi.org/10.1016/B978-0-12-819001-2.00022-X

Chellasamy, G., Kiriyanthan, R.M., Maharajan, T., Radha, A. and Yun, K., 2022. Remediation of microplastics using bionanomaterials: A review. *Environmental Research*, 208, p. 112724. https://doi.org/10.1016/j.envres.2022.112724

Chen, Y.J., Chen, Y., Miao, C., Wang, Y.R., Gao, G.K., Yang, R.X., Zhu, H.J., Wang, J.H., Li, S.L. and Lan, Y.Q., 2020. Metal–organic framework-based foams for efficient microplastics removal. *Journal of Materials Chemistry A*, 8(29), pp. 14644–14652. https://doi.org/10.1039/D0TA04891G

Corona, E., Martin, C., Marasco, R. and Duarte, C.M., 2020. Passive and active removal of marine microplastics by a mushroom coral (*Danafungia scruposa*). *Frontiers in Marine Science*, 7, p. 128. https://doi.org/10.3389/fmars.2020.00128

Cunha, C., Silva, L., Paulo, J., Faria, M., Nogueira, N. and Cordeiro, N., 2020. Microalgal-based biopolymer for nano-and microplastic removal: A possible biosolution for wastewater treatment. *Environmental Pollution*, 263, p. 114385. https://doi.org/10.1016/j.envpol.2020.114385

Cunningham, E.M., Rico Seijo, N., Altieri, K.E., Audh, R.R., Burger, J.M., Bornman, T.G., Fawcett, S., Gwinnett, C., Osborne, A.O. and Woodall, L.C., 2022. The transport and fate of microplastic fibres in the Antarctic: The role of multiple global processes. *Frontiers in Marine Science*, 9, p. 2353. https://doi.org/10.3389/fmars.2022.1056081

Ding, J., Sun, C., He, C., Zheng, L., Dai, D. and Li, F., 2022. Atmospheric microplastics in the Northwestern Pacific Ocean: Distribution, source, and deposition. *Science of The Total Environment*, 829, p. 154337. http://dx.doi.org/10.1016/j.scitotenv.2022.154337

Dris, R., Gasperi, J., Saad, M., Mirande, C. and Tassin, B., 2016. Synthetic fibers in atmospheric fallout: A source of Microplastics in the environment? *Marine Pollution Bulletin*, 104(1–2), pp. 290–293. https://doi.org/10.1016/j.watres.2022.119129

Duan, Y., Zhao, J., Qiu, X., Deng, X., Ren, X., Ge, W. and Yuan, H., 2022. Coagulation performance and floc properties for synchronous removal of reactive dye and polyethylene terephthalate microplastics. *Process Safety and Environmental Protection*, 165, pp. 66–76. https://doi.org/10.1016/j.psep.2022.07.010

Duis, K. and Coors, A., 2016. Microplastics in the aquatic and terrestrial environment: Sources (with a specific focus on personal care products), fate and effects. *Environmental Sciences Europe*, 28(1), pp. 1–25. https://doi.org/10.1186/s12302-015-0069-y

Enders, K., Tagg, A.S. and Labrenz, M., 2020. Evaluation of electrostatic separation of microplastics from mineral-rich environmental samples. *Frontiers in Environmental Science*, p. 112. https://doi.org/10.3389/fenvs.2020.00112

Esfandiari, A. and Mowla, D., 2021. Investigation of microplastic removal from greywater by coagulation and dissolved air flotation. *Process Safety and Environmental Protection*, 151, pp. 341–354. https://doi.org/10.1016/j.psep.2021.05.027

Fadare, O.O. and Okoffo, E.D., 2020. Covid-19 face masks: A potential source of Microplastic fibers in the environment. *The Science of the Total Environment*, 737, p. 140279. https://doi.org/10.1016/j.dibe.2023.100188

Fadare, O.O., Wan, B., Guo, L.H. and Zhao, L., 2020. Microplastics from consumer plastic food containers: Are we consuming it? *Chemosphere*, 253, p. 126787. https://doi.org/10.1016/j.chemosphere.2020.126787

Ferrero, L., Scibetta, L., Markuszewski, P., Mazurkiewicz, M., Drozdowska, V., Makuch, P., Jutrzenka-Trzebiatowska, P., Zaleska-Medynska, A., Andò, S., Saliu, F. and Nilsson, E.D., 2022. Airborne and marine microplastics from an oceanographic survey at the Baltic Sea: An emerging role of air-sea interaction? *Science of the Total Environment*, 824, p. 153709. http://dx.doi.org/10.1016/j.scitotenv.2022.154337

Gorji, M., Bagherzadeh, R. and Fashandi, H., 2017. Electrospun nanofibers in protective clothing. In *Electrospun nanofibers* (pp. 571–598). Woodhead Publishing. https://doi.org/10.1016/B978-0-08-100907-9.00021-0

Hidayaturrahman, H. and Lee, T.G., 2019. A study on characteristics of microplastic in wastewater of South Korea: Identification, quantification, and fate of microplastics during treatment process. *Marine Pollution Bulletin*, 146, pp. 696–702. https://doi.org/10.1016/j.marpolbul.2019.06.071

Horton, A.A. and Dixon, S.J., 2018. Microplastics: An introduction to environmental transport processes. *Wiley Interdisciplinary Reviews: Water*, 5(2), p. e1268. https://doi.org/10.1002/wat2.1268

Huang, Y., He, T., Yan, M., Yang, L., Gong, H., Wang, W., Qing, X. and Wang, J., 2021. Atmospheric transport and deposition of microplastics in a subtropical urban environment. *Journal of Hazardous Materials*, 416, p. 126168. https://doi.org/10.1016/j.jhazmat.2021.126168

Iyare, P.U., Ouki, S.K. and Bond, T., 2020. Microplastics removal in wastewater treatment plants: A critical review. *Environmental Science: Water Research & Technology*, 6(10), pp. 2664–2675. https://doi.org/ 10.1039/d0ew00397b

Jenner, L.C., Sadofsky, L.R., Danopoulos, E. and Rotchell, J.M., 2021. Household indoor microplastics within the Humber region (United Kingdom): Quantification and chemical characterisation of particles present. *Atmospheric Environment*, 259, p. 118512. https://doi.org/10.1016/j.atmosenv.2021.118512

Karbalaei, S., Hanachi, P., Walker, T.R. and Cole, M., 2018. Occurrence, sources, human health impacts and mitigation of Microplastic pollution. *Environmental Science and Pollution Research*, 25, pp. 36046–36063. https://doi.org/10.1007/s10311-023-01593-31213.

Kim, S., Sin, A., Nam, H., Park, Y., Lee, H. and Han, C., 2022. Advanced oxidation processes for microplastics degradation: A recent trend. *Chemical Engineering Journal Advances*, 9, p. 100213. https://doi.org/10.1016/j.ceja.2021.100213

Klein, M. and Fischer, E.K., 2019. Microplastic abundance in atmospheric deposition within the Metropolitan area of Hamburg, Germany. *Science of the Total Environment*, 685, pp. 96–103. https://doi.org/10.1016/j.scitotenv.2019.05.405

Kole, P.J., Löhr, A.J., Van Belleghem, F.G. and Ragas, A.M., 2017. Wear and tear of tyres: A stealthy source of microplatics in the environment. *International Journal of Environmental Research and Public Health*, 14(10), p. 1265. https://doi.org/10.3390/ijerph14101265

Li, J., Shan, E., Zhao, J., Teng, J. and Wang, Q., 2023. The factors influencing the vertical transport of microplastics in marine environment: A review. *Science of the Total Environment*, 870, p. 161893. http://dx.doi.org/10.1016/j.scitotenv.2023.161893

Liao, Z., Ji, X., Ma, Y., Lv, B., Huang, W., Zhu, X., Fang, M., Wang, Q., Wang, X., Dahlgren, R. and Shang, X., 2021. Airborne microplastics in indoor and outdoor environments of a coastal city in Eastern China. *Journal of Hazardous Materials*, 417, p. 126007. https://doi.org/10.1016/j.jhazmat.2021.126007

Liu, F., Nord, N.B., Bester, K. and Vollertsen, J., 2020. Microplastics removal from treated wastewater by a biofilter. *Water*, 12, p. 1085. https://doi.org/10.3390/w12041085

Liu, K., Wang, X., Fang, T., Xu, P., Zhu, L. and Li, D., 2019a. Source and potential risk assessment of suspended atmospheric Microplastics in Shanghai. *Science of the Total Environment*, 675, pp. 462–471. https://doi.org/10.1016/j.watres.2022.119129

Liu, K., Wang, X., Wei, N., Song, Z. and Li, D., 2019b. Accurate quantification and transport estimation of suspended atmospheric microplastics in megacities: Implications for human health. *Environment International*, 132, p. 105127. https://doi.org/10.1016/j.envint.2019.105127

Liu, X., Yuan, W., Di, M., Li, Z. and Wang, J., 2019c. Transfer and fate of microplastics during the conventional activated sludge process in one wastewater treatment plant of China. *Chemical Engineering Journal*, 362, pp. 176–182. https://doi.org/10.1016/j.cej.2019.01.033

Liu, S.Y., Leung, M.M.L., Fang, J.K.H. and Chua, S.L., 2021. Engineering a microbial 'trap and release' mechanism for microplastics removal. *Chemical Engineering Journal*, 404, p. 127079. https://doi.org/10.1016/j.cej.2020.127079

Lwanga, E.H., Van Roshum, I., Munhoz, D.R., Meng, K., Rezaei, M., Goossens, D., Bijsterbosch, J., Alexandre, N., Oosterwijk, J., Krol, M. and Peters, P., 2023. Microplastic appraisal of soil, water, ditch sediment and airborne dust: The case of agricultural systems. *Environmental Pollution*, 316, p. 120513. https://doi.org/10.1016/j.envpol.2022.120513

Ma, B., Xue, W., Hu, C., Liu, H., Qu, J. and Li, L., 2019. Characteristics of microplastic removal via coagulation and ultrafiltration during drinking water treatment. *Chemical Engineering Journal*, 359, pp. 159–167. https://doi.org/10.1016/j.cej.2018.11.155

Martin, C., Corona, E., Mahadik, G.A. and Duarte, C.M., 2019. Adhesion to coral surface as a potential sink for marine microplastics. *Environmental Pollution*, 255, p. 113281. https://doi.org/10.1016/j.envpol.2019.113281

Microbead Free Water Act, 2015. *US*. www.fda.gov/cosmetics/cosmetics-laws-regulations/microbead-free-waters-act-faqs

Misra, A., Zambrzycki, C., Kloker, G., Kotyrba, A., Anjass, M.H., Franco Castillo, I., Mitchell, S.G., Güttel, R. and Streb, C., 2020. Water purification and microplastics removal using magnetic polyoxometalate-supported ionic liquid phases (magPOM-SILPs). *Angewandte Chemie International Edition*, 59(4), pp. 1601–1605. https://doi.org/10.1002/anie.201912111

Mitrano, D.M. and Wohlleben, W., 2020. Microplastic regulation should be more precise to incentivize both innovation and environmental safety. *Nature Communications*, 11(1), p. 5324. https://doi.org/10.3389/fmars.2020.00128

Munyaneza, J., Jia, Q., Qaraah, F.A., Hossain, M.F., Wu, C., Zhen, H. and Xiu, G., 2022. A review of atmospheric microplastics pollution: In-depth sighting of sources, analytical methods, physiognomies, transport and risks. *Science of the Total Environment*, 822, p. 153339. http://dx.doi.org/10.1016/j.scitotenv.2022.153339

NOAA, 2023. https://oceanservice.noaa.gov/facts/microplastics.html

Oprea, S. and Doroftei, F., 2011. Biodegradation of polyurethane acrylate with acrylated epoxidized soybean oil blend elastomers by *Chaetomium globosum*. *International Biodeterioration & Biodegradation*, 65(3), pp. 533–538. https://doi.org/10.1016/j.ibiod.2010.09.011

Orr, I.G., Hadar, Y. and Sivan, A., 2004. Colonization, biofilm formation and biodegradation of polyethylene by a strain of Rhodococcus ruber. *Applied Microbiology and Biotechnology*, 65, pp. 97–104. https://doi.org/10.1007/s00253-004-1584-8

Osman, A.I., Hosny, M., Eltaweil, A.S., Omar, S., Elgarahy, A.M., Farghali, M., Yap, P.S., Wu, Y.S., Nagandran, S., Batumalaie, K. and Gopinath, S.C., 2023. Microplastic sources, formation, toxicity and remediation: A review. *Environmental Chemistry Letters*, 21(4), pp. 2129–2169. https://doi.org/10.1007/s10311-023-01593-3

Paço, A., Duarte, K., da Costa, J.P., Santos, P.S., Pereira, R., Pereira, M.E., Freitas, A.C., Duarte, A.C. and Rocha-Santos, T.A., 2017. Biodegradation of polyethylene microplastics by the marine fungus *Zalerion maritimum*. *Science of the Total Environment*, 586, pp. 10–15. http://dx.doi.org/10.1016/j.scitotenv.2017.02.017

Parashar, N. and Hait, S., 2023. Plastic rain-atmospheric microplastics deposition in urban and peri-urban areas of Patna City, Bihar, India: Distribution, characteristics, transport, and source analysis. *Journal of Hazardous Materials*, 458(5), p. 131883. https://doi.org/10.1016/j.jhazmat.2023.131883

Park, J.W., Lee, S.J., Hwang, D.Y. and Seo, S., 2021. Removal of microplastics via tannic acid-mediated coagulation and in vitro impact assessment. *RSC Advances*, 11(6), pp. 3556–3566. https://doi.org/10.1039/d0ra09645h

Perera, K., Ziajahromi, S., Bengtson Nash, S., Manage, P.M. and Leusch, F.D., 2022. Airborne microplastics in indoor and outdoor environments of a developing country in south asia: Abundance, distribution, morphology, and possible sources. *Environmental Science & Technology*, 56(23), pp. 16676–16685. https://doi.org/10.1021/acs.est.2c05885

Pizzichetti, A.R.P., Pablos, C., Álvarez-Fernández, C., Reynolds, K., Stanley, S. and Marugán, J., 2021. Evaluation of membranes performance for microplastic removal in a simple and low-cost filtration system. *Case Studies in Chemical and Environmental Engineering*, 3, p. 100075. https://doi.org/10.1016/j.cscee.2020.100075

Prasittisopin, L., Ferdous, W. and Kamchoom, V., 2023. Microplastics in construction and built environment. *Developments in the Built Environment*, p. 100188. https://doi.org/10.1016/j.dibe.2023.100188

Prata, J.C., 2018. Airborne microplastics: Consequences to human health? *Environmental Pollution*, 234, pp. 115–126. https://doi.org/10.1016/j.envpol.2017.11.043

Purwiyanto, A.I.S., Prartono, T., Riani, E., Naulita, Y., Cordova, M.R. and Koropitan, A.F., 2022. The deposition of atmospheric microplastics in Jakarta-Indonesia: The coastal urban area. *Marine Pollution Bulletin*, 174, p. 113195. https://doi.org/10.1016/j.marpolbul.2021.113195

Qian, Z.H.O.U., ChongGuo, T.I.A.N. and YongMing, L.U.O., 2017. Various forms and deposition fluxes of microplastics identified in the coastal urban atmosphere. *Chinese Science Bulletin*, 62(33), pp. 3902–3909. https://doi.org/10.1360/N972017-00956

Rajala, K., Grönfors, O., Hesampour, M. and Mikola, A., 2020. Removal of microplastics from secondary wastewater treatment plant effluent by coagulation/flocculation with iron, aluminum and polyamine-based chemicals. *Water Research*, 183, p. 116045. https://doi.org/10.1016/j.watres.2020.116045

Ruffell, H., Pantos, O., Northcott, G. and Gaw, S., 2021. Wastewater treatment plant effluents in New Zealand are a significant source of microplastics to the environment. *New Zealand Journal of Marine and Freshwater Research*, pp. 1–17. https://doi.org/10.1080/00288330.2021.1988647

Sarmah, P. and Rout, J., 2019. Cyanobacterial degradation of low-density polyethylene (LDPE) by *Nostoc carneum* isolated from submerged polyethylene surface in domestic sewage water. *Energy, Ecology and Environment*, 4, pp. 240–252. https://doi.org/10.1007/s40974-019-00133-6

Shahi, N.K., Maeng, M., Kim, D. and Dockko, S., 2020. Removal behavior of microplastics using alum coagulant and its enhancement using polyamine-coated sand. *Process Safety and Environmental Protection*, 141, pp. 9–17. https://doi.org/10.1016/j.psep.2020.05.020

Shen, M., Hu, T., Huang, W., Song, B., Zeng, G. and Zhang, Y., 2021. Removal of microplastics from wastewater with aluminosilicate filter media and their surfactant-modified products: Performance, mechanism and utilization. *Chemical Engineering Journal*, 421, p. 129918. https://doi.org/10.1016/j.cej.2021.129918

Siipola, V., Pflugmacher, S., Romar, H., Wendling, L. and Koukkari, P., 2020. Low-cost biochar adsorbents for water purification including microplastics removal. *Applied Sciences*, 10(3), p. 788. https//doi.org/10.3390/app10030788

Sturm, M.T., Herbort, A.F., Horn, H. and Schuhen, K., 2020. Comparative study of the influence of linear and branched alkyltrichlorosilanes on the removal efficiency of polyethylene and polypropylene-based microplastic particles from water. *Environmental Science and Pollution Research*, 27(10), pp. 10888–10898. https://doi.org/10.1007/s11356-020-07712-9

Sun, C., Wang, Z., Chen, L. and Li, F., 2020. Fabrication of robust and compressive chitin and graphene oxide sponges for removal of microplastics with different functional groups. *Chemical Engineering Journal*, 393, p. 124796. https://doi.org/10.1016/j.cej.2020.124796

Sun, J., Peng, Z., Zhu, Z.R., Fu, W., Dai, X. and Ni, B.J., 2022. The atmospheric microplastics deposition contributes to microplastic pollution in urban waters. *Water Research*, 225, p. 119116. https://doi.org/10.1016/j.watres.2022.119116

Szewc, K., Graca, B. and Dołęga, A., 2021. Atmospheric deposition of microplastics in the coastal zone: Characteristics and relationship with meteorological factors. *Science of the Total Environment*, 761, p. 143272. https://doi.org/10.1016/j.scitotenv.2020.143272

Talvitie, J., Mikola, A., Koistinen, A. and Setälä, O., 2017. Solutions to microplastic pollution–Removal of microplastics from wastewater effluent with advanced wastewater treatment technologies. *Water Research*, 123, pp. 401–407. http://dx.doi.org/10.1016/j.watres.2017.07.005

Tian, L., Jinjin, C., Ji, R., Ma, Y. and Yu, X., 2022. Microplasticss in agricultural soils: Sources, effects, and their fate. *Current Opinion in Environmental Science & Health*, 25, p. 100311. https://doi.org/10.1016/j.coesh.2021.100311

Tofa, T.S., Kunjali, K.L., Paul, S. and Dutta, J., 2019. Visible light photocatalytic degradation of microplastic residues with Zinc Oxide nanorods. *Environmental Chemistry Letters*, 17, pp. 1341–1346. https://doi.org/10.1007/s10311-019-00859-z

Uogintė, I., Pleskytė, S., Skapas, M., Stanionytė, S. and Lujanienė, G., 2023. Degradation and optimization of microplastic in aqueous solutions with graphene oxide-based nanomaterials. *International Journal of Environmental Science and Technology*, 20(9), pp. 9693–9706. https://doi.org/10.1007/s13762-022-04657-z

Vimala, P.P. and Mathew, L., 2016. Biodegradation of polyethylene using *Bacillus subtilis*. *Procedia Technology*, 24, pp. 232–239. https://doi.org/10.1016/j.protcy.2016.05.031

Wang, C., O'Connor, D., Wang, L., Wu, W.M., Luo, J. and Hou, D., 2022a. Microplastics in urban runoff: Global occurrence and fate. *Water Research*, p. 119129. https://doi.org/10.1016/j.watres.2022.119129

Wang, Z., Zhang, Y., Kang, S., Yang, L., Luo, X., Chen, P., Guo, J., Hu, Z., Yang, C., Yang, Z. and Gao, T., 2022b. Long-range transport of atmospheric microplastics deposited onto glacier in southeast Tibetan Plateau. *Environmental Pollution*, 306, p. 119415. https://doi.org/10.1016/j.envpol.2022.119415

Wang, X., Liu, K., Zhu, L., Li, C., Song, Z. and Li, D., 2021. Efficient transport of atmospheric microplastics onto the continent via the East Asian summer monsoon. *Journal of Hazardous Materials*, 414, p. 125477. https://doi.org/10.1016/j.jhazmat.2021.125477

Wang, Y., Liu, L., Cao, S., Yu, J., Li, X., Su, Y., Li, G., Gao, H. and Zhao, Z., 2023. Spatio-temporal variation of soil microplastics as emerging pollutant after long-term application of plastic mulching and organic compost in apple orchards. *Environmental Pollution*, 328, p. 121571. https://doi.org/10.1016/j.envpol.2023.121571

Wang, Z., Sedighi, M. and Lea-Langton, A., 2020. Filtration of microplastic spheres by biochar: Removal efficiency and immobilisation mechanisms. *Water Research*, 184, p. 116165. https://doi.org/10.1016/j.watres.2020.116165

Wright, S.L., Ulke, J., Font, A., Chan, K.L.A. and Kelly, F.J., 2020. Atmospheric microplastic deposition in an urban environment and an evaluation of transport. *Environment International*, 136, p. 105411. https://doi.org/10.1016/j.envint.2019.105411

Yao, Y., Glamoclija, M., Murphy, A. and Gao, Y., 2022. Characterization of microplastics in indoor and ambient air in northern New Jersey. *Environmental Research*, 207, p. 112142. https://doi.org/10.1016/j.envres.2021.112142

Zhang, M., Yang, J., Kang, Z., Wu, X., Tang, L., Qiang, Z., Zhang, D. and Pan, X., 2021. Removal of micron-scale microplastic particles from different waters with efficient tool of surface-functionalized microbubbles. *Journal of Hazardous Materials*, 404, p. 124095. https://doi.org/10.1016/j.jhazmat.2020.124095

Zhang, Q., Zhao, Y., Du, F., Cai, H., Wang, G. and Shi, H., 2020. Microplastic fallout in different indoor environments. *Environmental Science & Technology*, 54(11), pp. 6530–6539. https://dx.doi.org/10.1021/acs.est.0c00087

Zhou, G., Wang, Q., Li, J., Li, Q., Xu, H., Ye, Q., Wang, Y., Shu, S. and Zhang, J., 2021. Removal of polystyrene and polyethylene microplastics using PAC and $FeCl_3$ coagulation: Performance and mechanism. *Science of the Total Environment*, 752, p. 141837. https://doi.org/10.1016/j.scitotenv.2020.141837

5 Microplastic Pollution in the Soil Ecosystem and Emerging Recycling Techniques

Satyakirti Belwal, Anargha P. Nambiar, and Uma Hapani

5.1 INTRODUCTION

Plastic particles with a sizes less than 5 mm are generally distinguished as microplastics. These have been shown to pose a potential risk for living organisms, though the extent of effect is not known yet (Rillig *et al.*, 2017). In the past decade, concerns among researchers have been aroused toward the occurrence and impacts of microplastics worldwide. These types of microplastics in the form of small granules are used as cleaning agents in the areas of cosmetics and air blowing agents. Also, these are formed as a result of the disintegration and decomposition of microplastic materials (Ryan *et al.*, 2009). Hence it can be deduced that control on microplastics demands immediate attention where future research will strive a prominent place.

5.1.1 Microplastic Pollution

Interestingly, microplastic pollution is not manufactured purposefully; however, it results from the breakdown products of the commercial plastics that are used daily. Moreover, they have now originated to be a major cause of pollution from various sources that have already demonstrated the detrimental effects on marine organisms (do Sul and Costa, 2014). This can be attributed to the persistent nature of plastics and the presence of potentially toxic chemicals in it. As the matter of fact, the inappropriate use of plastics and their disposal has resulted in direct contact with aquatic organisms, which often consume them. These are small-sized particles, commonly known as plastic debris that enters the environment, disturbing its balance and subsequently affecting human health adversely. Such particles are termed microplastic pollution (Kirstein *et al.*, 2021).

DOI: 10.1201/9781032684574-5

5.1.2 Types of Microplastics

Microplastics are classified based on its properties like size, shape, and origin. Table 5.1 describes it in details (Zhu *et al.*, 2019).

5.1.3 Microplastics into the Soil Ecosystem

Irrespective of the origin, microplastics, being widespread when they reach the soil surface, have significant hazardous effects on the soil biota. For instance, the oxygen levels in the soil decline, and survival becomes difficult for the microorganisms living under the soil environment (Ren *et al.*, 2018). Research has proved the negative impacts of microplastics on plant growth and soil microbial activity, thereby altering the soil chemistry (Leed and Smithson, 2019). This change in soil conditions can be specifically attributed to the release of monomers, oligomers, impurities, and additive chemicals from the microplastics that exert selective pressure on the soil microbes (Iqbal *et al.*, 2020). Moreover, this also impacts the soil microbiota severely due to the changes that occur in soil conditions, namely, water holding capacity and bulk density (Boots *et al.*, 2019; de Souza Machado *et al.*, 2019; Iqbal *et al.*, 2020). It has been observed that microplastics are absorbed by some plants from the soil; however, the extent and implications of this absorption are still under scrutiny. This raises a serious concern pertaining to the transmission of microplastics into the human body via their diet, very likely due to the accumulation of them in the edible part of crop components (Sajjad *et al.*, 2022). It is important to note that Sajjad *et al.* (2022) confirmed the adsorption of potentially toxic elements (PTE) by microplastics that have long-lasting effect and the ability to bioaccumulate. Nevertheless, this observation was also affirmed by the research carried out by Nawab *et al.* (2021). To establish this conclusion, several reports have validated large surface area and high hydrophobic nature of microplastics that can act as a potential adsorbent for PTEs (Abbasi *et al.*, 2019; Akhbarizadeh *et al.*, 2017; Wang *et al.*, 2023).

5.2 MAJOR SOURCES OF MICROPLASTICS IN AIR, WATER, AND SOIL

The presence of microplastics in the air, water, and soil is a cause of major environmental and health-related issues. As discussed earlier, the extent of damage caused, likely due to microplastics, is still not established; however, the effects so far are found to be detrimental.

5.2.1 Sources of Microplastics in the Air

The presence of microplastics in the air has been paid little attention. The main contributors to airborne microplastic particulate matter (MPM) are urban dust, simulated textile material, and rubber tire erosion (Chen *et al.*, 2020). The lightweight nature of microplastics (MPs) facilitates the transportation of them from one location to another by air. These become harmful, and their negative effects become prominent when they enter the respiratory tract of humans. They may stimulate

TABLE 5.1
Classification of Microplastics

Sr no.	Based on	Category	Characteristics	Reference
1.	Size	Nanoplastics	Less than 1 μm in size	Kokilathasan and Dittrich, 2022
		Microplastics	Between 1 μm and 5 μm in size	
2.	Shape	Granule	Granules are small plastic particles used as raw materials in the production of plastic products and can be transformed into microplastics through the release of these particles into the environment in the process of manufacturing, transporting, or processing plastic materials.	Tanaka, and Takada, 2016; Zhu *et al.*, 2019
		Fragment	Fragments are the broken-down components of larger plastic objects that have deteriorated as a result of environmental factors such as weathering, ultraviolet radiation, and mechanical stresses.	
		Microbead	Microbeads are small, rounded plastic particles used in cosmetics and exfoliating products	
		Fiber	Fibers are plastic particles that are thin and thread-like, commonly found in textile, synthetic, and clothing materials.	
		Foam	Foam microplastics are expanded polystyrene beads (EPS) that decompose into smaller components. They are commonly employed in packing and disposable products like cups and food containers.	
3.	Origin	Polyethylene (PE)	$0.91 - 0.93\ g\ cm^{-3}$ Low-density PE $0.93 - 0.97\ g\ cm^{-3}$ High-density PE	Andrady, 2011; Avio *et al.*, 2017; Scheurer and Bigalke, 2018; Zhu *et al.*, 2019
		Polypropylene (PP)	$0.85 - 0.95\ g\ cm^{-3}$	
		Polystyrene (PS)	$1.04 - 1.11\ g\ cm^{-3}$	
		Polyester/polyethylene terephthalate (PET)	$1.37 - 1.45\ g\ cm^{-3}$	
		Polyvinyl chloride (PVC)	$1.16 - 1.58\ g\ cm^{-3}$	
		Nylon or polyamide (PA)	$1.08\ g\ cm^{-3}$ Nylon- 6 $1.31\ g\ cm^{-3}$ Nylon- 6,6	
		Others	Any other density	

chemotactic factors impeding the migration of macrophages and increase the permeability, thereby causing chronic inflammation due to dust load. Numerous respiratory disorders have been reported among workers employed in the textile industry and the polyvinyl chloride industry (Sajjad *et al.*, 2022; Sridharan *et al.*, 2021).

5.2.2 Sources of Microplastics in Water

The confirmed presence of microplastics in the oceans and the marine ecosystem pose a serious threat to aquatic biodiversity. The primary sources of their entry into the marine environment are from land-based sources, such as the accidental or mishandling of plastics, dumped plastic waste, and mismanagement of landfills such as fence surrounding landfill and non-availability of suitable synthetic material to cover the waste (Duis and Coors, 2016). The presence of microplastics in drinking water has become increasingly prominent and also a potential source of contamination. Agricultural plastics and other contaminants on the soil surface can penetrate deep into the land through the action of earthworms and leaching groundwater (Eerkes-Medrano *et al.*, 2019). So far, no studies have shown concrete evidence of microplastics in water as a source of pollution.

5.2.3 Sources of Microplastics in the Soil

Researchers have confirmed the presence of microplastics in soil by monitoring the movement of particles due to its property in the porous media and comparing it with the soil's physical characteristics (Rillig *et al.*, 2017). The microscopic movement of particles can be attributed to the biopores formed in the soil biota, ploughing and harvesting of the fields, and lastly soil cracking in the summers. Biopores are expected to be responsible for the downward movement of microplastics in the soil, while ploughing and harvesting are supposed to distribute the particles horizontally (Guo *et al.*, 2020; Yang *et al.*, 2021). Depending on the type of crop, microplastics enter the soil by harvesting portions of the plant that lie beneath the surface of the soil, such as potatoes and carrots. It is understood that the microplastic particles aggregates within the soil; however, their degree of aggregation is not known yet. The macroaggregation found in the soil is arranged hierarchically along with microplastic particles and microaggregates (<0.250 mm) other than organic matter, microorganisms, primary soil particles, and other soil constituents that are usually present in the soil (Tisdall and Oades, 1982; Rillig *et al.*, 2017). Figure 5.1 shows the major sources of microplastics in air, water, and soil.

5.3 EFFECTS OF MICROPLASTICS ON THE SOIL ECOSYSTEM

In soil ecosystem, the presence of microplastics can have a variety of impacts on soil's physical characteristics, microbial populations, plant health, soil organisms, and so on. However, research focusing on the impacts of microplastics in terrestrial ecosystems is still limited in comparison to marine ecosystems. Hence, extensive studies are essential to determine the magnitude of these impacts, their underlying mechanisms, and potential mitigation strategies on soil ecology due to microplastics.

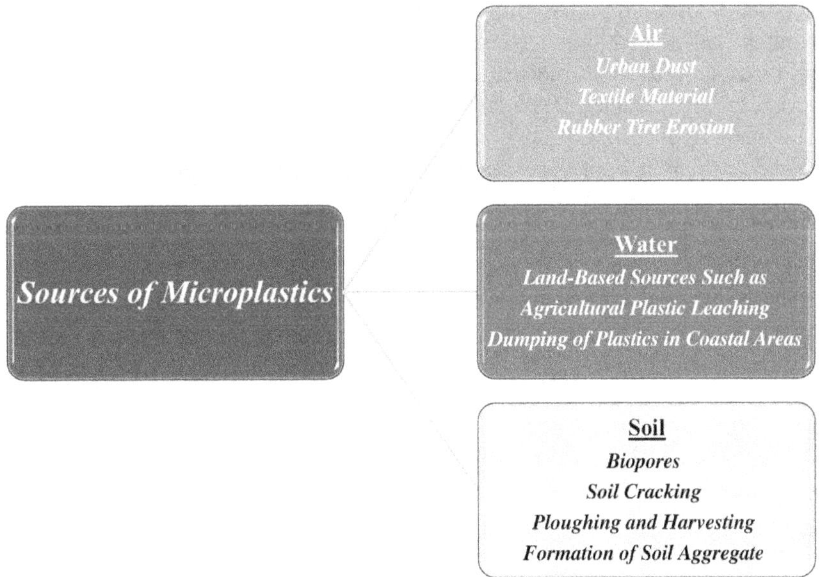

FIGURE 5.1 Sources of microplastics in air, water, and soil.

5.3.1 On Soil Microorganisms

The relationship between microplastics and soil microbial communities remains largely unknown. To date, very few studies have been conducted to evaluate the impact of microplastics on soil systems, particularly in terms of microbial activity, bacterial migration, and the dissemination of antibiotic-resistance genes (ARGs) (Ding *et al.*, 2022; Zhang *et al.*, 2021). The presence of microorganisms in the soil is essential; they are actively involved in the decomposition of organic matter, cycling of nutrients in the soil, and various other chemical reactions beneath the soil matrix. Soil microbial activity is highly dependent on the degradation process of organic matter. These include nitrification, denitrification, methane oxidation, sulfur oxidation, and mineralization of organic matter (Castellano-Hinojosa *et al.*, 2022; Joos and De Tender, 2022). Moreover, the organic matter decomposition from plant/animal residues maintains the soil fertility through the nutrient cycling and recycling process and carbon sequestration in the soil. It is important to note that the amount of microbial biomass present in the soil mixture is directly associated with soil health and complex chemical bonds cleavage. Additionally, they are also considered a source, sink, and regulator of soil in energy and nutrient transformations (Murphy *et al.*, 2007).

The presence of microplastics in soil invertebrates like nematodes, Oligochaeta (earthworms), Collembola, and Isopods is a less explored research area to date. This has created a research gap, and no information pertaining to the ramifications originated by microplastics on soil creatures' health has been reported yet, which is in contrast to the significant findings witnessed for aquatic creatures. Deliberate incorporation of microplastics into liquid media, food, and the soil matrix by a few researchers has helped in understanding the negative effects on soil organisms. This study focused on the soil microorganism's survival, growth, reproductivity, inflammatory response, metabolic functions, feeding behavior, neurotoxicity, and gut microbiota (Chang *et al.*, 2022; Huang *et al.*, 2023). As anticipated, significant alterations were observed in the biomass: tissue elemental composition, inhibition of organism's growth, and the approach to the lethal concentration of 1–2% (w/w). Subsequently, these findings suggest hinderance in their mobility and accidental ingestion when the microplastics adhere to its external surface. Moreover, when microplastics are ingested by the soil organisms, they issue false satiation, reduced carbon biomass accumulation, depleting energy and finally halting the organism's growth or even resulting in death. However, the effect of microplastics on living beings is still a point of discussion that is needed in order to attain clarity on their repercussions (Guo *et al.*, 2020; Boots *et al.*, 2019).

It was confirmed from one study that an artificially generated soil environment containing *E. fetida*, a class of earthworms, can directly absorb microplastic particles at very high rates. The study reported elevated peroxidase (POD) and catalase (CAT) activities and a falloff of super oxide dismutase (SOD) and thione S-transferase (GST) activity, causing tissue damage and oxidative stress. Furthermore, the bioavailability of hydrophobic organic contaminants (HOCs) in soil porewaters was found to decrease drastically. The model-based predictions for microplastic-sorbed earthworms demonstrated a complete withdrawal of HOCs from the soil matrix. It could be understood that the microplastic particles interact with HOCs based on their physicochemical properties, most importantly the hydrophobicity of HOCs. Fascinatingly, the negligible effect of microplastics on soil health, soil domesticated invertebrates, and bioaccumulation of hygroscopic contaminants are known. Such observations can be attributed to soil's so-called buffering effect, which includes dilution, competitive sorption, and various mitigating interactions. Hence, it becomes essential to draw a conclusion of microplastic risks on a large scale by assessing it for real samples, i.e., a real ground field, and by considering a wide variety of organisms and scenarios. Despite having relevant findings and risk factors associated with microplastics for aquatic organisms, the accurate determination of detrimental potentials on terrestrial ecosystems cannot be obtained, focusing mainly on agricultural soils (Wang *et al.*, 2019; Hurley et al., 2017; Lusher *et al.*, 2015).

5.3.2 On Plants

Impacts of microplastics on plants, their growth, potential risks, and food quality have not been thoroughly investigated. It could be possibly related to difficult screening of these particles in plant tissues and lack of attention paid to this part. Even with very little information available on microplastics, it endorses that both polypropylene microplastics (PP-MPs) and polyester microplastics (PES-MPs) have been found to exert a negative influence on crop productivity. It has been found that incorporation of PP-MPs in plants reduces the soil bulk density for *Glycine max* (GM, dicotyledon) and *Arachis hypoaea* (AH, dicotyledon). Also, the soil pH was observed to decrease

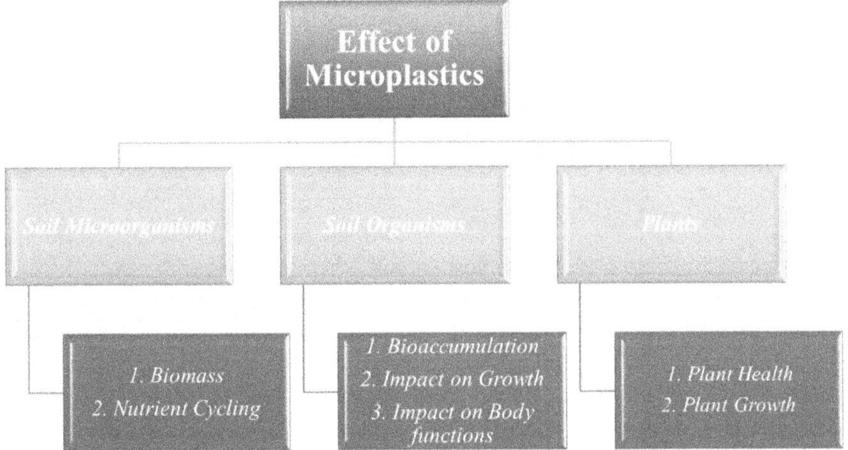

FIGURE 5.2 Effects of microplastics on the soil ecosystem.

the presence of PP-MPs; instead it increases with PES-MP. AH being is the mostly commonly used parameter and measures plant height, culm diameters, total biomass, shoot biomass, root biomass. These parameters have been significantly reduced in the presence of PP-MPs, addressing an urgent attention on food chain and threats on human health allied due to such decline (Zhou *et al.*, 2023; Colzi *et al.*, 2022; Gao *et al.*, 2019).

In addition, the phytotoxic properties of microplastics are directly and indirectly connected to plant performance. More importantly, the nanoscale microplastic particles enter the plant root system through a crack-entry mechanism and then spread out to various plant organs under the pressure of transpiration (Li *et al.*, 2020).

Zhang *et al.* (2022) confirmed microplastics exert strain on antioxidant systems, disturbing plant morphology and affecting the process of photosynthesis depressingly. This study identified negative, positive, and non-neutral interactions of microplastics altering the plant morphologies and the process of photosynthesis that are highly dependent on the type of polymer present in the microplastics. Therefore, the accumulation of microplastics in plants further migrates toward the upper trophic levels and biomagnify themselves. Figure 5.2 shows the effects of microplastics on the soil ecosystem.

5.4 EMERGING RECYCLING TECHNIQUES FOR MICROPLASTICS

A microplastic recycling is a complex process. This is due to the breakdown process of large particles into smaller ones. However, the concept is considered to be novel whereby microplastics can be extracted and reprocessed for future use. These are accepted to be present in a wide range of aquatic, soil, and airborne habitats. Moreover, traditional recycling methods step back certainly due to its small size and wide distribution in various territories. Thus several approaches have been established and put into practice to mitigate this type of pollution.

The methods include:

1. Filtration and collection.
2. Monitoring, identification, and extraction from the environment.
3. Microplastic to energy conversion.

5.4.1 Filtration and Collection

This process of filtration is the simplest among other recycling techniques. It is also a widely implemented method in the wastewater treatment plants to remove the sludge where microplastics are generally present. The use of advanced water filtration systems not only controls but also prevents microplastics penetration into the water bodies. A wastewater treatment study was conducted by Lofty *et al.* (2022) for microplastics from a primary settler tank on European agricultural land for sewage sludge. The work was able to successfully quantify microplastics at Nash Wastewater Plant, South Wales, United Kingdom (Lofty *et al.*, 2022).

5.4.2 Monitoring, Identification, and Extraction from the Environment

To analyze microplastic particles, the primary requirement is to isolate these particles from the matrix for a specimen under the study. The analysis should be done in such a way that microplastic particles are not destroyed or tampered with in any manner, but the focus of the adopted method for soil aggregate dispersion should be that on matrix.

5.4.3 Manual Extraction

The extraction of microplastics can be easily attained by the process of sifting and manually sorting the sample under a stereomicroscope. By doing this, unambiguous minerals or biogenic particulate matter and even the visible cellular particles can be separated. Later, this process was followed by a 'hot needle test' to decoct the potential number of microplastic particles in the sample. However, this method calls out an alternative for the fact related to various discrepancies. The inconsistency is accompanied by cumbersome manual sorting and visual identification (Martin *et al.*, 2018; Ivleva, 2021; Woo *et al.*, 2021).

5.4.4 Electrostatic Precipitation

Electrostatic precipitation of microplastics from the solid matter is a modified separation technique contained in a small device. This is a commonly used instrument in recycling industries that works for microplastics of sizes between 63 μm and 5 mm. Noticeably, the sample throughput and recovery rate were observed near to 99% and 90–100%, respectively (Felsing *et al.*, 2018). It is also recommended that the device can be efficiently used to analyze wet samples; however, for cohesive soil specimens, the technique is not advised. The suggested inappropriateness is its aggregation due to the adhesive nature toward the metal drum and scrapper of the electrostatic precipitation vessel, instigating loss of microplastics fraction in the final sample (Enders *et al.*, 2020; Rani *et al.*, 2023).

5.4.5 Visual Identification

Although a visual identification test is supposed to be the basic and fundamental approach for microplastic analysis on a light microscope, the method exhibits certain biasness and a 70% error ratio, approximately. To overcome this inaccuracy, few research works reviewed the thermal properties of synthetic polymers and proposed the hot needle test (Ruggero *et al.*, 2020; Syafina *et al.*, 2022). Subsequently, this experiment was thoroughly investigated by Zhang *et al.* (2021), and, based on the examination, a new process was developed exclusively for low-density soil polymers. In this method, initially the supernatant similar to the density of water is separated by heating the sample for 3–5 minutes at 130°C; later, the microscopic image is captured for both samples (before and after heat treatment). The melted particle here is determined to be a thermoplastic type of polymer. Nevertheless, the technique is not suitable for high-density polymers and does not support the melting of natural materials such as wax. Also, the protocol seems to be destructive, the and accurate marking of polymer becomes difficult. Polarized light microscope substitutes for this technique for the detection of synthetic particles (Sierra *et al.*, 2020).

5.4.6 Chromatography

Sophisticated extraction techniques, when hyphenated with chromatography, gives qualitative and quantitative information for polymers. These include high-temperature gel permeation chromatography (HT-GPC), 96 liquid extraction (LSE) with size exclusion chromatography (SEC) and pyrolysis gas chromatography mass spectrum (Py-GC-MS). One such example is the determination of polyolefins (polyols) in cosmetics on HT-GPC. It is important to understand that the bioavailability of polymer particles depends on the size and shape on other hand, the number of bioavailable particles is again subject to the environmental impact. Also, the morphology of these particles is compatible with the soil health, structure, and functionalities (de Souza Machado *et al.*, 2018).

5.4.7 Thermogravimetric Analysis

Thermal analysis is a destructive analytical technique. In this, the microplastics in soil and complex matrices can be detected; however, successive analyses are a concern. The drawbacks associated with it are the size, shape, and number of particles in the microplastics that are related to the environmental impacts. Samanta *et al.* (2022) carried out microbial analysis on wastewater by combining thermogravimetric analysis (TGA) and differential thermal calorimetry (DSC); the analysis could only reveal polyethylene (PE) and polypropylene (PP) in the sample. On the contrary, TGA and MS were hyphenated by David et al. (2018) to determine polyethylene terephthalate (PET) from soil samples with no prior sample preparation. Even though the method was found to be successful, a few researchers still suggest the necessity of further developing the method and restricting the PET analysis using the same (Huppertsberg and Knepper, 2018).

5.4.8 In-Situ Identification

A hypothesis was framed by Paul and his team, who proposed a technique combining near infrared (NIR) spectroscopy and chemometrics for microplastic determination in soil samples. The method was developed based on the understanding that NIR has an ability to penetrate beyond mid-infrared range. This will help in analyzing the samples even when they are coated by thick biofilm. Moreover, the water sensitiveness for the method supports the hypothesis. However, the experimental point of view rejected it and proved to be unsuccessful for the microplastic detection by this method (Paul et al., 2019; Vidal and Pasquini, 2021).

Thereafter, Corradini *et al.* (2019) put forward a proposal to examine the plastic composition and its concentration directly from the samples. Here, a portable visible near infrared spectroscope (vis-NIR) was designed to determine the polymer concentration for the given field in a short time, by passing the extraction procedure and detection processes. With this method, low-dispersive polymers like low-density polyethylene (LDPE), polyethylene terephthalate (PET), and polyvinyl chloride (PVC) were successfully detected from the designed simulated polluted soil samples. This has demonstrated an accuracy of 10 g/kg and sensitivity of 15 g/kg for the model, and these values are viewed as limitations with the model method. In addition, the method only discloses the concentration of the polymer particles and does not provide any information regarding its size and shape (Möller *et al.*, 2020; Corradini *et al.*, 2019).

Another unique and important technique is surface-enhanced raman scattering (SERS) where samples are exposed to a plasmonic environment. SERS works by amplifying the Raman signal of a sample molecule to plasmonic nanostructures. These nanostructures are generally metals such as gold or silver and are further amplified up to several orders of magnitude. SERS can be used to detect trace amounts of agripollutants in the field that offers appreciable sensitivity and selectivity. To eliminate the complex sample preparation or sample transfer to laboratories, portable detection systems (PDS) and spectrometers are employed to facilitate real-time analysis in the field (Yılmaz *et al.*, 2022).

5.4.9 Vibrational Spectroscopy or Raman Spectroscopy

A conjunction of microspectroscopic techniques, Raman (RM), and Fourier transform infrared spectroscopy (FTIR) can be used to acquire information on the spectral and spatial properties of particles that are less than 100 μm in size. This is a suitable technique for the analysis of particulate matter that is less than 100 μm in size, such as microplastics. Notably, the smallest particulate specimen found in bottled mineral water is 1 mm in size. Moreover, with RM analysis, polymer identification and the microplastics abundance can be known accurately for the sample under scrutiny (Anger *et al.*, 2018).

For agricultural pollutants, various detection methods are available; however, these are inbuilt for laboratories that lack onsite outdoor detection viability. To make it practicable, surface-enhanced Raman scattering (SERS) can be opted as a convenient choice for a portable detection system. These portable meters can be used in combination with a portable spectrometer to aid in a superior identification for the samples. SERS enables a high-speed detection, real-time tracking of agricultural contaminants on the ground or in processing plants.

It is known that the physical pollutant/microplastics can be determined by surface-enhanced Raman scattering (SERS). Specifically, SERS can detect the presence of microplastics in non-pretreated water samples. Here, the samples were treated with a sponge-supported layer containing Au nanoparticles that allows detection and concentration of the microplastics. Moreover, portable Raman spectrometer was able to detect 40 $\mu g/ml^{-1}$ at 100 nm, which is considered a good selectivity value for the detection of microplastics released in aquatic environments (Yılmaz *et al.*, 2022; Yin *et al.*, 2021; Kundu *et al.*, 2021).

5.4.10 Microplastic-to-Energy Conversion

Due to the extensive and unregulated usage of plastic items, their unfettered disposal at the end of their useful lives has resulted in microplastics posing a global risk for the environment. Triboelectric nanogenerator (TENG) is a technique where electricity is produced when two different types of materials come into contact and get separated. This creates a charge imbalance between the two materials. Later, when these materials are mechanically stimulated, by either touch or movement, the electrostatic energy generated is converted into useful electrical energy. By utilizing this energy, wearable devices and sensors can be charged efficiently. TENG promises a feasible mechanism to produce an output that is based on the inherent triboelectric properties of the microplastic material. The structural design parameter was evaluated to optimize the structure of the TENG and its resulting electrical output. Hence it is verified that TENG technique microplastics can be used as a renewable source to generate electricity (Cho *et al.*, 2023). Figure 5.3 showing different emerging recycling techniques for microplastics.

5.5 SUMMARY

The breakdown of commercial plastic products to small sizes, to be precise less than 5 mm particles, is known as microplastics. The source for microplastics is not one but many that can easily penetrate to the soil, microorganisms, marine organisms, plants, and so on. However, the research pertaining to the toxic effects of microplastics for the environment, living organisms, and human beings has not been studied extensively. This chapter discusses the reports published by various researchers where they have focused on gathering the data that states the major sources of microplastic pollution in soil, air, and water. Moreover, the chronic damages that target the soil and aquatic organisms' health and survival rate are disclosed to some extent. These particles also disturb the balance of the plant ecosystem that eventually reaches human beings via the food chain. They also alter soil composition and affect microbial communities in the soil. However, the negative effects on human beings have not been reported yet in this area of research. Additionally, the recycling methods are discussed that confirms that there is room for eradication and shows a sign of relief

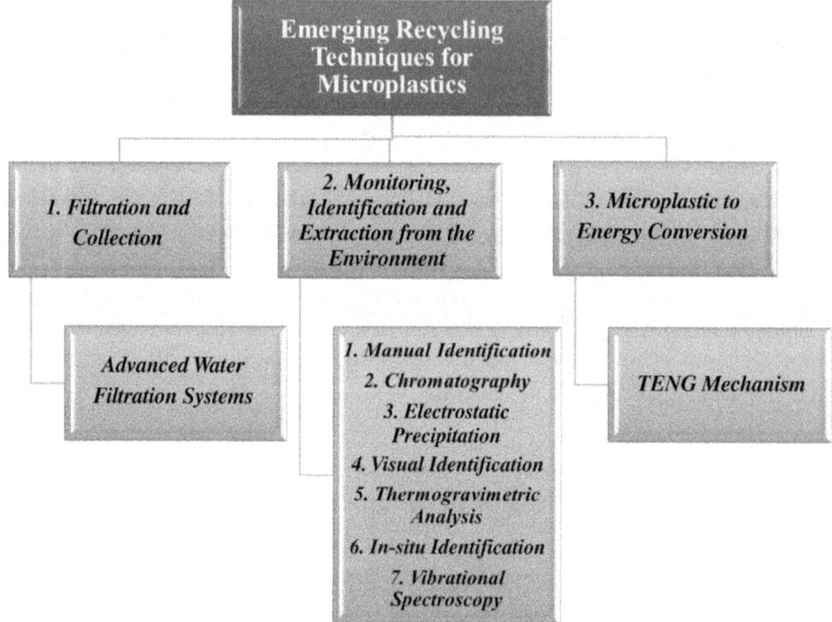

FIGURE 5.3 Emerging recycling techniques for microplastics.

to all the species that are threatened due to microplastics. Interestingly, researchers have come up with a technique in which the successful formation of electrical energy from microplastic pollution can be achieved. The implementation of these emerging recycling techniques helps in achieving the ultimate goal of sustainable development as well as sustaining the environment. However, the impacts of microplastics known from the research carried out so far suggest detrimental effects on soil and water habitats. It can be concluded that the research in this area has to be focused on filling the research gaps that are still unknown in order to evade a serious threat that is coming into alignment in the near future.

REFERENCES

Abbasi, S., Keshavarzi, B., Moore, F., Shojaei, N., Sorooshian, A., Soltani, N. and Delshab, H., 2019. Geochemistry and environmental effects of potentially toxic elements, polycyclic aromatic hydrocarbons and microplastics in coastal sediments of the Persian Gulf. *Environmental Earth Sciences*, 78, pp. 1–15.

Akhbarizadeh, R., Moore, F., Keshavarzi, B. and Moeinpour, A., 2017. Microplastics and potentially toxic elements in coastal sediments of Iran's main oil terminal (Khark Island). *Environmental Pollution*, 220, pp. 720–731.

Andrady, A.L., 2011. Microplastics in the marine environment. *The Marine Pollution Bulletin,* 62(8), pp. 1596–1605.

Anger, P.M., von der Esch, E., Baumann, T., Elsner, M., Niessner, R. and Ivleva, N.P., 2018. Raman microspectroscopy as a tool for microplastic particle analysis. *TrAC Trends in Analytical Chemistry*, 109, pp. 214–226. https://doi.org/10.1016/j.trac.2018.10.010

Avio, C.G., Gorbi, S. and Regoli, F., 2017. Plastics and microplastics in the oceans: From emerging pollutants to emerged threat. *Marine Environmental Research*, 128, pp. 2–11. https://doi.org/10.1016/j.marenvres.2016.05.012

Boots, B., Russell, C.W. and Green, D.S., 2019. Effects of microplastics in soil ecosystems: Above and below ground. *Environmental Science & Technology*, 53(19), pp. 11496–11506.

Castellano-Hinojosa, A., Boyd, N.S. and Strauss, S.L., 2022. Impact of fumigants on non-target soil microorganisms: A review. *Journal of Hazardous Materials*, 427, p. 128149.

Chang, J., Fang, W., Liang, J., Zhang, P., Zhang, G., Zhang, H., Zhang, Y. and Wang, Q., 2022. A critical review on interaction of microplastics with organic contaminants in soil and their ecological risks on soil organisms. *Chemosphere*, 306, p. 135573.

Chen, G., Feng, Q. and Wang, J., 2020. Mini-review of microplastics in the atmosphere and their risks to humans. *Science of the Total Environment*, 703, p. 135504. https://doi.org/10.1016/j.scitotenv.2019.135504

Cho, S., Cha, K., Kim, B., Lee, J., Park, K., Chung, S.H., Song, M., Heo, D., Son, J.H., Choi, M. and Lin, Z.H., 2023. Sustainable utilization of aging-deteriorated microplastics as triboelectric nanogenerator. *Chemical Engineering Journal*, p. 144283. https://doi.org/10.1016/j.cej.2023.144283

Colzi, I., Renna, L., Bianchi, E., Castellani, M.B., Coppi, A., Pignattelli, S., Loppi, S. and Gonnelli, C., 2022. Impact of microplastics on growth, photosynthesis and essential elements in Cucurbita pepo L. *Journal of Hazardous Materials*, 423, p. 127238.

Corradini, F., Bartholomeus, H., Lwanga, E.H., Gertsen, H. and Geissen, V., 2019. Predicting soil microplastic concentration using vis-NIR spectroscopy. *Science of the Total Environment*, 650, pp. 922–932. https://doi.org/10.1016/j.scitotenv.2018.09.101

David, J., Steinmetz, Z., Kučerík, J. and Schaumann, G.E., 2018. Quantitative analysis of poly (ethylene terephthalate) microplastics in soil via thermogravimetry–mass spectrometry. *Analytical Chemistry*, 90(15), pp. 8793–8799.

de Souza Machado, A.A., Lau, C.W., Kloas, W., Bergmann, J., Bachelier, J.B., Faltin, E., Becker, R., Görlich, A.S. and Rillig, M.C., 2019. Microplastics can change soil properties and affect plant performance. *Environmental Science & Technology*, 53(10), pp. 6044–6052.

de Souza Machado, A.A., Lau, C.W., Till, J., Kloas, W., Lehmann, A., Becker, R. and Rillig, M.C., 2018. Impacts of microplastics on the soil biophysical environment. *Environmental Science & Technology*, 52(17), pp. 9656–9665.

Ding, L., Huang, D., Ouyang, Z. and Guo, X., 2022. The effects of microplastics on soil ecosystem: A review. *Current Opinion in Environmental Science & Health*, 26, p. 100344.

do Sul, J.A.I. and Costa, M.F., 2014. The present and future of microplastic pollution in the marine environment. *Environmental Pollution*, 185, pp. 352–364.

Duis, K. and Coors, A., 2016. Microplastics in the aquatic and terrestrial environment: Sources (with a specific focus on personal care products), fate and effects. *Environmental Sciences Europe*, 28(1), pp. 1–25.

Eerkes-Medrano, D., Leslie, H.A. and Quinn, B., 2019. Microplastics in drinking water: A review and assessment. *Current Opinion in Environmental Science & Health*, 7, pp. 69–75.

Enders, K., Tagg, A.S. and Labrenz, M., 2020. Evaluation of electrostatic separation of microplastics from mineral-rich environmental samples. *Frontiers in Environmental Science*, p. 112.

Felsing, S., Kochleus, C., Buchinger, S., Brennholt, N., Stock, F. and Reifferscheid, G., 2018. A new approach in separating microplastics from environmental samples based on their electrostatic behavior. *Environmental Pollution*, 234, pp. 20–28.

Gao, M., Liu, Y. and Song, Z., 2019. Effects of polyethylene microplastic on the phytotoxicity of di-n-butyl phthalate in lettuce (Lactuca sativa L. var. ramosa Hort). *Chemosphere*, 237, p. 124482.

Guo, J.J., Huang, X.P., Xiang, L., Wang, Y.Z., Li, Y.W., Li, H., Cai, Q.Y., Mo, C.H. and Wong, M.H., 2020. Source, migration and toxicology of microplastics in soil. *Environment International*, 137, p. 105263. https://doi.org/10.1016/j.envint.2019.105263

Huang, M., Zhu, Y., Chen, Y. and Liang, Y., 2023. Microplastics in soil ecosystems: Soil fauna responses to field applications of conventional and biodegradable microplastics. *Journal of Hazardous Materials*, 441, p. 129943.

Huppertsberg, S. and Knepper, T.P., 2018. Instrumental analysis of microplastics—benefits and challenges. *Analytical and Bioanalytical Chemistry*, 410, pp. 6343–6352.

Hurley, R.R., Woodward, J.C. and Rothwell, J.J., 2017. Ingestion of microplastics by freshwater tubifex worms. *Environmental Science & Technology*, 51(21), pp. 12844–12851.

Iqbal, S., Atique, U., Mahboob, S., Haider, M.S., Iqbal, H.S., Al-Ghanim, K.A., Al-Misned, F., Ahmed, Z. and Mughal, M.S., 2020. Effect of supplemental selenium in fish feed boosts growth and gut enzyme activity in juvenile tilapia (Oreochromis niloticus). *Journal of King Saud University-Science*, 32(5), pp. 2610–2616.

Ivleva, N.P., 2021. Chemical analysis of microplastics and nanoplastics: Challenges, advanced methods, and perspectives. *Chemical Reviews*, 121(19), pp. 11886–11936.

Joos, L. and De Tender, C., 2022. Soil under stress: The importance of soil life and how it is influenced by (micro) plastic pollution. *Computational and Structural Biotechnology Journal*, 20, pp. 1554–1566.

Kirstein, I.V., Gomiero, A. and Vollertsen, J., 2021. Microplastic pollution in drinking water. *Current Opinion in Toxicology*, 28, pp. 70–75. https://doi.org/10.1016/j.cotox.2021.09.003

Kokilathasan, N. and Dittrich, M., 2022. Nanoplastics: Detection and impacts in aquatic environments–a review. *Science of the Total Environment*, 849, p. 157852. https://doi.org/10.1016/j.scitotenv.2022.157852

Kundu, A., Shetti, N.P., Basu, S., Reddy, K.R., Nadagouda, M.N. and Aminabhavi, T.M., 2021. Identification and removal of micro-and nano-plastics: Efficient and cost-effective methods. *Chemical Engineering Journal*, 421, p. 129816.

Leed, R. and Smithson, M., 2019. Ecological effects of soil microplastic pollution. *Science Insights*, 30, pp. 70–84.

Li, R., Yu, L., Chai, M., Wu, H. and Zhu, X., 2020. The distribution, characteristics and ecological risks of microplastics in the mangroves of Southern China. *Science of The Total Environment, 708*, 135025. https://doi.org/10.1016/j.scitotenv.2019.135025

Lofty, J., Muhawenimana, V., Wilson, C.A.M.E. and Ouro, P., 2022. Microplastics removal from a primary settler tank in a wastewater treatment plant and estimations of contamination onto European agricultural land via sewage sludge recycling. *Environmental Pollution*, 304, p. 119198. https://doi.org/10.1016/j.envpol.2022.119198

Lusher, A.L., Hernandez-Milian, G., O'Brien, J., Berrow, S., O'Connor, I. and Officer, R., 2015. Microplastic and macroplastic ingestion by a deep diving, oceanic cetacean: The True's beaked whale Mesoplodon mirus. *Environmental Pollution*, 199, pp. 185–191.

Martin, K.M., Hasenmueller, E.A., White, J.R., Chambers, L.G. and Conkle, J.L., 2018. Sampling, sorting, and characterizing microplastics in aquatic environments with high suspended sediment loads and large floating debris. *JoVE (Journal of Visualized Experiments)*, (137), p. e57969.

Möller, J.N., Löder, M.G. and Laforsch, C., 2020. Finding microplastics in soils: A review of analytical methods. *Environmental Science & Technology*, 54(4), pp. 2078–2090.

Murphy, D.V., Stockdale, E.A., Brookes, P.C. and Goulding, K.W., 2007. Impact of microorganisms on chemical transformations in soil. *Soil Biological Fertility: A Key to Sustainable Land Use in Agriculture*, pp. 37–59.

Nawab, J., Din, Z.U., Ahmad, R., Khan, S., Zafar, M.I., Faisal, S., Raziq, W., Khan, H., Rahman, Z.U., Ali, A. and Khan, M.Q., 2021. Occurrence, distribution, and pollution indices of potentially toxic elements within the bed sediments of the riverine system in Pakistan. *Environmental Science and Pollution Research*, 28(39), pp. 54986–55002. https://doi.org/10.1007/s11356-021-14783-9

Paul, A., Wander, L., Becker, R., Goedecke, C. and Braun, U., 2019. High-throughput NIR spectroscopic (NIRS) detection of microplastics in soil. *Environmental Science and Pollution Research*, 26, pp. 7364–7374. https://doi.org/10.1007/s11356-018-2180-2

Rani, M., Ducoli, S., Depero, L.E., Prica, M., Tubić, A., Ademovic, Z., Morrison, L. and Federici, S., 2023. A complete guide to extraction methods of microplastics from complex environmental matrices. *Molecules*, 28(15), p. 5710.

Ren, X., Tang, J., Yu, C. and He, J., 2018. Advances in research on the ecological effects of microplastic pollution on soil ecosystems. *Journal of Agro-Environment Science*, 37(6), pp. 1045–1058.

Rillig, M.C., Ingraffia, R. and de Souza Machado, A.A., 2017. Microplastic incorporation into soil in agroecosystems. *Frontiers in Plant Science*, 8, p. 1805. https://doi.org/10.3389/fpls.2017.01805

Ruggero, F., Gori, R. and Lubello, C., 2020. Methodologies for microplastics recovery and identification in heterogeneous solid matrices: A review. *Journal of Polymers and the Environment*, 28, pp. 739–748.

Ryan, P.G., Moore, C.J., Van Franeker, J.A. and Moloney, C.L., 2009. Monitoring the abundance of plastic debris in the marine environment. *Philosophical Transactions of the Royal Society B: Biological Sciences*, 364(1526), pp. 1999–2012. https://doi.org/10.1098/rstb.2008.0207

Sajjad, M., Huang, Q., Khan, S., Khan, M.A., Liu, Y., Wang, J., Lian, F., Wang, Q. and Guo, G., 2022. Microplastics in the soil environment: A critical review. *Environmental Technology & Innovation*, 27, p. 102408. https://doi.org/10.1016/j.eti.2022.102408

Samanta, P., Dey, S., Kundu, D., Dutta, D., Jambulkar, R., Mishra, R., Ghosh, A.R. and Kumar, S., 2022. An insight on sampling, identification, quantification and characteristics of microplastics in solid wastes. *Trends in Environmental Analytical Chemistry*, p. e00181.

Scheurer, M. and Bigalke, M., 2018. Microplastics in Swiss floodplain soils. *Environmental Science & Technology*, 52(6), pp. 3591–3598. http://dx.doi.org/10.1016/j.marenvres.2016.05.012

Sierra, I., Chialanza, M.R., Faccio, R., Carrizo, D., Fornaro, L. and Pérez-Parada, A., 2020. Identification of microplastics in wastewater samples by means of polarized light optical microscopy. *Environmental Science and Pollution Research*, 27, pp. 7409–7419.

Sridharan, S., Kumar, M., Singh, L., Bolan, N.S. and Saha, M., 2021. Microplastics as an emerging source of particulate air pollution: A critical review. *Journal of Hazardous Materials*, 418, p. 126245.

Syafina, P.R., Yudison, A.P., Sembiring, E., Irsyad, M. and Tomo, H.S., 2022. Identification of fibrous suspended atmospheric microplastics in Bandung Metropolitan Area, Indonesia. *Chemosphere*, 308, p. 136194.

Tanaka, K. and Takada, H., 2016. Microplastic fragments and microbeads in digestive tracts of planktivorous fish from urban coastal waters. *Scientific Reports*, 6(1), p. 34351.

Tisdall, J.M. and Oades, J.M., 1982. Organic matter and water-stable aggregates in soils. *Journal of Soil Science*, 33(2), pp. 141–163. https://doi.org/10.1111/j.1365-2389.1982.tb01755.x

Vidal, C. and Pasquini, C., 2021. A comprehensive and fast microplastics identification based on near-infrared hyperspectral imaging (HSI-NIR) and chemometrics. *Environmental Pollution*, 285, p. 117251.

Wang, J., Coffin, S., Sun, C., Schlenk, D. and Gan, J., 2019. Negligible effects of microplastics on animal fitness and HOC bioaccumulation in earthworm Eisenia fetida in soil. *Environmental Pollution*, 249, pp. 776–784. https://doi.org/10.1016/j.envpol.2019.03.102

Wang, Y., Zhang, G., Zhang, F. and Wang, H., 2023. Diagnostic strategy for the combined effects of microplastics and potentially toxic elements on microbial communities in catchment scale. *Science of the Total Environment*, 860, p. 160499.

Woo, H., Seo, K., Choi, Y., Kim, J., Tanaka, M., Lee, K. and Choi, J., 2021. Methods of analyzing microsized plastics in the environment. *Applied Sciences*, 11(22), p. 10640.

Yang, C., Yin, L., Guo, Y., Han, T., Wang, Y., Liu, G., Maqbool, F., Xu, L. and Zhao, J., 2023. Insight into the absorption and migration of polystyrene nanoplastics in *Eichhornia crassipes* and related photosynthetic responses. *Science of the Total Environment*, 892, p. 164518.

Yang, L., Zhang, Y., Kang, S., Wang, Z. and Wu, C., 2021. Microplastics in soil: A review on methods, occurrence, sources, and potential risk. Science of the Total Environment, 780, p. 146546.

Yılmaz, H., Yilmaz, D., Taskin, I.C. and Culha, M., 2022. Pharmaceutical applications of a nanospectroscopic technique: Surface-enhanced Raman spectroscopy. *Advanced Drug Delivery Reviews*, 184, p. 114184.

Yin, R., Ge, H., Chen, H., Du, J., Sun, Z., Tan, H. and Wang, S., 2021. Sensitive and rapid detection of trace microplastics concentrated through Au-nanoparticle-decorated sponge on the basis of surface-enhanced Raman spectroscopy. *Environmental Advances*, 5, p. 100096.

Zhang, X., Li, Y., Ouyang, D., Lei, J., Tan, Q., Xie, L., Li, Z., Liu, T., Xiao, Y., Farooq, T.H. and Wu, X., 2021. Systematical review of interactions between microplastics and microorganisms in the soil environment. *Journal of Hazardous Materials*, 418, p. 126288.

Zhang, Y., Cai, C., Gu, Y., Shi, Y. and Gao, X., 2022. Microplastics in plant-soil ecosystems: A meta-analysis: Uncited references. *Environmental Pollution*, p. 119718. https://doi.org/10.1016/j.envpol.2022.119718

Zhou, W., Wang, Q., Wei, Z., Jiang, J. and Deng, J., 2023. Effects of microplastic type on growth and physiology of soil crops: Implications for farmland yield and food quality. *Environmental Pollution*, 326, p. 121512. https://doi.org/10.1016/j.envpol.2023.121512

Zhu, F., Zhu, C., Wang, C. and Gu, C., 2019. Occurrence and ecological impacts of microplastics in soil systems: A review. *Bulletin of Environmental Contamination and Toxicology*, 102, pp. 741–749.

6 Impact of Microplastics in Terrestrial Food Production

Arghya Mandal and Apurba Ratan Ghosh

6.1 INTRODUCTION

There's the growing threat of ocean acidification, overfishing, and, of course, climate change. More and more clear is that this contamination extends over the ocean but with an increasing understanding of all the impacts that plastic has on the environment around the globe. In terrestrial environments, microplastics, *i.e.*, tiny particles produced when larger plastic items decompose, are an increasingly ubiquitous contaminant, as has been emphasized in a recent paper on their presence all along the food chain and in agricultural production systems (Jin et al.,2002). The question is then whether the two types of actions are complementary or inversive with reference to a particular issue like terrorism wherein both (terrorists and govt. armed forces) involve the use of armed force but have differing long-term outcomes for that issue. Plastic particles smaller than 5 mm in size—known as microplastics—will enter terrestrial ecosystems through various pathways. Microplastics can be generated through the breakdown of bigger materials, for example, bundling materials, horticultural films, and customer rubbishes (Wojnowska-Baryła et al., 2022). However, there remains a huge discrepancy in the availability of basic healthcare facilities across different regions of India. Microplastics enter water systems via seepage of wastewater into groundwater or when they are dispersed throughout water bodies as aerosols that eventually fall into river, lake, and ocean ecosystems. Furthermore, they can traverse continents through atmospheric transport and deposition. When microplastics do finally reach the soil, they can interfere with the delicate network of organisms and the processes needed for plant growth and food production. According to Wang et al. (2023), microplastics could accumulate in plant systems by reaching the roots and later getting transported to different plant parts including leaves, stems, fruit, and so on. These factors include the size, shape, and composition of microplastics, the properties of the soil (e.g., pH, mineral composition), and the traits of the specific plant species (Porter et al., 2023). As such, there's worry that the pollutants might be introduced into the human food chain through microplastics in vegetation (Khan et al., 2023). In addition to just polluting crops, microplastics affect the terrestrial food produced. These particles also affect the soil's fertility and quality that in turn affects agriculture produce overall. This is also known as strawing. Furthermore, the presence of microplastics disrupts important microbial ecosystems needed for

DOI: 10.1201/9781032684574-6

nutrient cycling in the soil according to Zhou and Zhou (2023). Such changes in soil dynamics pose a major challenge for the production of sustainable food as they could yield reduced agricultural produce of poorer quality and greater susceptibility to pests and diseases. Moreover, anxieties are expressed concerning the potential impacts on human health from the movement of microplastics through our terrestrial ecosystems (Richard et al., 2022). Plastic dust collected on plants could then be consumed by primary consumers such as sheep or cows, which could become contaminated with microplastics. The contaminated herbivores could become prey for predators like lions or foxes or eaten by humans; as a result, humans could come into contact with microplastics and the substances attached to or absorbed by them (Nor, 2022). Meanwhile questions are being raised as to whether some of these possible dangers, such as inflammation, oxidative stress, and disruption of both endocrine and immune systems in relation to long-term microplastics exposure, are known or potential threats.

Solving the issue of microplastics in the land-based food supply demands a multipronged approach. We need an overhaul of the whole system by reducing overall production and use of plastics, promoting sustainable waste management practices, and introducing tougher norms for the utilization of plastic in the agricultural sector. Additionally, further studies are required to develop effective methods of detecting microplastics in soils and vegetation, evaluating their human health risks via toxicology tests, and researching potential solutions for the cleanup.

This new challenge is that microplastics are affecting the agricultural yield on land. As we strive for more resilient and sustainable food systems, understanding and reducing, where possible, the risks posed by microplastics in our environment are critical. By coordinating all these measures like scientific research, how do we even measure this problem? Using legislative initiatives regulating plastic use and public awareness campaigns, we'll manage to reduce the level of microplastics in our soils and in our crops.

6.2 SOURCES AND PATHWAYS OF MICROPLASTICS

Microplastics, which are defined as tiny plastic particles smaller than 5 mm, are a persistent and alarming form of pollution that could have a negative impact on terrestrial food production systems. These minute particles have a variety of origins and entry points into the environment. Understanding microplastics' sources and migration routes is essential to understanding their effects on the growth of food crops and the possible threats to human health they may pose. We will examine the main sources and distribution channels for microplastics in terrestrial food production in this section.

6.2.1 Fragmentation of Larger Plastic Items

The disintegration of bigger plastic products is a significant source of microplastics in terrestrial ecosystems. Plastics used in a variety of industries, such as agricultural films, consumer goods, and packaging materials, can deteriorate over time as a result of exposure to environmental conditions such heat, light, and mechanical

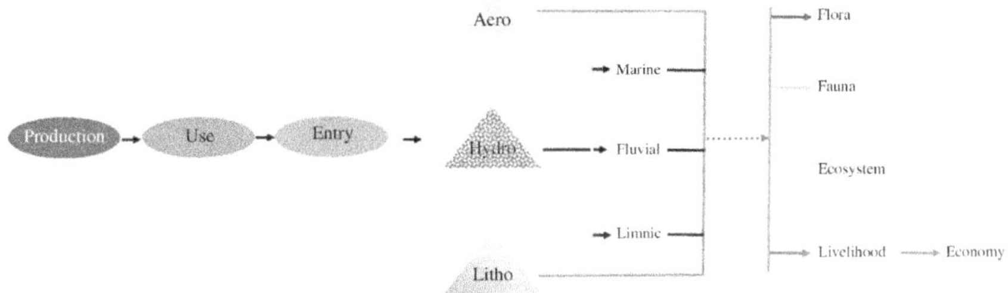

FIGURE 6.1 Pathways of microplastics.

stress (Ainali et al., 2022). These bigger plastic objects consequently disintegrate into tiny fragments, eventually forming microplastics.

6.2.2 Application of Plastic Mulches

The disintegration of bigger plastic products is a significant source of microplastics in terrestrial ecosystems. Plastics used in a variety of industries, such as agricultural films, consumer goods, and packaging materials, can deteriorate over time as a result of exposure to environmental conditions such heat, light, and mechanical stress (Ainali et al., 2022). These bigger plastic objects consequently disintegrate into tiny fragments, eventually forming microplastics.

6.2.3 Contaminated Compost and Organic Amendments

Compost, which is often used as a soil amendment to improve fertility, can contain microplastics if it includes contaminated organic waste with plastic residues (Braun et al., 2023). Similarly, other organic amendments like sewage sludge or biosolids, which are used as fertilizers in agriculture, can also introduce microplastics into the soil if the waste contains plastic particles (Hooge et al., 2023).

6.2.4 Atmospheric Deposition

Microplastics can be transported over long distances through the atmosphere and deposited onto agricultural lands. Airborne microplastics can originate from various sources, including urban areas, industrial activities, and atmospheric emissions (Jahandari, 2023). These particles settle onto the soil surface and can be subsequently incorporated into the soil profile through natural processes such as wind and rainfall.

6.2.5 Irrigation Water and Wastewater

Water used for irrigation purposes, especially in areas with high plastic pollution, can be a potential source of microplastics. Microplastics may be present in rivers, lakes, and reservoirs that collect runoff from cities or industrial facilities; these microplastics may then be transported onto agricultural lands via irrigation techniques (Jiang

TABLE 6.1
Sources of Microplastics

Sources	Descriptions
Single-use plastics	These are disposable plastic items like straws, bottles, bags, and cutlery that break down into microplastics over time.
Synthetic fabrics	Clothing made from synthetic materials like polyester and nylon release microfibers during washing and use.
Tire wear	As tires wear down on roads, they shed tiny particles containing microplastics.
Personal care products	Microbeads found in cosmetics, toothpaste, and exfoliating products can contribute to microplastic pollution.
Industrial processes	Plastic pellets and powders used in manufacturing processes can escape into the environment.
Microplastic litter	Larger plastic debris breaks down into smaller fragments, becoming microplastics in the environment.
Fishing gear	Discarded or lost fishing nets, lines, and gear release microplastics into oceans and water bodies.
Agricultural runoff	Microplastics from plastic mulches, films, and agricultural products can enter waterways through runoff.
Landfills	Plastics disposed of in landfills can break down into microplastics over time, leaching into the soil and water.
Sewage treatment effluents	Wastewater treatment plants may not capture all microplastics, leading to their release into rivers and oceans.

et al., 2023). Additionally, improper wastewater treatment when used for irrigation, in particular, can cause the introduction of microplastics into the soil.

6.2.6 Agricultural Practices and Plastic Debris

Plastic debris may accumulate on agricultural areas as a result of improper waste management techniques, including the disposal of plastic products like bags, containers, and agricultural machinery (Narayanan, 2023). These plastic objects have the potential to degrade over time, which would increase the amount of microplastics in the soil.

It is significant to highlight that, based on geographic location, agricultural techniques, and proximity to urban or industrial areas, the precise origins and paths of microplastics may differ. Understanding the regional context and carrying out research that is specific to a given area can help to shed light on the sources and migration routes of microplastics in terrestrial food production systems.

6.3 UPTAKE OF MICROPLASTICS BY PLANTS

Microplastics are easily absorbed by plants once they have settled in the soil. Because it raises concerns about the entry of these particles into the terrestrial food chain and the potential hazards to human health, the uptake of microplastics by plants is

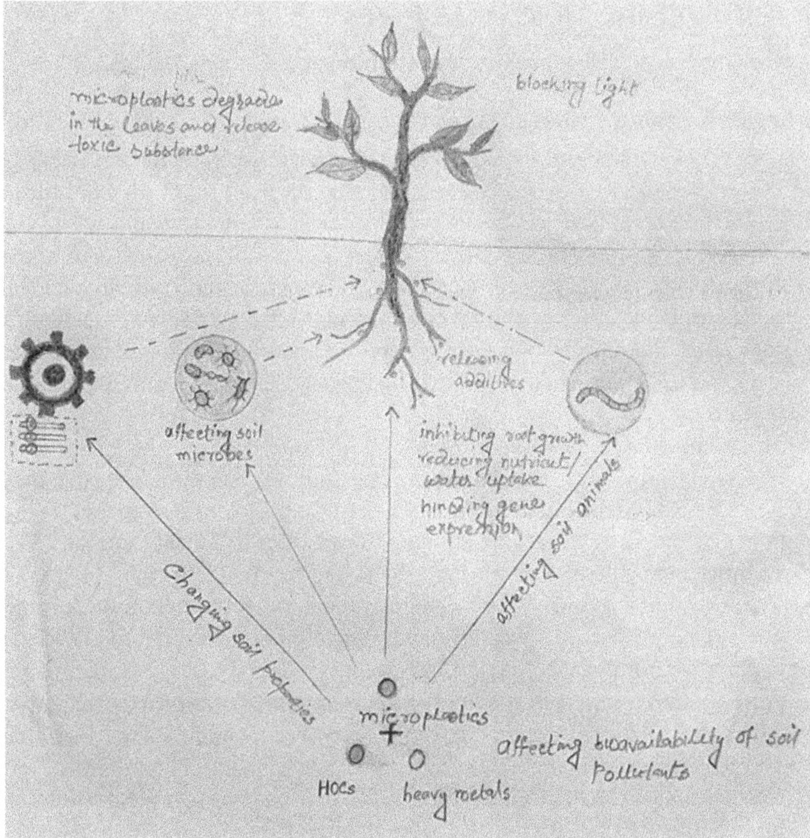

FIGURE 6.2 Pathways of microplastic uptake.

a cause for concern. We have looked at the processes and elements affecting how plants absorb microplastics in this part.

6.3.1 Mechanisms of Uptake

There are two main ways for plants to absorb microplastics through their roots: directly and indirectly.

- **Direct Uptake:** Plant roots have the ability to directly absorb microplastics that are present in the soil-pore water. These particles can enter the plant's vascular system through the root tissues and then travel to different areas of the plant, such as the stems, leaves, and fruits (Wang et al., 2023).
- **Indirect Uptake:** Indirect plant uptake of microplastics is also possible through relationships with mycorrhizal fungus. Mycorrhizal fungi collaborate with plant roots to improve nutrient intake. According to studies, soil-found microplastics can stick to mycorrhizal fungal hyphae, and as a result, the plants can absorb the microplastics through the hyphal networks.

6.4 FACTORS INFLUENCING UPTAKE

Several elements can affect how plants absorb microplastics, including:

- **Particle Size and Form:** Plant uptake of microplastic particles can be influenced by their size and shape. Larger particles are typically more difficult to absorb than smaller ones. According to studies, plant roots are more likely to absorb microplastics of a size between 1 and 10 µm (Song et al., 2023).
- **Chemical Structure:** The chemical makeup of microplastics may affect how well they are absorbed. According to certain research, some chemical compounds included in plastics, such as plasticizers and flame retardants, may make it easier for plants to absorb microplastics. To fully comprehend the impact of chemical composition on absorption pathways, more research is necessary.
- **Root Characteristics:** The morphology and surface features of plant roots can affect how well they absorb microplastics. Microplastic uptake may be more likely in plants with larger, more established root systems (Huang et al., 2023). The interaction between microplastics and plant roots may also be impacted by the presence of root hairs and root exudates.
- **Properties of the Soil:** The availability and mobility of microplastics in the soil can be influenced by the texture, organic matter concentration, and pH of the soil. For instance, microplastics may stick to clay or organic matter particles in the soil, which would make them less accessible to plant roots. Microplastics can adsorb and desorb differently depending on the pH of the soil (Dissanayake et al., 2022).
- **Species and Growth of Plants Stage:** The ability of various plant species to absorb microplastics may vary. Different plant species absorb microplastics differently, according to certain studies. Younger plants often have higher uptake rates, and the stage of the plant's growth can also have an impact on microplastic uptake (Singh et al., 2023).

6.5 CONSEQUENCES AND IMPLICATIONS

The food chain and the health of plants may both be affected by plants absorbing microplastics. After being ingested by plants, microplastics have the ability to build up in plant tissues and negatively impact growth, development, and general health (Chang et al., 2022). Furthermore, consumption of microplastic-contaminated plant-based food products could pose health risks due to human exposure through the passage of microplastics via the food chain (Al Mamun et al., 2023).

It is crucial to comprehend the mechanisms and elements affecting the uptake of microplastics by plants in order to develop mitigation methods for their presence in the food chain. To fully understand the long-term effects of microplastic absorption on plant physiology, crop productivity, and the possible repercussions on human health, more research is required. Additionally, minimizing the entry of microplastics into terrestrial food production systems can be achieved by investigating ways to limit microplastic contamination in agricultural activities and by putting sustainable waste management strategies into practice.

TABLE 6.2
Effects and Impacts of Microplastics on Soil Health

Effects	Impact
Reduced water infiltration	Microplastics can clog soil pores, leading to reduced water infiltration.
Nutrient depletion	Microplastics can adsorb and sequester nutrients, making them less available to plants and soil organisms.
Altered microbial community	Microplastics can disrupt the balance of soil microbial communities.
Soil compaction	Accumulation of microplastics can lead to soil compaction, reduced aeration, and root growth.
Toxicity to soil organisms	Microplastics may release toxic chemicals, affecting soil organisms such as earthworms and beneficial microorganisms
Transfer to food chain	Microplastics can be ingested by soil-dwelling organisms, potentially transferring up the food chain
Overall soil degradation	Cumulative effects of microplastics can lead to overall soil degradation, reducing its fertility and resilience

6.6 EFFECTS ON SOIL HEALTH

The health of the soil is seriously threatened by microplastic pollution, which could have an impact on terrestrial food production systems. (See Table 6.2.) Microplastics can interact with soil elements and organisms when they penetrate the soil ecosystem, having a variety of negative impacts. In this part, we will attempt to determine how microplastics affect soil health and what that means for food production.

- **Physical Effects:** The physical characteristics of soil can be changed by microplastics. They can limit soil porosity and fill soil pores because of their small size, which results in compacted soil and lower rates of water infiltration. This may reduce soil aeration, obstruct root development, and block the passage of water and nutrients across the soil profile. Additionally, the buildup of microplastics in the soil might have an impact on its composition, increasing soil erosion and decreasing soil stability (Liu et al., 2022).
- **Chemical Effects:** Microplastics may alter the chemical composition of soil. Heavy metals, herbicides, and organic pollutants are just a few of the chemical components that they can adsorb and accumulate. The toxins that have been adsorbed into the soil may be discharged, which would impair soil fertility and the availability of vital nutrients for plants. In addition to affecting soil health and crop nutrient availability, microplastics can also alter soil pH levels and nutrient cycling mechanisms (Wijesooriya et al., 2023).
- **Biological Effects:** Microplastics may negatively affect soil microbes and other species that live in the soil, disrupting soil ecosystems. Microplastics have the potential to change the diversity and abundance of soil bacteria,

fungus, and protozoa populations (Li et al., 2023). Important soil processes like organic matter breakdown, nitrogen cycling, and soil nutrient availability may be impacted by this change in microbial communities. In addition, earthworms and other beneficial invertebrates that are essential for the establishment of soil structure and nutrient cycling may be adversely affected by microplastics (Qiu et al., 2022).

- **Soil Contamination:** Microplastics have the ability to pollute the soil with hazardous and enduring materials. Chemical additives like plasticizers and flame retardants may be present in some microplastics or may adsorb to them. When these additions are dispersed in the soil, they may have a negative impact on soil quality and pose dangers to soil ecosystem organisms. Additionally, the buildup of microplastics in the soil can lead to long-term contamination, which lingers in the ecosystem and may have an impact on future crop development and food safety (Zhou et al., 2023; Onoja et al., 2022).
- **Crop Productivity and Quality:** Crop productivity and quality may ultimately be impacted by the detrimental effects of microplastics on soil health. Reduced agricultural yields and poor plant health can result from compacted soil, limited nutrient availability, and altered soil microbial activity (Shah et al., 2023). Additionally, the buildup of microplastics in plant tissues may have an impact on the safety and nutritional value of food crops, which may have an impact on human health (Huang et al., 2023; Iqbal et al., 2023).

Microplastic contamination of soil can have a substantial impact on soil health and ultimately have an impact on terrestrial food production systems. Creating mitigation solutions for the effects of microplastics on soil's physical, chemical, and biological qualities is crucial. We may endeavor to lessen microplastic pollution and protect soil health for sustainable food production by adopting sustainable practices, encouraging responsible plastic usage, and putting in place appropriate waste management practices.

6.7 TRANSFER TO THE FOOD CHAIN

Microplastics have the capacity to go up the food chain once they enter the environment, including the soil and water bodies. The accumulation of microplastics in species at higher trophic levels, such as animals and maybe humans, raises questions concerning their transmission through the food chain. We will examine the causes and effects of microplastic transmission to the food chain in this section.

- **Transfer to Herbivores:** Herbivorous species that consume contaminated plant matter, like insects, worms, and grazing mammals, are susceptible to ingesting microplastics. Microplastics can build up in a variety of plant tissues, such as leaves, stems, and fruits, when they are ingested by plants from the soil (Zhang et al., 2022; Graf et al., 2023). The microplastics in their diet may then be consumed by herbivores that eat these plants.
- **Transfer to Carnivores and Omnivores:** Through the predation and eating of herbivorous species, microplastics can climb the food chain. Microplastics

can build up in the tissues of carnivorous and omnivorous animals that eat prey contaminated with them (D'Costa, 2022). Higher trophic levels may be affected by this shift, which could include apex predators.

- **Bioaccumulation and Biomagnification:** Microplastics are difficult for organisms to metabolize or excrete, which allows them to bioaccumulate in their tissues. Microplastics can build up over time and reach larger concentrations in creatures' bodies. In addition, as they ascend the food chain, microplastics can biomagnify (Covernton et al., 2022; Chen et al., 2022). Microplastic concentrations in the bodies of organisms that eat prey contaminated with microplastics may be higher than in their food sources.
- **Potential Human Exposure:** Microplastic contamination in the food chain raises questions about possible human exposure. Consuming contaminated food products can expose people to microplastics, especially if they come from animals or plant-based meals that were grown in areas with microplastic contamination (Siddiquia et al., 2023). For instance, seafood should be avoided because marine species have a high risk for accumulating microplastic.
- **Health Risks:** Concerns regarding possible human exposure are raised by the movement of microplastics through the food chain. Consuming contaminated food products, especially those made from animals or plant-based meals grown in microplastic-affected areas, might expose humans to the particles (Siddiquia et al., 2023). Due to the significant likelihood that microplastic may accumulate in marine species—seafood, for instance—is particularly concerning.
- **Human Health Implications:** There are worries about the possible negative consequences of microplastic exposure on human health, despite the fact that their long-term impacts are still not fully understood. When consumed, harmful compounds and heavy metals contained in microplastics may leak into the gastrointestinal tract (Liao and Yang, 2022). Inflammation, oxidative stress, and the disturbance of the endocrine and immunological systems are just a few of the health effects that these drugs may have on people (Sangkham et al., 2022). Further worries are raised by the possibility that human tissues and organs could be impacted by microplastics due to their small size.

6.8 MITIGATION AND FUTURE DIRECTIONS

Researchers, policymakers, farmers, and consumers must all be involved in a comprehensive strategy to reduce the impact of microplastics in terrestrial food production. Next are some potential strategies and future directions for addressing the issue.

- **Reduce Plastic Use and Waste:** One of the fundamental steps in mitigating microplastic pollution is reducing the overall use and consumption of plastics. Governments, industries, and individuals should work together to promote alternative materials, sustainable packaging solutions, and recycling initiatives. By minimizing plastic waste, we can reduce the potential thrust of microplastics entering into agricultural systems.

- **Enhance Waste Management Practices:** Proper waste management is crucial to prevent the release of microplastics into the environment. Implementing efficient recycling and waste disposal systems, including improved wastewater treatment, can help to reduce the amount of plastic waste that reaches agricultural lands. Additionally, it is crucial to educate consumers and farmers on the need of appropriate plastic waste management.
- **Develop Plastic-Free Agricultural Practices:** Microplastic contamination in food production systems can be considerably reduced by agricultural techniques that minimize or eliminate the use of plastics. For instance, researching substitutes for plastic mulches, which are a major source of microplastic pollution, such as biodegradable films or organic matter, can help eliminate their use. Similarly, lessening the use of fertilizers made of plastic, like those made from sewage sludge, can help reduce contamination.
- **Enhance Soil Monitoring and Remediation:** For detecting and quantifying microplastics in soil, accurate and efficient methods must be developed in order to monitor their existence and comprehend their behavior. To increase the dependability of the data, researchers should continue to standardize and enhance analytical methods. Furthermore, investigating remediation methods to get rid of microplastics from polluted soils, like soil washing or bioremediation procedures, shows promise for reducing their effects.
- **Analyze the Effects on Ecology and Human Health:** More research is required to comprehend the ecological and health effects of microplastics on terrestrial food production. To ensure risk evaluations and policy choices are accurate, long-term research addressing the accumulation, destiny, and potential harmful effects of microplastics in plants, animals, and humans is required. This entails assessing the consequences of various microplastic properties (size, shape, composition), as well as possible thresholds for negative effects.
- **Create Regulatory Frameworks:** Governments and regulatory organizations are essential in reducing the contamination of microplastics. The release of microplastics into the environment can be reduced by establishing thorough laws and guidelines for plastic use in agriculture, including standards for plastic mulches, fertilizers, and packaging. Promoted labeling guidelines for goods containing microplastics can also increase customer choice and awareness.
- **Encourage Education and Awareness:** It's critical to educate people about the problem of microplastics in terrestrial food production. Promoting sustainable practices, responsible plastic use, and educated decision making can be accomplished with the aid of educational programs aimed at farmers, consumers, and policymakers. We can promote behavior change and aid in the shift to a plastic-free, sustainable food system by arming people with knowledge.

Microplastics' impact on terrestrial food production has security implications.

- **Detection Methods Standardization:** The absence of standardized and effective detection procedures presents one of the major obstacles in research on the effects of microplastics on terrestrial food production. The development of standardized techniques for the extraction, identification, and quantification of microplastics in soils, plants, and food products should be the main goal of future study. This would make it possible to estimate the quantities of microplastic contamination more precisely and make comparisons between research projects easier.

- **Extensive Ecological Research:** More long-term studies are required to better understand the long-term ecological effects of microplastics on terrestrial food production. Such research should look into the long-term impacts of microplastic exposure on ecosystem functioning, plant development, and soil health. Insightful information about the persistence, destiny, and transit of microplastics in agricultural systems can be gained from longitudinal studies.
- **Regulations and Risk Analysis:** For the purpose of informing regulatory actions addressing microplastic contamination in terrestrial food production, detailed risk assessment frameworks must be developed. Future studies should concentrate on determining the possible dangers connected to ingesting microplastics by people and animals while taking into account the particle size, chemical makeup, and bioaccumulation potential of various microplastics. The development of appropriate thresholds and regulations to protect food safety and human health can be guided by these analyses.
- **Strategies for Mitigating:** It should be investigated how to lessen the effect of microplastics on the production of terrestrial food. Limiting plastic contamination in soils entails examining agricultural methods that reduce the use of plastic mulches, encouraging substitute biodegradable materials, and putting sustainable waste management techniques into practice. Research on the creation of environmentally friendly additives for plastics used in agriculture can also lessen the amount of microplastics released into the environment.
- **Consumer Education and Awareness:** It is imperative to raise consumer awareness of the problem of microplastic contamination in terrestrial food production. Consumers can make educated decisions by choosing items with less plastic packaging, supporting sustainable farming methods, and properly recycling plastic waste with the help of educational campaigns and initiatives. We can all work together to minimize the demand for plastics and lessen their negative effects on the environment and human health by establishing a culture of sustainability and encouraging environmentally conscious consumer behavior.
- **International Cooperation and Governmental Initiatives:** Microplastics in terrestrial food production must be addressed through international cooperation and regulatory changes. International partnerships can make it easier to share information, exchange data, and harmonize rules. To establish comprehensive policies and standards for plastic use in agriculture, waste management, and food safety, governments and policymakers need to collaborate. International conventions and agreements can also promote sensible plastic production, consumption, and disposal methods.

The effect of microplastics on the production of terrestrial food is a complicated and dynamic topic that necessitates continual investigation and teamwork. We can work to reduce the negative effects of microplastics on our food systems and protect human health and the environment by focusing on the standardization of detection methods, conducting long-term ecological studies, putting risk assessment frameworks into place, exploring mitigation strategies, raising consumer awareness, and encouraging international cooperation. In conclusion, reducing the effect of microplastics on terrestrial food production necessitates a multifaceted strategy that includes source reduction, enhanced waste management, alternative farming methods, scientific improvements, regulation, and awareness-building initiatives.

6.9 CONCLUSION

The effect of microplastics on the production of terrestrial food is a critical issue that requires our attention. The effects of these tiny particles on the environment and public health are becoming increasingly clear as our knowledge of microplastic contamination expands. Microplastics can enter terrestrial ecosystems in a number of ways, such as when bigger plastic objects are broken up or used as mulch or fertilizer. Microplastics can be ingested by plants after being buried in the soil and then enter the food chain. This raises questions about the likelihood that eating tainted food could expose people to microplastics. Microplastics have severe negative effects on the health of the soil. These particles may change the soil's physicochemical characteristics, which may have an effect on the soil's fertility and nutrient availability. Microplastics can also have a negative impact on soil microbes, which are essential for soil health and the cycling of nutrients. Therefore, these alterations in soil health may lead to poorer crop yields and nutrient quality. Additional difficulties are related to the movement of microplastics through the food chain. Carnivores and omnivores may consume microplastics that build up in herbivore tissues, potentially increasing exposure and accumulation of these particles in higher trophic levels. Due to the possibility that tiny particles could release harmful compounds and heavy metals when consumed, the presence of microplastics in the human food chain raises questions about the potential effects on long-term health. A multifaceted strategy is needed to reduce the effect of microplastics in terrestrial food production. The use of plastic in agriculture must be strictly regulated, waste management procedures must be improved, and consumption of plastic must be decreased. Furthermore, the discovery of microplastics in cow's milk is a troubling problem that highlights how widespread plastic pollution is in our environment. Research has proven the presence of these microscopic plastic particles in milk, but because of the normally low amounts, no immediate health concerns have been brought up. Our knowledge of the long-term effects of consuming microplastics through food is very limited, thus further study is required to fully evaluate any potential health effects. To successfully address the problem, it is also necessary to develop detection techniques, conduct research on toxicological effects, and consider remediation options. In order to develop a sense of shared responsibility for minimizing microplastic pollution, public awareness and education are essential. We can help to lessen the effect of microplastics on terrestrial food production by encouraging sustainable methods and exercising informed consumer choice.

In conclusion, the effect of microplastics in the production of terrestrial food is a complex and pressing issue that needs to be addressed right away. By addressing the causes, distribution channels, and effects of microplastic pollution, we can work toward resilient and sustainable food systems that put the health of the environment and people first.

REFERENCES

Ainali, N.M., Kalaronis, D., Evgenidou, E., Kyzas, G.Z., Bobori, D.C., Kaloyianni, M., Yang, X., Bikiaris, D.N. and Lambropoulou, D.A., 2022. Do poly (lactic acid) microplastics instigate a threat? A perception for their dynamic towards environmental pollution and toxicity. *Science of the Total Environment*, *832*, p. 155014.

Al Mamun, A., Prasetya, T.A.E., Dewi, I.R. and Ahmad, M., 2023. Microplastics in human food chains: Food becoming a threat to health safety. *Science of the Total Environment*, *858*, p. 159834.

Braun, M., Mail, M., Krupp, A.E. and Amelung, W., 2023. Microplastic contamination of soil: Are input pathways by compost overridden by littering? *Science of the Total Environment*, *855*, p. 158889.

Chang, X., Fang, Y., Wang, Y., Wang, F., Shang, L. and Zhong, R., 2022. Microplastic pollution in soils, plants, and animals: A review of distributions, effects and potential mechanisms. *Science of the Total Environment*, *850*, p. 157857.

Chen, X.J., Ma, J.J., Yu, R.L., Hu, G.R. and Yan, Y., 2022. Bioaccessibility of microplastic-associated heavy metals using an in vitro digestion model and its implications for human health risk assessment. *Environmental Science and Pollution Research*, *29*(51), pp. 76983–76991.

Covernton, G.A., Cox, K.D., Fleming, W.L., Buirs, B.M., Davies, H.L., Juanes, F., Dudas, S.E. and Dower, J.F., 2022. Large size (> 100-μm) microplastics are not biomagnifying in coastal marine food webs of British Columbia, Canada. *Ecological Applications*, *32*(7), p. e2654.

D'Costa, A.H., 2022. Microplastics in decapod crustaceans: Accumulation, toxicity and impacts, a review. *Science of the Total Environment*, *832*, p. 154963.

Dissanayake, P.D., Kim, S., Sarkar, B., Oleszczuk, P., Sang, M.K., Haque, M.N., Ahn, J.H., Bank, M.S. and Ok, Y.S., 2022. Effects of microplastics on the terrestrial environment: A critical review. *Environmental Research*, *209*, p. 112734.

Graf, M., Greenfield, L.M., Reay, M.K., Bargiela, R., Williams, G.B., Onyije, C., Lloyd, C.E., Bull, I.D., Evershed, R.P., Golyshin, P.N. and Chadwick, D.R., 2023. Increasing concentration of pure micro-and macro-LDPE and PP plastic negatively affect crop biomass, nutrient cycling, and microbial biomass. *Journal of Hazardous Materials*, p. 131932.

Hooge, A., Hauggaard-Nielsen, H., Heinze, W.M., Lyngsie, G., Ramos, T.M., Sandgaard, M.H., Vollertsen, J. and Syberg, K., 2023. Fate of microplastics in sewage sludge and in agricultural soils. *TrAC Trends in Analytical Chemistry*, p. 117184.

Huang, F., Hu, J., Chen, L., Wang, Z., Sun, S., Zhang, W., Jiang, H., Luo, Y., Wang, L., Zeng, Y. and Fang, L., 2023. Microplastics may increase the environmental risks of Cd via promoting Cd uptake by plants: A meta-analysis. *Journal of Hazardous Materials*, *448*, p. 130887.

Iqbal, B., Zhao, T., Yin, W., Zhao, X., Xie, Q., Khan, K.Y., Zhao, X., Nazar, M., Li, G. and Du, D., 2023. Impacts of soil microplastics on crops: A review. *Applied Soil Ecology*, *181*, p. 104680.

Jahandari, A., 2023. Microplastics in the urban atmosphere: Sources, occurrences, distribution, and potential health implications. *Journal of Hazardous Materials Advances*, p. 100346.

Jiang, J.J., Hanun, J.N., Chen, K.Y., Hassan, F., Liu, K.T., Hung, Y.H. and Chang, T.W., 2023. Current levels and composition profiles of microplastics in irrigation water. *Environmental Pollution, 318*, p. 120858.

Jin, T., Tang, J., Lyu, H., Wang, L., Gillmore, A.B. and Schaeffer, S.M., 2022. Activities of microplastics (MPs) in agricultural soil: A review of MPs pollution from the perspective of agricultural ecosystems. *Journal of Agricultural and Food Chemistry, 70*(14), pp. 4182–4201.

Khan, A., Jie, Z., Wang, J., Nepal, J., Ullah, N., Zhao, Z.Y., Wang, P.Y., Ahmad, W., Khan, A., Wang, W. and Li, M.Y., 2023. Ecological risks of microplastics contamination with green solutions and future perspectives. *Science of The Total Environment*, p. 165688.

Li, H., Luo, Q.P., Zhao, S., Zhou, Y.Y., Huang, F.Y., Yang, X.R. and Su, J.Q., 2023. Effect of phenol formaldehyde-associated microplastics on soil microbial community, assembly, and functioning. *Journal of Hazardous Materials, 443*, p. 130288.

Liao, Y.L. and Yang, J.Y., 2022. The release process of Cd on microplastics in a ruminant digestion in-vitro method. *Process Safety and Environmental Protection, 157*, pp. 266–272.

Liu, B., Li, W., Pan, X. and Zhang, D., 2022. The persistently breaking trade-offs of three-decade plastic film mulching: Microplastic pollution, soil degradation and reduced cotton yield. *Journal of Hazardous Materials, 439*, p. 129586.

Nor, N.H.B.M., 2022. *Microplastics' journey into the gut: Human exposure to microplastics and associated chemicals* (Doctoral dissertation, Wageningen University and Research).

Onoja, S., Nel, H.A., Abdallah, M.A.E. and Harrad, S., 2022. Microplastics in freshwater sediments: Analytical methods, temporal trends, and risk of associated organophosphate esters as exemplar plastics additives. *Environmental Research, 203*, p. 111830.

Porter, A., Barber, D., Hobbs, C., Love, J., Power, A.L., Bakir, A., Galloway, T.S. and Lewis, C., 2023. Uptake of microplastics by marine worms depends on feeding mode and particle shape but not exposure time. *Science of the Total Environment, 857*, p. 159287.

Qiu, Y., Zhou, S., Zhang, C., Zhou, Y. and Qin, W., 2022. Soil microplastic characteristics and the effects on soil properties and biota: A systematic review and meta-analysis. *Environmental Pollution*, p. 120183.

Richard, B., Qi, A. and Fitt, B.D., 2022. Control of crop diseases through integrated crop management to deliver climate-smart farming systems for low-and high-input crop production. *Plant Pathology, 71*(1), pp. 187–206.

Sangkham, S., Faikhaw, O., Munkong, N., Sakunkoo, P., Arunlertaree, C., Chavali, M., Mousazadeh, M. and Tiwari, A., 2022. A review on microplastics and nanoplastics in the environment: Their occurrence, exposure routes, toxic studies, and potential effects on human health. *Marine Pollution Bulletin, 181*, p. 113832.

Shah, S., Ilyas, M., Li, R., Yang, J. and Yang, F.L., 2023. Microplastics and nanoplastics effects on plant–pollinator interaction and pollination biology. *Environmental Science & Technology, 57*(16), pp. 6415–6424.

Siddiquia, S.A., Khanc, S., Tariqd, T., Sameend, A., Nawaze, A., Walayath, N., Oboturovai, N.P., Ambartsumovi, T.G. and Nagdaliani, A.A., 2023. Potential risk assessment and toxicological impacts of nano/micro-plastics on human health through food products. *Nano/Micro-Plastics Toxicity on Food Quality and Food Safety, 103*, p. 361.

Singh, N., Abdullah, M.M., Ma, X. and Sharma, V.K., 2023. Microplastics and nanoplastics in the soil-plant nexus: Sources, uptake, and toxicity. *Critical Reviews in Environmental Science and Technology*, pp. 1–30.

Song, U., Kim, J. and Rim, H., 2023. Assessing phytotoxicity of microplastics on aquatic plants using fluorescent microplastics. *Environmental Science and Pollution Research*, pp. 1–10.

Wang, X., Xie, H., Wang, P. and Yin, H., 2023. Nanoparticles in plants: Uptake, transport and physiological activity in leaf and root. *Materials, 16*(8), p. 3097.

Wijesooriya, M., Wijesekara, H., Sewwandi, M., Soysa, S., Rajapaksha, A.U., Vithanage, M. and Bolan, N., 2023. Microplastics and soil nutrient cycling. *Microplastics in the Ecosphere: Air, Water, Soil, and Food*, pp. 321–338.

Wojnowska-Baryła, I., Bernat, K. and Zaborowska, M., 2022. Plastic waste degradation in landfill conditions: The problem with microplastics, and their direct and indirect environmental effects. *International Journal of Environmental Research and Public Health, 19*(20), p. 13223.

Zhang, Z., Cui, Q., Chen, L., Zhu, X., Zhao, S., Duan, C., Zhang, X., Song, D. and Fang, L., 2022. A critical review of microplastics in the soil-plant system: Distribution, uptake, phytotoxicity and prevention. *Journal of Hazardous Materials, 424*, p. 127750.

Zhou, J., Jia, R., Brown, R.W., Yang, Y., Zeng, Z., Jones, D.L. and Zang, H., 2023. The long-term uncertainty of biodegradable mulch film residues and associated microplastics pollution on plant-soil health. *Journal of Hazardous Materials, 442*, p. 130055.

Zhou, Y. and Zhou, S., 2023. Role of microplastics in microbial community structure and functions in urban soils. *Journal of Hazardous Materials*, p. 132141.

7 Microplastics Degradation and Remediation Techniques

Niladri Sekhar Mondal, Atanu Patra, and Apurba Ratan Ghosh

7.1 INTRODUCTION

Microplastic pollution has become a significant environmental concern, requiring urgent attention in today's ever-changing landscape of environmental challenges. These tiny plastic particles, known as microplastics, with sizes ranging from 0.1 μm to 5 mm, have infiltrated every corner of our planet, from the depths of the oceans to the most remote wilderness areas (Zhang *et al.*, 2021). This widespread distribution highlights the immense impact of human consumption and waste generation on the environment. In 1892, cellulosic fiber production started commercially and has since rapidly expanded. The market for synthetic fibers was projected to reach 60.4 million tons in 2014, with polyester production surpassing cotton in 2002 and expected to rise to 50 million tons by 2020, growing at a significantly faster rate compared to other types of fibers (Woodings, 2001; Gschwandtner, 2022). In addition to revolutionizing industries and enhancing product longevity, plastics' non-reactive nature also makes them persistent long after their intended utilization. A lot of plastic waste enters the environment through various pathways after excessive use, poor management, and disposal techniques (Ncube *et al.*, 2021). This plastic waste can degrade slowly under the influence of physical, chemical, and biological processes, and this process may take a few decades and produce millions of smaller plastic particles called microplastics (MPs) (Lusher *et al.*, 2017; Li *et al.*, 2021; Zhang *et al.*, 2021).

Understanding the fate, consequences, and effective control strategies of these miniature contaminants has become more important as these small pollutants become a point of concern in current environmental issues and pose a multifaceted threat (Abbasi *et al.*, 2020; Iroegbu *et al.*, 2021). The greater variety and complexity of microplastics found in real environments are believed to be linked to their toxicity. Smaller microplastics, particularly those in fiber form, are generally considered more hazardous due to prolonged exposure time and different mechanisms of harm (Pirsaheb *et al.*, 2020, Iroegbu *et al.*, 2021). Microplastics have a global impact, being transported by rivers from civilized areas to remote regions of the ocean where they accumulate in gyres. They can persist there for centuries, breaking down into smaller particles that enter ecosystems and food chains, potentially posing risks

DOI: 10.1201/9781032684574-7

to human health (Lusher *et al.*, 2017, Yuan *et al.*, 2022). This process affects marine organisms through the food web and ultimately impacts seafood safety and human health (Pirsaheb *et al.*, 2020, Yuan *et al.*, 2022). Their influence extends to terrestrial environments, where microplastics can also alter soil properties and nutrient cycles that impact plant growth. It has spread everywhere, even the air we breathe and in the seas from the Mariana Trench to the Arctic (Peng *et al.*, 2018; Bergmann *et al.*, 2022).

While mitigation efforts should prioritize reducing the production and release of plastic waste, addressing the vast quantities of existing microplastics necessitates targeted and innovative remediation techniques. The endeavor to improve the burden of microplastic pollution is fraught with challenges, across the scientific, technological, and regulatory domains. Hence a comprehensive understanding is required for available methods, their strengths, limitations, and potential unintended consequences.

This chapter embarks on a comprehensive exploration of the intricate web of microplastic degradation and the emerging strategies for their effective remediation. By delving into the intricacies of both natural and technological approaches, we aim to provide a holistic understanding of the challenges posed by microplastics and the diverse solutions available to mitigate their impact.

7.2 DEGRADATION OF MICROPLASTICS

Microplastics have emerged as a pervasive and concerning environmental issue in recent years, drawing much attention due to their extensive dispersion and potential ecological effects. These tiny plastic particles, measuring between a few micrometers to millimeters in size, can be found in every corner of our planet, from the depths of the ocean to the highest mountain peaks. Despite their different origins, microplastics are commonly produced from the breakdown of larger plastic materials like bottles, packaging containers, synthetic bags, etc. (Iroegbu *et al.*, 2021).

Understanding how microplastics degrade is crucial because it holds the key to reducing their harmful effects on ecosystems and public health. Depending on the properties of the plastic and the surrounding environmental variables, plastic materials can persist for hundreds or even thousands of years (Chamas *et al.*, 2020; Iroegbu *et al.*, 2021). This degradation of MPs occurs through different processes, including biological degradation, photodegradation, chemical degradation, and physical degradation.

7.2.1 Biological Degradation

The process of biological degradation, or biodegradation, occurs in nature by the help of microorganisms with the enzymatic ability to break down complex organic molecules, such as the polymer chains in plastics, into simpler compounds. These enzymes, such as lipases and esterases, hydrolyze the ester bonds of plastic polymers, which results in depolymerization and finally splits into smaller fragments (Lai *et al.*, 2023). Numerous species of bacteria, fungi, and even algae have been found to have the potential to break down microplastics. Utilizing microbial activity

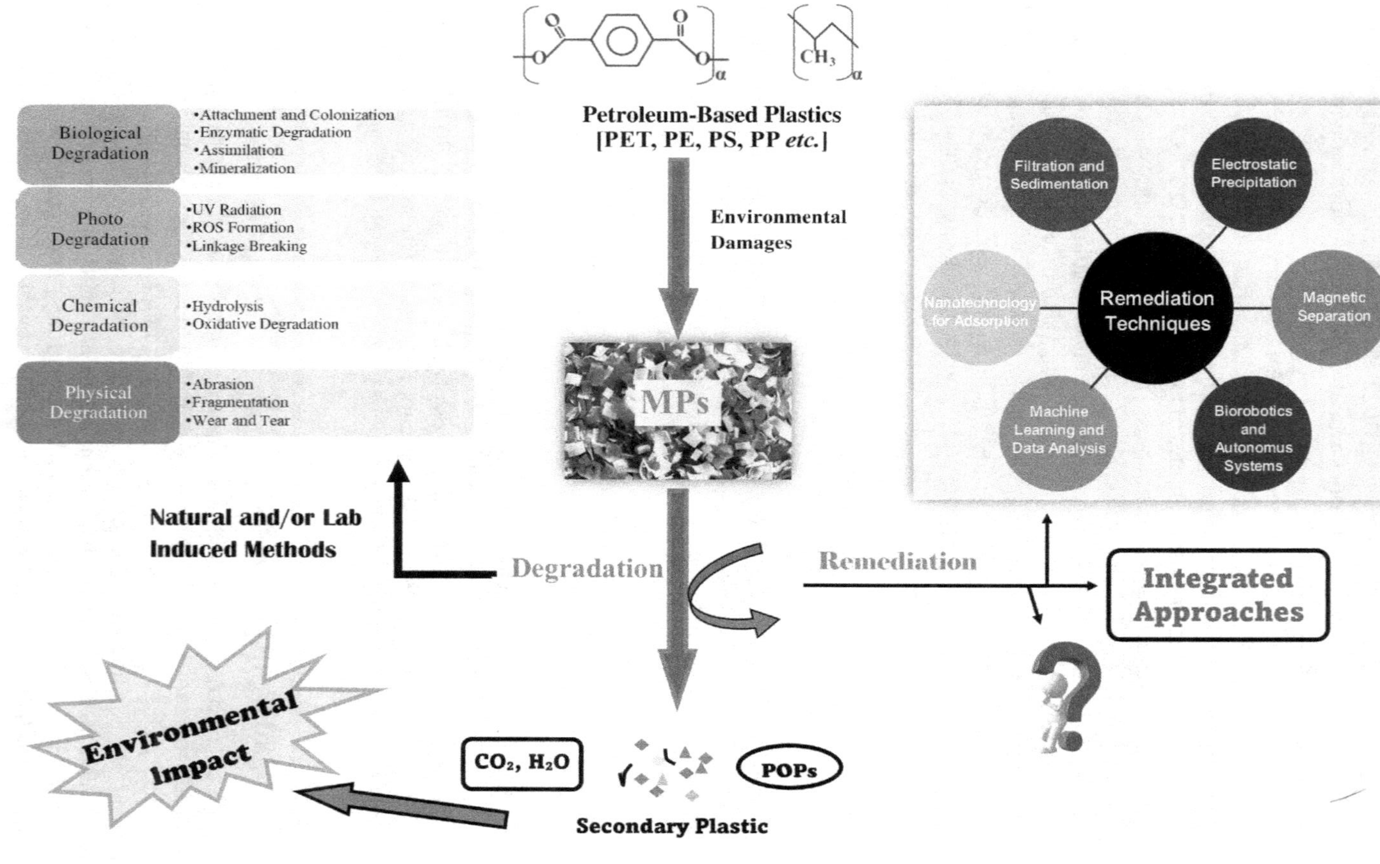

FIGURE 7.1 Microplastic degradation and techniques for remediation.

to reduce microplastic pollution has drawn a significant level of interest because of its eco-friendliness and potential for long-term effect reduction (Debroy *et al.*, 2022).

The researchers are particularly interested in microbial communities that persist in plastic-contaminated environments like sediments and wastewater. From that contaminated environment, researchers try to isolate the microbes with the best potential of degrading the MPs. Initially, a biological consortium with a significant capability for degrading microplastics was identified, and later the most productive species within that microbial consortium were identified using DNA mapping (Yuan *et al.*, 2020). Most typically, a variety of microbe strains are introduced on different types of microplastic materials (e.g., PET, PE, PS, PP, etc.) and find out the most efficient plastic degrading strain, some of which are listed in Table 7.1. However, the efficiency of the biological degradation process is influenced by a number of variables, such as the type of plastic, microbial activity, temperature, pH, and the availability of nutrients present in the belonging environment (Chamas *et al.*, 2020; Auta *et al.*, 2022). The challenges of extending those processes and ensuring their efficiency in varied environmental conditions are presently being studied.

7.2.1.1 Biodegradation Pathways

The process of biological degradation involves several stages:

- **Attachment and Colonization:** Microorganisms initially attach to the surface of microplastics, facilitated by their hydrophobic nature (Debroy *et al.*, 2022). This attachment is the pioneer stage that will allow the remaining enzymatic action to occur.
- **Enzymatic Degradation:** Once microbes are attached to the surface of plastic materials, they release enzymes that enhance the hydrolysis of ester bonds present in plastic polymers. Enzymes released by microorganisms, which include urease, protease, and lipases, influence the enzymatic degradation of microplastics both internally and externally (Chandra and Singh, 2020; Saravanan *et al.*, 2021). These enzymes cut the polymer chains into oligomers and monomers, reducing the size of the polymer chain as well as creating microplastic particles.
- **Assimilation:** The broken-down parts of polymer chains from plastic materials can be used as food by microorganisms for growth and metabolism in the form of carbon and energy. This step is essential for the complete elimination of plastics from the environment without producing toxic byproducts (Kjeldsen *et al.*, 2018). Microbial cells have difficulties in absorbing some plastic monomers because of the semipermeable cell membrane; therefore the monomers are absorbed by microbial cells through the processes of biotransformation and then assimilation (Thakur *et al.*, 2023).
- **Mineralization:** According to certain studies (Shah *et al.*, 2008; Kjeldsen *et al.*, 2018), microbes can further degrade the absorbed breakdown products into simpler forms such as carbon dioxide, CH_4, water, and biomass. The conversion of plastic waste into the fundamental building blocks of life is referred to as mineralization.

TABLE 7.1
Degradation of Different Types of Microplastic by Bacteria Fungi and Algae

Type of Microplastic	Microorganism	Favorable Conditions	Efficiency	Days	References
Polyethylene tetraphthalate (PET)	**Bacteria**				
	Vibrio	Temp: 50°C ± 2, nutrient broth	35%	45	Sarkhel *et al.*, 2020
	Bacillus gottheilii	Temp: 37°C, nutrient broth	3%	30	Auta *et al.*, 2017
	Thermobifida fusca	Temp: 55°C ± 2	54%	21	Wei *et al.*, 2019
	Pseudomonas sps.	Temp: 30°C ± 2, Bushnell Hass broth	5%	100	Taghavi *et al.*, 2021
	Algae				
	Phaeodactylum tricornutum	Temp: 21°C ± 2, culture agar media		54	Moog *et al.*, 2019
	Spirulina sp.		48.61%	112	Khoironi *et al.*, 2019
	Fungi				
	Penicillium funiculosum	Temp: 30°C ± 2, Czapek-Doxa broth medium	0.21%	84	Nowak *et al.*, 2011
	Aspergillus sp.	Temp: 37°C ± 2, potato dextrose agar medium	22%	42	Sarkhel *et al.*, 2020
Polyethylene (PE)	**Bacteria**				
	Pseudomonas fluorescens	Temp: 27°C, nutrient broth	18%	270	Thomas *et al.*, 2015
	Pseudomonas citronellolis	Temp: 25°C, nutrient medium	17.8%	1	Bhatia *et al.*, 2014
	Bacillus brevies	Temp: 30 °C, mineral salt medium		60	Watanabe *et al.*, 2009
	Aneurinibacillus sp.	Salt-free media	45.7%	140	Skariyachan *et al.*, 2017
	Algae				
	Oscillatoria subbrevis	Temp: 27°C	30%	42	Sarmah and Rout, 2019
	Scenedesmus dimorphous	Temp: 27°C, diatomic medium Bold basal medium	3.74%	30	Kumar *et al.*, 2017
	Navicula popular	Temp: 27°C, bold basal medium	4.44%	30	Kumar *et al.*, 2017

	Fungi				
	Aspergillus niger	Temp: 25°C	17.4%	126	Nandi and Joshi, 2013
	Aspergillus glaucus	Agar media	6.63%	30	Kathiresan, 2003
	Penicillium oxalicum NS4	Temp: 27°C, Czapek-Doxa broth	36.60%	30	Ojha *et al.*, 2017
	Cephalosporium sp.	Temp: 28°C, mineral salt medium	7.18%	20	Chaudhary and Vijayakumar, 2020
Polystyrene (PS)	**Bacteria**				
	Bacillus subtilis	Temp: 37°C, nutrient broth	20%	28	Asmita *et al.*, 2015
	Rhodococcus ruber C208	Temp: 35°C, mineral medium, synthetic medium	0.8%	56	Mor and Sivan, 2008
	Pseudomonas sp.	Temp: 20°C ± 2, salt-free medium	5%	28	Taghavi *et al.*, 2021
	Algae				
	Chlamydomonas reinhardtii				Barone *et al.*, 2020
	Pseudokirchneriella subcapita				Padervand *et al.*, 2020
	Fungi				
	Cepahlosporium sps.	Temp: 28°C, mineral salt medium	2.17%	56	Chaudhary and Vijayakumar, 2020
	Penicillium sps.	Mineral liquid medium	8.4%		Oviedo-Anchundia *et al.*, 2021
	Geomices sp.	Temp: 18°C, mineral salt medium	6.8%	90	Oviedo-Anchundia *et al.*, 2021
Polypropylene (PP)	**Bacteria**				
	Pseudomonas sp.	Temp: 25°C, Bushnell Hass broth	9%	40	Jeon *et al.*, 2021
	Bacillus sp.	Bushnell Hass broth	6%	40	Auta *et al.*, 2018
	Rhodococcus sps.	Temp: 25°C, Bushnell Hass broth	6%	40	Auta *et al.*, 2018
	Algae				
	Spirulina		36.7%	112	Khoironi *et al.*, 2019
	Fungi				
	Aspergillus sps.	Temp: 27°C, Rose Bengal medium	12%	90	Raaman *et al.*, 2012
	Aspergillus niger	Temp: 27°C, Rose Bengal medium	53.09%	90	Williams and Osahon, 2021

However, the effectiveness of this biological degradation process is influenced by several factors: (1) The type of plastic greatly affects its susceptibility to biodegradation. Polyesters like polyethylene terephthalate (PET) and polyhydroxyalkanoates (PHA) are more biodegradable compared to polyolefins like polyethylene and polypropylene (Siracusa *et al.*, 2008). (2) Environmental conditions like temperature, pH, nutrient availability, and oxygen levels in the environment impact microbial activity (Paredes *et al.*, 2007; D'Alò *et al.*, 2021). Certain microorganisms thrive under specific conditions, so optimizing these factors can enhance biodegradation rates. (3) Microplastics found in natural environments often develop surface modifications due to exposure to sunlight, pollutants, and biological interactions (Chandra and Singh, 2020; Auta *et al.*, 2022). These modifications can influence microbial attachment and degradation rates. (4) Enzymatic capacities differ among microorganisms. A diversified microbial population with complementing enzymatic capabilities can improve the microplastic degradation process as a whole, for example, PETase and MHETase. These two enzymes enhance PET degradation (Blair *et al.*, 2021; Cui *et al.*, 2021).

Recent research actively explores strategies to enhance biological degradation rates, such as genetically engineering microbes to improve their plastic-degrading capabilities. Microbiome manipulation, where specific microbial communities are introduced to target plastic pollution, is also being investigated. Collaborative efforts among microbiologists, material scientists, and environmental scientists are crucial to harnessing the full potential of biological degradation as a microplastic mitigation strategy.

7.2.2 Photodegradation

Photodegradation is a most prominent and natural abiotic process that plays a pivotal role in mediating the persistence of microplastics in the environment. These substances are found everywhere in the water and land ecosystems. As ultraviolet (UV) rays from sunlight penetrate them, a process called photodegradation comes into play and causes chemical bonds within the polymer chains of these microplastics to break down. When this occurs, the structural strength of these microplastics is lost, causing them to be easily damaged by mechanical and biological degradation processes (Zhang *et al.*, 2021).

Factors such as plastic composition, UV intensity, exposure length, and the inclusion of additives that absorb or scatter UV radiation all influence the rate of photodegradation. Plastics having pigments or stabilizers may have slower photodegradation rates (Yousif and Haddad, 2013). While photodegradation offers a natural and passive approach to microplastic mitigation, its efficacy is influenced by environmental conditions, and not all plastics are equally susceptible to this process. Photodegradation alters the fate of microplastics in several ways; for example, this process leads to the physical breakdown of microplastics into smaller particles, facilitating their transport through ecosystems and, due to photodegradation-generated smaller fragments of microplastics becoming more accessible to microbial enzymes, potentially enhancing subsequent biodegradation. There is some negative implication that photodegraded microplastics can enter the food web at various trophic levels, affecting aquatic and terrestrial organisms in nanoplastic form, which would show a divergent way of toxicity.

7.2.2.1 Photodegradation Mechanisms

The photodegradation process begins with the absorption of UV radiation by plastic polymers. This energy absorption leads to the excitation of electrons within the polymer chains, causing bond dissociation and the formation of highly reactive free radicals; the Norrish mechanism (Figure 7.2) is involved in this phenomenon (Torikai *et al.*, 1983; Al-Malaika *et al.*, 2017). These free radicals initiate a cascade of chemical reactions, including chain scission, cross-linking, and oxidation.

$$-CH_3-\overset{O}{\overset{\|}{c}}-CH_2- \xrightarrow{\text{UV radiation}/h\nu} -CH_2 \underset{\cdot CH_2+CO}{\overset{-}{\downarrow}} \overset{O}{\overset{\|}{c^{\bullet}}}+{}^{\bullet}CH_2$$

7.2.2.2 Effects on Polymer Structure

As the photodegradation process unfolds, microplastics experience several structural changes:

- **Chain Scission:** Bond cleavage weakens polymer chains, leading to the fragmentation of microplastics into smaller particles (Song *et al.*, 2017). This increase in surface-area-to-volume ratio exposes more polymer bonds to UV rays.
- **Cross-Linking:** Simultaneously, the interlinking of polymer chains can also take place. These cross-links have the capability to change the mechanical attributes of the plastic, rendering it more fragile and vulnerable to external forces (Zhu *et al.*, 2021).
- **Oxidation:** Oxidation-driven reactions cause the incorporation of functional oxygenated groups (e.g., hydroxyl, carboxyl, etc.) onto the polymer's exterior. These groupings can additionally enhance the sensitivity of microplastics to microbial degradation processes.

7.2.2.3 Factors Influencing Photodegradation

The efficiency of photodegradation of microplastic is influenced by a range of factors:

- **Plastic Composition:** Like the biodegradation of plastic materials, the susceptibility of various plastic polymers to ultraviolet (UV) radiation also varies. Due to the presence of UV-absorbing carbonyl chromophores present in aromatic polymers like polyethene terephthalate (PET), they are more susceptible to photodegradation (Wiles and Carlsson, 1980).
- **UV Intensity and Wavelength:** UV radiation intensity and wavelength significantly affect the rate of photodegradation. Shorter wavelengths, such as UV-C, are more energy rich ($E = \frac{hc}{\lambda}, \lambda =$ wavelength) and lead to faster degradation (Torikai *et al.*, 1983).
- **Exposure Time:** The degree of photodegradation is directly proportional to the exposure durations. However, long-term exposure can lead to the generation of photoproducts that may have different environmental consequences.

- **Presence of Additives:** The sensitivity of microplastics to photodegradation can be influenced by additives used in plastic, such as stabilizers and colors. Some additives like titanium dioxide (TiO_2), act as photocatalysts when exposed to UV radiation and can accelerate the degradation of microplastics by generating reactive oxygen species (ROS) that break down the plastic (Nabi *et al.*, 2021).

7.2.3 Chemical Degradation

A potential solution for the enduring problem of microplastic pollution is chemical degradation. Chemical degradation is the process of exposing microplastics to various chemical reactions that disintegrate the polymer structures inside them, ultimately leading to depolymerization and the breakdown of these synthetic substances. The hydrolysis reaction is a common chemical degradation process that involves the reaction of plastic polymer with water molecules, which results in the cleavage of polymer chains (Chamas *et al.*, 2020; Lastovina and Budnyk, 2021). On the other hand, oxidative degradation involves treating microplastics with oxidizing agents, such as ozone, hydrogen peroxide, or other reactive chemicals to induce degradation (Lastovina and Budnyk, 2021; Kim *et al.*, 2022). Although chemical degradation can break down plastic materials into smaller fragments, it often requires extremely controlled conditions and can produce potentially harmful byproducts. Researchers are exploring ways to optimize degradation conditions and identify environmentally safe chemical agents, although integration with other remediation techniques could offer a more comprehensive approach to microplastic (MPs) removal.

Box 7.1: Chemical Degradation of MPs

Hydrolysis: Breaking Polymer Bonds

Hydrolysis is a phenomenon in which water molecules cut the chemical bonds holding the polymer chains together. It is one of the basic mechanisms of chemical degradation. Temperature, pH, and the presence of catalysts can accelerate this process. The polymer chains gradually break down into smaller fragments as a result of hydrolysis, making them more vulnerable to further degradation by other mechanisms. (Deshoulles *et al.*, 2022).

Oxidative Degradation: Introducing Reactive Groups

Microplastics are subject to oxidative destruction when they come into contact with oxidizing substances like ozone (O_3), hydrogen peroxide (H_2O_2), and other reactive compounds. These substances apply hydroxyl, carbonyl, and carboxyl functional groups to the surface of the polymer (Klein *et al.*, 2018). These entities render the plastic more vulnerable to future processes of degradation, such as microbial action and mechanical fragmentation.

While chemical degradation offers a potential solution to the microplastics problem, several challenges and considerations need to be addressed:

- **Selectivity:** Different plastics require specific conditions and reagents for efficient chemical degradation. Finding a universal method to degrade all types of plastic components is highly challenging work.

- **Byproduct Formation:** Byproducts from chemical degradation may include some hazardous or enduring substances. A critical life cycle assessment is required for those byproducts to confirm environmental safety.
- **Reaction Conditions:** To achieve effective degradation without causing additional environmental impact, the reaction variables like temperature, pressure, and reaction time must be optimized crucially.
- **Scale-Up and Practicality:** Moving from lab-scale experiments to large-scale applications presents significant engineering and logistical issues that must be resolved before practical implementation.

With investigations into new reagents, catalysts, and reaction conditions, the chemical degradation of microplastics is advancing rapidly. Biodegradable plastics, designed to be more vulnerable to chemical degradation, may be a potential solution. Additionally, the integration of chemical degradation with other remediation techniques, such as mechanical methods or biological degradation, could be a more effective and sustainable strategy for microplastic removal.

7.2.4 Physical Degradation

Physical degradation methods involve mechanical processes that break down larger plastic pieces into smaller fragments physically. These methods include grinding, milling, and shredding (Hu *et al.*, 2021). Fragmentation of MPs through mechanical forces increases their surface area, potentially facilitating faster biodegradation and microbial colonization (Lucas *et al.*, 2008; Kumar *et al.*, 2017; Sarkhel *et al.*, 2020).

Mechanical methods are commonly used in combination with other techniques such as filtration and sedimentation. For example, wastewater treatment plants often employ screens and filters to capture larger microplastic particles, which are subsequently subjected to mechanical breakdown for further size reduction (Ziajahromi *et al.*, 2017). However, managing the produced microplastic fragments is challenging work, and it also has to be ensured that they do not contribute to any other secondary pollution.

The mechanical forces applied during abrasion, fragmentation, and wear transform plastic particles into smaller sizes. This may have an effect on how they move, are transported, and interact with living organisms (Pelegrini *et al.*, 2023). Larger plastic items fragment into microplastic particles and/or nanoplastic that are more easily transported by wind and water currents (Chamas *et al.*, 2020; Zhang *et al.*, 2021; Pelegrini *et al.*, 2023). These fragments may then gather in different dimensions of the environment and exacerbate pollution in both terrestrial and marine environments.

7.2.4.1 Mechanical Degradation Processes

Microplastics are subjected to experience a variety of mechanical forces in their environment:

- **Abrasion:** In the natural environment, where microplastic interacts with sediments, waves, and currents, plastic material degradation happens slowly due to friction between microplastic surfaces and abrasive materials such as rocks, sand, and other plastics.

- **Fragmentation:** Repeated mechanical stress causes fragmentation of larger plastic objects, such as bottles, large plastic sheets, polybags, and synthetic textiles (Acharya *et al.*, 2021). As a result, smaller microplastic particles are produced, with increased surface-area.
- **Wear and Tear:** The continuous exposure of microplastics to mechanical agitation, as in the roiling action of waves, can lead to the gradual wearing down of their edges and surfaces.

7.2.4.2 Ecological Implications

Physical degradation has multifaceted implications for ecosystems:

- **Transport and Distribution:** Smaller microplastic fragments resulting from degradation can be transported over longer distances by water currents and wind, leading to their widespread distribution (Eerkes-Medrano *et al.*, 2015; Hale *et al.*, 2020).
- **Ingestion and Trophic Transfer:** The smaller particle size resulting from physical degradation can increase the possibility of microplastics being ingested by a wider range of organisms, from zooplankton to filter-feeding organisms (Hale *et al.*, 2020; Pelegrini *et al.*, 2023).
- **Chemical Interactions:** Increased surface area resulting from degradation might increase the adsorption of persistent organic pollutants such as DDT onto microplastics, potentially altering contaminant bioavailability in the environment (Bakir *et al.*, 2012). Partitioning, surface sorption (hydrogen bonding, electrostatic interaction, and van der Waals force), and pore filling are the key processes involved in the sorption of organic pollutants to NPs/MPs (Wang *et al.*, 2020).

7.3 REMEDIATION OF MICROPLASTICS

7.3.1 Filtration and Sedimentation

Filtration and sedimentation are basic physical separation techniques widely employed in water and wastewater treatment to remove suspended particles including microplastics. Filtration involves passing water through porous surfaces that trap microplastics while allowing clean water to leach out (Tang and Hadibarata, 2021). Sedimentation relies on the principle of gravity to allow microplastics to settle down in water over time, typically in sedimentation basins or tanks.

Filtration methods range from simple sand filters to sophisticated membrane filtration systems. Particle size distribution, filter pore size, flow rates, and water quality are only a few examples of the variables that affect the efficiency of filtration and sedimentation systems. Although these methods can effectively remove larger microplastic particles, in the case of smaller particles additional treatment processes may be required. According to reports, the primary filtration of WTPs can eliminate 16.5–98.4% of microplastics from wastewater (Tang and Hadibarata, 2021). Overall MP removal efficiencies for secondary treatment in wastewater treatment facilities range from 78.1 to 100%, and stage-by-stage efficiencies range from 7% (activated sludge) to 99.9% (membrane reactor) (Ziajahromi *et al.*, 2017). The tertiary treatment process can eliminate 87.3–99.9% of microplastics in total from the wastewater. Typically, the drinking water treatment plants initially remove 1.8–54.5% of microplastic through the process of coagulation and sedimentation, but advanced treatment increases that removal to 88.6% overall. In one trial, utilizing a higher amount of flocculant than normal increased removal of MP by up to 62% Carr *et al.*, 2016; Tang and Hadibarata, 2021).

7.3.1.1 Filtration of Microplastics

Filtration involves the passage of a fluid through a porous medium, which captures particles larger than the pore size while allowing smaller particles and the fluid to pass through. The selection of the filter medium and pore size is important to effectively target MPs. Pore size, flow velocity, filter material, and microplastic properties are just a few examples of the variables that affect how effectively microplastics are filtered. Due to their ability to pass through the currently used filters, smaller microplastics pose a threat to outdated traditional filtration systems. Jams of the filter pores by organic matter or particles can also reduce effectiveness and necessitate regular maintenance.

Various filtration systems can be adapted for microplastic removal:

- **Depth Filtration:** In this process, porous media with different pore sizes are utilized. Smaller particles are retained deeper in the medium than larger ones, which are collected on the surface. For instance, turbidities in the PET-MP and Arizona test dust suspensions were reduced to 80.1 ± 1.5 and 91.9 ± 0.3% NTU, respectively, by an oyster-inspired adsorption system with 4000 stimulated compression/relaxation cycles; the filtration efficiency is proportional to the depth of the chitosan NF sponges (Risch and Adlhart, 2021).
- **Membrane Filtration:** This filtration system uses synthetic membranes with precise pore sizes, capable of removing particles selectively based on their size. For instance, a composite membrane made of graphene oxide (GO) and polyvinyl alcohol (PVA) can separate MPs made of high-density polyethene (HDPE) from wastewater within 15 seconds utilizing electrostatic repulsion mechanisms (Dey *et al.*, 2023).
- **Screen Filtration:** This technique physically prevents the passage of microplastics by using screens with various mesh sizes. Microscreen filtration and disc filters (DF) were installed at the Viikinmaki Waste Water Treatment Plant (WWTP) in Finland's Helsinki metropolis city for the removal of microplastic from wastewater (Talvitie *et al.*, 2017).

7.3.1.2 Sedimentation of Microplastics

Sedimentation relies on gravity to settle suspended particles out of a fluid. Larger and denser microplastics tend to settle faster than smaller particles, facilitating their removal from the water column. Sedimentation basins at water treatment facilities allow particles, particularly microplastics, to settle because of lower flow velocities that give particles enough retention time to do so. The settling speed of microplastics depends on size, shape, density, and interactions with other suspended matter. Microplastics can change their sedimentation behavior when they are grouped with other particles (Elagami *et al.*, 2022). Commercial coagulants worked effectively for the quantitative removal of relatively large MPs, such as 20–90 μm polystyrene (PS) and 15–140 μm polyethene (PE) (Lapointe *et al.*, 2020; Na *et al.*, 2021). Lower removal efficiencies were observed, for example, 33% for 10 μm PS, $< 2\%$ for 10–20 μm PE, and $< 0.1\%$ for 1 μm PS, despite the fact that there were relatively few studies on the coagulation/sedimentation of smaller microplastics (Zhang *et al.*, 2020; Na *et al.*, 2021). In some circumstances, the addition and modification of coagulants may help in the microplastic aggregate formation and speed up the sedimentation process. The production of tiny, stable particles must be prevented; therefore, optimization must be done with extra care.

7.3.2 Emerging Technologies

Microplastic pollution is a global scourge that has triggered a wave of research and invention, leading to the creation of cutting-edge technologies that are targeted at this widespread environmental concern through improved microplastic degradation and removal efficiency. These technologies are environmentally friendly and effective in removing microscopic particles that standard technology finds difficult to remove. The electrostatic method creates electric fields that attract and capture microplastics from water or air (Feilin and Mingwei, 2022). Similarly, magnetic separation exploits the magnetic properties of certain plastics, allowing them to be separated from water using magnetic fields (Grbic *et al.*, 2019; Mato and Noguchi, 2021). Microplastic segregation potentiality is anticipated to improve with the use of second-generation high-temperature superconducting (HTS) magnets that can produce fields greater than 15 T (Mato and Noguchi, 2021). Nanomaterials are gaining importance due to their ability to absorb and degrade microplastics. Nanoparticles with large surfaces attract and bind microplastic particles, allowing them to be easily removed from soil or water (Sajid *et al.*, 2023). However, the impact of these nanoparticles on the environment and human health needs to be thoroughly investigated prior to their widespread use in real environments.

7.3.2.1 Nanotechnology for Adsorption and Degradation

Nanoparticles possess high surface-area-to-volume ratios, enabling efficient adsorption of microplastics from water and sediment matrices. Functionalized nanoparticles can specifically target microplastics due to their surface properties, enhancing removal efficiency. The high surface areas and strong adsorption properties of nanoparticles like graphene oxide, carbon nanotubes, and magnetic nanoparticles are capable of efficiently absorbing microplastics from water (Ouda *et al.*, 2023; Sajid *et al.*, 2023). A material that binds to microplastics can be coated on magnetic nanoparticles like iron oxide nano. When these magnetic nanoparticles are added to water that has been polluted with microplastics, the pollutants can be removed from the water successfully utilizing magnets. Some of the complex nanoparticles like magnetic polyoxometalate-supported ionic liquid phases (magPOM-SILPs) that are composed of magnetic Fe_2O_3/SiO_2 core–shell nanoparticles and that are functionalized with a polyoxometalate ionic liquid, can effectively remove microplastic and also other pollutants like heavy metals, organic pollutants from the wastewater (Misra *et al.*, 2020).

Additionally, nanomaterials with catalytic properties can facilitate the degradation of adsorbed microplastics through advanced oxidation processes, producing smaller and potentially less harmful breakdown products. Nanoparticles, for instance titanium dioxide (TiO_2) and zinc oxide (ZnO), can be used as photocatalysts to degrade microplastics faster under UV light. Therefore, when microplastic particles interact with these catalysts, they go through a photodegradation process, transforming into smaller, less toxic compounds (Nabi *et al.*, 2021; Ouda *et al.*, 2023).

7.3.2.2 Electrostatic Precipitation

These technologies obey the principles of electrostatic attraction to capture microplastics from air and water. Electrostatic precipitators utilize charged plates or wires to create electric fields that attract and collect microplastics suspended in the air. Electrostatic forces can also be employed in water treatment systems, where charged electrodes attract and aggregate microplastics, facilitating their removal through sedimentation or filtration (Padervand *et al.*, 2020).

7.3.2.3 Magnetic Separation Techniques

Magnetic separation harnesses the magnetic properties of certain plastics, enabling their efficient removal from environmental matrices. Microplastics infused with magnetic nanoparticles (e.g., magPOM-SILPs, superparamagnetic $Fe_3O_4/\gamma\text{-}Fe_2O_3$ NPs, etc.) can be selectively captured using magnetic fields (Misra *et al.*, 2020; Gunatilake *et al.*, 2022). This strategy has benefits in terms of scalability and environment adaptability. However, effective separation and tagging of magnetic nanoparticles to various types of plastic material need precise control of magnetic characteristics and particle size distribution, which is a subject of extensive research.

7.3.2.4 Biorobotics and Autonomous Systems

Advancements in robotics and autonomous systems have paved the way for innovative approaches to microplastic removal (Urso and Pumera, 2022). Biorobots, inspired by aquatic organisms like jellyfish and manta rays, can navigate water bodies to capture microplastics while minimizing ecological disruption (Zhang *et al.*, 2022). They can be equipped with specialized arms or appendages capable of collecting microplastics from various environments, specifically from remote areas including the seabed or from the deep water column. These robots can work autonomously or under remote human control. Autonomous vessels equipped with sensors and collection mechanisms can prospect marine environments and deploy targeted removal techniques (Beladi-Mousavi *et al.*, 2021; Urso and Pumera, 2022).

7.3.2.5 Machine Learning and Data Analytics

In managing microplastic pollution, machine learning and data analytics are essential. Machine learning can be used to evaluate the ecological risks that MPs pose utilizing big data research. Machine learning algorithms can identify and categorize microplastics using images derived from environmental materials such as sediment or water samples. Convolutional neural networks (CNNs) are commonly used for this purpose. Data-driven machine learning can boost the control of dangerous MPs and reduce the impact, on both local and global scales as compared to traditional models (Su *et al.*, 2023). Algorithms can be created to identify microplastics from hyperspectral or multispectral data collected from remote sensing data, drone imagery, and spectroscopic measurements. These methods improve the effectiveness and accuracy of microplastic detection and monitoring when combined with artificial intelligence, enabling immediate decision making and adaptive solutions (Xiouras *et al.*, 2022; Yu and Hu, 2022). Machine learning can empower autonomous robots to detect and collect microplastics from aquatic environments using computer vision and sensor data.

7.4 CHALLENGES AND CONSIDERATIONS

The necessity to quickly address microplastic pollution has prompted the development of numerous degradation and remediation methods. The pollution caused by microplastics can be reduced with the help of these methods, but they are not without challenges and the possibility of unexpected consequences. This scientific literature reviewed the intricate web of issues and considerations that researchers and other stakeholders must take into account while formulating strategies to minimize the pollution caused by microplastics. An additional concern is the potential for the production of harmful byproducts and releases to the environment during degradation processes. For instance, when plastics break down, chemicals and additives that were previously incorporated into the plastic matrix may be released into the environment creating new environmental hazards. It is critical to ensure that the degradation products are safe and nontoxic.

In addition, unforeseen effects on non-target organisms and ecosystems must be taken into account. Even though removing microplastic is important, the techniques utilized shouldn't harm or disturb helpful organisms or change ecosystem dynamics. It is crucial to evaluate the overall environmental impact of remediation processes in order to prevent exacerbating already existing environmental issues.

- **Complexity of Microplastics:** A problem arises from the wide variety of microplastics in terms of polymer types, sizes, and surface modification. Based on different plastics' distinct degradation characteristics, they react to different techniques (Lucas *et al.*, 2008; Zhang *et al.*, 2021). Due to their complex composition and degrading behaviors, it is difficult to design an all-encompassing solution that tackles the complete range of microplastics.
- **Generation of Harmful Byproducts:** During the process of degradation of microplastics, the output product may be harmful to the environment and as well as to human health. For instance, during the chemical and biological degradation process of microplastics, smaller plastic fragments and other chemical compounds like additives, plasticizers, and additional toxic compounds may be released to the environment as residual products. Although the transformation of microplastics into smaller particles is not necessarily

parallel to the reduction of their environmental impact, these fragments might still carry potential risks (Pelegrini *et al.*, 2023).

- **Ecological Disruption:** Microplastic degradation and removal techniques can unintentionally alter crucial processes running in the ecosystem. For example, mechanical methods such as filtration can trap other natural materials, nutrients, and small microbes along with the microplastic particles and disrupt the balance of natural dynamics within aquatic systems. Furthermore, removing microplastics can alter microbial communities in unexpected ways that could affect the microbes dependent on biogeochemical cycles.
- **Resuspension and Redistribution:** Mechanical removal techniques, such as sedimentation and skimming, may lead to the resuspension of microplastics in the medium. In future, these resuspended MP particles may be translocated through the water current to another new location, exacerbating the pollution rather than mitigating it (Carr *et al.*, 2016). So, it's essential to examine the sediment dynamics and water currents before assessing the efficiency of these techniques.
- **Microplastics in Remote Environments:** A controlled environment with sufficient logistical support is required for the utilization of several degradation and removal methods. However, extreme MP-contaminated places like deep sea sediments and polar regions remain substantially untouched by these approaches (Peng *et al.*, 2018; Bergmann *et al.*, 2022) because technique adoption and implementation in these locations are difficult due to logistical constraints and potential environmental consequences.
- **Unintended Consequences on Non-Target Organisms:** The application of certain techniques may affect non-target organisms. A wide range of creatures, such as beneficial microorganisms, nutrient cycles, and non-plastic materials, may be impacted by chemical degradation techniques. It is difficult to selectively target microplastics in an open system while limiting collateral damage.
- **Sustainability and Resource Consumption:** The environmental sustainability of microplastic degradation and remediation strategies should be taken into account throughout the development and implementation process. If not properly controlled, the energy consumption, utilization of resources, and waste generation associated with these strategies may have the opposite effect.

7.5 INTEGRATED APPROACHES

Considering the complexity of microplastic pollution, no single technique can address its different issues comprehensively. Integrated approaches that combine multiple techniques may offer more effective solutions. For example, combining biological degradation with physical methods like filtration can enhance microplastic removal rates. By utilizing the synergetic strengths of various methods, these integrated strategies are capable of achieving more efficient, holistic, and sustainable mitigation outcomes.

7.5.1 Synergistic Advantages of Integrated Approaches

Integrated approaches sum up the unique strengths of different techniques, resulting in synergistic advantages. The advantages are as follows.

- **Comprehensive Removal:** Combining physical, chemical, and biological methods addresses microplastics at various stages of degradation and from different sources, ensuring more comprehensive removal.
- **Enhanced Efficiency:** Individual techniques may have limitations or specific operational requirements. Integrating methods can compensate for these limitations and enhance overall efficiency. It reduces the drawback of one technique by involving another process of microplastic remediation.
- **Reduced Byproducts:** Integrated approaches can potentially minimize the generation of harmful residual products that are generated from some individual techniques. For instance, physical degradation, photodegradation, and chemical degradation produce smaller particles of plastic (nanoplastics) and harmful chemical composite during degradation, but, by integrating biodegradation, these residuals can be minimized.

7.5.1.1 Integrated Approaches: Example

Some typical and feasible integrated approaches are as follows.

- **Filtration plus Biological Degradation:** Filtration methods such as membrane filters or disk filters can segregate microplastics from liquid substances quite efficiently. With the integration of the additional method like the biological degradation method, microbes capable of releasing plastic-degrading enzymes enhance the breakdown process of the segregated microplastics (Chaudhary and Vijayakumar, 2020; Padervand *et al.*, 2020). The combination not only reduces the initial load but also speeds up the breakdown process, preventing plastic waste from accumulating in the filtration system and minimizing the requirement of the frequent replacement of filter membrane.
- **Photodegradation plus Mechanical Removal:** UV exposure on microplastics initiate photodegradation, which breaks down larger polymer chains of the plastic materials to the smaller parts. These pieces can subsequently be captured more efficiently by incorporating mechanical methods like sedimentation or filtration (Wiles and Carlsson, 1980; Lapointe *et al.*, 2020). The photodegradation-induced reduction in particle size of microplastic boosts the efficiency of mechanical removal methods, which are less effective against larger particles.
- **Nanomaterials plus Electrostatic Separation:** Nanomaterials with adjustable surface charges and high surface area can adsorb microplastic particles. Using this capability of the nanoparticles and clubbing with the electrostatic separation process, microplastics can be removed quite effectively. Electrostatic separation uses electric fields to attract and remove the charged microplastics, with the nanomaterials acting as efficient adsorbents that

also attract neutral microplastics but attached with charged nanomaterials, leading to improved microplastic removal rates (Feilin and Mingwei, 2022; Ouda *et al.*, 2023).

- **Nanomaterials plus Magnetic Separation:** Similar to the previous example, microplastics can be adsorbed by nanomaterials because of their high surface-to-volume ratios and magnetic characteristics. By utilizing the magnetism and adsorption capability of these nanoparticles, the efficacy of the magnetic separation process can be increased. Furthermore by employing a strong magnetic field to attract and remove nanotagged microplastics, the nanomaterials serve as effective adsorbents, and the removal rates of microplastics from the environment are improved (Misra *et al.*, 2020; Gunatilake *et al.*, 2022).

Although integrated approaches offer immense potential, they also pose challenges. Interlinking of different techniques requires complex engineering and optimization to ensure seamless integration. The combined effect of multiple techniques must be critically assessed to prevent unintended consequences on ecosystems. To overcome this issue, it requires to collaborate with the experts from various fields including environmental science, engineering, physics, microbiology, and material science. Vigorous research will be required to understand the interactions among the different techniques and predict their combined outcomes precisely.

7.6 CONCLUSION

Microplastic pollution is a significant global concern, requiring scientific investigation and innovative solutions. Microplastic degradation involves biodegradation, photodegradation, physical degradation, and chemical degradation. Biodegradation uses microorganisms' enzymatic capabilities to break down polymer chains into simpler compounds. Photodegradation, driven by UV radiation, involves chain scission, cross-linking, and oxidation. Chemical degradation involves hydrolysis and oxidation reactions, breaking down plastic polymers into smaller fragments. Physical degradation processes occur through abrasion and then fragmentation. The effectiveness of these processes depends on factors like plastic composition, environmental conditions, and microbial communities. Innovative techniques like filtration and sedimentation, electrostatics separation, nanomaterials, and magnetic separation offer improved removal efficiency. Bioremediation harnesses microbial communities to degrade plastics. Integrated approaches combining mechanical, biological, and chemical techniques can increase removal efficiency and minimize potential drawbacks. An integrated approach also minimizes the secondary waste produced through a single technique and safeguards the ecosystem from undesired consequences. Collaboration among scientists, engineers, policymakers, and industry is crucial for developing holistic strategies. The potential for toxic byproducts, unintended consequences on ecosystems, and long-term effects of degraded microplastics is still being investigated. Collective action is needed to bridge the gap between scientific insights and real-world application.

REFERENCES

Abbasi, S.A., Khalil, A.B., and Arslan, M., 2020. Extensive use of face masks during COVID-19 pandemic: (micro-) plastic pollution and potential health concerns in the Arabian Peninsula. *Saudi Journal of Biological Sciences*, 27 (12), 3181–3186.

Acharya, S., Rumi, S.S., Hu, Y., and Abidi, N., 2021. Microfibers from synthetic textiles as a major source of microplastics in the environment: A review. *Textile Research Journal*, 91 (17–18), 2136–2156.

Al-Malaika, S., Axtell, F., Rothon, R., and Gilbert, M., 2017. Additives for plastics. In: *Brydson's plastics materials*. Elsevier, 127–168.

Asmita, K., Shubhamsingh, T., and Tejashree, S., 2015. Isolation of plastic degrading micro-organisms from soil samples collected at various locations in Mumbai, India. *International Research Journal of Environment Sciences*, 4 (3), 77–85.

Auta, H.S., Abioye, O.P., Aransiola, S.A., Bala, J.D., Chukwuemeka, V.I., Hassan, A., Aziz, A., and Fauziah, S.H., 2022. Enhanced microbial degradation of PET and PS microplastics under natural conditions in mangrove environment. *Journal of Environmental Management*, 304, 114273.

Auta, H.S., Emenike, C.U., and Fauziah, S.H., 2017. Screening of Bacillus strains isolated from mangrove ecosystems in Peninsular Malaysia for microplastic degradation. *Environmental Pollution*, 231, 1552–1559.

Auta, H.S., Emenike, C.U., Jayanthi, B., and Fauziah, S.H., 2018. Growth kinetics and biodeterioration of polypropylene microplastics by *Bacillus* sp. and *Rhodococcus* sp. isolated from mangrove sediment. *Marine Pollution Bulletin*, 127, 15–21.

Bakir, A., Rowland, S.J., and Thompson, R.C., 2012. Competitive sorption of persistent organic pollutants onto microplastics in the marine environment. *Marine Pollution Bulletin*, 64 (12), 2782–2789.

Barone, G.D., Ferizović, D., Biundo, A., and Lindblad, P., 2020. Hints at the applicability of microalgae and cyanobacteria for the biodegradation of plastics. *Sustainability*, 12 (24), 10449.

Beladi-Mousavi, S.M., Hermanova, S., Ying, Y., Plutnar, J., and Pumera, M., 2021. A maze in plastic wastes: Autonomous motile photocatalytic microrobots against microplastics. *ACS Applied Materials & Interfaces*, 13 (21), 25102–25110.

Bergmann, M., Collard, F., Fabres, J., Gabrielsen, G.W., Provencher, J.F., Rochman, C.M., van Sebille, E., and Tekman, M.B., 2022. Plastic pollution in the Arctic. *Nature Reviews Earth & Environment*, 3 (5), 323–337.

Bhatia, M., Girdhar, A., Tiwari, A., and Nayarisseri, A., 2014. Implications of a novel Pseudomonas species on low density polyethylene biodegradation: An in vitro to in silico approach. *SpringerPlus*, 3, 1–10.

Blair, E.M., Dickson, K.L., and O'Malley, M.A., 2021. Microbial communities and their enzymes facilitate degradation of recalcitrant polymers in anaerobic digestion. *Current Opinion in Microbiology*, 64, 100–108.

Carr, S.A., Liu, J., and Tesoro, A.G., 2016. Transport and fate of microplastic particles in wastewater treatment plants. *Water Research*, 91, 174–182.

Chamas, A., Moon, H., Zheng, J., Qiu, Y., Tabassum, T., Jang, J.H., Abu-Omar, M., Scott, S.L., and Suh, S., 2020. Degradation rates of plastics in the environment. *ACS Sustainable Chemistry & Engineering*, 8 (9), 3494–3511.

Chandra, P., and Singh, D.P., 2020. Microplastic degradation by bacteria in aquatic ecosystem. In: *Microorganisms for sustainable environment and health*. Elsevier, 431–467.

Chaudhary, A.K., and Vijayakumar, R.P., 2020. Effect of chemical treatment on biological degradation of high-density polyethylene (HDPE). *Environment, Development and Sustainability*, 22 (2), 1093–1104.

Cui, Y., Chen, Y., Liu, X., Dong, S., Tian, Y., Qiao, Y., Mitra, R., Han, J., Li, C., and Han, X., 2021. Computational redesign of a PETase for plastic biodegradation under ambient condition by the GRAPE strategy. *ACS Catalysis*, 11 (3), 1340–1350.

D'Alò, F., Odriozola, I., Baldrian, P., Zucconi, L., Ripa, C., Cannone, N., Malfasi, F., Brancaleoni, L., and Onofri, S., 2021. Microbial activity in alpine soils under climate change. *Science of the Total Environment*, 783, 147012.

Debroy, A., George, N., and Mukherjee, G., 2022. Role of biofilms in the degradation of microplastics in aquatic environments. *Journal of Chemical Technology & Biotechnology*, 97 (12), 3271–3282.

Deshoulles, Q., Le Gall, M., Benali, S., Raquez, J.M., Dreanno, C., Arhant, M., Priour, D., Cerantola, S., Stoclet, G., and Le Gac, P.Y., 2022. Hydrolytic degradation of biodegradable poly (butylene adipate-co-terephthalate) (PBAT)-towards an understanding of microplastics fragmentation. *Polymer Degradation and Stability*, 205, 110122.

Dey, T.K., Jamal, M., and Uddin, M.E., 2023. Fabrication and performance analysis of graphene oxide-based composite membrane to separate microplastics from synthetic wastewater. *Journal of Water Process Engineering*, 52, 103554.

Eerkes-Medrano, D., Thompson, R.C., and Aldridge, D.C., 2015. Microplastics in freshwater systems: A review of the emerging threats, identification of knowledge gaps and prioritisation of research needs. *Water Research*, 75, 63–82.

Elagami, H., Ahmadi, P., Fleckenstein, J.H., Frei, S., Obst, M., Agarwal, S., and Gilfedder, B.S., 2022. Measurement of microplastic settling velocities and implications for residence times in thermally stratified lakes. *Limnology and Oceanography*, 67 (4), 934–945.

Feilin, H. and Mingwei, S., 2022. Ecofriendly removing microplastics from rivers: A novel air flotation approach crafted with positively charged carrier. *Process Safety and Environmental Protection*, 168, 613–623.

Grbic, J., Nguyen, B., Guo, E., You, J.B., Sinton, D., and Rochman, C.M., 2019. Magnetic extraction of microplastics from environmental samples. *Environmental Science & Technology Letters*, 6 (2), 68–72.

Gschwandtner, C., 2022. Outlook on global fiber demand and supply 2030. *Growth*, 65, 113.

Gunatilake, U.B., Morales, R., Basabe-Desmonts, L., and Benito-Lopez, F., 2022. Magneto twister: Magneto deformation of the water–air interface by a superhydrophobic magnetic nanoparticle layer. *Langmuir*, 38 (11), 3360–3369.

Hale, R.C., Seeley, M.E., La Guardia, M.J., Mai, L., and Zeng, E.Y., 2020. A global perspective on microplastics. *Journal of Geophysical Research: Oceans*, 125 (1), e2018JC014719.

Hu, K., Zhou, P., Yang, Y., Hall, T., Nie, G., Yao, Y., Duan, X., and Wang, S., 2021. Degradation of microplastics by a thermal Fenton reaction. *ACS ES&T Engineering*, 2 (1), 110–120.

Iroegbu, A.O.C., Ray, S.S., Mbarane, V., Bordado, J.C., and Sardinha, J.P., 2021. Plastic pollution: A perspective on matters arising: Challenges and opportunities. *ACS Omega*, 6 (30), 19343–19355.

Jeon, J.-M., Park, S.-J., Choi, T.-R., Park, J.-H., Yang, Y.-H., and Yoon, J.-J., 2021. Biodegradation of polyethylene and polypropylene by Lysinibacillus species JJY0216 isolated from soil grove. *Polymer Degradation and Stability*, 191, 109662.

Kathiresan, K., 2003. Polythene and plastics-degrading microbes from the mangrove soil. *Revista de biologia tropical*, 51 (3–4), 629–633.

Khoironi, A., Anggoro, S., and Sudarno, 2019. Evaluation of the interaction among microalgae *Spirulina* sp, plastics polyethylene terephthalate and polypropylene in freshwater environment. *Journal of Ecological Engineering*, 20 (6), 161–173.

Kim, S., Sin, A., Nam, H., Park, Y., Lee, H., and Han, C., 2022. Advanced oxidation processes for microplastics degradation: A recent trend. *Chemical Engineering Journal Advances*, 9, 100213.

Kjeldsen, A., Price, M., Lilley, C., Guzniczak, E., and Archer, I., 2018. A review of standards for biodegradable plastics. *Industrial Biotechnology Innovation Centre*, 33 (1).

Klein, S., Dimzon, I.K., Eubeler, J., and Knepper, T.P., 2018. Analysis, occurrence, and degradation of microplastics in the aqueous environment. *Freshwater Microplastics: Emerging Environmental Contaminants?* 51–67.

Kumar, R.V., Kanna, G.R., and Elumalai, S., 2017. Biodegradation of polyethylene by green photosynthetic microalgae. *Journal of Bioremediation & Biodegradation*, 8 (381), 2.

Lai, J., Huang, H., Lin, M., Xu, Y., Li, X., and Sun, B., 2023. Enzyme catalyzes ester bond synthesis and hydrolysis: The key step for sustainable usage of plastics. *Frontiers in Microbiology*, 13, 1113705.

Lapointe, M., Farner, J.M., Hernandez, L.M., and Tufenkji, N., 2020. Understanding and improving microplastic removal during water treatment: Impact of coagulation and flocculation. *Environmental Science & Technology*, 54 (14), 8719–8727.

Lastovina, T.A., and Budnyk, A.P., 2021. A review of methods for extraction, removal, and stimulated degradation of microplastics. *Journal of Water Process Engineering*, 43, 102209.

Li, Y., Sun, Y., Li, J., Tang, R., Miu, Y., and Ma, X., 2021. Research on the influence of microplastics on marine life. In: *IOP conference series: Earth and environmental science*. IOP Publishing, 012006.

Lucas, N., Bienaime, C., Belloy, C., Queneudec, M., Silvestre, F., and Nava-Saucedo, J.-E., 2008. Polymer biodegradation: Mechanisms and estimation techniques–a review. *Chemosphere*, 73 (4), 429–442.

Lusher, A., Hollman, P., and Mendoza-Hill, J., 2017. *Microplastics in fisheries and aquaculture: Status of knowledge on their occurrence and implications for aquatic organisms and food safety*. FAO.

Mato, T., and Noguchi, S., 2021. Microplastic collection with ultra-high magnetic field magnet by magnetic separation. *IEEE Transactions on Applied Superconductivity*, 32 (4), 1–5.

Misra, A., Zambrzycki, C., Kloker, G., Kotyrba, A., Anjass, M.H., Franco Castillo, I., Mitchell, S.G., Güttel, R., and Streb, C., 2020. Water purification and microplastics removal using magnetic polyoxometalate-supported ionic liquid phases (magPOM-SILPs). *Angewandte Chemie International Edition*, 59 (4), 1601–1605.

Moog, D., Schmitt, J., Senger, J., Zarzycki, J., Rexer, K.-H., Linne, U., Erb, T., and Maier, U.G., 2019. Using a marine microalga as a chassis for polyethylene terephthalate (PET) degradation. *Microbial Cell Factories*, 18 (1), 1–15.

Mor, R., and Sivan, A., 2008. Biofilm formation and partial biodegradation of polystyrene by the actinomycete Rhodococcus ruber: Biodegradation of polystyrene. *Biodegradation*, 19, 851–858.

Na, S.-H., Kim, M.-J., Kim, J.-T., Jeong, S., Lee, S., Chung, J., and Kim, E.-J., 2021. Microplastic removal in conventional drinking water treatment processes: Performance, mechanism, and potential risk. *Water Research*, 202, 117417.

Nabi, I., Ahmad, F., and Zhang, L., 2021. Application of titanium dioxide for the photocatalytic degradation of macro-and micro-plastics: A review. *Journal of Environmental Chemical Engineering*, 9 (5), 105964.

Nandi, R.G., and Joshi, M., 2013. Biodegradation of pretreated polythene by different species of Aspergillus isolated from garbage soil. *Bioscience Biotechnology Research Communications*, 6, 199–201.

Ncube, L.K., Ude, A.U., Ogunmuyiwa, E.N., Zulkifli, R., and Beas, I.N., 2021. An overview of plastic waste generation and management in food packaging industries. *Recycling*, 6 (1), 12.

Nowak, B., Pająk, J., Drozd-Bratkowicz, M., and Rymarz, G., 2011. Microorganisms participating in the biodegradation of modified polyethylene films in different soils under laboratory conditions. *International Biodeterioration & Biodegradation*, 65 (6), 757–767.

Ojha, N., Pradhan, N., Singh, S., Barla, A., Shrivastava, A., Khatua, P., Rai, V., and Bose, S., 2017. Evaluation of HDPE and LDPE degradation by fungus, implemented by statistical optimization. *Scientific Reports*, 7 (1), 39515.

Ouda, M., Banat, F., Hasan, S.W., and Karanikolos, G.N., 2023. Recent advances on nanotechnology-driven strategies for remediation of microplastics and nanoplastics from aqueous environments. *Journal of Water Process Engineering*, 52, 103543.

Oviedo-Anchundia, R., del Castillo, D.S., Naranjo-MorÃ, J., Francois, N., AlarcÃ³n, A., Villafuerte, J.S., and Barcos-Arias, M., 2021. Analysis of the degradation of polyethylene, polystyrene and polyurethane mediated by three filamentous fungi isolated from the Antarctica. *African Journal of Biotechnology*, 20 (2), 66–76.

Padervand, M., Lichtfouse, E., Robert, D., and Wang, C., 2020. Removal of microplastics from the environment. A review. *Environmental Chemistry Letters*, 18, 807–828.

Paredes, D., Kuschk, P., Mbwette, T.S.A., Stange, F., Müller, R.A., and Köser, H., 2007. New aspects of microbial nitrogen transformations in the context of wastewater treatment–a review. *Engineering in Life Sciences*, 7 (1), 13–25.

Pelegrini, K., Pereira, T.C.B., Maraschin, T.G., Teodoro, L.D.S., Basso, N.R.D.S., De Galland, G.L.B., Ligabue, R.A., and Bogo, M.R., 2023. Micro-and nanoplastic toxicity: A review on size, type, source, and test-organism implications. *Science of the Total Environment*, 878, 162954.

Peng, X., Chen, M., Chen, S., Dasgupta, S., Xu, H., Ta, K., Du, M., Li, J., Guo, Z., and Bai, S., 2018. Microplastics contaminate the deepest part of the world's ocean. *Geochemical Perspectives Letters*, 9 (1), 1–5.

Pirsaheb, M., Hossini, H., and Makhdoumi, P., 2020. Review of microplastic occurrence and toxicological effects in marine environment: Experimental evidence of inflammation. *Process Safety and Environmental Protection*, 142, 1–14.

Raaman, N., Rajitha, N., Jayshree, A., and Jegadeesh, R., 2012. Biodegradation of plastic by Aspergillus spp. isolated from polythene polluted sites around Chennai. *Journal of Academia and Industrial Research*, 1 (6), 313–316.

Risch, P., and Adlhart, C., 2021. A chitosan nanofiber sponge for oyster-inspired filtration of microplastics. *ACS Applied Polymer Materials*, 3 (9), 4685–4694.

Sajid, M., Ihsanullah, I., Tariq Khan, M., and Baig, N., 2023. Nanomaterials-based adsorbents for remediation of microplastics and nanoplastics in aqueous media: A review. *Separation and Purification Technology*, 305, 122453.

Saravanan, A., Kumar, P.S., Vo, D.-V.N., Jeevanantham, S., Karishma, S., and Yaashikaa, P.R., 2021. A review on catalytic-enzyme degradation of toxic environmental pollutants: Microbial enzymes. *Journal of Hazardous Materials*, 419, 126451.

Sarkhel, R., Sengupta, S., Das, P., and Bhowal, A., 2020. Comparative biodegradation study of polymer from plastic bottle waste using novel isolated bacteria and fungi from marine source. *Journal of Polymer Research*, 27, 1–8.

Sarmah, P., and Rout, J., 2019. Cyanobacterial degradation of low-density polyethylene (LDPE) by Nostoc carneum isolated from submerged polyethylene surface in domestic sewage water. *Energy, Ecology and Environment*, 4, 240–252.

Shah, A.A., Hasan, F., Hameed, A., and Ahmed, S., 2008. Biological degradation of plastics: A comprehensive review. *Biotechnology Advances*, 26 (3), 246–265.

Siracusa, V., Rocculi, P., Romani, S., and Dalla Rosa, M., 2008. Biodegradable polymers for food packaging: A review. *Trends in Food Science & Technology*, 19 (12), 634–643.

Skariyachan, S., Setlur, A.S., Naik, S.Y., Naik, A.A., Usharani, M., and Vasist, K.S., 2017. Enhanced biodegradation of low and high-density polyethylene by novel bacterial consortia formulated from plastic-contaminated cow dung under thermophilic conditions. *Environmental Science and Pollution Research*, 24, 8443–8457.

Song, Y.K., Hong, S.H., Jang, M., Han, G.M., Jung, S.W., and Shim, W.J., 2017. Combined effects of UV exposure duration and mechanical abrasion on microplastic fragmentation by polymer type. *Environmental Science & Technology*, 51 (8), 4368–4376.

Su, J., Zhang, F., Yu, C., Zhang, Y., Wang, J., Wang, C., Wang, H., and Jiang, H., 2023. Machine learning: Next promising trend for microplastics study. *Journal of Environmental Management*, 344, 118756.

Taghavi, N., Singhal, N., Zhuang, W.-Q., and Baroutian, S., 2021. Degradation of plastic waste using stimulated and naturally occurring microbial strains. *Chemosphere*, 263, 127975.

Talvitie, J., Mikola, A., Koistinen, A., and Setälä, O., 2017. Solutions to microplastic pollution–Removal of microplastics from wastewater effluent with advanced wastewater treatment technologies. *Water Research*, 123, 401–407.

Tang, K.H.D., and Hadibarata, T., 2021. Microplastics removal through water treatment plants: Its feasibility, efficiency, future prospects and enhancement by proper waste management. *Environmental Challenges*, 5, 100264.

Thakur, B., Singh, J., Singh, J., Angmo, D., and Vig, A.P., 2023. Biodegradation of different types of microplastics: Molecular mechanism and degradation efficiency. *Science of the Total Environment*, 877, 162912.

Thomas, B.T., Olanrewaju-Kehinde, D.S.K., Popoola, O.D., and James, E.S., 2015. Degradation of plastic and polythene materials by some selected microorganisms isolated from soil. *World Applied Sciences Journal*, 33 (12), 1888–1891.

Torikai, A., Suzuki, K., and Fueki, K., 1983. Photodegradation of polypropylene and polypropylene containing pyrene. *Polymer Photochemistry*, 3 (5), 379–390.

Urso, M., and Pumera, M., 2022. Nano/microplastics capture and degradation by autonomous nano/microrobots: A perspective. *Advanced Functional Materials*, 32 (20), 2112120.

Wang, F., Zhang, M., Sha, W., Wang, Y., Hao, H., Dou, Y., and Li, Y., 2020. Sorption behavior and mechanisms of organic contaminants to nano and microplastics. *Molecules*, 25 (8), 1827.

Watanabe, T., Ohtake, Y., Asabe, H., Murakami, N., and Furukawa, M., 2009. Biodegradability and degrading microbes of low-density polyethylene. *Journal of Applied Polymer Science*, 111 (1), 551–559.

Wei, R., Breite, D., Song, C., Gräsing, D., Ploss, T., Hille, P., Schwerdtfeger, R., Matysik, J., Schulze, A., and Zimmermann, W., 2019. Biocatalytic degradation efficiency of post-consumer polyethylene terephthalate packaging determined by their polymer microstructures. *Advanced Science*, 6 (14), 1900491.

Wiles, D.M., and Carlsson, D.J., 1980. Photostabilisation mechanisms in polymers: A review. *Polymer Degradation and Stability*, 3 (1), 61–72.

Williams, J.O., and Osahon, N.T., 2021. Assessment of microplastic degrading potential of fungal isolates from an estuary in rivers state, Nigeria. *South Asian Journal of Research in Microbiology*, 9 (2), 11–19.

Woodings, C., 2001. A brief history of regenerated cellulosic fibers. *Regenerated Cellulose Fibers*, 1–21.

Xiouras, C., Cameli, F., Quilló, G.L., Kavousanakis, M.E., Vlachos, D.G., and Stefanidis, G.D., 2022. Applications of artificial intelligence and machine learning algorithms to crystallization. *Chemical Reviews*, 122 (15), 13006–13042.

Yousif, E., and Haddad, R., 2013. Photodegradation and photostabilization of polymers, especially polystyrene. *Springer Plus*, 2 (1), 1–32.

Yu, F., and Hu, X., 2022. Machine learning may accelerate the recognition and control of microplastic pollution: Future prospects. *Journal of Hazardous Materials*, 432, 128730.

Yuan, J., Ma, J., Sun, Y., Zhou, T., Zhao, Y., and Yu, F., 2020. Microbial degradation and other environmental aspects of microplastics/plastics. *Science of the Total Environment*, 715, 136968.

Yuan, Z., Nag, R., and Cummins, E., 2022. Human health concerns regarding microplastics in the aquatic environment-from marine to food systems. *Science of the Total Environment*, 823, 153730.

Zhang, C.W., Zou, W., Yu, H.C., Hao, X.P., Li, G., Li, T., Yang, W., Wu, Z.L., and Zheng, Q., 2022. Manta Ray inspired soft robot fish with tough hydrogels as structural elements. *ACS Applied Materials & Interfaces*, 14 (46), 52430–52439.

Zhang, K., Hamidian, A.H., Tubić, A., Zhang, Y., Fang, J.K.H., Wu, C., and Lam, P.K.S., 2021. Understanding plastic degradation and microplastic formation in the environment: A review. *Environmental Pollution*, 274, 116554.

Zhang, Y., Diehl, A., Lewandowski, A., Gopalakrishnan, K., and Baker, T., 2020. Removal efficiency of micro- and nanoplastics (180 nm–125 µm) during drinking water treatment. *Science of the Total Environment*, 720, 137383.

Zhu, C.N., Zheng, S.Y., Qiu, H.N., Du, C., Du, M., Wu, Z.L., and Zheng, Q., 2021. Plastic-like supramolecular hydrogels with polyelectrolyte/surfactant complexes as physical cross-links. *Macromolecules*, 54 (17), 8052–8066.

Ziajahromi, S., Neale, P.A., Rintoul, L., and Leusch, F.D.L., 2017. Wastewater treatment plants as a pathway for microplastics: Development of a new approach to sample wastewater-based microplastics. *Water Research*, 112, 93–99.

8 Exploring Microplastic and Natural Fiber Emissions from Fabrics and Textiles

Sirat Sandil, Vineet Kumar, and Gyula Zaray

8.1 INTRODUCTION

Since the 1940s, when plastic production commenced, the rate of plastic production has rapidly enhanced, reaching 465 million metric tons globally in 2019 (Carney Almroth et al., 2018; Geyer et al., 2022). Plastics have a multitude of applications due to properties like durability and resistance to degradation; however, the same features also contribute to their persistence in nature (Galvão et al., 2020). Plastics, including microplastics (MPs), have become ubiquitous in our natural environment (Cesa et al., 2020), to the point where they are now considered a defining characteristic of the Anthropocene (Belzagui et al., 2019) and have been underlined as contaminants of global concern (Coppock et al., 2019). Microplastics comprise plastic particles <5 mm in size arising from the degradation of larger plastics (secondary MPs), as well as those intentionally manufactured in the concerned size (primary MPs) (Geyer et al., 2022). In the environment, MPs are encountered in an array of morphologies that include fragments, films, pellets, beads, foams, and fibers (Mahbub and Shams, 2022). Of these, microplastic fibers (MPFs) comprise the largest fraction of MPs in the marine, freshwater, and terrestrial environments (Coppock et al., 2019; Cai et al., 2020). MPFs are easily ingested by organisms and thereby pose significant adverse effects in microorganisms, animals, plants, and humans and thus affect ecosystem functioning (Zhang et al., 2022).

The textile industry plays a substantial role in the global economic landscape, providing a wide range of products from clothing to technical textiles. This sector accounts for the third largest share (15%) of worldwide plastic consumption as a result of utilizing single-use plastics in textiles (Berruezo et al., 2021; Sudheshna et al., 2022). Over the last ten years, the increased usage of synthetic fabrics has skyrocketed their demand, and their compound annual growth rate (CAGR) is predicted to rise from 4% to more than 7% within a decade. However, this industry also has a notable environmental footprint, particularly concerning microfibers. Microfibers (MFs), both synthetic (MPFs) and natural, are shed from synthetic and natural fabrics, respectively, during various stages of their life cycle, particularly laundry

 DOI: 10.1201/9781032684574-8

(Zhang et al., 2021; Raja Balasaraswathi and Rathinamoorthy, 2022). Laundry, using household washing machines, is regarded as the principal contributor of MFs in the environment. Currently, about 840 million units of washing machines are in use globally, and this number is expected to rise (Cesa et al., 2020). The laundry effluents are guided to the wastewater treatment plants (WWTPs), but it has been discovered that WWTPs inadequately retain MFs (Bayo et al., 2021). Hence a significant quantity of MFs are released into the environment through WWTP effluents and, additionally, through sludge (Cesa et al., 2020; Cai et al., 2020). The presence of MFs in the environment has been documented to have detrimental impacts on aquatic and terrestrial organisms, plants, soil, and human beings (Qiao et al., 2019; De Souza Machado et al., 2019; Yee et al., 2021). This chapter delves into the world of MPFs and natural MFs, examining their characteristics, sources, emissions, impacts, and potential solutions.

8.2 SOURCES OF MICROFIBERS

In the textile industry, three primary types of fibers are employed: (1) natural fibers stemming mainly from plants (cotton, linen) and animals (wool, silk), (2) regenerated fibers derived from dissolved plant cellulosic materials and shaped into fibers through extrusion (viscose, lyocell, modal, rayon), and (3) synthetic textiles (plastic) derived from petrochemical products (e.g., polyester, polypropylene, polyamide, acrylic) (Napper and Thompson, 2016; Stanton et al., 2019; Zambrano et al., 2021). In 2018, worldwide fiber production amounted to 111 million tons, with synthetic fibers encompassing 65% of the entirety, natural fibers making up 29%, and regenerated fibers comprising 6% (Dalla Fontana et al., 2021; Zhang et al., 2021). Among synthetic textiles, polyester dominates the market, constituting 52% of worldwide fiber production with an annual output of 57 million tonnes (Sudheshna et al., 2022). Cotton is an important natural textile fiber with the second highest rate of production and consumption globally. Fibers find usage in the production of garments, bed linen, blankets, towels, carpets, upholstery, etc. (Napper and Thompson, 2016). Tragically, washing these textiles releases colossal amounts of MPFs and natural MFs into the environment, circumventing washing machine filters and WWTPs (Dalla Fontana et al., 2021).

The domestic washing of textiles is a significant driver of MPFs and natural MFs in the environment (Cai et al., 2020). Galvão et al., 2020 estimated the release of about 18,000,000 MPFs from a washing load of 6 kg, whereas Belzagui et al. (2019) appraised that 1.4×10^{17} MFs/year were reaching the oceans solely from domestic laundering. Additionally, De Falco et al. (2020) established that the discharge of MFs from clothing to the atmosphere as a result of wear constituted an equally important route. While contemporary research has largely focused on MP contamination spurred by MPFs, there are scarcely any investigations on the detrimental influences of MFs stemming from natural textiles or regenerated ones. This is despite several studies reporting the dominance of natural MFs in the MFs encountered in the environment (Zambrano et al., 2021). Stanton et al. (2019) observed that natural MFs constituted 93.8% of the MFs collected in the River Trent, UK. Additionally, in a majority of studies, the natural MFs are ignored and only the number of MPFs are reported. This is primarily because natural fibers are deemed biodegradable and therefore innocuous, due to which any probable risk posed by them has been disregarded (Zambrano et al., 2021). However, even the natural and regenerated fabrics are treated with chemicals like dyes and finishing materials along with surface modifications by chemicals, similar to synthetic textiles (Zambrano et al., 2021). They can thus act as vectors of the contaminants and readily release the toxic chemicals into the environment due to their shorter lifespan, posing a risk to the biota (Stanton et al., 2019). It is therefore crucial to probe the adverse consequences of natural MFs as well.

Another noteworthy contributor of MPs in the environment is WWTPs, with MPFs comprising the majority of MPs encountered in WWTPs' effluent and sludge (79.65–82%)(Lares et al., 2018; Bayo et al., 2021). The MFs arise from the use and care of synthetic and natural clothing materials and laundering activities (Ziajahromi et al., 2017; Zhang et al., 2022), and WWTPs effluent and sludge are prominent sources of MFs in the aquatic and terrestrial environments (Ziajahromi et al., 2017; Cesa et al., 2020).

8.3 EMISSIONS OF MICROFIBERS FROM FABRICS AND TEXTILES

The vast majority of MFs pollution in the environment can be attributed to the textile industry, domestic washing of clothes, household effluent, and disposal of clothes (Mishra et al., 2022). The MFs comprise MPFs and natural MFs emanating from synthetic fabrics and natural textiles, respectively (Sudheshna et al., 2022). These MFs are shed by garments and apparel throughout their life cycle from textile production, clothing manufacturing, usage, laundry, and disposal (Belzagui et al., 2019).

The term 'shedding' refers to the weakening, extrication, and removal of fibers from the fabric, mainly during the washing process when the textile is agitated (Carney Almroth et al., 2018; Geyer et al., 2022). This shedding is prompted by the centrifugal force, mechanical tension, and chemical stressors that textiles experience in the washing machine (Volgare et al., 2021). These stresses cause furring or pilling of the unprotected surfaces. The formation of MFs involves the following stages: fuzzing, fibrillation, pill formation, and pill detachment. The amount of MFs liberated by the fabric depends on the properties of the fabrics like type, chemical composition, compactness, and age (Napper and Thompson, 2016; Gaylarde et al., 2021).

Cai et al. (2020) investigated the release of MPFs during the manufacturing process and confirmed that MPFs were released during manufacture depending on the techniques employed. During yarn production and textile manufacturing, rotor spinning and surface treatment were essential steps regulating MPFs formation, as rotor yarn and surface processed textiles (fleece, microfiber, plain brushed) resulted in a higher number of MPFs. Similarly, the cutting methods also had a critical effect on MPFs emissions, with scissor-cut textiles releasing more MPFs (3–31 times) than laser-cut ones. Microfiber release during the manufacturing process was quantified by Grillo et al. (2023), wherein a denim garment underwent industrial washing, and the MFs released were 10.95 times more than observed in household washing.

During domestic laundry, a multitude of factors have been reported to regulate the release of MFs (Figure 8.2), mainly the physicochemical characteristics of the

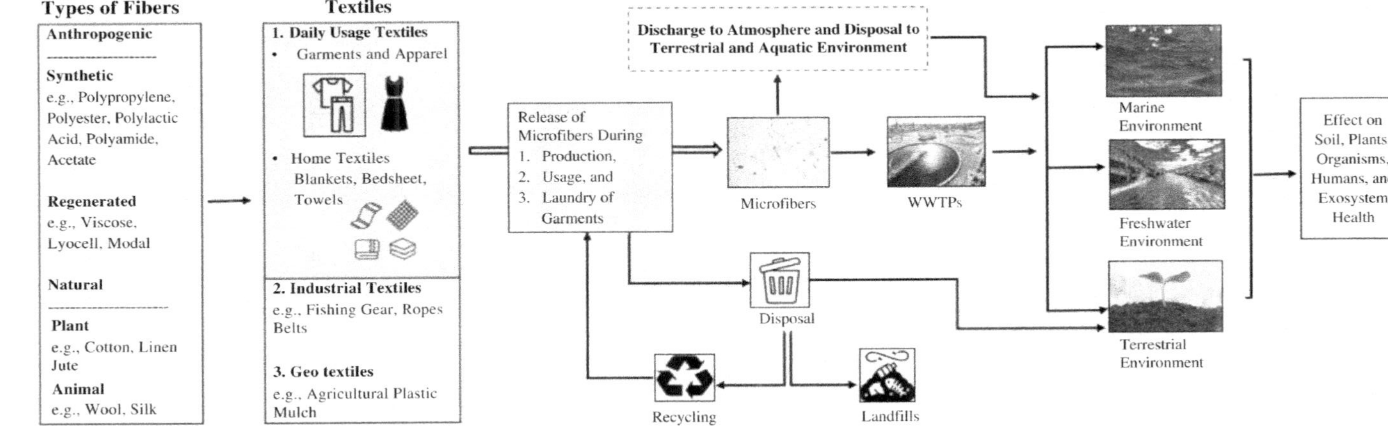

FIGURE 8.1 Types of fibers, textiles, emission of microfibers, and their environmental fates.

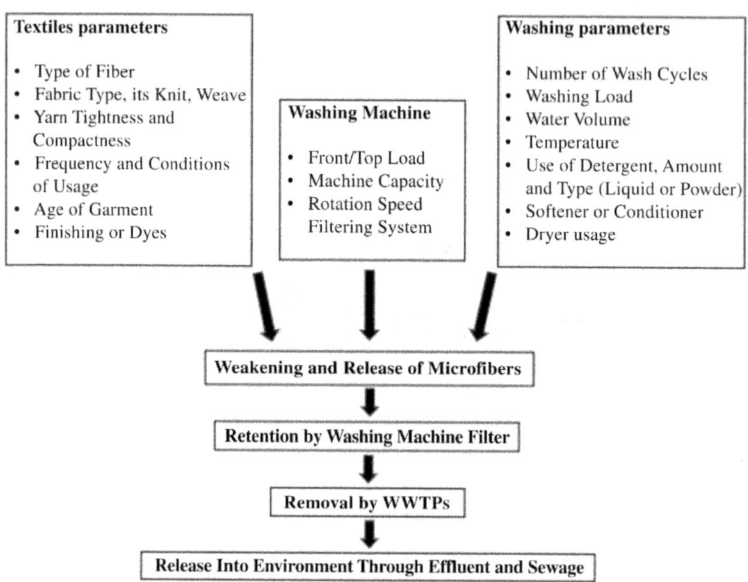

FIGURE 8.2 Parameters influencing microfibers release during laundry.

fiber, type of fabric (woven, knit), texture (dense, loose), the knit/weave/stitch of fabric, yarn type (Napper and Thompson, 2016; Zambrano et al., 2019), and the usage of finishing materials or dyes (Hernandez et al., 2017). Other laundry parameters that influence MFs release include washing machine design (Sudheshna et al., 2022), number of wash cycles (Belzagui et al., 2019), washing time (Hernandez et al., 2017), volume of washing load (Galvão et al., 2020), volume of water used (Rathinamoorthy and Balasaraswathi, 2022), water temperature (Zambrano et al., 2019), detergent type and volume (De Falco et al., 2018), use of softener (Zambrano et al., 2021), and usage of dryer (Pirc et al., 2016).

8.3.1 Microfibers from Domestic Laundering

The domestic washing of fabrics is the key driver of MFs in the environment (Geyer et al., 2022). The act of washing can lead to greater fabric damage through abrasion than regular wear and tear, resulting in the release of more MFs (Yang et al., 2019). Abrasion on the fabric's surface intensifies mechanical stress, disrupts the structure of MFs, and consequently releases a greater quantity of MFs (Rathinamoorthy and Balasaraswathi, 2022). Numerous studies (see Table 8.1) have narrated the release of MFs during the domestic laundering procedure. The quantification of MFs arising from the washing process is a tedious task due to the enormous quantity of MFs released. This task is further complicated due to their small sizes, lack of standard methodology, and the presence of MPFs as well as natural MFs. While the studies have documented the MFs released from different types of fabrics, various washing techniques, and laundering conditions, the employment of varying

TABLE 8.1
Microfibers Detected in Different Studies

Sample Material	Laundry Method	Laundry Parameters (Time, Temperature, Wash Cycles, Detergent Usage)	Filter Size (μm)	Detection Method	Amount of MFs	Other observations	References
PE, acrylic, PE-cotton	Washing machine	75 min; 30, 40°C; detergent and conditioner	25	SEM	PE-cotton: 137,951 MFs/6 kg*; PE: 496,030 MFs/6 kg; Acrylic: 728,789 MFs/6 kg	MFs dimension: 11.9–17.7 μm diameter and 5–7.8 μm length; Significant effect of wash conditions on MFs release	Napper and Thompson (2016)
PE fleece	Front-load washing machine and tumble dryer	15 min; 30°C; 10 cycles; detergent and softener	200	FTIR, SEM, Stereomicroscope	0.0012 wt%/wash	MFs dimension: 20–200 μm; 3.5 times higher MFs during tumble drying	Pirc et al. (2016)
PE, PE-Spandex	Gyrowash	45 min; 40°C; 5 wash cycles; detergent; and 1, 2, 4, 8 h; 25, 40, 60, 80°C; detergent	0.45	Digital microscope	0.025 mg MPFs/g textile in every wash	MFs dimension: <1000 μm; no significant effect of washing temperature or duration on MFs release	Hernandez et al. (2017)
PE, Cotton	Front-load washing machine	75 min; 40°C; 5 wash cycles; detergent	0.7	Stereoscopic microscope	PE: 2.23×10^5 particles/wash or 340 mg/wash Cotton: 9.73×10^5/wash or 809 mg/wash	Yearly release of PE and cotton MFs from domestic washing machines was 154,000 kg and 411,000 kg	Sillanpää and Sainio (2017)
Acrylic, nylon, PE	Gyrowash	30 min, 60°C; 2, 5, 20 wash; liquid detergent	1.2	Light microscope	PE fleece: 7360 MPFs/m^{2}L/wash cycle	MFs dimension: 25–3000 μm; decrease in MFs with increasing wash cycles; higher MFs release with detergent usage	Carney Almroth et al. (2018)

(Continued)

TABLE 8.1 (*Continued*)
Microfibers Detected in Different Studies

Sample Material	Laundry Method	Laundry Parameters (Time, Temperature, Wash Cycles, Detergent Usage)	Filter Size (μm)	Detection Method	Amount of MFs	Other observations	References
PE, PP	Gyrowash	45, 60 mins; 40, 75°C; 5 wash cycles; liquid and powder detergent	5	FTIR, stereomicroscope, SEM	6,000,000 MPFs/5 kg wash load	MFs dimension: 340 ± 292 μm length and 14 ± 3 μm diameter; increased MFs release in case of detergent usage, comparatively higher in case of powder detergent	De Falco et al. (2018)
PE, PE-elastane, acrylic-PA	Front-load washing machine	15 min; ambient temp.; detergent	20	Stereomicroscope, SEM	PE-elastane: 175 MPFs/g/wash; acrylic-PA: 560 MPFs/g/wash	MFs dimension: 200–400 μm; high correlation between MFs release and textile density	Belzagui et al. (2019)
Cotton, PE, rayon	Gyrowash	16 min; 25 and 44°C; detergent	1.2	Fiber quality analyzer	Cellulose-based: 0.2–4 mg/g fabric PE: 0.1–1 mg/g fabric	MFs dimension: >200 μm; Higher MFs shedding with detergent usage and high water temperature	Zambrano et al. (2019)
PE, recycled PE, PE-cotton-modal	Washing machine	107 min; 40°C; 10 wash cycles; detergent	5, 20, 60	SEM	PE: 640,000 MFs/garment/wash; PE-cotton-modal*: 1,500,000 MFs/garment/wash	Type of fiber (yarn and twist) influenced MFs release	De Falco et al. (2019)
PE, PA, acetate	Platen and pulsator washing machine	15 min; 30, 40, 60°C; detergent	5	Stereomicroscope	Acetate: 74,816 MPFs/m^2/wash (with detergent); 72,130 MPFs/m^2/wash (without detergent)	Both washing machines generated huge amounts of MFs; washing temperature and detergent enhanced MFs shedding	Yang et al. (2019)

PE	Ultrasonic extraction	90 min; 23–27°C; 9 extractions; detergent	0.45	SEM	15 MPFs/g for a filament and 45,400 MPFs/g for a scissor-cut microfiber textile	Fiber dimension: 100–800 μm; large amount of MFs produced during manufacturing	Cai et al. (2020)
PE, PE-cotton	Washing machine	107 min; 40°C; detergent	5, 20, 50, 400	Optical microscope and SEM	PE: 709–1747 MFs/g/wash; PE-cotton: 3898 MFs/g/wash	Garment with compact weave, highly twisted yarn, and continuous filament released less MFs	De Falco et al. (2020)
Cotton, acrylic, PE, PA	Washing machine	40 min; 24°C; 10 wash; detergent	450	Biological microscope	Cotton and synthetic textiles: 285.65 and 112.35–267.90 mg/10 cycles	Detergent reduced MFs emission significantly; 10 sequential wash cycles reduced MFs release	Cesa et al. (2020)
Mixture of cotton, wool, PE, viscose, elastane, acrylic (clothing material and home textile items)	Top load and front load washing machine	In-built setting; detergent	0.7 and 2.7	Stereomicroscope, SEM, FTIR	Front-load: 7453635 MF/7 kg* and 7375500 MFs/6 kg; Top load: 10692255 MFs/7 kg and 7,589,017 MFs/6.2 kg	Synthetic fiber emission rate was 19%; 48.64% of MFs length was in the range <5 μm	Sudheshna et al. (2022)

* MFs: PE-polyester, PA-polyamide, PP-polypropylene, FTIR-Fourier transform infrared spectroscopy, SEM-scanning electron microscopy

fiber characterization techniques and units of quantification render any comparison impossible (Lares et al., 2018; Gaylarde et al., 2021).

8.3.2 Factors Influencing Microfibers Release

The amount of MFs emitted during laundry is contingent mainly on the type of textile (polymer) and the structure of fibers in the fabric. Carney Almroth et al. (2018) determined the release of MPFs from three synthetic textiles (acrylic, nylon, and polyester) with differing fiber structure (knit, microfleece, and fleece) and observed that polyester fleece fabric discharged the maximum amount of MPFs (7360 MPFs/m^2/L/wash) as compared to polyester fabric (87 MPFs/m^2/L/wash). Similarly, Yang et al. (2019) reported that among polyester, polyamide, and acetate fabrics, the maximum MFs (74,816 MPFs/m^2/wash) were shed by acetate fibers. However, Zambrano et al. (2019) observed significantly more MFs in the case of cotton, polyester-cotton, and rayon (0.2–4 mg/g fabric) as compared to pure polyester fabrics (0.1–1 mg/g fabric). Likewise, Sillanpää and Sainio (2017) noted that cotton garments released a higher number of MFs as compared to polyester.

Fabric physical properties have been noted to be an important parameter determining MFs shedding rate. The fabrics with greater compactness, higher stitch density, higher tightness, lower loop length, and lesser thickness have been reported to lose fewer MFs (Raja Balasaraswathi and Rathinamoorthy, 2022). Berruezo et al. (2021) and De Falco et al. (2020) reported that the loss of MFs was minimized by increased density of yarn (fibers' compactness) and the presence of amply twisted yarns fashioned of continuous filaments. Additionally, loose and aged textiles shed more MFs throughout the washing procedure (Carney Almroth et al., 2018). Other fabric characters influencing MFs release are lock stitch, fiber cohesion, tenacity, and size, interlacing, and yarn twist (Cesa et al., 2020; Raja Balasaraswathi and Rathinamoorthy, 2022). On the other hand, Dalla Fontana et al (2021) mentioned that sewing thread and edge effect had a stronger influence on MFs release, masking even the effect of textile geometry and piling tendency.

The emission of MFs has been observed to decrease with increasing wash cycles (Cesa et al., 2020). Belzagui et al (2019) reported that the MF detachment rate was reduced with the number of wash cycles until it stabilized during the fifth cycle. The larger quantity of MFs in the initial wash cycles was probably due to the existence of MFs left over during the manufacturing process or due to MFs attached, entwined, or torn from the fiber structure as a result of the mechanical force placed upon the garment during laundering. On the other hand, an increase in the washing and drying time has been found to increase the release of MFs due to prolonged mechanical stress and rotational force on the fabric (Mahbub and Shams, 2022).

The washing load profoundly impacts the emission of MFs with the amount of MFs increasing by over five times with decreasing washing load. In the case of a lower load, the mechanical stress on the garment is enhanced and a synergistic effect occurs between water volume and fabric (Volgare et al., 2021). The amount of water utilized during washing is another major factor influencing MFs release; a higher volume of water intensifies the impact of hydrostatic force and viscous drag force, which hinders the fabric's free movement within the liquid and results in heightened shedding (Rathinamoorthy and Balasaraswathi, 2022). The application of detergent during laundry significantly enhances the shedding of MFs (Carney Almroth et al., 2018; De Falco et al., 2018). Yang et al. (2019) noted that washing with detergents boosted the MFs release from fabrics by 1.2–5.6-fold. An augmented MFs loss due to detergent occurs because detergents diminish surface tension and act as dispersion agents, carrying away shed MFs (Carney Almroth et al., 2018). The use of softeners and durable press also causes an increase in the emission of MFs by altering mechanical properties of the fabric such as abrasion resistance and friction coefficient. Cotton fabric treated with softener and durable press released 1.3–1.63 mg/g fabric while untreated fabric released 0.73 mg/g fabric (Zambrano et al., 2021). Contrarily, De Falco et al. (2018) witnessed a diminution in the number of MFs due to softener usage. An increase in the temperature of the water used for washing also enhances MFs release (Zambrano et al., 2019). An increase in temperature from 25 to 44°C significantly enhanced the release of MFs by a cotton garment (Zambrano et al., 2019). This is because an increase in water temperature encourages the expansion of cellulosic fibers, creating a space for the movement of broken strands and the development of fuzz (Zambrano et al., 2019). However, the emission of polyester MFs was also greatly enhanced while washing in hot water 54% as opposed to cold water (30°C) (Yang et al., 2019).

The emission of MFs is affected by the type of washing machine (top-load or front-load) because of their varying levels of agitation. Top-load machines, in particular, release over 10% more MFs compared to front-load machines (Sudheshna et al., 2022). While the washing process has been documented to aid in the release of large amounts of MFs, studies have reported an enhanced emission of MFs during the drying process. Pirc et al. (2016) observed that tumble drying caused the discharge of 3.5 times more MFs as opposed to laundry. The clothing undergoes more mechanical stress when tumble drying than during washing, hence the MFs emission would be higher. This is further facilitated by the loosening or weakening of fibers on the cloth surface during washing (Kärkkäinen and Sillanpää, 2021).

In literature, a broad spectrum of MF concentrations have been documented due to the use of filters with varying pore sizes. Applying a pore size of 60 μm, De Falco et al. (2019) conveyed the discharge of $640,000 - 1,500,000$ MFs/kg of fabric. Contrarily, Pirc et al. (2016) and Cesa et al. (2020) employed filters with larger pore sizes (200 and 450 μm) and noted much lower amounts of MFs. Due to the dearth of appropriate filters in the washing machine, MFs are able to bypass them and are subsequently channeled into either WWTPs or, in numerous instances, directly into the water source (Mishra et al., 2022).

8.4 TRANSPORT, ENVIRONMENTAL FATE, AND ECOTOXICITY

The majority of MFs arising from domestic or industrial sources are discharged to the WWTPs, while a smaller proportion enters the environment directly without any treatment (Bayo et al., 2021). However, WWTPs also serve as a source of MFs to the aquatic and terrestrial environments as a significant amount of MFs evade treatment and end up in the environment through the discharge of effluent and sewage sludge (Bayo et al., 2021). The WWTPs are able to eliminate up to 90–98% of

MPs; however, the removal efficiency of MPFs is poor (56%) (Lares et al., 2018; Bayo et al., 2021). An estimated 1.6×10^7 MPs make their way to the aquatic bodies every year through effluent (Bayo et al., 2021), despite the retention of up to 70–99% in the sludge (Carney Almroth et al., 2018). Since sludge is utilized as fertilizer in agricultural fields, the MFs retained in the sludge can also be delivered to the terrestrial environment, resulting in the dispersion of MFs in the terrestrial environment and consequent runoff into the water bodies (Cesa et al., 2020). Geyer et al. (2022) appraised that in 2019, 2.2 kilotons of MPFs were released from laundry in California, of which 95% was diverted by the WWTPs from the aquatic environment to the terrestrial one through the agricultural application of biosolids.

Microfibers are prevalent all over the world in dissimilar environmental matrices, such as water, sediment, soil, and organisms (Carney Almroth et al., 2018). They are actively ingested by organisms such as fish, crab, lobster, zooplankton, and soil nematodes (Table 8.2) due to their small sizes comparable to that of the planktons (food) and sediments (Au et al., 2015; Woods et al., 2020). Among MPs of similar sizes but different shapes, MPFs are uptaken most readily and are found in the largest quantities in the gut of organisms. Qiao et al. (2019) reported the shape-dependent accretion of MPs in the gut of zebrafish in the order: MPFs (8 μg/mg) > fragments (1.7 μg/mg) > beads (0.5 μg/mg). While MPFs have been documented as the predominant form of MPs in the environment in a large number of studies (Horn et al., 2020), the examination of the influences of MFs is scarce. MPFs have been documented to adversely affect organisms in multiple ways, such as reduction of feeding activity, modification of feeding behavior, alteration of prey selection, damage to the intestines of organisms, and negative effect on growth, reproduction, embryo development, and egg survival (Ziajahromi et al., 2017; Coppock et al., 2019; Qiao et al., 2019) (Table 8.2). MPFs, due to their shape and flexibility, also exert more severe toxicity effects as compared to other MP shapes like fragments or beads (Ziajahromi et al., 2017; Qiao et al., 2019; De Souza Machado et al., 2019). In the freshwater environment, Au et al. (2015) compared the effect of polyethylene particles and polypropylene fibers on the amphipod *Hyalella azteca* and reported that the MPFs exhibited greater toxicity to the organism due to longer residence time in the gut. The extended residence time of MPFs in the gut probably affected the organism's ability to intake food, resulting in lower energy and thus less growth. In zebrafish, the accretion of polypropylene MPFs (25 μm) in the gut resulted in an array of toxicity effects in the intestine, including impairment of the mucosal layer, augmented permeability and inflammation, and disturbed metabolism (Qiao et al., 2019). Moreover, the MPFs transformed the gut microbiota, altering the quantity of certain bacteria (Qiao et al., 2019). Polypropylene fibers were noted to enhance the mortality of adult crabs, reduce the retention of egg clutches, and affect embryo development rates (Horn et al., 2020). Nylon fibers were observed to cause a 6% decline in the intake of feed (algae) by the marine copepod *Calanus helgolandicus*. Further, the copepod's feeding behavior was also modified, and the organism avoided the consumption of algae comparable in shape and size to MPs in order to avoid consuming plastic (Coppock et al., 2019). The uptake of MPFs by organisms is dependent on MPF concentration, the growth stage of the organism, and the availability of food (Woods et al., 2020). In the terrestrial environment, MPFs alter soil properties, such as soil bulk density, soil aggregation,

TABLE 8.2
Effect of Microplastic Fibers (MPFs) on Organisms, Soil, and Plants

Organism	Type of Microfibers, Length and Diameter, Duration of Experiment, Concentration	Observed Effects	Reference
Marine environment			
Copepod (*Calanus helgolandicus*)	Nylon MPFs; 40 μm and 10 μm, 23 μm and 100 μm; 24 h; 100 MPFs/mL	6% decline in consumption of similar algae; alteration of feeding activity; modification of prey selection	Coppock et al. (2019)
Pacific mole crab (*Emerita analoga*)	PP MPFs; 1 mm; 71 days; 3 MPFs/L	Enhanced adult crab mortality, diminished retention of egg clutches, variation in embryo development	Horn et al. (2020)
American lobster larvae (*Homarus americanus*)	PET MPFs; 459 μm; 5 and 10 days; 1, 10, 25 MPFs/mL	Decreased survival of early larvae; lower oxygen consumption in late larval stages	Woods et al. (2020)
Freshwater environment			
Amphipod (*Hyalella azteca*)	PE MP particles; 10–27 μm; 10 days; 0–100,000 MPs/mL PP MPFs; 20–75 μm and 20 μm; 10 days; 0–90 MPFs/mL	More toxicity of PP MPFs than PE particles; significantly less growth on exposure to MPFs	Au et al. (2015)
Waterflea (*Ceriodaphnia dubia*)	PES MPFs; 100–400 μm; 48 h; Acute (0.125–4 mg/L) and chronic (31.25–1000 μg/L)	Dose-dependent effect on mortality during acute exposure and on growth and reproduction during chronic exposure; reduced reproductive capacity; carapace and antenna deformities	Ziajahromi et al. (2017)

(Continued)

TABLE 8.2 (*Continued*)
Effect of Microplastic Fibers (MPFs) on Organisms, Soil, and Plants

Organism	Type of Microfibers, Length and Diameter, Duration of Experiment, Concentration	Observed Effects	Reference
Zebrafish (*Danio rerio*)	PP MPFs; 25 μm; 24 h; 20 mg/L	Damage to mucosal layer, inflammation, enhanced permeability, alteration of metabolism and gut microbiota	Qiao et al. (2019)
Terrestrial environment			
Spring onion (*Allium fistulosum*)	PES MPFs; 5000 μm and 8 μm; 2 months; 0.2%	Transformed soil properties (bulk density, soil aggregation, water holding capacity); increase in root biomass, change in leaf C and N content and C-N ratio	De Souza Machado et al. (2019)
Nematode (*Caenorhabditis elegans*)	Polyacrylicnitrile MPFs; 630 μm; 24 hours; 0.001, 0.01, 0.1%	High toxicity of MPFs on *C. elegans* in soil, especially after a long duration of wet and dry cycles	Kim et al. (2020)
7 plants (*Festuca brevipila, Holcus lanatus, Calamagrostis epigejos, Achillea millefolium, Hieracium pilosella, Plantago lanceolata, Potentilla argentea*	PES MPFs; 1280 μm and 30 μm; 2 months; 0.4%	MPFs reduced soil bulk density and improved aeration; increased root and shoot mass; effect on plant community structure and species dominance	Lozano and Rillig (2020)

aeration, and water-holding capacity (De Souza Machado et al., 2019; Lozano and Rillig, 2020) and cause toxicity to soil nematodes (Kim et al., 2020). In plants, polyester MPFs were found to alter the root structure, increase root biomass, and change the content of C and N as well as the C-N ratio in leaves (De Souza Machado et al., 2019). They were also found to alter the plant community structure, wherein MPFs increased the dominance of invasive species (*Calamagrsotis*) and decreased the biomass of species (*Holcus*) facilitating the growth of other species (Lozano and Rillig, 2020). The detrimental effects of MPFs on organisms have additional consequences for other organisms higher up in the trophic chain (Horn et al., 2020). The presence of MPFs in the trophic chain can pose health risks to human beings as well. These adverse effects include physical stress, inflammation, oxidative stress, altered immune response, and apoptosis (Yee et al., 2021). The existing limited literature has focused on the toxicity of only MPFs despite both MPFs and natural MFs posing similar toxicities due to their small sizes.

Microfibers in the environment can result in several other negative consequences. MPs contain chemicals added in the form of additives to improve their properties and also tend to adsorb contaminants from the surroundings owing to their high surface-area-to-volume ratio. MPFs can thus transport, magnify, and leach contaminants inside the organism's body and the surrounding environment (Lares et al., 2018; Belzagui et al., 2019). They act as vectors for certain microorganisms, allowing them to flourish and form a distinct ecosystem with an altered abundance of microorganisms termed 'plastisphere' (Yee et al., 2021). The abundance and potential hosts of extracellular antibiotic resistance genes (eARGs) in the sewage sludge were observed to increase in response to exposure to MPFs. The ARGs are accountable for the pervasiveness of antibiotic resistance bacteria (ARB), which renders antibiotics ineffective and presents health hazards including pathogenic bacterial infections (Zhang et al., 2022).

8.5 MITIGATION AND CONCLUSION

A number of strategies have been proposed to mitigate the release of MFs at the different stages of the textile life cycle:

- Improvement of manufacturing practices (production of durable textiles, application of appropriate finishing materials, prewashing at point source), and development of sustainable textiles (Carney Almroth et al., 2018; Belzagui et al., 2019; Cai et al., 2020; Le et al., 2022)
- Upgrade of the washing machine with better filters to reduce MFs release (in-drum devices like GuppyFriend washing bag, Cora Ball, and Fourth Element washing bag can trap 54%, 31%, and 21% MFs, respectively, while external filters like XFiltra, Lint LUV-R, and Planet Care can remove 78%, 29%, ad 25% MFs, respectively) (Napper et al., 2020; Kärkkäinen and Sillanpää, 2021).
- Optimization of washing parameters (lower detergent use, low washing temperature, appropriate washing load, and water volume)
- Improved removal by WWTPs (by applying membrane bioreactors, sand filters, and disc filters) (Gaylarde et al., 2021)

- Regulatory measures and industry initiatives (such as implementation of policies to minimize MFs emission) (Gaylarde et al., 2021)
- Consumer awareness (reduction of synthetic textiles usage and enhancement of natural textiles use) (Le et al., 2022)

The shedding of MPFs and natural MFs from fabrics is a multifaceted challenge with far-reaching implications for the environment and society. Hence solving this problem requires a holistic approach involving the entire life cycle of textiles, beginning from the textile industry to consumers and to WWTPs. As we navigate this complex landscape, understanding the sources, impacts, and potential solutions associated with MFs becomes crucial for promoting sustainable practices within the textile industry and minimizing the environmental footprint of our clothing and textiles.

ACKNOWLEDGMENT

This research was supported by the National Laboratory for Water Science and Water Security (RRF-2.3.1-21-2022-00008).

REFERENCES

Au, S.Y., Bruce, T.F., Bridges, W.C., Klaine, S.J., 2015. Responses of Hyalella azteca to acute and chronic microplastic exposures. *Environ. Toxicol. Chem.* 34, 2564–2572. https://doi.org/10.1002/etc.3093

Bayo, J., Olmos, S., López-Castellanos, J., 2021. Assessment of microplastics in a municipal wastewater treatment plant with tertiary treatment: Removal efficiencies and loading per day into the environment. *Water (Switzerland).* 13. https://doi.org/10.3390/w13101339

Belzagui, F., Crespi, M., Álvarez, A., Gutiérrez-Bouzán, C., Vilaseca, M., 2019. Microplastics' emissions: Microfibers' detachment from textile garments. *Environ. Pollut.* 248, 1028–1035. https://doi.org/10.1016/j.envpol.2019.02.059

Berruezo, M., Bonet-Aracil, M., Montava, I., Bou-Belda, E., Díaz-García, P., Gisbert-Payá, J., 2021. Preliminary study of weave pattern influence on microplastics from fabric laundering. *Text. Res. J.* 91, 1037–1045. https://doi.org/10.1177/0040517520965708

Cai, Y., Mitrano, D.M., Heuberger, M., Hufenus, R., Nowack, B., 2020. The origin of microplastic fiber in polyester textiles: The textile production process matters. *J. Clean. Prod.* 267, 121970. https://doi.org/10.1016/j.jclepro.2020.121970

Carney Almroth, B.M., Åström, L., Roslund, S., Petersson, H., Johansson, M., Persson, N.K., 2018. Quantifying shedding of synthetic fibers from textiles; a source of microplastics released into the environment. *Environ. Sci. Pollut. Res.* 25, 1191–1199. https://doi.org/10.1007/s11356-017-0528-7

Cesa, F.S., Turra, A., Checon, H.H., Leonardi, B., Baruque-Ramos, J., 2020. Laundering and textile parameters influence fibers release in household washings. *Environ. Pollut.* 257. https://doi.org/10.1016/j.envpol.2019.113553

Coppock, R.L., Galloway, T.S., Cole, M., Fileman, E.S., Queirós, A.M., Lindeque, P.K., 2019. Microplastics alter feeding selectivity and faecal density in the copepod, Calanus helgolandicus. *Sci. Total Environ.* 687, 780–789. https://doi.org/10.1016/j.scitotenv.2019.06.009

Dalla Fontana, G., Mossotti, R., Montarsolo, A., 2021. Influence of sewing on microplastic release from textiles during washing. *Water. Air. Soil Pollut.* 232. https://doi.org/10.1007/s11270-021-04995-7

De Falco, F., Cocca, M., Avella, M., Thompson, R.C., 2020. Microfiber release to water, via laundering, and to air, via everyday use: A comparison between polyester clothing with differing textile parameters. *Environ. Sci. Technol.* 54, 3288–3296. https://doi.org/10.1021/acs.est.9b06892

De Falco, F., Di Pace, E., Cocca, M., Avella, M., 2019. The contribution of washing processes of synthetic clothes to microplastic pollution. *Sci. Rep.* 9, 1–11. https://doi.org/10.1038/s41598-019-43023-x

De Falco, F., Gullo, M.P., Gentile, G., Di Pace, E., Cocca, M., Gelabert, L., Brouta-Agnésa, M., Rovira, A., Escudero, R., Villalba, R., Mossotti, R., Montarsolo, A., Gavignano, S., Tonin, C., Avella, M., 2018. Evaluation of microplastic release caused by textile washing processes of synthetic fabrics. *Environ. Pollut.* 236, 916–925. https://doi.org/10.1016/j.envpol.2017.10.057

De Souza Machado, A.A., Lau, C.W., Kloas, W., Bergmann, J., Bachelier, J.B., Faltin, E., Becker, R., Görlich, A.S., Rillig, M.C., 2019. Microplastics can change soil properties and affect plant performance. *Environ. Sci. Technol.* 53, 6044–6052. https://doi.org/10.1021/acs.est.9b01339

Galvão, A., Aleixo, M., De Pablo, H., Lopes, C., Raimundo, J., 2020. Microplastics in wastewater: Microfiber emissions from common household laundry. *Environ. Sci. Pollut. Res.* 27, 26643–26649. https://doi.org/10.1007/s11356-020-08765-6

Gaylarde, C., Baptista-Neto, J.A., da Fonseca, E.M., 2021. Plastic microfibre pollution: How important is clothes' laundering? *Heliyon.* 7, e07105. https://doi.org/10.1016/j.heliyon.2021.e07105

Geyer, R., Gavigan, J., Jackson, A.M., Saccomanno, V.R., Suh, S., Gleason, M.G., 2022. Quantity and fate of synthetic microfiber emissions from apparel washing in California and strategies for their reduction. *Environ. Pollut.* 298, 118835. https://doi.org/10.1016/j.envpol.2022.118835

Grillo, J.F., López-Ordaz, A., Hernández, A.J., Catarí, E., Sabino, M.A., Ramos, R., 2023. Synthetic microfiber emissions from denim industrial washing processes: An overlooked microplastic source within the manufacturing process of blue jeans. *Sci. Total Environ.* 884. https://doi.org/10.1016/j.scitotenv.2023.163815

Hernandez, E., Nowack, B., Mitrano, D.M., 2017. Polyester textiles as a source of microplastics from households: A mechanistic study to understand microfiber release during washing. *Environ. Sci. Technol.* 51, 7036–7046. https://doi.org/10.1021/acs.est.7b01750

Horn, D.A., Granek, E.F., Steele, C.L., 2020. Effects of environmentally relevant concentrations of microplastic fibers on Pacific mole crab (Emerita analoga) mortality and reproduction. *Limnol. Oceanogr. Lett.* 5, 74–83. https://doi.org/10.1002/lol2.10137

Kärkkäinen, N., Sillanpää, M., 2021. Quantification of different microplastic fibres discharged from textiles in machine wash and tumble drying. *Environ. Sci. Pollut. Res.* 28, 16253–16263. https://doi.org/10.1007/s11356-020-11988-2

Kim, S.W., Waldman, W.R., Kim, T.Y., Rillig, M.C., 2020. Effects of different microplastics on nematodes in the soil environment: Tracking the extractable additives using an ecotoxicological approach. *Environ. Sci. Technol.* 54, 13868–13878. https://doi.org/10.1021/acs.est.0c04641

Lares, M., Ncibi, M.C., Sillanpää, M., Sillanpää, M., 2018. Occurrence, identification and removal of microplastic particles and fibers in conventional activated sludge process and advanced MBR technology. *Water Res.* 133, 236–246. https://doi.org/10.1016/j.watres.2018.01.049

Le, L.T., Nguyen, K.Q.N., Nguyen, P.T., Duong, H.C., Bui, X.T., Hoang, N.B., Nghiem, L.D., 2022. Microfibers in laundry wastewater: Problem and solution. *Sci. Total Environ.* 852. https://doi.org/10.1016/j.scitotenv.2022.158412

Lozano, Y.M., Rillig, M.C., 2020. Effects of microplastic fibers and drought on plant communities. *Environ. Sci. Technol.* 54, 6166–6173. https://doi.org/10.1021/acs.est.0c01051

Mahbub, M.S., Shams, M., 2022. Acrylic fabrics as a source of microplastics from portable washer and dryer: Impact of washing and drying parameters. *Sci. Total Environ.* 834, 155429. https://doi.org/10.1016/j.scitotenv.2022.155429

Mishra, S., Dash, D., Das, A.P., 2022. Detection, characterization and possible biofragmentation of synthetic microfibers released from domestic laundering wastewater as an emerging source of marine pollution. *Mar. Pollut. Bull.* 185, 114254. https://doi.org/10.1016/j.marpolbul.2022.114254

Napper, I.E., Barrett, A.C., Thompson, R.C., 2020. The efficiency of devices intended to reduce microfibre release during clothes washing. *Sci. Total Environ.* 738. https://doi.org/10.1016/j.scitotenv.2020.140412

Napper, I.E., Thompson, R.C., 2016. Release of synthetic microplastic plastic fibres from domestic washing machines: Effects of fabric type and washing conditions. *Mar. Pollut. Bull.* 112, 39–45. https://doi.org/10.1016/j.marpolbul.2016.09.025

Pirc, U., Vidmar, M., Mozer, A., Kržan, A., 2016. Emissions of microplastic fibers from microfiber fleece during domestic washing. *Environ. Sci. Pollut. Res.* 23, 22206–22211. https://doi.org/10.1007/s11356-016-7703-0

Qiao, R., Deng, Y., Zhang, S., Wolosker, M.B., Zhu, Q., Ren, H., Zhang, Y., 2019. Accumulation of different shapes of microplastics initiates intestinal injury and gut microbiota dysbiosis in the gut of zebrafish. *Chemosphere*. 236, 124334. https://doi.org/10.1016/j.chemosphere.2019.07.065

Raja Balasaraswathi, S., Rathinamoorthy, R., 2022. Effect of fabric properties on microfiber shedding from synthetic textiles. *J. Text. Inst.* 113, 789–809. https://doi.org/10.1080/00405000.2021.1906038

Rathinamoorthy, R., Balasaraswathi, S.R., 2022. Investigations on the interactive effect of laundry parameters on microfiber release from polyester knitted fabric. *Fibers Polym.* 23, 2052–2061. https://doi.org/10.1007/s12221-022-4929-y

Sillanpää, M., Sainio, P., 2017. Release of polyester and cotton fibers from textiles in machine washings. *Environ. Sci. Pollut. Res.* 24, 19313–19321. https://doi.org/10.1007/s11356-017-9621-1

Stanton, T., Johnson, M., Nathanail, P., MacNaughtan, W., Gomes, R.L., 2019. Freshwater and airborne textile fibre populations are dominated by 'natural', not microplastic, fibres. *Sci. Total Environ.* 666, 377–389. https://doi.org/10.1016/j.scitotenv.2019.02.278

Sudheshna, A.A., Srivastava, M., Prakash, C., 2022. Characterization of microfibers emission from textile washing from a domestic environment. *Sci. Total Environ.* 852, 158511. https://doi.org/10.1016/j.scitotenv.2022.158511

Volgare, M., De Falco, F., Avolio, R., Castaldo, R., Errico, M.E., Gentile, G., Ambrogi, V., Cocca, M., 2021. Washing load influences the microplastic release from polyester fabrics by affecting wettability and mechanical stress. *Sci. Rep.* 11, 1–12. https://doi.org/10.1038/s41598-021-98836-6

Woods, M.N., Hong, T.J., Baughman, D., Andrews, G., Fields, D.M., Matrai, P.A., 2020. Accumulation and effects of microplastic fibers in American lobster larvae (Homarus americanus). *Mar. Pollut. Bull.* 157, 111280. https://doi.org/10.1016/j.marpolbul.2020.111280

Yang, L., Qiao, F., Lei, K., Li, H., Kang, Y., Cui, S., An, L., 2019. Microfiber release from different fabrics during washing. *Environ. Pollut.* 249, 136–143. https://doi.org/10.1016/j.envpol.2019.03.011

Yee, M.S.L., Hii, L.W., Looi, C.K., Lim, W.M., Wong, S.F., Kok, Y.Y., Tan, B.K., Wong, C.Y., Leong, C.O., 2021. Impact of microplastics and nanoplastics on human health. *Nanomaterials*. 11, 1–23. https://doi.org/10.3390/nano11020496

Zambrano, M.C., Pawlak, J.J., Daystar, J., Ankeny, M., Cheng, J.J., Venditti, R.A., 2019. Microfibers generated from the laundering of cotton, rayon and polyester based fabrics and their aquatic biodegradation. *Mar. Pollut. Bull.* 142, 394–407. https://doi.org/10.1016/j.marpolbul.2019.02.062

Zambrano, M.C., Pawlak, J.J., Daystar, J., Ankeny, M., Venditti, R.A., 2021. Impact of dyes and finishes on the microfibers released on the laundering of cotton knitted fabrics. *Environ. Pollut.* 272, 115998. https://doi.org/10.1016/j.envpol.2020.115998

Zhang, L., Sun, J., Zhang, Z., Peng, Z., Dai, X., Ni, B.J., 2022. Polyethylene terephthalate microplastic fibers increase the release of extracellular antibiotic resistance genes during sewage sludge anaerobic digestion. *Water Res.* 217, 118426. https://doi.org/10.1016/j.watres.2022.118426

Zhang, Y.Q., Lykaki, M., Alrajoula, M.T., Markiewicz, M., Kraas, C., Kolbe, S., Klinkhammer, K., Rabe, M., Klauer, R., Bendt, E., Stolte, S., 2021. Microplastics from textile origin-emission and reduction measures. *Green Chem.* 23, 5247–5271. https://doi.org/10.1039/d1gc01589c

Ziajahromi, S., Kumar, A., Neale, P.A., Leusch, F.D.L., 2017. Impact of microplastic beads and fibers on Waterflea (*Ceriodaphnia dubia*) survival, growth, and reproduction: Implications of single and mixture exposures. *Environ. Sci. Technol.* 51, 13397–13406. https://doi.org/10.1021/acs.est.7b03574

9 Effect of Microplastic Pollution on Soil, Plants, and Soil Microbes

Rahul Kandpal and Nishu Goyal

9.1 INTRODUCTION

The overutilization of plastic has emerged as a prevalent issue, giving rise to substantial environmental challenges. Plastic particles measuring less than 5 mm are identified as microplastics. The transformation of larger plastic items into microplastics transpires via physical, biological, and chemical mechanisms. (Yu et al., 2021). Mulching films (thin sheets or films placed on the soil surfaces in gardening and agriculture to suppress weeds, conserve moisture, and regulate soil temperature), which are used to improve crop growth by enhancing soil conditions and water efficiency, have become a main producer of microplastics in agricultural biospheres (Serrano-Ruiz et al., 2023). The leftover debris of plastic waste generated from mulching activities undergoes a gradual transformation into increasingly smaller particles. Eventually, these particles evolve into microplastics due to a combination of physical, chemical, and biological processes. These processes include exposure to ultraviolet radiation, erosion caused by water or air, and the involvement of organisms like earthworms (Sajjad et al., 2022; Wright & Kelly, 2017).

Microplastics (MPs) have shown adverse effects on the microbial community, potentially leading to changes in soil's physical and chemical characteristics (Bandopadhyay et al., 2018; Dong et al., 2021). In a scientific investigation, it was determined that microplastics have adverse effects on the organic carbon content in the soil and nitrogen, causing disturbances in the nutrient cycle and influencing the overall development of plants. (Napper, 2015). The main type of plastic comprises polystyrene (PS) polyethylene (PE), polypropylene (PP), polyvinylchloride (PVC), and polyethylene terephthalate (PET) (Geyer et al., 2017; He et al., 2018; Qi et al., 2018; Sajjad et al., 2022). In various studies, it has been shown that the excessive accumulation of microplastics in agricultural soil could result in the following outcomes. The piling up of residual plastic waste can disrupt the natural arrangement of soil particles, inducing change in the soil's physical and chemical properties. This disruption affects the soil's capacity to retain moisture, nutrients, and air. Microplastics impact the soil environment by impacting the soil structure, notably altering its porosity. The phenomenon of bioturbation arises from the downward movement of microplastic within the soil matrix. These microplastic residues accumulate in substantial quantities, and they obstruct and occupy soil pores, resulting in

 DOI: 10.1201/9781032684574-9

a gradual reduction in the soil's capacity for infiltration due to which natural nutrient cycle within the soil gets disturbed and brings about the modifications in the microbial composition, ultimately influencing the growth of crops (Sajjad et al., 2022). A substantial amount of microplastics is disposed of in landfills, leading to adverse effects on both terrestrial and wetland ecosystems. These microplastics can be categorized into further primary microplastics and secondary microplastics. Primary microplastics have commercial importance, whereas secondary microplastics are produced by the degradation of large plastics (Ceschin et al., 2023). Microplastics can have detrimental impacts on the existing microbial community, potentially leading to changes in the soil's physical and chemical characteristics.

This interference can also extend to nutrient cycling, as microplastics have the potential to bind to nutrients, leading to a reduced uptake of nutrients by plants. The presence of microplastics can have a detrimental effect on seed germination. Seeds may encounter difficulties when trying to establish roots and undergo initial sprouting. Additionally, microplastics might interfere with soil aeration and microbial activities. Consequently, this interference could lead to the release of gases like carbon dioxide and methane from the soil into the atmosphere (Sajjad et al., 2022). MPs disperse within the soil matrix due to the influence of alternating dryness and moisture, various soil management techniques, and disruptions caused by biological factors. This dispersion consequently leads to alteration in soil physicochemical attributes including the carbon shift (C) nitrogen (N) and phosphorous (P) content as well as changes in pH (Yu et al., 2022). Many researchers have suggested that the harmful consequences of plastic involve restricting photosynthesis and impacting the growth of roots and shoots. The consequences appear to arise from the attachment of microplastic particles to the outer tissue of the plants. This attachment of microplastics creates a physical barrier that blocks light and air, subsequently impeding the processes of photosynthesis and respiration (Ceschin et al., 2023). In this chapter, we will explore how microplastics impact the soil and soil fauna living in it and how plant growth and development are impacted by the microplastic present in the environment. MPs are known for their resistance to degradation in the environment, and this persistence can lead to long-term availability of plastic particles in the ecosystem. Bioaccumulation refers to the process by which substances, including MPs, accumulate in tissues of the organism over time. Due to their small size, MPs can be ingested by the organisms living in the lower trophic level and then move up the food chain as predators consume contaminated prey. Toxicological data have shown that MPs can cause various types of toxicity in target organisms, including reproductive, growth, and behavioral toxicities. The impact of microplastics on freshwater ecosystems, specifically on *Daphnia* in which MPs (Figure 9.1) lead to changes in the behavior and morphology of *Daphnia*, which are small aquatic organisms often used in ecotoxicology studies (Y. Liu et al., 2022). Studies have shown that microplastics can have adverse effects on the health of fish including damage to their gut, liver, and kidneys. These particles accumulate in the tissues and organs, leading to various adverse health outcomes. Studies have shown that excess consumption of these microplastics leads to substantial liver harm, marked by the level of lipid apoptosis (cell death)(Subaramaniyam et al., 2023). In many experimental studies, it was proved that the ecological impact of microplastics

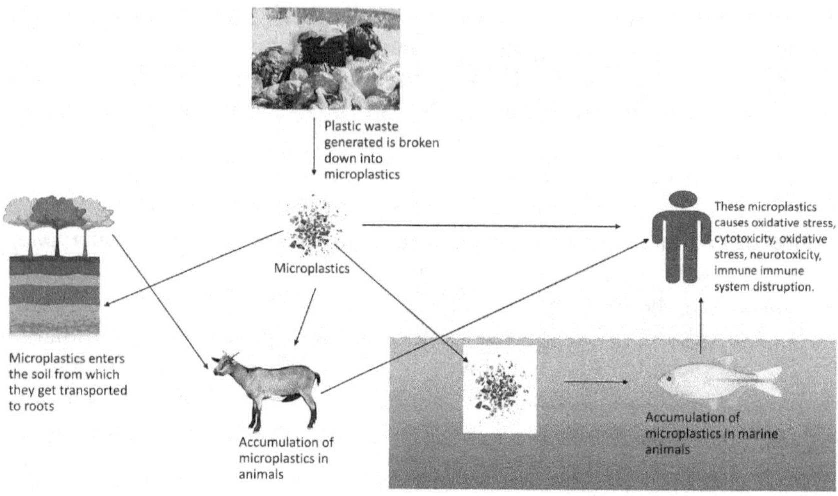

FIGURE 9.1 Bioaccumulation of microplastics.

on higher plants, particularly microplastics brings alterations in morphofunctional traits across various crop species (Gentili et al., 2022).

9.2 ABUNDANCE OF MICROPLASTIC AND ITS CLASSIFICATION BASED ON SHAPE AND SIZE

9.2.1 Sources of Microplastics

Microplastics in terrestrial environments originate from various sources, with three major contributors being agricultural practices, runoff effects, and the improper disposal of large plastics. In agricultural environments, various procedures can introduce microplastics into the soil, such as the application of sewage sludge as a nutrient source. Research findings indicate that a substantial proportion of microplastics is concentrated in the solid phase of sludge. Diverse origins contribute to the existence of microplastics in sludge, encompassing microbeads derived from cosmetics and industrial products, fibers released during the laundering of synthetic garments, tire debris, and broken plastics transported by urban runoff. Additionally, specific agricultural methodologies aimed at enhancing crop yields and mitigating pest concerns might also contribute to the presence of microplastics. For instance, the widespread practice of plastic mulching often results in the persistence of plastic films in soils. Over time, these plastics undergo fragmentation and degradation, leading to the generation of plastic particles of various sizes, ranging from micro- to nanoplastic (Martinho et al., 2022). Microplastic origins can be classified into primary and secondary categories. Deliberately manufactured for distinct functions, primary microplastics serve specific purposes like abrasive components in cosmetics, carriers for drugs, and applications in industries such as air blasting. On the other hand, secondary microplastics emerge as particles resulting from the fragmentation of larger objects, including bottles and bags (Ceschin et al., 2023).

Primary microplastics are deliberately created small plastic particles or products that are intentionally manufactured to be less than 5 mm in size. They are incorporated into various products or systems in their small form from the outset. Examples of common primary microplastics include:

- **Microbeads:** Minuscule plastic beads added to personal care products like exfoliating scrubs and toothpaste
- **Microfibers:** Tiny plastic fibers that are shed during the washing of synthetic textiles such as polyester and nylon clothing
- **Nurdles:** Small plastic pellets used as raw materials in manufacturing processes, which can inadvertently escape into the environment during transportation or production
- **Plastic Pellets:** Tiny plastic particles used in the production of a diverse array of plastic goods
- **Microplastic Particles:** Incorporated into abrasive cleaning products, paints, and coatings

Secondary microplastics are the result of larger plastic items breaking down over time due to ecological influences such as sunlight, wind, and wave action, or through mechanical processes like abrasion (Guo et al., 2020). These smaller plastic fragments are generated when:

- Plastic bottles and containers break into pieces.
- Plastic bags degrade and fragment.
- Weathered fishing gear produces small particles.
- Larger plastic debris in oceans, rivers, and other environments deteriorates into tiny plastic fragments.

Secondary microplastics are a significant concern because they can endure in the environment for extended periods, accumulate in ecosystems, and be ingested by marine life. This accumulation can potentially lead to their entry into the food chain, posing risks to both wildlife and human health (Akdogan & Guven, 2019; Lehtiniemi et al., 2018). Moreover, observations highlighted that atmospheric microplastic particles exhibit significantly diminished dimensions when compared to aquatic microplastics. This suggests a lower concentration of minute particles within the cryosphere as opposed to the atmosphere (Y. Zhang et al., 2020).

9.2.2 Distribution of Microplastic

There is very little understanding regarding the prevalence of microplastics in snow and ice due to limited available data. Humans generate millions of tons of plastic annually for their own benefit. In one report, total plastic manufacturing across the globe was found to reach 335 million tons in 2016, with an average increase of 8.6% since the 1950s. Every year there is a huge release of microplastics by different areas of the world; refer to Figure 9.2 for the regional release of microplastics in kilotons (Boucher & Friot, n.d.). This was all the result of excessive plastic production, improper disposal practices, and population growth, all of which have contributed to

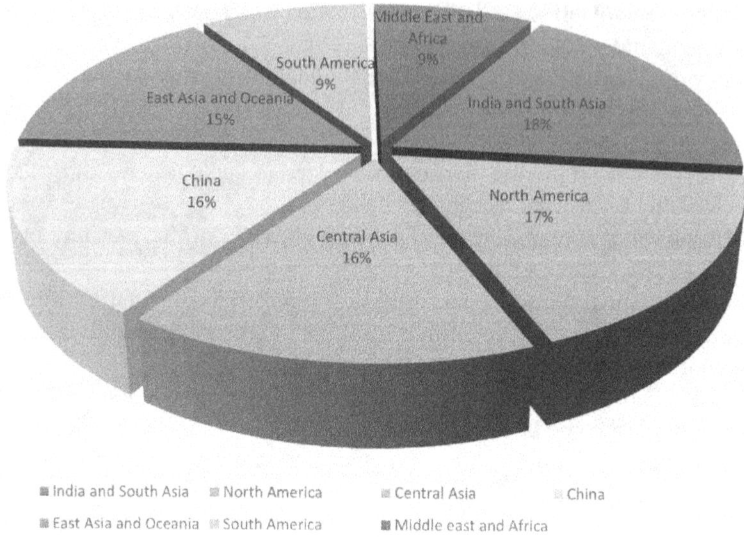

FIGURE 9.2 Release of microplastics across the globe (Boucher & Friot, n.d.).

such a great increment of waste generated by plastic (Rai et al., 2023). Apart from the primary microplastics produced from tires in the Americas, Europe, and Central Asia, another important note is that the emission of primary microplastic surpasses those of the secondary microplastic resulting from poorly managed waste in Europe and North America (Hale et al., 2020)

The distribution of microplastics in different regions is as follows:

- **Marine Environments:** Microplastics are extensively found in oceans and seas. They can be found at the surface, in the water column, and even in deep-sea sediments. Marine microplastics originate from various sources, including maritime activities, land-based runoff, and the breakdown of larger plastic debris (Andrady, 2011).
- **Freshwater Ecosystems:** Microplastics are also present in rivers, lakes, and freshwater bodies. They can enter these environments through urban runoff, wastewater discharge, and atmospheric deposition. River systems can transport microplastics from upstream sources to downstream areas (Wagner et al., 2014)
- **Soil:** Microplastics have been identified in soil, with notable concentrations found in regions employing plastic film mulching for agriculture. These microplastics enter soil ecosystems through the degradation of plastic films, the use of fertilizers containing plastic components, and deposition from the atmosphere. Their presence in soil has implications for soil health and raises concerns about the possibility of microplastics entering the food chain through plant uptake (Rillig et al., 2017)
- **Airborne Plastics**: Studies have revealed the presence of microplastics in the atmosphere, predominantly in the form of airborne fibers and particles.

These airborne microplastics can travel considerable distances and eventually settle on both terrestrial and aquatic surfaces. This deposition plays a role in contaminating the environment with microplastics (Dris et al., 2017).

- **Food and Drinking Water:** Microplastics have been identified in a variety of food products, encompassing seafood, drinking water, and even table salt. The increasing concern revolves around human exposure to microplastics through dietary consumption and drinking water, as these tiny particles have the potential to enter the human body, giving rise to health-related possibilities (Schwabl et al., 2019).

9.3 EFFECT OF MICROPLASTICS ON SOIL

The soil also functions as a repository for MPs, and, in recent times, numerous researchers have investigated the implications of MPs on the soil ecosystem (Sajjad et al., 2022; Yang et al., 2018). The introduction of microplastics into the soil can lead to changes in the physicochemical and hydrological characteristics, potentially impacting the enzymatic activities and the composition of the soil microbial community. In a research study, fresh soil samples were taken, followed by drying them with air, and then processing them into a ground form for the detection of enzymatic activity in it. The analysis of soil enzymes such as urease, sucrase, and catalase was conducted with commercially available kits by a company named Solarbio, a company based in China. All these are necessary for soil nutrient cycling and the ecosystem's overall makeup. These soil enzymes consequently serve as crucial and sensitive indicators for the alterations in the soil environment caused by the presence of contaminants (Yu et al., 2021). MPs can be understood as physical contaminants within the soil, as indicated by the research of de Souza Machado et al. (2018). Microfibers have been associated with a decrease in soil bulk density (de Souza Machado et al., 2018). Research conducted by Yu et al. (2021) analyzed how microplastics affect various soil properties, including soil organic matter, total phosphorus, total nitrogen, and potassium content, with different types of microplastics. The control group consists of soil samples or experimental units that were not exposed to any microplastics. It serves as the baseline or reference group for comparing the effects of various types of microplastics (MPs). Here's what was observed:

1. **Soil Organic Matter:** Except PVCs and MPs, all other categories of microplastics demonstrated a notable augmentation in soil organic matter, showing an increase ranging from 72 to 324% when compared to the control group. This implies that the majority of microplastic varieties exerted a favorable influence on soil organic matter.
2. **Total Nitrogen:** Except for the amalgamated category, all alternative interventions involving microplastics PP and PE led to a reduction in the overall nitrogen content within the soil. This decline varied from 17 to 49% when contrasted with the control group. This implies a predominantly adverse influence of microplastics on the overall nitrogen concentrations in the soil.
3. **Potassium Content:** Applications involving microplastics (specifically, PP, PE, and mixed varieties) exhibited notably diminished soil potassium levels

in contrast to the control cohort. This shows us the existence of these microplastics resulted in a depletion of potassium content in the soil.

4. **Total Phosphorus:** The presence of both PS and PVC MPs negatively affects the overall phosphorus content within the soil. PS MPs resulted in a 17% reduction, whereas PVC MPs caused a 14% decrease compared to the control group. This suggests that these microplastic variants contribute to a decline in the total phosphorus levels within the soil (Yu et al., 2021).

While some types of microplastics may increase soil organic matter, they can also negatively impact other essential soil nutrients like phosphorus, nitrogen, and potassium. These findings suggest that the ability of microplastics in the soil can have complex and potentially detrimental effects on soil quality and nutrient levels (Yu et al., 2021). When the wastewater is not treated properly, it has very a significant quantity of microplastic. The origin of these microplastics spans diverse outlets, including laundry machines, as well as personal care items like hair cleansers and exfoliants. When this untreated wastewater is used directly for the irrigation of agricultural fields, it is most likely to introduce microplastic into the soil, leading to plastic contamination in the farmland environment and causing further harm to the soil, soil microbial community, and crops (He et al., 2018). The effect of microplastic can cause alterations in the soil composition and the soil texture. These changes can further lead to more adverse effects on water circulation and the overall working of the terrestrial ecosystems, which includes the interaction between the plants and the soil (de Souza Machado et al., 2018). Microplastic may impact the soil properties by changing the level of pH, bulk density, nutrient retention, and bioaccumulation (Lozano & Rillig, 2022). (S. Zhang et al., 2019). Due to the limitations of existing techniques, alternative techniques are being worked on for the removal of microplastic (MPs) from the soil environment. The alternative techniques that are being used currently are (1) air flotation, (2) density suspension, (3) heating (3–5 seconds at 130°Celsius. Although these methods are being used, they have significant challenges:

1. These processes are inefficient and slow in retrieving microplastics from the soil.
2. Difficulty arises in accurately capturing the three-dimensional heterogeneity present in the soil plastic pollution, and the results and the analysis would be biased and would not represent the proper distribution of microplastic in the soil matrix.
3. These existing technologies are not capable of extracting plastic particles in the size range of nano- and picometers, which are the chief sources of plastic pollution.
4. The density suspension method is a widely used method for extracting MPs, but its suitability for soil extraction remains uncertain (Sajjad et al., 2022)

9.4 EFFECT OF MICROPLASTICS ON SOIL MICROBES

Microplastics offer a distinctive ecological niche for microorganisms, establishing a conducive setting for their proliferation and establishment, all while functioning as a reservoir of carbon. The degradation of microplastics involves three successive stages: biodeterioration, fragmentation, and assimilation (Kumari et al., 2022). In one study (Rillig et al., 2017), it was found that the presence of earthworms had a significant positive impact on the movement of microplastic particles away from the soil surface. Further, it was clear that the overall distribution of microplastics within different soil depths relied on the existence of earthworms and the dimensions of the particles. In the absence of earthworms, microplastics predominately remained in the topsoil layer. However, in the presence of earthworms, microplastic particles of all sizes were observed to reach the middle and bottom soil layers within the 21-day experimental period. Bacteria are known for their extensive distribution across diverse environments, including soil, water, and the atmosphere, and these microorganisms have been the focus of extensive research. Numerous studies have explored bacterial species, recognizing their crucial involvement in the decomposition of intricate polymers and the remediation of environmental contaminants such as crude oil, metal compounds, antibiotics, plastics, and various other ecological pollutants concerns (Omidoyin & Jho, 2023)

In a few research studies conducted, it was found that the occurrence of microplastics has been shown to have several impacts on microbial population. In farmland and soil, the use of plastic mulching has been found to increase the abundance of specific plastic-associated microorganisms, including *Actinobacteria*, *Bacteroidetes*, and *Proteobacteria*, that are involved in the degradation of polyethylene. However, the presence of microplastic (MP) residues in soil was associated with a negative impact on soil bacteria's motility. Bacterial motility is recognized to have a pivotal role in maintaining biodiversity in micro-ecosystems, degrading soil pollutants, and cycling biochemicals. Additionally, certain algae, such as *Chlamydomonas*, were more abundant in treatments involving polystyrene (PS) microplastics, and the presence of microplastics altered the structures of biological soil crusts. As a result, microplastics not only affect the variety of microorganisms but also exert an influence on their metabolic functions and contributions to nutrient cycling (Yu et al., 2021). After amassing, microplastics are conveyed to different internal organs via the circulatory system, instigating disturbances in metabolic routes and the endocrine system. This leads to oxidative stress and cell necrosis, ultimately culminating in cell apoptosis and contributing to the organism's demise (Yu et al., 2022). The temperature, presence of roots, vapor movement, the level of moisture in the soil all together influence the microbial activities in the soil, so the ecosystem of the soil could be altered by the extensive use of microplastics, which would affect the microbes living in the soil. Additionally, synthetic mulching materials that are made of plastic could also lead to the loss of the soil invertebrate population (Bandopadhyay et al., 2018).

The properties of microplastics can have adverse impacts on the structure, diversity, and richness of the microbial flora (refer to Table 9.1). The dimensions of the microplastics play a vital role in that microbes attach to them and colonize them, and the variety and distribution of the microbial communities all depend on MP size. Some studies suggested that the size of MPs did not greatly impact the bacteria at the genus level, while some showed they affected the microbial communities. In a study, it was reported that actinobacteria thrived on certain types of microplastic, but a lot of research is still needed to be conducted for more understanding of the consequences of MPs on soil microbe communities (Li et al., 2021; Omidoyin & Jho, 2023).

TABLE 9.1
Effect of Microplastsics

Types of Polymers	Species Name	Impact	References
PVC (polyvinyl chloride)	*Daphnia magna*	Has detrimental effects on the reproductive cycle, as well as psychological aspects.	Y. Liu et al., 2022
PE (polyethylene)	Earthworm, *Lumbricus terrestris*	Helps in the transportation and degradation of microplastics.	Rillig et al., 2017
PP (polypropylene)	*Oreochromis mossambicus*	There was the death of the liver cells, with significant histological changes, damaged DNA.	Jeyavani et al., 2023
PET (polyethylene terephthalate)	*Achatina fulica*	There was reduced food intake, excretion, gastrointestinal villi damage, and a fall in the level of glutathione peroxidase.	K. Liu et al., 2019
PS (polystyrene)	Larval zebrafish *(Danio rerio)*	It affects the microbiome and metabolism.	Wan et al., 2019

Of the existing 200 types of plastic material, each exhibits its own distinct properties, such as their responsiveness and interactions within the soil surroundings. So while classifying microplastics, it becomes important to not only take their dimensions into account but also their chemical characteristics like their hydrophobic tendencies and their physical qualities like their shape. These affect both the biochemical attributes of the soil, as well as its biological constituents (S. Zhang et al., 2019).

The presence of microplastics (MP) in soil has shown diverse effects on the soil microbiome. These consequences encompass alterations in microbial functions, nutrient cycling, bacterial movement, and the breakdown of organic substances. Microplastics can also disrupt microbiome composition, trigger immune responses, alter enzyme activities, and induce gene expression changes in soil microorganisms. Furthermore, researchers have indicated that MP materials serve as selective niches for bacteria and fungi, with microbial biofilms forming on MP surfaces. These biofilms impact the physical and chemical properties of MPs, influencing their bioavailability, degradability, and mobility in soil (Kublik et al., 2022).

9.5 EFFECT OF MICROPLASTICS ON PLANTS

Crops play a crucial role both in human nutrition and in the well-being of ecosystems. Nevertheless, various environmental and biological challenges, including microplastic pollution, can have severe impacts on the financial return of these plants. The two main concerns that arise are the capability of the plants to absorb and accumulate MPs and the consequences of MPs on plant growth and the quality of food they produce (Kumari et al., 2022). MPs can have far-reaching detrimental impacts on various biological and chemical processes in plants.

The consequences of MPs on plants have been studied extensively in recent years, and research papers have provided valuable insights into the specific impacts. Here are detailed effects of microplastics on plants, based on research findings:

1. **Seed Germination Inhibition:** Research has shown that microplastics can hinder seed germination. MPs can create physical barriers around seeds, preventing the water absorption and gas exchange essential for germination. This inhibition can delay or reduce the number of seeds that successfully germinate (Qi et al., 2018; Yu et al., 2022).
2. **Root Growth:** Microplastics can detrimentally impact root growth and development. As microplastics attach to root surfaces, they induce alterations in root morphology, potentially leading to deformities. This, in turn, hinders the root's capacity to effectively explore the soil for water and essential nutrients, ultimately resulting in compromised root growth and a reduction in overall plant development (Kumari et al., 2022)
3. **Photosynthesis Disruption:** Microplastics have the potential to disrupt photosynthesis, a vital process for plant energy production. They can obstruct sunlight from reaching chloroplasts or interfere with gas exchange in leaves, resulting in decreased rates of photosynthesis and a decline in overall plant vitality (Kumari et al., 2022)
4. **Oxidative Stress:** The presence of microplastics within plant tissues can trigger oxidative stress. This occurs as a result of the buildup of reactive oxygen species (ROS) caused by microplastic-induced damage. In reaction, plants engage antioxidant defense mechanisms, which could potentially result in changes in cellular processes. (Subaramaniyam et al., 2023)
5. **Genetic Damage:** Microplastics can potentially cause genetic damage in plants. Physical interactions between MPs and plant cells may result in mutations or other genetic abnormalities, impacting plant reproduction and health (Kumari et al., 2022).
6. **Microbial Community Alterations:** Microplastics present in soil can modify the composition and functionality of microbial flora. Shifts in microbial diversity and activity may influence nutrient cycling and soil well-being, thereby indirectly affecting plant growth and nutrition (Rillig et al., 2019).

MPs can affect the development as well as the functioning of plants through a variety of mechanisms, which include:

1. Direct detriment to plants, primarily instigated by nanoplastics (NPs).
2. Indirectly impacting plant growth by altering soil characteristics and the makeup of microbial communities.
3. Direct harm stemming from pollutants (including substances like plastic softeners, fire inhibitors, anti-oxidants, pigments, and more) found within MPs (Rillig et al., 2019).

The consequences of MPs on fauna can be attributed to adsorption. For instance, MPs can adhere to plant roots, altering their characteristics and the uptake of

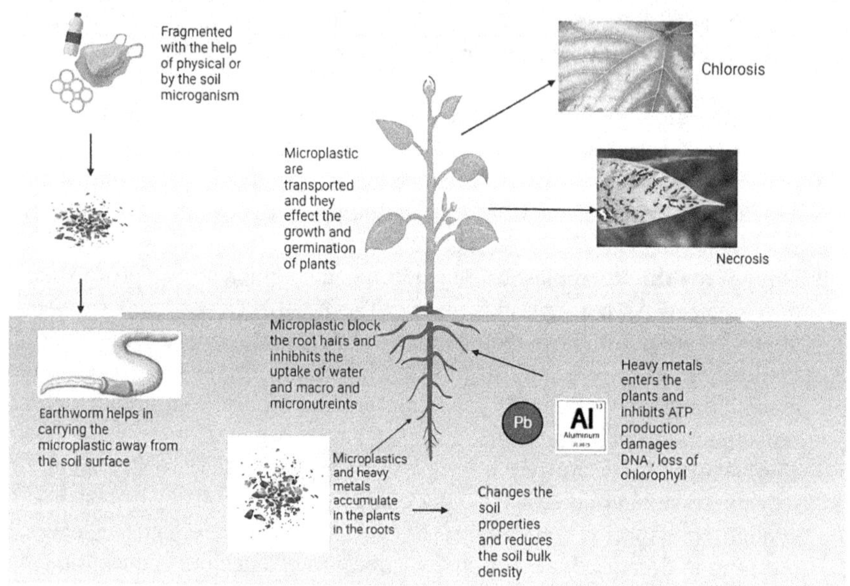

FIGURE 9.3 Consequences of microplastics on plants and soil microbes.

nutrients and water, with smaller MPs demonstrating higher toxicity to plants (Yu et al., 2022)

Plastic film mulching is widely used in agriculture because it helps to control the temperature of the soil, improves water efficiency, and enhances growth and quality. However, it has also been a chief contributor to pollution caused by microplastics in the soil. In a study, it was found that in the areas where there was extensive use of plastic films, the concentration in 5, 15, and 25 years was found to be significantly heightened over time. Plastic film mulching is considered a root cause of microplastic in terrestrial ecosystems due to excessive widespread use and improper disposal. This represents the effect of microplastic and heavy metals in plants and how they cause the loss of chlorophyll and DNA damage to ATP production (refer to Figure 9.3 for the effect of microplastics).

Additionally, the dustproof net that is used extensively in China in agricultural lands is also contributing to soil microplastic contamination (Yu et al., 2022). Nanoparticles of microplastic were discovered amassed in the plant roots, potentially impeding the transportation of nutrients through cellular linkages or pores in the cell wall. (Gentili et al., 2022). Many research studies have suggested potential ways in which microplastics can affect plants through physical, chemical, and biological mechanisms. In the soil ecosystem, the microplastics are broken down into smaller particles through fragmentation. Experimental studies were done using fluorescent nanoplastic, and it was found that these particles can stick to plant surfaces, particularly seeds and roots, when the plant is growing (Guo et al., 2020). Their physical presence can block the seed pores and the root structures, which can lead to reduced

TABLE 9.2
Consequences of Microplastics on Some Species

Type of Polymer	Species Name	Impact	Reference
PVC (poly vinyl chloride)	*Phaseolus vulgaris l*	PVC microparticles had dual effects on both root morphology and development, as well as photosynthesis, exhibiting contrasting impacts on chlorophyll and carotenoid content.	Gentili et al., 2022
PE (polyethylene)	(Glycine max) soybean	It affects growth and germination.	Wang et al., 2021
PET (polyethylene terephthalate)	Spring onion (*Allium istulosum*)	Microplastics can influence the characteristics of leaves, plant roots, and overall biomass.	Rai et al., 2023
PS (polystyrene)	*Chlamydomonas*	It alters the structures of biological soil crusts.	Yu et al., 2021
PP (polypropylene)	*Solanum lycopersicum*	There was no significant effect on seed germination but positive effects on shoot and root elongation.	Shorobi et al., 2023

water and nutrient uptake, and it could even hamper germination and have adverse impacts on plant growth. Nanoscopic and micrometric plastic particles spanning dimensions between 20 μm and 5 mm can even accumulate in the plant tissues' intracellular uptake in various terrestrial species. The absorption has been associated with varied oxidative stress patterns in both root and shoot systems, inducing the formation of reactive oxygen species (ROS) and instigating antioxidant responses.

All these things collectively contribute to a decline in plant growth and performance (Subaramaniyam et al., 2023). It has been found that the impact of microplastics is very harmful in many species (refer to Table 9.2 for the consequences of microplastics on plants).

In some studies, it was found that MPs potentially impact root growth in both monocots and dicots and that the presence of NMPs on the root surface and seed capsule may account for this accumulation. Some positive effects were found in the root biomass of a few plant species, which can be a sign of stress-induced growth. In short, there was very little impact compared to those observed on seed sprouting (Bosker et al., 2019; Jiang et al., 2019; Zantis et al., 2023). Microplastic (MP) stress has a distinct impact on plants, with direct interactions occurring at the root level (refer Figure 9.4).

9.6 CONCLUSION AND DISCUSSION

The correlation between the size of microplastics and the duration for which they are present in the atmosphere as well as their deposition characteristics, and the speed at which various microplastics get degraded is not clear. Proper and consistent sampling

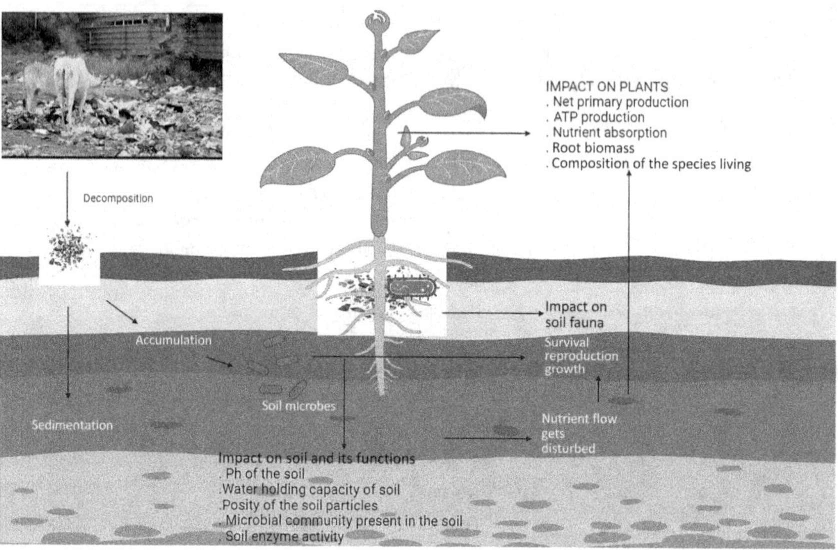

FIGURE 9.4 Impact of microplastics on soil and plants. Microplastic (MP) stress has a distinct impact on plants, with direct interactions occurring at the root level. MP particles can adhere to root hairs, leading to alterations in root growth. MP stress is more likely to induce oxidative stress, resulting in diminished leaf growth and a reduction in photosynthetic activity (Jia et al., 2023).

and pretreatment procedures encompass matching filters with appropriate pore sizes for density separation techniques and analytical methodologies to ensure a precise global comparison of microplastic discovered in ice and snow (Y. Zhang et al., 2022). The absorption and accumulation of microplastics by flora in the terrestrial ecosystem can lead to many ecological imbalances and may also have adverse effects on the environment. The primary issue is the potential transfer of these microplastics to various trophic levels, posing a threat to food security. To date, scientists are still researching the potential threats of this challenging pollution (Kumari et al., 2022). The use of environmentally friendly plastic mulches looks like a better alternative instead of using plastic mulches. But still we do not know clearly the long-term effects of these decomposable plastic mulches or the effect of PE mulches on how it affects the soil ecosystem and the microbial community living in it (Bandopadhyay et al., 2018). The temperature, presence of the roots, vapor movement, the level of moisture in the soil all together influence the microbial activities in the soil, so the ecosystem of the soil could be altered by the extensive use of microplastic, which would affect the microbes living in the soil. Additionally, synthetic mulching materials made of plastic could also lead to the loss of the soil invertebrates population (Bandopadhyay et al., 2018). Microplastic particles that are present in facial scrubs will be continuously entering the sewage system as these are direct products that are extensively used by consumers. These particles are so small that they even pass through sewage treatment filters. The effluent carrying this microplastic would subsequently be released into the inland water bodies and from there they can enter irrigated lands, estuaries, and oceans (Napper et al., 2015).

REFERENCES

Akdogan, Z., & Guven, B. (2019). Microplastics in the environment: A critical review of current understanding and identification of future research needs. *Environmental Pollution, 254*, 113011. https://doi.org/10.1016/j.envpol.2019.113011

Andrady, A. L. (2011). Microplastics in the marine environment. *Marine Pollution Bulletin, 62*(8), 1596–1605. https://doi.org/10.1016/j.marpolbul.2011.05.030

Bandopadhyay, S., Martin-Closas, L., Pelacho, A. M., & DeBruyn, J. M. (2018). Biodegradable plastic mulch films: Impacts on soil microbial communities and ecosystem functions. In *Frontiers in Microbiology* (Vol. 9, Issue APR). Frontiers Media S.A. https://doi.org/10.3389/fmicb.2018.00819

Bosker, T., Bouwman, L. J., Brun, N. R., Behrens, P., & Vijver, M. G. (2019). Microplastics accumulate on pores in seed capsules and delay germination and root growth of the terrestrial vascular plant Lepidium sativum. *Chemosphere, 226*, 774–781. https://doi.org/10.1016/j.chemosphere.2019.03.163

Boucher, J. and Friot D. (2017). *Primary Microplastics in the Oceans: A Global Evaluation of Sources* (p. 43). IUCN. https://dx.doi.org/10.2305/IUCN.CH.2017.01.en

Ceschin, S., Mariani, F., Di Lernia, D., Venditti, I., Pelella, E., & Iannelli, M. A. (2023). Effects of microplastic contamination on the aquatic plant Lemna minuta (least duckweed). *Plants, 12*(1), 207. https://doi.org/10.3390/plants12010207

de Souza Machado, A. A., Kloas, W., Zarfl, C., Hempel, S., & Rillig, M. C. (2018). Microplastics as an emerging threat to terrestrial ecosystems. In *Global Change Biology* (Vol. 24, Issue 4, pp. 1405–1416). Blackwell Publishing Ltd. https://doi.org/10.1111/gcb.14020

Dong, Y., Gao, M., Qiu, W., & Song, Z. (2021). Effect of microplastics and arsenic on nutrients and microorganisms in rice rhizosphere soil. *Ecotoxicology and Environmental Safety, 211*, 111899. https://doi.org/10.1016/j.ecoenv.2021.111899.

Dris, R., Gasperi, J., Mirande, C., Mandin, C., Guerrouache, M., Langlois, V., & Tassin, B. (2017). A first overview of textile fibers, including microplastics, in indoor and outdoor environments. *Environmental Pollution, 221*, 453–458. https://doi.org/10.1016/j.envpol.2016.12.013

Gentili, R., Quaglini, L., Cardarelli, E., Caronni, S., Montagnani, C., & Citterio, S. (2022). Toxic impact of soil microplastics (PVC) on two weeds: Changes in growth, phenology and photosynthesis efficiency. *Agronomy, 12*(5). https://doi.org/10.3390/agronomy12051219

Geyer, R., Jambeck, J. R., & Law, K. L. (2017). *Production, Use, and Fate of All Plastics Ever Made*. www.science.org

Guo, J. J., Huang, X. P., Xiang, L., Wang, Y. Z., Li, Y. W., Li, H., Cai, Q. Y., Mo, C. H., & Wong, M. H. (2020). Source, migration, and toxicology of microplastics in soil. In *Environment International* (Vol. 137). Elsevier Ltd. https://doi.org/10.1016/j.envint.2019.105263

Hale, R. C., Seeley, M. E., La Guardia, M. J., Mai, L., & Zeng, E. Y. (2020). A global perspective on microplastics. In *Journal of Geophysical Research: Oceans* (Vol. 125, Issue 1). Blackwell Publishing Ltd. https://doi.org/10.1029/2018JC014719

He, D., Luo, Y., Lu, S., Liu, M., Song, Y., & Lei, L. (2018). Microplastics in soils: Analytical methods, pollution characteristics, and ecological risks. In *TrAC—Trends in Analytical Chemistry* (Vol. 109, pp. 163–172). Elsevier B.V. https://doi.org/10.1016/j.trac.2018.10.006

Jeyavani, J., Sibiya, A., Stalin, T., Vigneshkumar, G., Al-Ghanim, K. A., Riaz, M. N., Govindarajan, M., & Vaseeharan, B. (2023). Biochemical, genotoxic and histological implications of polypropylene microplastics on freshwater fish oreochromis mossambicus: An aquatic eco-toxicological assessment. *Toxics*, *11*(3). https://doi.org/10.3390/toxics11030282

Jia, L., Liu, L., Zhang, Y., Fu, W., Liu, X., Wang, Q., Tanveer, M., & Huang, L. (2023). Microplastic stress in plants: Effects on plant growth and their remediations. In *Frontiers in Plant Science* (Vol. 14). Frontiers Media SA. https://doi.org/10.3389/fpls.2023.1226484

Jiang, C., Yin, L., Li, Z., Wen, X., Luo, X., Hu, S., Yang, H., Long, Y., Deng, B., Huang, L., & Liu, Y. (2019). Microplastic pollution in the rivers of the Tibet Plateau. *Environmental Pollution*, *249*, 91–98. https://doi.org/10.1016/j.envpol.2019.03.022

Kublik, S., Gschwendtner, S., Magritsch, T., Radl, V., Rillig, M. C., & Schloter, M. (2022). Microplastics in soil induce a new microbial habitat, with consequences for bulk soil microbiomes. *Frontiers in Environmental Science*, *10*. https://doi.org/10.3389/fenvs.2022.989267

Kumari, A., Rajput, V. D., Mandzhieva, S. S., Rajput, S., Minkina, T., Kaur, R., Sushkova, S., Kumari, P., Ranjan, A., Kalinitchenko, V. P., & Glinushkin, A. P. (2022). Microplastic pollution: An emerging threat to terrestrial plants and insights into its remediation strategies. In *Plants* (Vol. 11, Issue 3). MDPI. https://doi.org/10.3390/plants11030340

Lehtiniemi, M., Hartikainen, S., Näkki, P., Engström-Öst, J., Koistinen, A., & Setälä, O. (2018). Size matters more than shape: Ingestion of primary and secondary microplastics by small predators. *Food Webs*, *17*, e00097. https://doi.org/10.1016/j.fooweb.2018.e00097

Li, H.-Z., Zhu, D., Lindhardt, J. H., Lin, S.-M., Ke, X., & Cui, L. (2021). Long-term fertilization history alters effects of microplastics on soil properties, microbial communities, and functions in diverse farmland ecosystem. *Environmental Science & Technology*, *55*(8), 4658–4668. https://doi.org/10.1021/acs.est.0c04849

Liu, K., Wang, X., Wei, N., Song, Z., & Li, D. (2019). Accurate quantification and transport estimation of suspended atmospheric microplastics in megacities: Implications for human health. *Environment International*, *132*. https://doi.org/10.1016/j.envint.2019.105127

Liu, Y., Zhang, J., Zhao, H., Cai, J., Sultan, Y., Fang, H., Zhang, B., & Ma, J. (2022). Effects of polyvinyl chloride microplastics on reproduction, oxidative stress, and reproduction and detoxification-related genes in Daphnia magna. *Comparative Biochemistry and Physiology Part C: Toxicology & Pharmacology*, *254*, 109269. https://doi.org/10.1016/j.cbpc.2022.109269

Lozano, Y. M., & Rillig, M. C. (2022). Legacy effect of microplastics on plant–soil feedbacks. *Frontiers in Plant Science*, *13*. https://doi.org/10.3389/fpls.2022.965576

Martinho, S. D., Fernandes, V. C., Figueiredo, S. A., & Delerue-Matos, C. (2022). Microplastic pollution focused on sources, distribution, contaminant interactions, analytical methods, and wastewater removal strategies: A review. *International Journal of Environmental Research and Public Health, 19*(9). MDPI. https://doi.org/10.3390/ijerph19095610

Napper, I. E., Bakir, A., Rowland, S. J., & Thompson, R. C. (2015). Characterization, quantity, and sorptive properties of microplastics extracted from cosmetics. *Marine Pollution Bulletin*, *99*(1–2), 178–185. https://doi.org/10.1016/j.marpolbul.2015.07.029

Omidoyin, K. C., & Jho, E. H. (2023). Effect of microplastics on soil microbial community and microbial degradation of microplastics in soil: A review. *Environmental Engineering Research*, *28*(6), 220716–0. https://doi.org/10.4491/eer.2022.716

Qi, Y., Yang, X., Pelaez, A. M., Huerta Lwanga, E., Beriot, N., Gertsen, H., Garbeva, P., & Geissen, V. (2018). Macro- and micro-plastics in soil-plant system: Effects of plastic mulch film residues on wheat (Triticum aestivum) growth. *Science of the Total Environment*, *645*, 1048–1056. https://doi.org/10.1016/j.scitotenv.2018.07.229

Rai, M., Pant, G., Pant, K., Aloo, B. N., Kumar, G., Singh, H. B., & Tripathi, V. (2023). Microplastic pollution in terrestrial ecosystems and its interaction with other soil pollutants: A potential threat to soil ecosystem sustainability. *Resources*, *12*(6), 67. https://doi.org/10.3390/resources12060067

Rillig, M. C., Lehmann, A., de Souza Machado, A. A., & Yang, G. (2019). Microplastic effects on plants. In *New Phytologist* (Vol. 223, Issue 3, pp. 1066–1070). Blackwell Publishing Ltd. https://doi.org/10.1111/nph.15794

Rillig, M. C., Ziersch, L., & Hempel, S. (2017). Microplastic transport in soil by earthworms. *Scientific Reports*, *7*(1). https://doi.org/10.1038/s41598-017-01594-7

Sajjad, M., Huang, Q., Khan, S., Khan, M. A., Liu, Y., Wang, J., Lian, F., Wang, Q., & Guo, G. (2022). Microplastics in the soil environment: A critical review. In *Environmental Technology and Innovation* (Vol. 27). Elsevier B.V. https://doi.org/10.1016/j.eti.2022.102408

Schwabl, P., Köppel, S., Königshofer, P., Bucsics, T., Trauner, M., Reiberger, T., & Liebmann, B. (2019). Detection of various microplastics in human stools. *Annals of Internal Medicine*, *171*(7), 453–457. https://doi.org/10.7326/M19-0618

Serrano-Ruíz, H., Eras, J., Martín-Closas, L., & Pelacho, A. M. (2020). Compounds released from unused biodegradable mulch materials after contact with water. Polymer Degradation and Stability, 178, 109202. https://doi.org/10.1016/j.polymdegradstab.2020.109202

Shorobi, F. M., Vyavahare, G. D., Seok, Y. J., & Park, J. H. (2023). Effect of polypropylene microplastics on seed germination and nutrient uptake of tomato and cherry tomato plants. *Chemosphere*, *329*, 138679. https://doi.org/10.1016/j.chemosphere.2023.138679

Subaramaniyam, U., Allimuthu, R. S., Vappu, S., Ramalingam, D., Balan, R., Paital, B., Panda, N., Rath, P. K., Ramalingam, N., & Sahoo, D. K. (2023). Effects of microplastics, pesticides, and nano-materials on fish health, oxidative stress, and antioxidant defense mechanism. *Frontiers in Physiology*, *14*. https://doi.org/10.3389/fphys.2023.1217666

Wagner, M., Scherer, C., Alvarez-Muñoz, D., Brennholt, N., Bourrain, X., Buchinger, S., Fries, E., Grosbois, C., Klasmeier, J., Marti, T., Rodriguez-Mozaz, S., Urbatzka, R., Vethaak, A. D., Winther-Nielsen, M., & Reifferscheid, G. (2014). Microplastics in freshwater ecosystems: What we know and what we need to know. *Environmental Sciences Europe*, *26*(1), 1–9. https://doi.org/10.1186/s12302-014-0012-7

Wan, Z., Wang, C., Zhou, J., Shen, M., Wang, X., Fu, Z., & Jin, Y. (2019). Effects of polystyrene microplastics on the composition of the microbiome and metabolism in larval zebrafish. *Chemosphere*, *217*, 646–658. https://doi.org/10.1016/j.chemosphere.2018.11.070

Wang, L., Liu, Y., Kaur, M., Yao, Z., Chen, T., & Xu, M. (2021). Phytotoxic effects of polyethylene microplastics on the growth of food crops soybean (glycine max) and mung bean (Vigna radiata). *International Journal of Environmental Research and Public Health*, *18*(20), 10629. https://doi.org/10.3390/ijerph182010629

Wright, S. L., & Kelly, F. J. (2017). Plastic and human health: A micro issue? *Environmental Science and Technology*, *51*(12), 6634–6647. https://doi.org/10.1021/acs.est.7b00423

Yang, C., Potts, R., & Shanks, D. R. (2018). Enhancing learning and retrieval of new information: A review of the forward testing effect. In *NPJ Science of Learning* (Vol. 3, Issue 1). Springer Nature. https://doi.org/10.1038/s41539-018-0024-y

Yu, H., Qi, W., Cao, X., Hu, J., Li, Y., Peng, J., Hu, C., & Qu, J. (2021). Microplastic residues in wetland ecosystems: Do they truly threaten the plant-microbe-soil system? *Environment International*, *156*. https://doi.org/10.1016/j.envint.2021.106708

Yu, H., Zhang, Y., Tan, W., & Zhang, Z. (2022). Microplastics as an Emerging Environmental Pollutant in Agricultural Soils: Effects on Ecosystems and Human Health. *Frontiers in Environmental Science, 10*. Frontiers Media S.A. https://doi.org/10.3389/fenvs.2022.855292

Zantis, L. J., Borchi, C., Vijver, M. G., Peijnenburg, W., Di Lonardo, S., & Bosker, T. (2023). Nano- and microplastics commonly cause adverse impacts on plants at environmentally relevant levels: A systematic review. *Science of the Total Environment*, *867*, 161211. https://doi.org/10.1016/J.SCITOTENV.2022.161211

Zhang, S., Wang, J., Liu, X., Qu, F., Wang, X., Wang, X., Li, Y., & Sun, Y. (2019). Microplastics in the environment: A review of analytical methods, distribution, and biological effects. *TrAC Trends in Analytical Chemistry*, *111*, 62–72. https://doi.org/10.1016/j.trac.2018.12.002

Zhang, Y., Gao, T., Kang, S., Shi, H., Mai, L., Allen, D., & Allen, S. (2022). Current status and future perspectives of microplastic pollution in typical cryospheric regions. *Earth-Science Reviews, 226*. Elsevier B.V. https://doi.org/10.1016/j.earscirev.2022.103924

Zhang, Y., Kang, S., Allen, S., Allen, D., Gao, T., & Sillanpää, M. (2020). Atmospheric microplastics: A review on current status and perspectives. *Earth-Science Reviews, 203*. Elsevier B.V. https://doi.org/10.1016/j.earscirev.2020.103118

10 Unveiling the Microplastic Menace

Exploring the Soil Microbiome and Ecological Consequences

Nilendu Basak, Atif Aziz Chowdhury, Ankita Chatterjee, Parama Das Gupta, and Ekramul Islam

10.1 INTRODUCTION

Plastics have been used in all aspects of life because of their versatility, durability, and cost-effectiveness. A large amount of plastic waste is dumped every day in the environment, and, thanks to its poor biodegradability and limited recycling facilities, that waste is increasing daily. However, plastics are often broken into small pieces because of physical, chemical, and biological impacts. Microplastics (MPs) (<5 mm in size) are small plastic particles often generated by such breaking down or byproducts such as facial cleansers and toothpaste, which have become ubiquitous environmental pollutants of emerging concern (Sajjad et al., 2022). Their small size makes MPs prone to being transferred through organisms, posing a threat to them as well as to the environment. In addition, they often contain additives like plasticizers and colorants, multiplying the deadly consequences (Guo et al., 2020). While the prevalence of MPs in marine environments has been widely studied, there is a growing recognition that soils represent a major global sink for MPs. Recent studies have detected MPs in soils from both the agricultural and the natural environments across the globe (He et al., 2018). The accumulation of MPs in soils has raised concerns about their potential impacts on soil health and functioning.

A significant portion of the world's biodiversity is found in soils, which support several essential ecological services. Functions such as air and water quality and composition, nutrient cycling, and plant growth are essential for the health and sustainability of our planet. Multiple elements of soil biodiversity drive ecosystem functions across biomes, and soil food web properties explain ecosystem services (Yu et al., 2022). The continuous ability of the soil to serve as an essential living ecosystem that supports humans, animals, and plants is known as soil health. Problems with environmental sustainability that are connected to soil include climate change, biodiversity loss, water security, energy security, and other issues. Soil microbes are

DOI: 10.1201/9781032684574-10

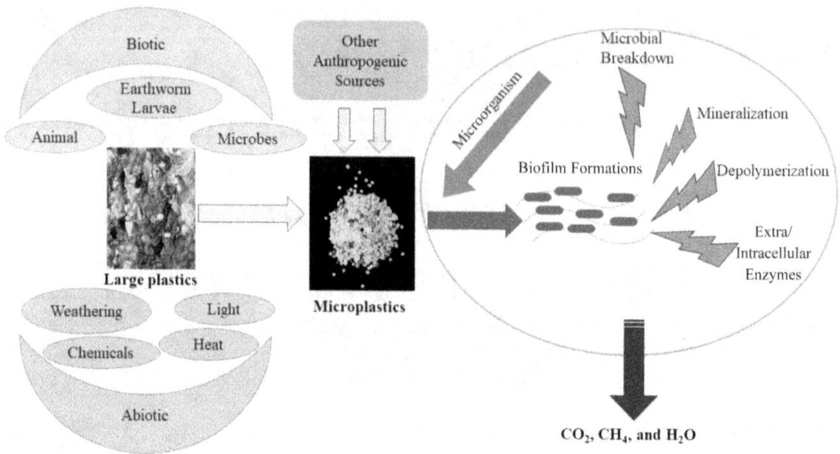

FIGURE 10.1 Schematic representation of soil microplastics flux.

the key element in sustaining these interactions by participating in nutrient (carbon, nitrogen, phosphorus, and sulfur) cycling, soil structure and aggregation, plant nutrition, and fertility, ecosystem functioning, and carbon storage (Yu et al., 2014, 2022).

Unfortunately, soil has become an important reservoir of MPs because they are generally retained in the soil. Various MPs present in soil affect the soil environment differently (Wen et al., 2022). Agricultural mulching, sludge operations, sewage irrigation, use of organic fertilizers, abrasion of tires, and personal care products are the principal sources of MP pollution in soil (Sajjad et al., 2022; Wen et al., 2022). Due to the bioturbation process, harvesting, and soil management methods, these MPs may disperse into soil particles (Lwanga et al., 2017). These can alter soil properties such as soil pH, soil organic matter, water holding capacity, soil erosion, and, most importantly, microbial activities and diversities. Moreover, any disruptions to soil microbiomes could therefore have far-reaching consequences. A few microbes are reported to biodegrade MPs through various mechanisms in association with abiotic factors (Figure 10.1). A handful of studies have analyzed the effects of MPs on soil microbial biomass and community structure. However, our understanding of MP-microbiome interactions remains limited.

Despite increased scientific attention, MP contamination in soils is still an emerging environmental concern, with major information gaps in understanding the magnitude, fate, and ecological implications of soil MPs. In this chapter, we have summarized the present level of knowledge on the origins, occurrence, and ecological hazards of MPs in soil. Additionally, we have explored the effects of MPs on the soil microbiome and how it changes microbial diversity. It is critical to enhance our knowledge in order to create solutions for lowering MP burdens and minimizing their effects on long-term soil health and ecosystem functioning.

10.2 ORIGIN OF MICROPLASTICS IN SOIL

Considering the vital characters of soil in sustaining biodiversity, regulating nutrient cycles, and providing food, it is crucial to understand the sources of MPs in soils

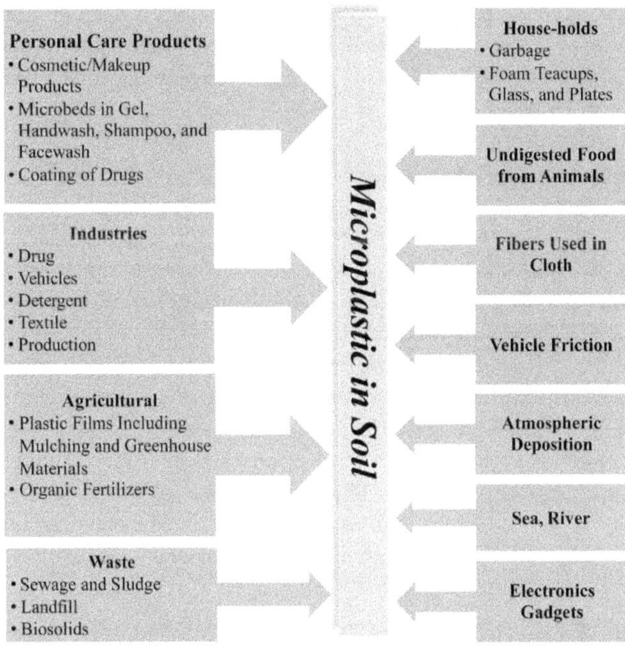

FIGURE 10.2 Anthropogenic sources leading to microplastics pollution in soil.

across different land use types. Due to its wide applicability, many manufactural industries in agro, textile, vehicle, personal care items, and drug production are some of the biggest suppliers of MPs in the terrestrial environment (Zhao et al., 2022). Figure 10.2 depicts the possible routes of entry of MPs in the soil system that includes sources such as sewage sludge, irrigation, plastic mulching, runoffs, compost, etc.

Microplastics primarily enter the lithosphere through the degradation of plastic mulches and open dumping. Sunlight breaks down these polymers into MPs over time. In wet agricultural systems like paddy fields or irrigation water systems, plastic-coated urea fertilizers (organic fertilizers contain MPs with >1 mm size) have been identified as major MP sources (Gao et al., 2021; Li et al., 2020). Plastic mulch helps suppress weeds, conserve moisture, and maintain optimal soil temperature, thereby promoting crop growth (Li et al., 2020). Sewage sludge is used in agriculture as a fertilizer and soil amendment because of its high organic matter content. This can provide nutrients and improve soil structure. In rain-fed farms, plastic mulch and sewage sludge application contribute significantly to soil MP loads. Though sewage treatment removes many MPs, sludge often contains high concentrations of micro- and macroplastics ranging from 1500 to 24,000 to 170,000 items/kg (Cydzik-Kwiatkowska et al., 2022; He et al., 2018). The average annual release of MPs from sludge into the environment in China is estimated at 1.56×10^{14} particles. Estimates also show total yearly inputs of up to 430,000 and 300,000 tons of MPs to farmlands in Europe and North America, respectively (He et al., 2018).

The use of untreated wastewater to irrigate farms is another source of MPs in agricultural soils. Irrigation wastewater has been found to contain many small plastic particles with diameters of 164–327 mm on average. High concentrations of MPs up to 627,000 items/m^3 have been detected in household wastewater. Total concentrations of 80–260 mg/m^3 of polyethylene (PE) and polypropylene (PP) MPs have also been found in wastewater (Hartline et al., 2016; Majewsky et al., 2016).

Atmospheric transport can move MPs over long distances, contributing to soil contamination. Studies show varied shapes, deposition fluxes, and seasonal variations of atmospheric MPs in coastal urban areas in China, indicating a key source for coastal soils. Other research found mean MP fallout rates of 29–280 items/m^2 per day in an urban area near Paris, with over 90% plastic fibers and 50% larger than 1000 mm. Atmospheric transport likely exchanges MPs and other pollutants between regions (He et al., 2018).

Other notable MP sources include municipal waste dumping, tire abrasion, and littering. Primary MPs like nurdles, plastic beads, fibers, and powders used in manufacturing and personal care products directly enter the soil (~4,500 to 94,500 microbeads). For example, facial cleansers contain microbeads that are released during use. Additional sources of MPs include littering, illegal dumping, and tire wear. Along with household ammonites, laundry and plastic cups and plates, are also contributing sources of MPs in the environment (Zhao et al., 2022).

10.3 EFFECT OF MICROPLASTICS ON SOIL PROPERTIES

The properties of soils are closely related to their composition. Several contaminants, like MPs from different sources entering the soil, may adversely affect the soil environment and result in an alteration in soil properties (Wang et al., 2022; Yu et al., 2021). The presence of MPs may alter the physicochemical properties of soil such as soil texture, soil aggregation, soil hydraulic conductivity, soil water holding capacity, soil temperature, soil organic matter, soil pH, soil sorption, soil electrical conductivity, soil nitrogen content, soil phosphorus and potassium levels, soil carbon and nitrogen mineralization, and soil enzyme activities (Chia et al., 2022). These may have a negative effect on nutrient availability, alter the soil microbial community, affect plant growth, etc. (de Souza Machado et al., 2018; Wang et al., 2021).

10.3.1 Effect on Soil pH and Physical Properties

Microplastics can alter soil pH through interactions with organic and inorganic components like cations and protons. Studies have shown PE microplastics increase pH in alkaline soils while decreasing pH in acidic soils; however, the underlying mechanisms are unclear (Dissanayake et al., 2022). More research across soil and plastic types is needed to understand the impact on soil pH. MPs also affect key physical properties of soils, including bulk density, water retention, and aggregate stability (Chia et al., 2022). Depending on the type of MPs, the impact on the soil properties varies. For example, polyester reduces the bulk density of soil, while polyester microfiber does not (Li et al., 2023). By changing soil porosity, MPs can increase evaporation leading to cracking. Different MPs have varying effects on water-holding capacity; for example, polyester fibers are impacted more than PE and polyacrylic (de Souza Machado et al., 2018). Some 72% of soil MPs reduce soil water aggregates that control soil structure, porosity, and stability (Zhang and Liu, 2018).

10.3.2 Effect on Fertility and Nutrient Cycles

Soil organic matter and mineral decomposition are the main sources of soil nutrients, and the soil nutrient cycle is an indicator of soil fertility. Because MPs change the soil's physical properties, they add up to a decrease in soil fertility. The presence of different types and forms of MPs affect the nutrient cycle—mainly the C and N in the soil.

10.3.2.1 Effect on Soil Nitrogen Content

Nitrogen is considered one of the most important limiting factors of soil (Liu et al., 2022). Nitrogen is present in soil in a form that is generally unavailable to plants. In soil, nitrogen is present as organic nitrogen, which is not utilized by plants unless it is mineralized to inorganic form (Du et al., 2020; Liu et al., 2022). Altered physical, chemical, biological properties of soils in the presence of MPs negatively impact nutrient availability, soil microbial community, plant growth, etc., with potential impact on nitrogen availability and mineralization (de Souza Machado et al., 2018; Liu et al., 2017; Wang et al., 2020). Obviously, the type, size, and concentration of MPs present in soil affect soil nitrogen transformation differentially (Shen et al., 2022). Although some researchers have studied the effect of MPs on nitrogen availability in soil, the results are rather conflicting or inconsistent. In one study, plasticizers containing polyvinyl chloride (PVC) MPs at 0.5% (w/w) significantly increased soil NH_4^+-N content and decreased NO_3^--N content by 91% (Zhu et al., 2022). This study also showed that, under this condition, nitrogen-fixing microorganisms, urea decomposers, and nitrate reducers are present in abundance but lowered the number of nitrifying microorganisms (Zhu et al., 2022). On the contrary, the addition of polylactic acid (PLA) increases the nitrate content while decreasing the ammonium concentration (Chen et al., 2020). Another study showed that, upon addition of 28% (w/w) PP-MPs in loess soil, a significant change in dissolved organic nitrogen (DON) and nitrate (NO_3^-) occurred (Liu et al., 2017). The addition of PE-MPs affected microorganisms related to the nitrogen cycle and also reduced nitrogen availability (Boots et al., 2019). PP directly reduced nitrogen availability due to the chelation of ammonium nitrogen present in the soil (Bandopadhyay et al., 2020). Another study indicated that PE can reduce the activities of the enzyme related to the soil nitrogen cycle and thus reduces its availability (Green et al., 2016).

10.3.2.2 Effect on Soil Carbon Content

Carbon turnover in soil refers to a process of the migration and transformation of carbon through plants and soil components and also through the atmosphere and water (Yu et al., 2014). Various factors like the physicochemical properties of soil, vegetation, microbial activity can affect the carbon turnover in soil (Yao et al., 2022; Yu et al., 2014). As MPs have an effect on the physical and chemical properties of soil and thus affect plant growth and microbial activity, they influence the carbon turnover of soil. MPs influence carbon turnover by inhibiting plant growth and thus reducing the primary productivity of plants and CO_2 fixation. Plant roots and litters serve as an important source of soil organic matter. MPs may affect this by altering the nutrient content and water-holding capacity of the soil, thus reducing plant biomass accumulation (Yao et al., 2022). Some microbial hydrolases can hydrolyze the various bonds in different MPs and utilize them as carbon sources, resulting in the production of greenhouse gas like CO_2 (Stubbins et al., 2021). It is observed that PHBV increases microbial activity that exhibit co-metabolism, resulting in an increase in carbon mineralization (Zhou et al., 2021). The addition of PE-MP also increases the mineralization of soil organic matter (Kuzyakov, 2010).

10.3.3 Effect on Soil Enzyme Activity

All six classes of enzymes—oxidoreductase, hydrolase, transferase, isomerase, lyase, and ligase—are present in soil. They are produced and secreted by plants and soil microorganisms as well as released during the decomposition of plant and animal remains (Huang et al., 2019). These enzymes are associated with various functions like the soil nutrient cycle, flow of soil energy, decomposition of soil organic matter, quality of soil, etc. (Cui et al., 2019). These enzymes are very sensitive to soil environment; any changes may lead to alterations in the activity of these enzymes. They are also involved in assessing soil fertility. Presence of MPs affect the soil enzyme activity. MPs act differentially on enzyme activity; studies showed the results are somehow conflicting. Some MPs increase or decrease the enzyme activity, whereas others have no effect at all. Generally, the type of MPs, concentration, structure, and exposer time determine the effect of MPs on soil enzyme activity (Yu et al., 2022). MPs reduce the activity of urease, catalase, and dehydrogenase enzymes significantly (Huang et al., 2019; Yi et al., 2021). Various enzymes involved in the nutrient cycle are also affected by MPs. Activities of leucine aminopeptidase, alkaline phosphatase, beta-glucosidase, and cellobiohydrolase, involved in the N, P, and C cycles respectively, are also reduced in the presence of MPs (Awet et al., 2018). Some studies showed that high concentrations of PLA-MPs, PBS-MPs, and PHB-MPs $(2\%\ w/w)$ increase the activities of urease, phosphatase, and catalase; low concentrations $(0.2\%\ w/w)$ of these MPs decrease the activity of phosphatase but do not significantly change the activities of urease and catalase (Feng et al., 2022). Another study showed that PE-MPs increase urease, catalase, and acid phosphatase activity and inhibit FDAse activity, while having no significant effect on sucrase (Fei et al., 2020; Huang et al., 2019). In some studies, it is observed that PA-MPs have no significant effect on catalase, urease, and beta-glucosidase activities during a 70-day incubation period but that they reduce soil enzyme activity in the first 20 days significantly (Cheng et al., 2021).

10.3.4 Effect of Compound Pollution with Other Pollutants

Due to their large surface area and hydrophobic nature, MPs can adsorb various potentially toxic compounds like heavy metals and persistent organic pollutants from the terrestrial environment (Sajjad et al., 2022). Together, they can cause pollution to a much larger extent. Many studies showed that MPs not only adsorb various heavy metals like Zn, Cu, Ni, Cd, Pb, Cr, and As but also affect the transformation and migration of these heavy metals, resulting in their bioavailability (Wen et al., 2022). MPs can also act as a vector to carry these toxic materials to different organisms (Koelmans et al., 2016). They also serve as transporters to increase the concentration of these heavy metals, facilitate their transport to the food chain and cause biomagnification (Tang et al., 2020). Due to differences in their physical and chemical nature like surface area and polarity, different MPs adsorb heavy metals differentially (Massos and Turner, 2017). Further, the capacity of adsorption of heavy metals by MPs increases with factors like aging of the MPs and UV exposure (Ding et al., 2020). MPs may exhibit some positive role by limiting the movement of heavy metals in soil. A study carried out by Dong et al. (2021) showed that MPs can reduce arsenic pollution in rice rhizospheric soil.

Apart from heavy metals, MPs can adsorb a large array of persistent organic pollutants (POPs). Several organic pollutants like polychlorinated bisphenols (PCBs), polycyclic aromatic hydrocarbons (PAHs), various pesticides, herbicides are adsorbed on the surface of MPs. MPs act as a vector for these organic pollutants and transport them to organisms that consume them. Thus the pollutants adsorbed by the MPs are bioavailable to animals and enter the food chain (Koelmans et al., 2016; Tang et al., 2020). MPs can change the properties of POPs like mobility, leaching, sorption ability (Sajjad et al., 2022). Together they can cause a more severe threat to the soil ecosystem. In one study, it was observed that the combined effect of MPs and phenanthrene reduced photosynthetic capability, antioxidant activities, and growth of wheat seedlings (Liu et al., 2021).

10.3.5 Antibiotic-Resistant Genes

In recent years, antibiotics have emerged as a potential pollutant. Due to the extensive application of antibiotics in human, veterinary medicine, and animal husbandry, antibiotic-resistant genes (ARGs) are widespread in soil. MPs also adsorb several antibiotics on their surface (Sun et al., 2018). The adsorption of different MPs (PA, PE, PS, PP, PVC) and antibiotics (ampicillin, ciprofloxacin, tetracycline, trimethoprim, sulfadiazine) were studied, and it was found that PA has the maximum adsorption ability to antibiotics (Li et al., 2018). Furthermore, MPs have a coexistence with an antibiotic-resistant microbiome that develops biofilm on the surface of the MPs (Yang et al., 2020). Horizontal gene transfer between phylogenetically distinct microorganisms that are present on surface of MPs is much faster than in free living microorganisms. Thus MPs have emerged as a potential health threat globally (Richard et al., 2019).

10.4 ECOLOGICAL CONSEQUENCES OF MP POLLUTION IN SOIL

10.4.1 Effect of Microplastics on Soil Animals

There are limited studies conducted to assess the toxic effects of MPs on soil animals. The effect of polybrominated diphenyl ether (PBDE) present in polyurethane (PU) foam was studied on earthworms by Gaylor et al. (2013). It was reported that the PBDE released from the PU forms aggregates in the earthworms' bodies. This study indicated that hazardous chemicals present in the MPs might be ingested by the soil organisms when they enter the soil system (Gaylor et al., 2013). Exposure to LDPE on earthworms at higher concentrations (28–60% w/w) affects the health of the earthworms, and the MPs are transferred to other organisms through the food chain (Lwanga et al., 2016). The presence of MPs in the soil reaches to the burrows created by the earthworms in the soil as well. This occurs due to the movement of the earthworms in the soil once they accumulate the MPs in their bodies (Lwanga et al., 2017). The burrowing of earthworms has also been observed to increase soil porosity as well as water infiltration in the soil, which in turn supports entry of MPs to the depth of the soil (Helmberger et al., 2020). Though MPs are reported to have been accumulated in the earthworms' bodies, no significant hazardous impacts are noticed on the growth, survival, or reproduction of the earthworms. However, immunological responses and histopathological alterations are induced by the MPs in the earthworms (Rodriguez-Seijo et al., 2017). The effect of MPs was further studied in species of springtails where the transport of urea-formaldehyde components was analyzed. The experimental analyses revealed that the type and size of particles along with the size of the organisms are the major factors for determining the transport rate of the MPs in the soil (Maaß et al., 2017). Soil arthropods are also capable of transferring MPs in the depth of the soil. Though not much research has been conducted regarding this subject, the conducted studies conclude that MPs can be accumulated by the soil organisms and cause certain chemical alterations and adsorption, thereby getting transferred throughout the food chain (Chae and An, 2018).

10.4.2 Effect on Plants

Researchers are keenly studying the impact of MPs on the plants over the past few years; however, the effect is very unclear and requires greater exploration. When exploring the effect of MPs in higher plants, it has been observed that two types of effects are found: stimulating effect and inhibitory effect. Though the MPs are pollutants in nature, they are observed to stimulate growth of higher plants in certain cases. Certain plastic components, being organic polymer in nature, act as a source of nitrogen for the plants. The nitrogen present in the MPs leaches into the soil and might enhance or aid in fertilization. Nitrogen-containing MPs have been observed to increase the nitrogen content of leaf and biomass growth in the plant *Allium fistulosum* (Rillig et al., 2019). Studies also revealed that non-nitrogenous polystyrene (PS) positively affected the presence of photosynthetic pigment in the plant leaf. The growth conditions and concentration of chlorophyll in wheat plants were enhanced with efficient rate post the exposure to PS at a rate of 0.01–1 mg/L (Lian et al., 2020). Scientists also hypothesized that PS induces the alpha-amylase activity and thereby enhances starch hydrolysis, ultimately increasing the availability of soluble sugar for the seedling of wheat plants to germinate. Polyester microfibers are able to enhance the mass of the shoot of *Calamagrostis* under watered conditions and *Hieracium* under drought environmental conditions (Lozano and Rillig, 2020).

Though the stimulating effects were found to be promising, the positive impact can only persist under a limited MP concentration in the environment. Studies across the world have stated that overexposure of MPs has a significant hazardous impact or inhibitory effect on plants such as cress, perennial ryegrass, spring onion, etc. (Bosker et al., 2019; Guo et al., 2020; Jiang et al., 2019). In studies related to the herb cress, the accumulation of MPs was observed in the pores of the seed capsules, which effectively reduces the rate of germination if exposed for more than 8 hours. The root lengths of the herbs are also impacted by hampering their growth in event the herbs are in contact with MPs for more than 24 hours (Bosker et al., 2019). The hydroponic variant of *Vicia faba* (commonly known as broad bean), when exposed to PS-MPs, was observed to accumulate the MPs in their root tips and thereby exhibited inhibition in growth, oxidative, and genotoxic damages (Jiang et al., 2019). The effect of six different MPs was observed on the growth of spring onions. The MPs could significantly halt the growth of the biomass and create imbalance in the carbon/nitrogen ratio, water content, and nitrogen content of the leaf. It is also reported that the root characteristics like root length, diameter, area, and tissue density significantly decreased in the presence of MPs in their surrounding (de Souza Machado et al., 2019). The accumulation of MPs in several parts of plants severely blocks the cellular connections or might clog the pores of the cell walls, which further pauses the nutrient transfer in the plant cells (Guo et al., 2020). PS-MPs aggregates in the leaf vessels of the higher plants block the uptake of nutrients and water, and thus it can be concluded that they impact the transport of mineral elements as well (Senavirathna et al., 2022; Sun et al., 2021). The plants, once they experience the lack of water and nutrients, would be inhibited in their growth. The exposure of lettuce leaves to PS MPs significantly affected the photosynthetic reactions as the accumulation of the MPs on the leaf surface blocks the trapping of sunlight (Li et al., 2022).

10.4.3 Effects on Soil Microbiome

Soil microbes are the drivers of soil ecosystem services. They are involved in maintaining soil fertility, plant growth, and productivity, as well as environmental biogeochemical cycles. But MP buildup in soil impedes soil microbial growth and activity. Yet there is still a void in terms of available studies that focus on the consequences of MPs in the soil microbiome. Short life and relatively small size of soil microbiota complicates the study. In recent years, various studies have been engaged to analyze the consequences of MPs in the environment as well as the microorganisms dwelling there. Since MP pollution is highly associated with the aquatic environment, most of the studies conducted focuses on the alarming effects of MPs among aquatic microorganisms. However, scientists now are focusing on the deposition of plastics and their impacts in agricultural land and landfill sludges. The studies conducted to derive the effect of MPs among soil microbes have reported mixed views on the final impact. Certain studies revealed that a significant amount of LDPE, HDPE, polyethylene terephthalate (PET), and PVC did not impact the microbial consortium present in the soil (Judy et al., 2019). To the contrary, various other studies have suggested that there are structural and functional changes in the microbial community when infected with MPs in the soil (Gao et al., 2021). Microbial communities are effectively hampered in the ample presence of PE and PP in the soil. It is reported that the count of *Acidobacteria* and *Bacteroidetes* increases in the presence of PE and PP, thereby reducing the existence of *Deinococcus-Thermus* and *Chloroflexi* (Yi et al., 2021). Moreover, the presence of LDPE and PET at a rate of 3% w/w and 0.2–0.4% w/w, respectively, affects the community of soil bacteria. Earlier studies reveal that PS and polyacrylic components impact the soil quality by affecting fluorescein diacetate, which reduces the microbial count in the soil in the long term (de Souza Machado et al., 2019). Presence of MP in the soil negatively impacts the bacterial transport as well as possibly having an effect on the elimination or destruction of antibiotic-resistant genes. MP also impacts the microbial count when present along with other contaminants, such as heavy metals and chemicals (W. Wang et al., 2020). A significant decrease of diverse bacterial community also occurs when biodegradable cigarette butts, consisting of cellulose acetate, are exposed in the soil (Koroleva et al., 2021). The change in soil quality by the MPs also impacts the aerobic and anaerobic microbial community, causing loss of habitats (Lozano et al., 2021). The increased concentration of PVC-MPs in the soil reduces the count of *Sphingomonadaceae*. PVC along with PE enhances the growth of nitrogen-fixing *Burkholderiaceae* as well as *Pseudomonadaceae*, indicating that MPs might enhance the biological nitrogen fixation in the soil (Fei et al., 2020). Soil microbial strains attach to the surface of MPs and causes formation of biofilms on the MPs. This concept led to the introduction of the term plastisphere (Zettler et al., 2013). Bacteria belonging to the groups *Pseudoalteromonadaceae* and *Vibrionaceae* were found to be enriched by the presence of fragmented plastics in their surroundings as they could colonize on the MPs surface (Feng et al., 2020).

In acidic soil, polyester and polyacrylic MPs are reported to decrease soil microbial activity and diversity, but membranous PE and fibrous PP increased microbial diversity (Li et al., 2023). MP accumulation in yeast and filamentous fungi can introduce MPs into the food chain and thereby risk biomagnification. *Aspergillus oryzae* and *Aspergillus nidulans* were reported to be affected by nanoplastics (He et al., 2018). Soil MPs increase the MP-degrading microbial community (Zhang et al., 2019). In the presence of Pb and Zn, MPs were shown to decrease microbial richness and diversity while enhancing special taxa (Feng et al., 2022). As MPs interfere with soil and water, they can alter water availability, resulting in altered microbial community structure, and decrease the functional genes involved in metabolism (Zhao et al., 2022). Illumina MiSeq analysis of soil indicated a higher succession rate in the presence of LDPE MPs. MPs can act as habitat of soil microbiomes; thus the presence of MPs can lead to increased microbial abundance in soil (Fei et al., 2020; Yi et al., 2021). For example, *Vibrio* spp. can grow on MP surfaces, suggesting MPs' role as a vector. Bacterial population on MPs differs from the surroundings. These can alter the microbial community of the soil. However, the toxic element absorbed on the MP surface can increase the persistence of toxic elements, thereby hampering microbial growth. MPs can provide the site of biofilm formation and can contribute to increased microbial survivability (Moyal et al., 2023). Moreover, in the presence of 0.007% LDPE, microbial species turnover tripled (J. Wang et al., 2020); a combination of 3% LDPE and 0.2% and 0.4% PET impacted the soil microbial richness and evenness (Ng et al., 2021).

10.5 BREAKDOWN PATHWAYS OF MICROPLASTIC BY MICROBES

PET is a widely used polymer, with global consumption exceeding 56 million tons annually (Liu et al., 2017). Its prevalence in single-use packaging means much of it ultimately contributes to MPs pollution of soils. Biodegrading PET is a more sustainable approach than recycling it. Microbial degradation of MPs like PET involves enzymes breaking the polymers down in steps like biodeterioration and mineralization. Studying the enzymatic mechanisms involved in PET MP degradation is therefore important. Enzymatic degradation has advantages over other methods, as enzymes require no expensive setups or external energy input (Khoonkari et al., 2015).

The degradation of PET is conducted by several enzymes commonly called PETase. The breakdown of PET occurs in two mechanisms: abiotic and biotic degradation. Abiotic mode of breakdown includes chemical and thermal degradation as well as hydrolysis, whereas biotic mechanism involves the application of the biological approach (Arhant et al., 2019). Biotic degradation is much more attainable due to the complete mineralization ability of the MPs, ending up in the production of nontoxic end products. The mechanism involved in the degradation mostly depends on the activity of two PETase enzymes. The first PETase primarily cleaves the PET into mono (2-hydroxyethyl) terephthalic acid (MHET) along with trace presence of bis (2-hydroxyethyl)-TPA (BHET) and terephthalic acid (TPA). After the primary breakdown, the second enzyme, called as MHETase, plays a role in the mineralization of MHET to form terephthalate and ethylene glycol. The ethylene glycol can further be converted into acetyl CoA, followed by acetate or direct conversion of isocitrate. Terephthalate continues to enter in a series of metabolic reactions to produce protocatechuate (PCA), which undergoes a metabolic pathway called as beta-ketoadipate pathway, as shown in Figure 10.3 (Salvador et al., 2019).

10.6 CONCLUSION AND FUTURE OUTLOOK

An imminent concern that deserves urgent attention is the emergence of MP pollution in soils. Research has uncovered that MPs stem from a range of sources, such as the application of sewage sludge and plastic mulching and through the use of wastewater for irrigation, which imparts contamination on both agricultural and natural soil worldwide. The initial signs indicate that MPs can have adverse effects on soil health

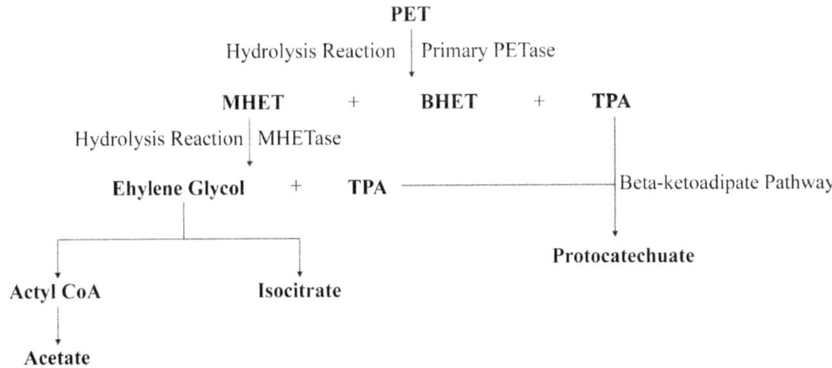

FIGURE 10.3 Degradation pathway of polyethylene terephthalate (PET) by microbial enzymes.

and function. It is believed that MPs will alter the physicochemical properties of soil, thus reducing overall fertility, disrupting biodiversity across trophic levels, and compromising the ecological functions that are responsible for maintaining productivity and sustainability. Despite the limited research done on this matter, several aspects, including the quantification of MP accumulation patterns, the analysis of effects on soil processes, and the elucidation of risks to organisms, continue to require further attention, especially in regard to interactions with other pollutants.

It is vital to undertake further inquiries to fill in the gaps in our knowledge. One way to achieve this is by measuring MP loads across different land uses, regions, and soil types, which can help detect areas with high pollution levels and devise targeted remedies accordingly. A thorough analysis of how various plastic qualities such as sizes, shapes, concentrations, and polymers influence soil properties (physical, chemical, and biological) can help unveil the effects of plastics on soil processes. Hence, a comprehensive investigation of toxicological and microbiome aspects is necessary to comprehend the repercussions on soil biodiversity. These findings can lead to taking preventive measures such as decreasing the use of plastic in the agriculture sector, upgrading wastewater treatments, developing degradable substitutes, and enforcing regulations to control plastic pollution. Reducing plastic usage and litter could be encouraged through public education. To curb the accumulation of MP, though, swift global solutions to limit plastic pollution would need to be implemented. Moreover, the environmental consequences of biodegradation intermediates of MPs should be access in future research to develop their sustainable management.

REFERENCES

Arhant, M., Le Gall, M., Le Gac, P.-Y., Davies, P., 2019. Impact of hydrolytic degradation on mechanical properties of PET—towards an understanding of microplastics formation. *Polymer Degradation and Stability* 161, 175–182. https://doi.org/10.1016/j.polymdegradstab.2019.01.021

Awet, T.T., Kohl, Y., Meier, F., Straskraba, S., Grün, A.-L., Ruf, T., Jost, C., Drexel, R., Tunc, E., Emmerling, C., 2018. Effects of polystyrene nanoparticles on the microbiota and functional diversity of enzymes in soil. *Environmental Sciences Europe* 30, 11. https://doi.org/10.1186/s12302-018-0140-6

Bandopadhyay, S., Sintim, H.Y., DeBruyn, J.M., 2020. Effects of biodegradable plastic film mulching on soil microbial communities in two agroecosystems. *PeerJ* 8, e9015. https://doi.org/10.7717/peerj.9015

Boots, B., Russell, C.W., Green, D.S., 2019. Effects of microplastics in soil ecosystems: Above and below ground. *Environmental Science & Technology* 53, 11496–11506. https://doi.org/10.1021/acs.est.9b03304

Bosker, T., Bouwman, L.J., Brun, N.R., Behrens, P., Vijver, M.G., 2019. Microplastics accumulate on pores in seed capsule and delay germination and root growth of the terrestrial vascular plant *Lepidium sativum*. *Chemosphere* 226, 774–781. https://doi.org/10.1016/j.chemosphere.2019.03.163

Chae, Y., An, Y.-J., 2018. Current research trends on plastic pollution and ecological impacts on the soil ecosystem: A review. *Environmental Pollution* 240, 387–395. https://doi.org/10.1016/j.envpol.2018.05.008

Chen, H., Wang, Y., Sun, X., Peng, Y., Xiao, L., 2020. Mixing effect of polylactic acid microplastic and straw residue on soil property and ecological function. *Chemosphere* 243, 125271. https://doi.org/10.1016/j.chemosphere.2019.125271

Cheng, Y., Song, W., Tian, H., Zhang, K., Li, B., Du, Z., Zhang, W., Wang, J., Wang, J., Zhu, L., 2021. The effects of high-density polyethylene and polypropylene microplastics on the soil and earthworm *Metaphire guillelmi* gut microbiota. *Chemosphere* 267, 129219. https://doi.org/10.1016/j.chemosphere.2020.129219

Chia, R.W., Lee, J.-Y., Jang, J., Kim, H., Kwon, K.D., 2022. Soil health and microplastics: A review of the impacts of microplastic contamination on soil properties. *Journal of Soils and Sediments* 22, 2690–2705. https://doi.org/10.1007/s11368-022-03254-4

Cui, Y., Fang, L., Guo, X., Wang, X., Wang, Y., Zhang, Y., Zhang, X., 2019. Responses of soil bacterial communities, enzyme activities, and nutrients to agricultural-to-natural ecosystem conversion in the Loess Plateau, China. *Journal of Soils and Sediments* 19, 1427–1440. https://doi.org/10.1007/s11368-018-2110-4

Cydzik-Kwiatkowska, A., Milojevic, N., Jachimowicz, P., 2022. The fate of microplastic in sludge management systems. *Science of the Total Environment* 848, 157466. https://doi.org/10.1016/j.scitotenv.2022.157466

de Souza Machado, A.A., Lau, C.W., Kloas, W., Bergmann, J., Bachelier, J.B., Faltin, E., Becker, R., Görlich, A.S., Rillig, M.C., 2019. Microplastics can change soil properties and affect plant performance. *Environmental Science & Technology* 53, 6044–6052. https://doi.org/10.1021/acs.est.9b01339

de Souza Machado, A.A., Lau, C.W., Till, J., Kloas, W., Lehmann, A., Becker, R., Rillig, M.C., 2018. Impacts of microplastics on the soil biophysical environment. *Environmental Science & Technology* 52, 9656–9665. https://doi.org/10.1021/acs.est.8b02212

Ding, L., Zhang, S., Wang, X., Yang, X., Zhang, C., Qi, Y., Guo, X., 2020. The occurrence and distribution characteristics of microplastics in the agricultural soils of Shaanxi Province, in north-western China. *Science of the Total Environment* 720, 137525. https://doi.org/10.1016/j.scitotenv.2020.137525

Dissanayake, P.D., Kim, S., Sarkar, B., Oleszczuk, P., Sang, M.K., Haque, M.N., Ahn, J.H., Bank, M.S., Ok, Y.S., 2022. Effects of microplastics on the terrestrial environment: A critical review. *Environmental Research* 209, 112734. https://doi.org/10.1016/j.envres.2022.112734

Dong, Y., Gao, M., Qiu, W., Song, Z., 2021. Effect of microplastics and arsenic on nutrients and microorganisms in rice rhizosphere soil. *Ecotoxicology and Environmental Safety* 211, 111899. https://doi.org/10.1016/j.ecoenv.2021.111899

Du, E., Terrer, C., Pellegrini, A.F.A., Ahlström, A., Van Lissa, C.J., Zhao, X., Xia, N., Wu, X., Jackson, R.B., 2020. Global patterns of terrestrial nitrogen and phosphorus limitation. *Nature Geoscience* 13, 221–226. https://doi.org/10.1038/s41561-019-0530-4

Fei, Y., Huang, S., Zhang, H., Tong, Y., Wen, D., Xia, X., Wang, H., Luo, Y., Barceló, D., 2020. Response of soil enzyme activities and bacterial communities to the accumulation of microplastics in an acid cropped soil. *Science of the Total Environment* 707, 135634. https://doi.org/10.1016/j.scitotenv.2019.135634

Feng, L., He, L., Jiang, S., Chen, J., Zhou, C., Qian, Z.-J., Hong, P., Sun, S., Li, C., 2020. Investigating the composition and distribution of microplastics surface biofilms in coral areas. *Chemosphere* 252, 126565. https://doi.org/10.1016/j.chemosphere.2020.126565

Feng, X., Wang, Q., Sun, Y., Zhang, S., Wang, F., 2022. Microplastics change soil properties, heavy metal availability and bacterial community in a Pb-Zn-contaminated soil. *Journal of Hazardous Materials* 424, 127364. https://doi.org/10.1016/j.jhazmat.2021.127364

Gao, B., Yao, H., Li, Y., Zhu, Y., 2021. Microplastic addition alters the microbial community structure and stimulates soil carbon dioxide emissions in vegetable-growing soil. *Environmental Toxicology and Chemistry* 40, 352–365. https://doi.org/10.1002/etc.4916

Gaylor, M.O., Harvey, E., Hale, R.C., 2013. Polybrominated diphenyl ether (PBDE) accumulation by earthworms (*Eisenia fetida*) exposed to biosolids-, Polyurethane foam microparticle-, and penta-BDE-amended soils. *Environmental Science & Technology* 47, 13831–13839. https://doi.org/10.1021/es403750a

Green, D.S., Boots, B., Sigwart, J., Jiang, S., Rocha, C., 2016. Effects of conventional and biodegradable microplastics on a marine ecosystem engineer (*Arenicola marina*) and sediment nutrient cycling. *Environmental Pollution* 208, 426–434. https://doi.org/10.1016/j.envpol.2015.10.010

Guo, J.-J., Huang, X.-P., Xiang, L., Wang, Y.-Z., Li, Y.-W., Li, H., Cai, Q.-Y., Mo, C.-H., Wong, M.-H., 2020. Source, migration and toxicology of microplastics in soil. *Environment International* 137, 105263. https://doi.org/10.1016/j.envint.2019.105263

Hartline, N.L., Bruce, N.J., Karba, S.N., Ruff, E.O., Sonar, S.U., Holden, P.A., 2016. Microfiber masses recovered from conventional machine washing of new or aged garments. *Environmental Science & Technology* 50, 11532–11538. https://doi.org/10.1021/acs.est.6b03045

He, D., Luo, Y., Lu, S., Liu, M., Song, Y., Lei, L., 2018. Microplastics in soils: Analytical methods, pollution characteristics and ecological risks. *TrAC Trends in Analytical Chemistry* 109, 163–172. https://doi.org/10.1016/j.trac.2018.10.006

Helmberger, M.S., Tiemann, L.K., Grieshop, M.J., 2020. Towards an ecology of soil microplastics. *Functional Ecology* 34, 550–560. https://doi.org/10.1111/1365-2435.13495

Huang, Y., Zhao, Y., Wang, J., Zhang, M., Jia, W., Qin, X., 2019. LDPE microplastic films alter microbial community composition and enzymatic activities in soil. *Environmental Pollution* 254, 112983. https://doi.org/10.1016/j.envpol.2019.112983

Jiang, X., Chen, H., Liao, Y., Ye, Z., Li, M., Klobučar, G., 2019. Ecotoxicity and genotoxicity of polystyrene microplastics on higher plant *Vicia faba*. *Environmental Pollution* 250, 831–838. https://doi.org/10.1016/j.envpol.2019.04.055

Judy, J.D., Williams, M., Gregg, A., Oliver, D., Kumar, A., Kookana, R., Kirby, J.K., 2019. Microplastics in municipal mixed-waste organic outputs induce minimal short to long-term toxicity in key terrestrial biota. *Environmental Pollution* 252, 522–531. https://doi.org/10.1016/j.envpol.2019.05.027

Khoonkari, M., Haghighi, A.H., Sefidbakht, Y., Shekoohi, K., Ghaderian, A., 2015. Chemical recycling of PET wastes with different catalysts. *International Journal of Polymer Science* 2015, e124524. https://doi.org/10.1155/2015/124524

Koelmans, A.A., Bakir, A., Burton, G.A., Janssen, C.R., 2016. Microplastic as a vector for chemicals in the aquatic environment: Critical review and model-supported reinterpretation of empirical studies. *Environmental Science & Technology* 50, 3315–3326. https://doi.org/10.1021/acs.est.5b06069

Koroleva, E., Mqulwa, A.Z., Norris-Jones, S., Reed, S., Tambe, Z., Visagie, A., Jacobs, K., 2021. Impact of cigarette butts on bacterial community structure in soil. *Environmental Science and Pollution Research* 28, 33030–33040. https://doi.org/10.1007/s11356-021-13152-w

Kuzyakov, Y., 2010. Priming effects: Interactions between living and dead organic matter. *Soil Biology and Biochemistry* 42, 1363–1371. https://doi.org/10.1016/j.soilbio.2010.04.003

Li, J., Yu, S., Yu, Y., Xu, M., 2022. Effects of microplastics on higher plants: A review. *Bulletin of Environmental Contamination and Toxicology* 109, 241–265. https://doi.org/10.1007/s00128-022-03566-8

Li, J., Zhang, K., Zhang, H., 2018. Adsorption of antibiotics on microplastics. *Environmental Pollution* 237, 460–467. https://doi.org/10.1016/j.envpol.2018.02.050

Li, W., Luo, Y., Pan, X., 2020. Microplastics in agricultural soils, in: He, D., Luo, Y. (Eds.), *Microplastics in Terrestrial Environments, The Handbook of Environmental Chemistry*. Springer International Publishing, Cham, pp. 63–76. https://doi.org/10.1007/698_2020_448

Li, Z., Yang, Y., Chen, X., He, Y., Bolan, N., Rinklebe, J., Lam, S.S., Peng, W., Sonne, C., 2023. A discussion of microplastics in soil and risks for ecosystems and food chains. *Chemosphere* 313, 137637. https://doi.org/10.1016/j.chemosphere.2022.137637

Lian, J., Wu, J., Xiong, H., Zeb, A., Yang, T., Su, X., Su, L., Liu, W., 2020. Impact of polystyrene nanoplastics (PSNPs) on seed germination and seedling growth of wheat (*Triticum aestivum* L.). *Journal of Hazardous Materials* 385, 121620. https://doi.org/10.1016/j.jhazmat.2019.121620

Liu, H., Yang, X., Liu, G., Liang, C., Xue, S., Chen, H., Ritsema, C.J., Geissen, V., 2017. Response of soil dissolved organic matter to microplastic addition in Chinese loess soil. *Chemosphere* 185, 907–917. https://doi.org/10.1016/j.chemosphere.2017.07.064

Liu, S., Wang, J., Zhu, J., Wang, J., Wang, H., Zhan, X., 2021. The joint toxicity of polyethylene microplastic and phenanthrene to wheat seedlings. *Chemosphere* 282, 130967. https://doi.org/10.1016/j.chemosphere.2021.130967

Liu, W., Cao, Z., Ren, H., Xi, D., 2022. Effects of microplastics addition on soil available nitrogen in farmland soil. *Agronomy* 13, 75. https://doi.org/10.3390/agronomy13010075

Lozano, Y.M., Aguilar-Trigueros, C.A., Onandia, G., Maaß, S., Zhao, T., Rillig, M.C., 2021. Effects of microplastics and drought on soil ecosystem functions and multifunctionality. *Journal of Applied Ecology* 58, 988–996. https://doi.org/10.1111/1365-2664.13839

Lozano, Y.M., Rillig, M.C., 2020. Effects of microplastic fibers and drought on plant communities. *Environmental Science & Technology* 54, 6166–6173. https://doi.org/10.1021/acs.est.0c01051

Lwanga, E.H., Gertsen, H., Gooren, H., Peters, P., Salánki, T., Van Der Ploeg, M., Besseling, E., Koelmans, A.A., Geissen, V., 2016. Microplastics in the terrestrial ecosystem: Implications for *Lumbricus terrestris* (Oligochaeta, Lumbricidae). *Environmental Science & Technology* 50, 2685–2691. https://doi.org/10.1021/acs.est.5b05478

Lwanga, E.H., Gertsen, H., Gooren, H., Peters, P., Salánki, T., Van Der Ploeg, M., Besseling, E., Koelmans, A.A., Geissen, V., 2017. Incorporation of microplastics from litter into burrows of *Lumbricus terrestris*. *Environmental Pollution* 220, 523–531. https://doi.org/10.1016/j.envpol.2016.09.096

Maaß, S., Daphi, D., Lehmann, A., Rillig, M.C., 2017. Transport of microplastics by two collembolan species. *Environmental Pollution* 225, 456–459. https://doi.org/10.1016/j.envpol.2017.03.009

Majewsky, M., Bitter, H., Eiche, E., Horn, H., 2016. Determination of microplastic polyethylene (PE) and polypropylene (PP) in environmental samples using thermal analysis (TGA-DSC). *Science of the Total Environment* 568, 507–511. https://doi.org/10.1016/j.scitotenv.2016.06.017

Massos, A., Turner, A., 2017. Cadmium, lead and bromine in beached microplastics. *Environmental Pollution* 227, 139–145. https://doi.org/10.1016/j.envpol.2017.04.034

Moyal, J., Dave, P.H., Wu, M., Karimpour, S., Brar, S.K., Zhong, H., Kwong, R.W.M., 2023. Impacts of biofilm formation on the physicochemical properties and toxicity of microplastics: A concise review. *Reviews of Environmental Contamination (Formerly: Residue Reviews)* 261, 8. https://doi.org/10.1007/s44169-023-00035-z

Ng, E.L., Lin, S.Y., Dungan, A.M., Colwell, J.M., Ede, S., Huerta Lwanga, E., Meng, K., Geissen, V., Blackall, L.L., Chen, D., 2021. Microplastic pollution alters forest soil microbiome. *Journal of Hazardous Materials* 409, 124606. https://doi.org/10.1016/j.jhazmat.2020.124606

Richard, H., Carpenter, E.J., Komada, T., Palmer, P.T., Rochman, C.M., 2019. Biofilm facilitates metal accumulation onto microplastics in estuarine waters. *Science of the Total Environment* 683, 600–608. https://doi.org/10.1016/j.scitotenv.2019.04.331

Rillig, M.C., Lehmann, A., De Souza Machado, A.A., Yang, G., 2019. Microplastic effects on plants. *New Phytologist* 223, 1066–1070. https://doi.org/10.1111/nph.15794

Rodriguez-Seijo, A., Lourenço, J., Rocha-Santos, T.A.P., Da Costa, J., Duarte, A.C., Vala, H., Pereira, R., 2017. Histopathological and molecular effects of microplastics in *Eisenia andrei Bouché*. *Environmental Pollution* 220, 495–503. https://doi.org/10.1016/j.envpol.2016.09.092

Sajjad, M., Huang, Q., Khan, S., Khan, M.A., Liu, Y., Wang, J., Lian, F., Wang, Q., Guo, G., 2022. Microplastics in the soil environment: A critical review. *Environmental Technology & Innovation* 27, 102408. https://doi.org/10.1016/j.eti.2022.102408

Salvador, M., Abdulmutalib, U., Gonzalez, J., Kim, J., Smith, A.A., Faulon, J.-L., Wei, R., Zimmermann, W., Jimenez, J.I., 2019. Microbial genes for a circular and sustainable bio-PET economy. *Genes* 10, 373. https://doi.org/10.3390/genes10050373

Senavirathna, M.D.H.J., Zhaozhi, L., Fujino, T., 2022. Short-duration exposure of 3-μm polystyrene microplastics affected morphology and physiology of watermilfoil (sp. *roraima*). *Environmental Science and Pollution Research* 29, 34475–34485. https://doi.org/10.1007/s11356-022-18642-z

Shen, M., Song, B., Zhou, C., Almatrafi, E., Hu, T., Zeng, G., Zhang, Y., 2022. Recent advances in impacts of microplastics on nitrogen cycling in the environment: A review. *Science of the Total Environment* 815, 152740. https://doi.org/10.1016/j.scitotenv.2021.152740

Stubbins, A., Law, K.L., Muñoz, S.E., Bianchi, T.S., Zhu, L., 2021. Plastics in the earth system. *Science* 373, 51–55. https://doi.org/10.1126/science.abb0354

Sun, H., Lei, C., Xu, J., Li, R., 2021. Foliar uptake and leaf-to-root translocation of nanoplastics with different coating charge in maize plants. *Journal of Hazardous Materials* 416, 125854. https://doi.org/10.1016/j.jhazmat.2021.125854

Sun, M., Ye, M., Jiao, W., Feng, Y., Yu, P., Liu, M., Jiao, J., He, X., Liu, K., Zhao, Y., Wu, J., Jiang, X., Hu, F., 2018. Changes in tetracycline partitioning and bacteria/phage-comediated ARGs in microplastic-contaminated greenhouse soil facilitated by sophorolipid. *Journal of Hazardous Materials* 345, 131–139. https://doi.org/10.1016/j.jhazmat.2017.11.036

Tang, S., Lin, L., Wang, X., Feng, A., Yu, A., 2020. Pb(II) uptake onto nylon microplastics: Interaction mechanism and adsorption performance. *Journal of Hazardous Materials* 386, 121960. https://doi.org/10.1016/j.jhazmat.2019.121960

Wang, W., Ge, J., Yu, X., Li, H., 2020. Environmental fate and impacts of microplastics in soil ecosystems: Progress and perspective. *Science of the Total Environment* 708, 134841. https://doi.org/10.1016/j.scitotenv.2019.134841

Wang, J., Huang, M., Wang, Q., Sun, Y., Zhao, Y., Huang, Y., 2020. LDPE microplastics significantly alter the temporal turnover of soil microbial communities. *Science of The Total Environment* 726, 138682. https://doi.org/10.1016/j.scitotenv.2020.138682

Wang, L., Wu, W.-M., Bolan, N.S., Tsang, D.C.W., Li, Y., Qin, M., Hou, D., 2021. Environmental fate, toxicity and risk management strategies of nanoplastics in the environment: Current status and future perspectives. *Journal of Hazardous Materials* 401, 123415. https://doi.org/10.1016/j.jhazmat.2020.123415

Wang, W., Ge, J., Yu, X., Li, H., 2020. Environmental fate and impacts of microplastics in soil ecosystems: Progress and perspective. *Science of the Total Environment* 708, 134841. https://doi.org/10.1016/j.scitotenv.2019.134841

Wang, Y., Wang, F., Xiang, L., Bian, Y., Wang, Z., Srivastava, P., Jiang, X., Xing, B., 2022. Attachment of positively and negatively charged submicron polystyrene plastics on nine typical soils. *Journal of Hazardous Materials* 431, 128566. https://doi.org/10.1016/j.jhazmat.2022.128566

Wen, X., Yin, L., Zhou, Z., Kang, Z., Sun, Q., Zhang, Y., Long, Y., Nie, X., Wu, Z., Jiang, C., 2022. Microplastics can affect soil properties and chemical speciation of metals in yellow-brown soil. *Ecotoxicology and Environmental Safety* 243, 113958. https://doi.org/10.1016/j.ecoenv.2022.113958

Yang, Y., Liu, W., Zhang, Z., Grossart, H.-P., Gadd, G.M., 2020. Microplastics provide new microbial niches in aquatic environments. *Applied Microbiology and Biotechnology* 104, 6501–6511. https://doi.org/10.1007/s00253-020-10704-x

Yao, Y., Lili, W., Shufen, P., Gang, L., Hongmei, L., Weiming, X., Lingxuan, G., Jianning, Z., Guilong, Z., Dianlin, Y., 2022. Can microplastics mediate soil properties, plant growth and carbon/nitrogen turnover in the terrestrial ecosystem? *Ecosystem Health and Sustainability* 8, 2133638. https://doi.org/10.1080/20964129.2022.2133638

Yi, M., Zhou, S., Zhang, L., Ding, S., 2021. The effects of three different microplastics on enzyme activities and microbial communities in soil. *Water Environment Research* 93, 24–32. https://doi.org/10.1002/wer.1327

Yu, H., Zhang, Y., Tan, W., Zhang, Z., 2022. Microplastics as an emerging environmental pollutant in agricultural soils: Effects on ecosystems and human health. *Frontiers in Environmental Science* 10, 855292. https://doi.org/10.3389/fenvs.2022.855292

Yu, H., Zhang, Z., Zhang, Y., Fan, P., Xi, B., Tan, W., 2021. Metal type and aggregate microenvironment govern the response sequence of speciation transformation of different heavy metals to microplastics in soil. *Science of the Total Environment* 752, 141956. https://doi.org/10.1016/j.scitotenv.2020.141956

Yu, J., Fang, L., Bian, Z., Wang, Q., Yu, Y., 2014. A review of the composition of soil carbon pool. *Acta Ecologica Sinica*, 34. https://doi.org/10.5846/stxb201301050036

Zettler, E.R., Mincer, T.J., Amaral-Zettler, L.A., 2013. Life in the "plastisphere": Microbial communities on plastic marine debris. *Environmental Science & Technology* 47, 7137–7146. https://doi.org/10.1021/es401288x

Zhang, G.S., Liu, Y.F., 2018. The distribution of microplastics in soil aggregate fractions in southwestern China. *Science of the Total Environment* 642, 12–20. https://doi.org/10.1016/j.scitotenv.2018.06.004

Zhang, M., Zhao, Y., Qin, X., Jia, W., Chai, L., Huang, M., Huang, Y., 2019. Microplastics from mulching film is a distinct habitat for bacteria in farmland soil. *Science of the Total Environment* 688, 470–478. https://doi.org/10.1016/j.scitotenv.2019.06.108

Zhao, S., Zhang, Z., Chen, L., Cui, Q., Cui, Y., Song, D., Fang, L., 2022. Review on migration, transformation and ecological impacts of microplastics in soil. *Applied Soil Ecology* 176, 104486. https://doi.org/10.1016/j.apsoil.2022.104486

Zhou, J., Gui, H., Banfield, C.C., Wen, Y., Zang, H., Dippold, M.A., Charlton, A., Jones, D.L., 2021. The microplastisphere: Biodegradable microplastics addition alters soil microbial community structure and function. *Soil Biology and Biochemistry* 156, 108211. https://doi.org/10.1016/j.soilbio.2021.108211

Zhu, F., Yan, Y., Doyle, E., Zhu, C., Jin, X., Chen, Z., Wang, C., He, H., Zhou, D., Gu, C., 2022. Microplastics altered soil microbiome and nitrogen cycling: The role of phthalate plasticizer. *Journal of Hazardous Materials* 427, 127944. https://doi.org/10.1016/j.jhazmat.2021.127944

11 Bioplastic
An Alternate for Plastic Materials and Its Application

Poushali Chakraborty, Venkatalakshmi Jakka, Shubhalakshmi Sengupta, and Papita Das

11.1 INTRODUCTION

Synthetic plastics derived from fossil fuel have various applications in the recent times. Until 2015, the record production of petroleum-based plastics was more than 300 million tons annually (Emadian et al., 2017). The usage of plastics has marked its effect in various fields, such as, packaging industry, vehicles, furniture, construction materials, household appliances, toys, medical equipment, etc. (Bayer et al., 2014). The wide range of plastics available in markets can be divided into two classes: low-density polyethylene plastics (LDPE) and high-density polyethylene plastics (HDPE). Excellent characteristic properties, like high mechanical strength, light weight, insulation properties, resistance to degradation from physical, mechanical, and microbial exposure, along with low production rate, have raised their demand in markets (Zeller et al., 2013). The major drawback of these types of polyethylene plastics is their non-biodegradability. These plastics need several to hundreds of years to decompose resulting in the huge production of solid waste and bursting landfills. Toxic leaches from these landfills interact with nearby water sources and pollutes those sources (Emadian et al., 2017). Moreover, the manufacturing of these plastics emits carbon and other toxic gases that are a threat for the environment (Jain and Tiwari, 2015). Multiples efforts have been taken for recycling these conventional synthetic plastics for reducing the risks related to one-time usage and non-biodegradability. But survey reports have shown that only 7% of the millions of tons of produced plastics can be recycled. Water bodies and land sites contaminated with these remaining plastics can cause serious ill threats to the living organisms (Pathak et al., 2014). Marine lives face multiple negative impacts like deformities, endocrine problems, skin problems, reduction in reproduction rate, and even death after swallowing these plastics (Delacuvellerie et al., 2019). Incineration is also not eco-friendly as, during the process, the emission of hazardous gases like CO_2, SO_2, CO, and NO_2 causes a high rate of air pollution.

Solving the problems caused by conventional plastics needs attention toward production of alternative bioplastics. As environmental concern is increasing in past decades, demands for alternative bioplastics have also risen tremendously in past years. Properties like biodegradability, nontoxic nature, favorable mechanical

DOI: 10.1201/9781032684574-11

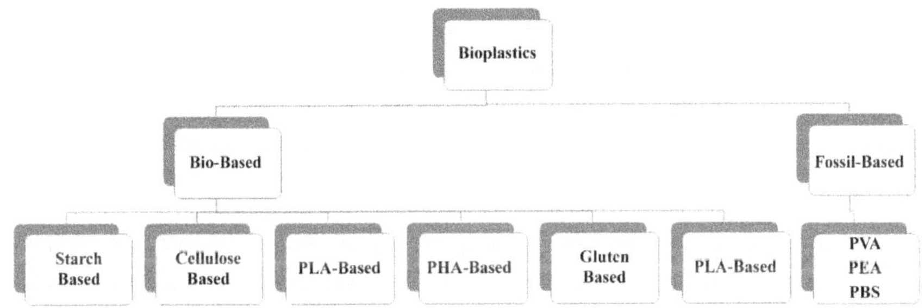

FIGURE 11.1 Different types of bioplastics based on their production.

strength, transparency, etc. have made them a suitable alternative for synthetic plastics (Hasan et al., 2018). Several sources have been explored as raw materials for the production of bioplastics. In this chapter, different types of bioplastics along with their uses in different fields are discussed.

11.2 BIOPLASTICS: DEFINITION

Bioplastics are the type of polymer synthesized from renewable sources and differ from conventional plastics in their comprising materials as they are made from bio-based or biodegradable materials. From the aspect of environment, biopolymers having the important properties of biodegradability and bio-based origin of raw materials can be labeled as bioplastics (Babu et al., 2013).

11.3 TYPES OF BIOPLASTICS

Biopolymers or bioplastics can be classified in a broad range as fossil-based biodegradable polymers and bio-based biodegradable polymers. Figure 11.1 summarizes the different types of bioplastics based on their production.

Based on material properties like strength and toughness, bioplastics can also be divided into thermosetting (comprised of hard and durable materials and impossible to remold) and thermoplastic (comprised of less rigid materials and can be remolded).

The property of biodegradability has categorized bioplastics into two groups: hydro-biodegradable and oxo-biodegradable (Iwata, 2015). Hydro-biodegradable plastics decompose hydrolytically and can also be a source of synthetic fertilizers. Starch-based bioplastics and polylactic-acid- (PLA-) based plastics are included in this category. Oxo-biodegradable plastics are produced by mixing prodegradant additives like iron or manganese salts with petroleum-based polymers. Additives enhance the biodegradation process of these polymers but are way slower than hydro-biodegradable plastics and take several months to years for degradation (Thomas et al., 2012; da Luz et al., 2013).

In the following sections, the major categories of bioplastics, the process of synthesis as well as their uses are discussed.

11.3.1 Fossil-Based Biopolymers

These type of bioplastics are produced from petrochemicals with a combination of starch or cellulose. Examples include polyvinyl alcohol (PVA), polybutylene succinate (PBS), polyethylene adipate (PEA), etc. (Song et al., 2011). PVA is the widely used water-soluble biopolymer for film formation and has properties like biodegradability and biocompatibility (Rudnik, 2019).

11.3.2 Bio-Based Biopolymers

Bio-based biopolymers, or bioplastics, can be classified into three subdivisions based on their source of production. First-generation bioplastics are comprised of starch-based bioplastics synthesized from food crops and plant (such as corn, sugarcane, plant oil, potato, etc.) and polylactic acid (PLA). Cellulose-based bioplastics produced from non-food crops or waste materials are known as second generation bioplastics. Microbial biomass derived bioplastics include the examples of third generation bioplastics (Tan et al., 2022)

11.3.2.1 Starch-Based Bioplastics

Starch is a type of carbohydrate molecule produced by plants as the form of their stored food (Le Corre et al., 2010). At present, it is the most widely used plant polysaccharide for producing bioplastics (Thakur et al., 2019). It consists of two units: amylose in a linear chain form and branched amylopectin. The crystallinity of starch comes from amylopectin units while the amorphous region is formed with amylose units. Starch with a higher amylose concentration forms films having more mechanical strength, gelling ability, elongation, and gas barrier properties, whereas increasing amylopectin concentration enhances the water-holding capacity in starch-based films (Ashok et al., 2016; Zia et al., 2015).

As native starch is brittle in nature, plasticizers and fillers are used to enhance the plasticizing property and strength of the bioplastic composite films, respectively (Ashok et al., 2016). Some examples of raw materials for extracting starch are corn, wheat, sugarcane, rice, potato, etc. The process of synthesizing starch-based films includes gelatinization of starch at first. In this step, starch is heated in the presence of water. As a result, the semicrystalline structure starts to loosen and amorphous amylose units start to swell and expands. At higher temperature, breaking down of intermolecular bonds of starch molecules facilitates the hydrogen bonding sites to engage with more water molecules. Amylose molecules then leach out in the surrounding water and forms a gel matrix (Waterschoot et al., 2015). The plasticizers help to irreversibly dissolve starch in the water by penetrating into them and enlarging the formed cavities by destroying the inner hydrogen bonds. Replacing starch–starch bonds with starch–plasticizer bonds increases the elasticity of the matrix, whereas adding suitable fillers forms a denser matrix, leading to strengthened bioplastics (Amri et al., 2018; Mallakpour, 2018). As an alternative of available polystyrene in the market, thermoplastic starch is utilized, and films produced by starch are now a widely used to form coating materials for food items (Geueke et al.,2014)

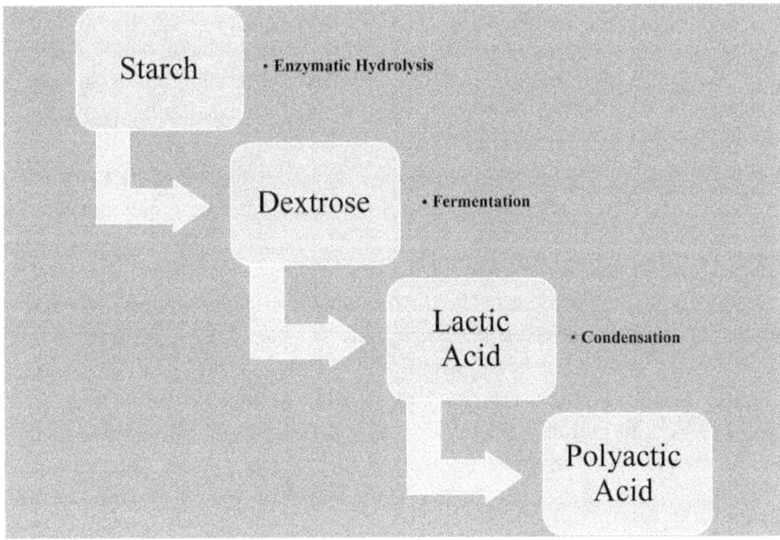

FIGURE 11.2 Schematic representation of PLA bioplastics.

11.3.2.2 PLA- (Polylactic Acid-) Based Bioplastics

Due to the industrial polycondensation of lactic acid as well as the ring-opening polymerization of lactide, PLA is produced primarily as an aliphatic polyester and biodegradable in nature (Jem and Tan, 2020). It is transparent, having the characteristics of tensile strength, elongation, etc. similar to synthetic plastics like polyethylene and polypropylene. However, properties like poor oxygen barrier and high brittleness have limited its usage in a wide field compared to conventional plastics (Reddy et al., 2013; Chieng et al., 2017). To overcome this problem, materials like nanoclay, polyethylene glycol (PEG), tributyl citrate (TBC), and oligomeric lactic acid (OLA) are associated with it (Chieng et al., 2017).

The procedure of synthesizing PLA bioplastics is described schematically in Figure 11.2. First, starch is hydrolyzed to form sugar (dextrose), which undergoes fermentation and produces lactic acid monomers. These monomers then polymerize to create polylactic acid having high molecular weights (Ashok et al., 2016). Corn, potato, sugarcane, etc. are examples of renewable carbohydrate sources from which lactic acid can be obtained. The application of this type of bioplastic can be seen in the food packaging industry mainly as it has the resistance toward oil-based products, including the properties of acting as a flavor and odor barriers for food materials (Ashok et al., 2016; Laadila et al., 2017).

11.3.2.3 Cellulose-Based Bioplastics

Cellulose is the most abundant biopolymer present in the earth obtained from different plants (such as cotton, jute, etc.), animals, algae, and bacterial species. The composition of cellulose includes linear chains of D-glucose that are linked by β-(1,4)-glyosidic bonds into a fibril-like structure surrounded by lignin and hemicellulose matrix

(Mood et al., 2013). Gopi et al. (2019) described the various characteristic features of cellulose like high degree of crystallinity, greater specific surface area, high polymerization rate, and high surface tension. Cellulose is insoluble to water and other organic solvents due to the presence of intra- and intermolecular hydrogen bonds in their structure. High immiscibility in water poses a major problem for bioplastic formation by cellulosic biomass along with the challenge caused by the presence of non-film-forming lignin and hemicellulose in a large amount (Chen and Shi, 2015).

11.3.2.4 Gluten-Based Bioplastics

Gluten is a plant-based protein derived from wheat, barley, rye, and their derivatives (Jasthi et al., 2020). Bioplastic production from gluten relies upon the most used process of extrusion like other conventional plastics. When heated, gluten undergoes cross-linking reactions that increases their viscoelastic properties, indicating the fact of extruding gluten in specific parametric conditions (Jiménez-Rosado et al., 2019). Gluten extracted from wheat is used to produce plastic materials for application in the food packaging field. However, based on recent studies, the usage of wheat-gluten-based bioplastics is restricted for food packaging applications as it is associated with celiac disease, a human autoimmune disorder (Mahadov and Green, 2011; Gómez-Heincke et al., 2017).

11.3.2.5 Polyglycolic Acid- (PGA-) Based Bioplastics

PGA can be synthesized by ring opening polymerization of glycolide with the help of a catalyst (metal salt) at low concentration. It is a type of biodegradable aliphatic polyester (Hill, 2005). Currently the usage of PGA-based plastic materials is mostly in the field of medical applications. Though these bioplastics have the resistance property toward most organic solvents, they are still subjected to hydrolysis in many instances (Song et al., 2011).

11.3.3 Microbial Bioplastics

The most widely produced microbial bioplastics include polyhydroxyalkanoates (PHA) and their derivatives. PHAs are a type of natural, intracellular polyesters with a biodegradable property and are synthesized by different bacteria in unfavorable condition for storing energy in their cells (Geueke et al., 2014). Unfavorable conditions are associated with nutrients like phosphorous and nitrogen depletion, low oxygen availability, as well as an insufficient presence of growth-promoting enzymes (Dietrich et al., 2019). However, certain groups of bacteria can synthesize PHAs without the condition of nutrient depletion (Albuquerque and Malafaia, 2018). Characteristics like chemically inertness, hydrophobicity, and high melting point have made it efficient as a sustainable alternative for petrochemical polymers. PHAs thus can be used as bioplastic wraps, shampoo bottles, and polyester fibers for clothing. Evidences have shown the degradation of PHA bioplastics naturally by marine microbes after decomposing into methane (Vigneswari et al., 2014). Several Gram-positive and -negative bacteria are responsible for synthesizing PHAs. Some examples are *Alcaligenes latus*, *Azotobacter vinelandii*, *Bacillus cereus*, *Enterobacter* sp.,

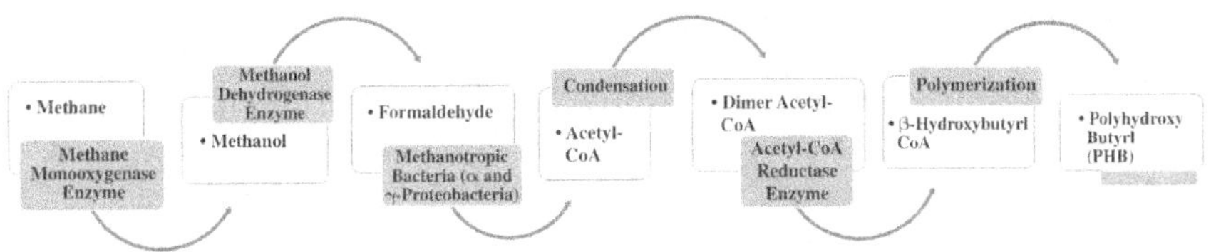

FIGURE 11.3 Schematic diagram of PHB production from methane.

Leptothrix sp., *Methylocystis* sp., *Pseudomonas* Sp., etc. Different raw materials are used as a substrate for producing PHAs. Organic wastes, peanut oil, methane, glycerol are used as a form of non-conventional carbon sources.

Depending on the growth media, polymers with desirable features can be produced having different side chains. Poly-3-hydroxybutyrate (PHB) is the most common type of PHA (Jost and Langowski, 2015). High brittleness, stiffness, crystallinity, and high melting temperature are some properties of pure PHB (Jost and Langowski, 2015). It is biodegradable, eco-friendly in nature, consisting of 14 monomeric units, and can be decomposed by microbes in CO_2 and water. Microorganisms capable of producing PHB from methane include *Methylobacterium rhodesianum*, *Cupriavidus nectar*, *Bacillus megaterium*, etc. PHB synthesis from methane is described schematically in Figure 11.3.

In industries, PHAs are cultivated using batch and fed-batch reactors. First, bacterial cells are grown up to a specific predetermined concentration, and then they are allowed to grow further in a nutrient-restricted environment with excess carbon sources. They will increase in bacterial size and weight as PHA accumulation occurs intracellularly (Tsang et al., 2019)

The applications of PHA biopolymers are implemented in various fields. The packaging industry, biomedical sector, cosmetics, agricultural field, etc., actively use PHAs nowadays; apart from these, they are also used in the textiles and building industries and have efficiently marked its role in drug delivery, artificial organ reconstruction, and surgical sutures (Lomas et al., 2013; Zhang et al., 2015). Despite all the advantages, the cost of production of PHA-based bioplastics is still high.

11.3.4 Composite Bioplastics

Nowadays most bioplastics are produced as biocomposites to improve their mechanical properties and cost-effectiveness. Soykeabkaew et al. (2012) described that more effective biocomposites can be made by extracting starch from corn, plasticized with glycerol, and incorporating bacterial cellulose into it rather than as in jute-fiber-based biocomposites. Incorporating cellulose nanocrystals (CNC) or chitosan improves the tensile strength. Filler-like silver nanoparticles induces the antimicrobial properties for making the films more effective for longer usage as food packaging material (Xu et al., 2018; Hasan et al., 2018; Abreu et al., 2015). In recent research fields, the major focus is on the production of composite bioplastics.

11.4 CONCLUSION

Replacing harmful conventional petroleum-derived plastics with biodegradable or bio-based plastic is a well-known solution for environmental issues. Plant and microbial-based bioplastics have been used widely in recent years due to its properties, which are mostly related to petroleum-derived plastics. However, the significant challenge for bioplastics is its cost and performance. But in recent years, bioplastic-derived composites have an improved mechanical property and effective performance at lower cost compared to petroplastics. Along with these improved properties, they have an advantage of being biocompatible and biodegradable. These bioplastics are the material for the future. If they can be made cost-effective, these materials will have a wider usage and benefit environmental sustainability immensely.

ACKNOWLEDGMENTS

VJ and SS would like to acknowledge DST (SEED div), Government of India for the financial support received from DST SEED Div (SYST) project (SP/YO/2019/1283) during the writing of this chapter.

REFERENCES

Abreu, A.S., Oliveira, M., de Sá, A., Rodrigues, R.M., Cerqueira, M.A., Vicente, A.A. and Machado, A.V., 2015. Antimicrobial nanostructured starch based films for packaging. *Carbohydrate Polymers*, *129*, pp. 127–134.

Albuquerque, P.B. and Malafaia, C.B., 2018. Perspectives on the production, structural characteristics and potential applications of bioplastics derived from polyhydroxyalkanoates. *International Journal of Biological Macromolecules*, *107*, pp. 615–625.

Amri, A., Ekawati, L., Herman, S., Yenti, S.R., Aziz, Y. and Utami, S.P., 2018, April. Properties enhancement of cassava starch based bioplastics with addition of graphene oxide. In *IOP Conference Series: Materials Science and Engineering* (Vol. 345, No. 1, p. 012025). IOP Publishing.

Ashok, A., Rejeesh, C. and Renjith, R., 2016. Biodegradable polymers for sustainable packaging applications: A review. *IJBB*, *1*(11).

Babu, R.P., O'Connor, K. and Seeram, R., 2013. Current progress on bio-based polymers and their future trends. *Progress in Biomaterials*, *2*, pp. 1–16.

Bayer, I. S., Guzman-Puyol, S., Heredia-Guerrero, J. A., Ceseracciu, L., Pignatelli, F., Ruffilli, R., . . . Athanassiou, A. (2014). Direct transformation of edible vegetable waste into bioplastics. *Macromolecules*, *47*(15), 5135–5143.

Chen, M.J. and Shi, Q.S., 2015. Transforming sugarcane bagasse into bioplastics via homogeneous modification with phthalic anhydride in ionic liquid. *ACS Sustainable Chemistry & Engineering*, *3*(10), pp. 2510–2515.

Chieng, B.W., Ibrahim, N.A., Then, Y.Y. and Loo, Y.Y., 2017. Epoxidized jatropha oil as a sustainable plasticizer to poly (lactic acid). *Polymers*, *9*(6), p. 204.

da Luz, J.M.R., Paes, S.A., Nunes, M.D., da Silva, M.D.C.S. and Kasuya, M.C.M., 2013. Degradation of oxo-biodegradable plastic by Pleurotus ostreatus. *PLoS One*, *8*(8), p. e69386.

Delacuvellerie, A., Cyriaque, V., Gobert, S., Benali, S. and Wattiez, R., 2019. The plastisphere in marine ecosystem hosts potential specific microbial degraders including Alcanivorax borkumensis as a key player for the low-density polyethylene degradation. *Journal of Hazardous Materials*, *380*, p. 120899.

Dietrich, K., Dumont, M.J., Del Rio, L.F. and Orsat, V., 2019. Sustainable PHA production in integrated lignocellulose biorefineries. *New Biotechnology*, *49*, pp. 161–168.

Emadian, S.M., Onay, T.T. and Demirel, B., 2017. Biodegradation of bioplastics in natural environments. *Waste Management*, *59*, 526–536.

Geueke, B., Wagner, C. C. and Muncke, J., 2014. Food contact substances and chemicals of concern: A comparison of inventories. *Food Additives & Contaminants: Part A*, *31*(8), pp. 1438–1450.

Gómez-Heincke, D., Martínez, I., Stading, M., Gallegos, C. and Partal, P., 2017. Improvement of mechanical and water absorption properties of plant protein based bioplastics. *Food Hydrocolloids*, *73*, pp. 21–29.

Gopi, S., Balakrishnan, P., Chandradhara, D., Poovathankandy, D. and Thomas, S., 2019. General scenarios of cellulose and its use in the biomedical field. *Materials Today Chemistry*, *13*, pp. 59–78.

Hasan, M., Rahmayani, R.F.I. and Munandar, 2018, March. Bioplastic from chitosan and yellow pumpkin starch with castor oil as plasticizer. In *IOP Conference Series: Materials Science and Engineering* (Vol. 333, p. 012087). IOP Publishing.

Hill, R.G., 2005. Biomedical polymers. In *Biomaterials, Artificial Organs and Tissue Engineering* (pp. 97–106). Woodhead Publishing.

Iwata, T., 2015. Biodegradable and bio-based polymers: Future prospects of eco-friendly plastics. *Angewandte Chemie International Edition*, *54*(11), pp. 3210–3215.

Jain, R. and Tiwari, A., 2015. Biosynthesis of planet friendly bioplastics using renewable carbon source. *Journal of Environmental Health Science and Engineering*, *13*, pp. 1–5.

Jasthi, B., Pettit, J. and Harnack, L., 2020. Addition of gluten values to a food and nutrient database. *Journal of Food Composition and Analysis*, *85*, p. 103330.

Jem, K.J. and Tan, B., 2020. The development and challenges of poly (lactic acid) and poly (glycolic acid). *Advanced Industrial and Engineering Polymer Research*, *3*(2), pp. 60–70.

Jiménez-Rosado, M., Zarate-Ramírez, L.S., Romero, A., Bengoechea, C., Partal, P. and Guerrero, A., 2019. Bioplastics based on wheat gluten processed by extrusion. *Journal of Cleaner Production*, *239*, p. 117994.

Jost, V. and Langowski, H.C., 2015. Effect of different plasticisers on the mechanical and barrier properties of extruded cast PHBV films. *European Polymer Journal*, *68*, pp. 302–312.

Laadila, M.A., Hegde, K., Rouissi, T., Brar, S.K., Galvez, R., Sorelli, L., Cheikh, R.B., Paiva, M. and Abokitse, K., 2017. Green synthesis of novel biocomposites from treated cellulosic fibers and recycled bio-plastic polylactic acid. *Journal of Cleaner Production*, *164*, pp. 575–586.

Le Corre, D., Bras, J. and Dufresne, A., 2010. Starch nanoparticles: A review. *Biomacromolecules*, *11*(5), pp. 1139–1153.

Lomas, A.J., Webb, W.R., Han, J., Chen, G.Q., Sun, X., Zhang, Z., El Haj, A.J. and Forsyth, N.R., 2013. Poly (3-hydroxybutyrate-co-3-hydroxyhexanoate)/collagen hybrid scaffolds for tissue engineering applications. *Tissue Engineering Part C: Methods*, *19*(8), pp. 577–585.

Mahadov, S. and Green, P.H., 2011. Celiac disease: A challenge for all physicians. *Gastroenterology & Hepatology*, *7*(8), p. 554.

Mallakpour, S., 2018. Fructose functionalized MWCNT as a filler for starch nanocomposites: Fabrication and characterizations. *Progress in Organic Coatings*, *114*, pp. 244–249.

Mood, S.H., Golfeshan, A.H., Tabatabaei, M., Jouzani, G.S., Najafi, G.H., Gholami, M. and Ardjmand, M., 2013. Lignocellulosic biomass to bioethanol, a comprehensive review with a focus on pretreatment. *Renewable and Sustainable Energy Reviews*, *27*, pp. 77–93.

Pathak, S., Sneha, C.L.R. and Mathew, B.B., 2014. Bioplastics: Its timeline based scenario & challenges. *Journal of Polymer and Biopolymer Physics Chemistry*, *2*(4), pp. 84–90.

Reddy, R.L., Reddy, V.S. and Gupta, G.A., 2013. Study of bio-plastics as green and sustainable alternative to plastics. *International Journal of Emerging Technology and Advanced Engineering*, *3*(5), pp. 76–81.

Rudnik, E., 2019. Compostable Polymer Materials (p. 410). Newne.
Song, J., Kay, M., and Coles, R., 2011. Bioplastics. In *Food and Beverage Packaging Technology* (pp. 295–319). Wiley-Blackwell.
Soykeabkaew, N., Laosat, N., Ngaokla, A., Yodsuwan, N. and Tunkasiri, T., 2012. Reinforcing potential of micro-and nano-sized fibers in the starch-based biocomposites. *Composites Science and Technology*, *72*(7), pp. 845–852.
Tan, S.X., Andriyana, A., Ong, H.C., Lim, S., Pang, Y.L. and Ngoh, G.C., 2022. A comprehensive review on the emerging roles of nanofillers and plasticizers towards sustainable starch-based bioplastic fabrication. *Polymers*, *14*(4), p. 664.
Thakur, R., Pristijono, P., Scarlett, C.J., Bowyer, M., Singh, S.P. and Vuong, Q.V., 2019. Starch-based films: Major factors affecting their properties. *International Journal of Biological Macromolecules*, *132*, pp. 1079–1089.
Thomas, N.L., Clarke, J., McLauchlin, A.R. and Patrick, S.G., 2012, August. Oxodegradable plastics: Degradation, environmental impact and recycling. In *Proceedings of the Institution of Civil Engineers-Waste and Resource Management* (Vol. 165, No. 3, pp. 133–140). ICE Publishing.
Tsang, Y.F., Kumar, V., Samadar, P., Yang, Y., Lee, J., Ok, Y.S., Song, H., Kim, K.H., Kwon, E.E. and Jeon, Y.J., 2019. Production of bioplastic through food waste valorization. *Environment International*, *127*, pp. 625–644.
Vigneswari, S., Bhubalan, K. and Amirul, A., 2014. Design and tailoring of polyhydroxyalkanoate-based biomaterials containing 4-hydroxybutyrate monomer. In *Biotechnology and Bioinformatics: Advances and Applications for Bioenergy, Bioremediation and Biopharmaceutical Research* (p. 281). Apple Academic Press.
Waterschoot, J., Gomand, S.V., Fierens, E. and Delcour, J.A., 2015. Production, structure, physicochemical and functional properties of maize, cassava, wheat, potato and rice starches. *Starch-Stärke*, *67*(1–2), pp. 14–29.
Xu, Y., Rehmani, N., Alsubaie, L., Kim, C., Sismour, E. and Scales, A., 2018. Tapioca starch active nanocomposite films and their antimicrobial effectiveness on ready-to-eat chicken meat. *Food Packaging and Shelf Life*, *16*, pp. 86–91.
Zeller, M.A., Hunt, R., Jones, A. and Sharma, S., 2013. Bioplastics and their thermoplastic blends from Spirulina and Chlorella microalgae. *Journal of Applied Polymer Science*, *130*(5), pp. 3263–3275.
Zhang, W., Chen, C., Cao, R., Maurmann, L. and Li, P., 2015. Inhibitors of polyhydroxyalkanoate (PHA) synthases: Synthesis, molecular docking, and implications. *ChemBioChem*, *16*(1), pp. 156–166.
Zia, F., Zia, K.M., Zuber, M., Kamal, S. and Aslam, N., 2015. Starch based polyurethanes: A critical review updating recent literature. *Carbohydrate Polymers*, *134*, pp. 784–798.

12 Ecological Impact of Microplastics in the Terrestrial Ecosystem

A Concise Review

Anirban Biswas and Saroni Biswas

12.1 INTRODUCTION

Since its invention, plastic has been an integral part of our society; however, the waste plastic creates a hazard for the ecosystems. The ever-increasing plastic waste is creating pollution in the environmental systems, most of which is expected to end up in oceans and landfills, and eventually their fragmentation and/or degradation will create microplastics (MPs) and finally nanoplastics (NPs) (Vaid et al. 2021). For the first time, Carpenter and Smith (1972) documented the microplastic in the Sargasso Sea. Later, the scientific community took interest and investigated the potential effects of micro- and nanoplastics on marine organisms after Thompson et al. (2004). Afterward, attention spread to the freshwater (Hoffman and Hittingaler 2017, Li et al. 2018) and terrestrial environments (de Souza Machado et al. 2018, Rillig and Lehmann 2020, Windsor et al. 2019).

The characteristics of MPs in terrestrial environments increase their propensity to produce ecological shocks because they are (1) highly persistent global pollutants that will hang around, (2) capable of complex interactions with the abiotic environment, (3) capable of having an immediate or indirect effect on terrestrial organisms, and (4) interact with the contaminants (Baho et al. 2021). Plastic powder used in cosmetics, paint, and coatings is the main source of MPs and NPs; waste plastics, tire abrasion, dust, and synthetic fabrics are secondary sources (Alexy et al. 2020; Toussaint et al. 2019). Plastics used in industry, fisheries, agriculture, and household use are additional sources of MPs and NPs (Toussaint et al. 2019; Ding et al. 2020). Terrestrial and aquatic plastic pollution is caused by both human-made and natural biopolymers (Watt et al. 2021). Some 95% of the microbeads used in personal care products had a size of less than 300 mm, and the amount of microbeads per unit of product ranged from 1.9 to 71.9 mg (Conkle et al. 2018). These minute particles are known to contain dangerous substances, but little is known about how they affect human health (Agamuthu 2018; Mendenhall 2018; Smith 2018). All habitats that contain degraded plastics experience negative effects on the soil, water, and air environments, which

DOI: 10.1201/9781032684574-12

negatively affect aquatic and terrestrial biota. Plastic particles (MPs, NPs, etc.) are slowly contaminating the food and feed chains. Current researches have focused on the source-sink-behavior chain of microplastics besides the research and development about the aquatic ecosystems (Nizzetto et al. 2016; Büks and Kaupenjohann 2020; Evangeliou et al. 2020). Current concise review has summarized the source-sink-behavior chain of microplastics into the terrestrial ecosystems.

12.2 METHODOLOGY

12.2.1 Search Selection

Published works were systematically searched in *pub med* (www.ncbi.nlm.nih.gov/pubmed), *science direct* (www.sciencedirect.com/), *web of science* (https://webofknowledge.com/), and *scopus* (www.scopus.com/). To search for the database, English terms representing microplastic, ecosystem, and terrestrial ecosystem were searched with the Boolean operator AND within the scientific data base. Articles with the following criteria were included in the review process: (1) studies published in journals, (2) studies conducted to identify the effects of microplastics, (3) microplastic effects on ecosystem, (4) microplastic effects on plants and animals, and (5) microplastic ecosystem structure and functions.

12.2.2 Documentation, Identification, Screening

Searched data were categorized (Biswas 2019) on the basis of 'selected criteria' to meet the article aim (Table 12.1). Article titles and keywords were checked thoroughly for this selection. The selection resulted from research and review articles on plastic waste and microplastic formation, microplastic contamination in soil and aquatic environments, microplastic uptake by plants and animals, direct and indirect health issues related to plant, animal, and human, etc. One limitation was applied concerning data acceptability and consideration for the review. We restricted the search and documentation as per Table 12.1.

TABLE 12.1
Database Search Criteria and Revision Including PUBMED, SCIENCE DIRECT, Web of Science, SCOPUS, and Open Source

Total records identified through database search	4509
Full text articles appended in the present review	119
Microplastic in ecosystem	30
Microplastic and soil contamination	20
Microplastic uptake in plants	14
Microplastic uptake in animals	22
Microplastic in aquatic environment	38

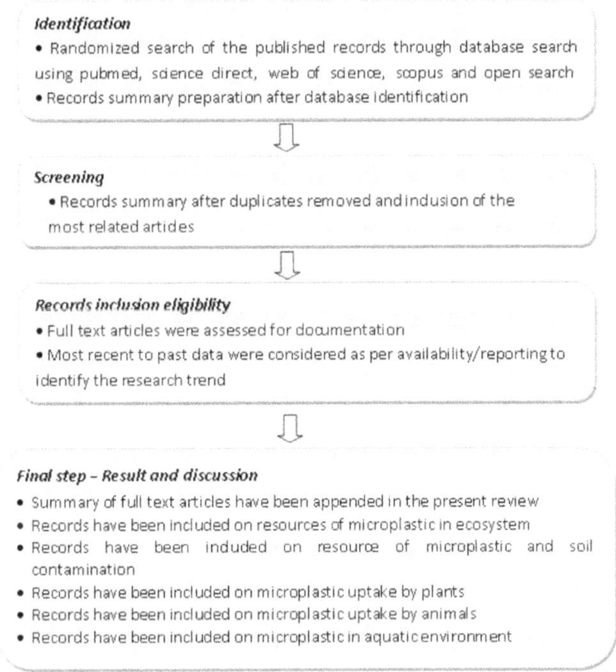

FIGURE 12.1 Review process flowchart and selection criteria of the databases.

12.3 RESULTS AND DISCUSSION

MPs that measure less than 5 mm in diameter are commonly associated with marine environments, but research has shown that MPs also have ecological impacts in terrestrial ecosystems. The review has summarized diverse trends of MPs' ecological impacts.

12.3.1 Soil Contamination

Micro- (<5 mm) and nano- (<0.1 μm) plastics have been ubiquitous in terrestrial soils, and soil has been a base and transport source for them (UNEP 2016; Nizzetto et al. 2016). MPs enter soil through various ways, leading to potential environmental and ecological impacts, such as the plastic mulches application, compost, and sewage sludge, which alters soil properties (porosity, aeration), nutrient cycling, water retention, and microbial communities (Boots et al. 2019; de Souza Machado et al. 2018; Liang et al. 2019; Liu et al. 2017; Zhang et al. 2020). The result is impaired plant growth, soil health, nutrient availability, and ecosystem health. MP contamination routes and consequences are summarized next.

- **Direct Deposition:** Abandoned MPs are directly deposited onto soil surfaces from the atmosphere and can be transported by wind or rainfall, resettling onto soil and vegetation (Huang et al., 2021). Reportedly up to 35 particle/m^3 may be suspended in MP concentrations in urban air (Xu et al. 2022).

- **Plastic Mulches:** Using plastic mulches in agricultural practices is very common for weed control and moisture retention, which degrade and release MPs into the soil in future (Bläsing and Amelung, 2018; Qi, R. et al. 2020).
- **Runoff and Erosion:** Municipal and industrial wastes sources contain enormous amounts of plastic materials, which are used in landfills, creating potential reservoirs of MPs for soil dumping due to runoff and erosion from landfill surfaces (Monkul and Özhan 2021; Kabir et al. 2023). These landfill leachates create secondary microplastics through abrasive activities (Iyare et al. 2020).

12.3.1.1 Consequences of Soil Contamination by Microplastics

- **Soil Physical Properties:** MPs cause changes in the soil's bulk density, aggregates, conductivity, water-holding capacity, cohesion, and pore structure once they get introduced into the environment (Yang et al. 2018). As per Jiang et al. (2017), MPs can degrade the soil structure, reduce air and water circulation, and produce a water-resistant layer. This may have an impact on how water moves through the soil, perhaps causing more runoff and erosion.
- **Nutrient Cycling:** In roadside soil, up to 7% of the soil weight has been reported the presence of MP by weight volume (Fuller and Gautam et al. 2016), and such a high amount of MPs affect soil chemistry by organic matter degradation (Liu et al. 2017). By affecting biophysical properties of the soil, MPs influence soil nutrient cycling processes affecting the nutrient availability and uptake by plants (Machado et al. 2019); the overall change of the soil properties are highly dependent on MP types (de Souza Machado et al. 2018).
- **Soil Microbial Communities:** Soil is important sink of MP particles due to the use of plastic mulches and the aggregation of sludge as fertilizer (Mahon et al. 2017). MP particles can be holding surfaces of microbes for colonization, affect the soil composition and microbial activities, and influence soil health and overall ecosystem functioning (Stabnikova et al. 2021). Low-density MPs (polyethylene, polypropylene, etc.) are concentrated in the neustonic layer together with dissolved organic matter, hydrophobic cells, and bacterial spores causing environmental and public health concerns as floating micro- and nanoplastic particles could be coated with bacterial biofilm (Fuller and Gautam 2016). The study about the effects of MP particles on the soil is still developing (Awet et al. 2018; de Souza Machado et al. 2018, 2019; Zhang et al. 2019; Sun et al. 2022).
- **Contaminant Transport:** MP particles, gathered within soil environment adsorb and concentrate chemical contaminants from the surroundings. Like a Trojan Horse, these particles act as carriers, facilitating the transportation of these contaminants onto different compartments of the ecosystems. A number of researches have highlighted that As, Cd, Cr, Cu, Mn, Ni, Pb, Zn, Cd, Cu, Fe, Co, and Al were found to be

adhering and cotransporting via different types of MPs (Allen et al. 2019; Verla et al. 2019).

- **Soil Organism:** MPs can be ingested by the soil invertebrates (earthworms, mites, springtails, and more) including while feeding on soil organic matters as the MP particles are mistaken for food particles due to their smaller size and similitude to organic material. These ingested MPs occupy space in the digestive tract, leading to reduced feeding efficiency and nutrient absorption (Rodriguez-Seijo et al. 2017). Earthworms (*Lumbricus terrestris*) could be found to consume MPs in the soil, affecting the growth ratio and effecting histopathological damage to these organisms (Cao et al. 2017; Lwanga et al. 2017). Anecic earthworms can physically move MP particles downward into the soil along with altering soil hydraulic properties (Edwards et al. 1990). MPs ingested by the soil invertebrates bear the implication of food chain contamination upon predators' consumption.

12.3.2 Plant Uptake and Impaired Plant Growth

Researchers have found that MPs have an impact on plants upon uptake of MPs, although the information is scarce (Azeem et al. 2021). Only three studies could be found to discuss the effects of MPs and NPs on nonvascular plants (Prata et al. 2019; Nolte et al. 2017; Sjollema et al. 2016), along with ten studies regarding vascular plants (Sun et al. 2020; Li et al. 2020, 2021a; Schwab et al. 2020; Giorgetti et al. 2020; Lian et al. 2020, 2021; Sun et al. 2021; Jiang et al. 2019; Bosker et al. 2019). Upon plant uptake, MPs can affect plant growth, development, and nutrient uptake (Lian et al. 2020).

- **Root Uptake:** MPs present in soil adhere to the plant root surfaces and are absorbed by the roots and transported to different plant parts. MP particles carry a charge on themselves that enhances their adsorption on plant roots due to electrostatic attraction (Lian et al. 2020). Upon root uptake, MPs move through the vascular system and accumulate in different tissues, including leaves and fruits (Roy et al. 2022). In the long run, those MPs can hinder nutrient immobilization and photosynthetic process (Lian et al. 2020). Adult plants can absorb the materials 3–4 nm in size; some studies have identified the plants to absorb NPs 10–12 times larger than that.
- **Foliar Uptake:** The suspended MPs are carried by wind or rainfall and can settle on plant surfaces, especially on leaves. This may be a potential route of MP intake by the plants, although the evidence is very, very limited (Sun et al. 2021; Lian et al. 2021).

12.3.2.1 Impaired Plant Growth and Development

- **Nutrient and Water Competition:** MPs block root surfaces and so reduce their ability to access soil nutrients and hygroscopic water, leading to deferred plant growth due to nutrient deficiencies. In addition, MPs accumulated within plant tissues cause physical damage to plant cells and tissues, affecting their structure and function (Pflugmacher et al. 2020).

- **Hormonal Disruption:** MPs interfere with hormonal regulation of plants, leading to impaired growth, flowering, and fruiting. But these abnormalities are plant species specific as ascribed from responses of hormones and amino acids (Bouaicha et al. 2022). Auxin and cytokinin govern the root system and regulate root hair formation, crown roots, and lateral roots. Studies have found abnormal concentrations IBA, OPA, and cis-zeatin riboside in roots indicating MP stress effect on root development (Li et al. 2021b; Neogy et al. 2021).
- **Oxidative Stress and Photosynthesis:** MP stress has been found to affect plant root growth by adsorbing MPs onto root hairs (Bosker et al. 2019). In leaves, MP stress causes oxidative stress and reduces photosynthesis and leaf growth (Lian et al. 2021; Colzi et al. 2022).

12.3.3 Ingestion by Wildlife

MPs are ingested by the terrestrial animals, insects, birds, and mammals because these particles are mistaken as food, leading to negative impacts on feeding behavior, digestion, and overall health. The fate of the ingested MPs is accumulation onto the digestive tract, causing potential physical damage and interruption in nutrient absorption. Ecotoxicological effects of MPs and NPs on marine animals have been documented for a long time, indicating their role in prey–predator relationships and possible harms (Piehl et al. 2018; Ferrante et al. 2022).

12.3.3.1 Sources of Microplastics for Wildlife Ingestion

- **Aquatic Environments:** MPs are originated from larger plastic materials by breaking into smaller particles over time. Surface runoff, streams, rivers, and wastewater treatment plant discharge are the main entry points for primary MPs into the aquatic environment. Secondary MPs are created physically (wind, wave, and current), chemically (UV radiation), and biologically (microbial) breakdown of meso- and macroplastic waste that has accumulated in the environment (Naidu 2019).
- **Terrestrial Environments:** MPs enter into any terrestrial ecosystem either directly as primary MPs or indirectly as secondary MPs (Waldman and Rillig 2020). Insects, birds, small mammals, and other terrestrial species ingest MPs when the MP particles are ingested due to their resemblance to food materials when come they into contact with plastic debris (Morey et al. 2007; Larson et al. 2015; Santana and Armstrong 2017).

12.3.3.2 Impacts of Microplastics Ingestion on Wildlife

Physical Harm and Toxicity

More than 250 marine species have been found to be impacted by MP ingestion (Laist 1997). Upon MP ingestion in the marine lives, the particles get accumulated within their digestive tract and block it, resulting in satiety, reduced feeding and energy reserves in the body, affecting overall growth (Haibo et al. 2016; Xiaoxia 2016). As has been tested with *Arenicola marina*, the polystyrene MPs significantly inhibit its growth where impaired growth was significantly correlated with MPs concentration (Qiang et al. 2017). MP-exposed bivalve molluscs were found to have decreased energy reserve (Zouxia et al. 2017). In the test animal *Mytilus galloprovincialis*, polyethylene MPs was found to disrupt the body balance, and the final result was increased energy consumption and decreased growth rate.

MPs build up in the intestine, gills, and liver, resulting in pathological alterations in these organs. Ultimately, MPs break down the gut's epithelial barrier, influencing inflammation, oxidative stress levels, gene expression (protein production), and the gut's microbiota. The livers of fish afflicted with MP have also been found to exhibit an imbalanced metabolism of lipids and carbohydrates. Certain contaminants, such as phenanthrene (Karami et al. 2016), mercury (Barboza et al. 2018), cadmium (Lu et al. 2018; Banaee et al. 2019), PCBs (Rainieri et al. 2018), gold ions (Lee et al. 2019), and antibiotics (Zhang et al. 2019), may have more toxic effects on fish when MPs are present.

12.3.3.3 Behavioral and Reproductive Effects

Recent studies have found MPs to impact the behavior and reproductive success of terrestrial animals by altering the nesting behavior of birds or affecting the reproduction of insects, which in turn affects ecosystem dynamics.

12.3.3.4 Behavioral Effects

- **Feeding Behavior and Prey–Predator Interaction:** For the aquatic invertebrates, MPs are mistaken as food materials due to their micro size (0.5–816 mm) and resemblance to natural prey. Ingested MPs lead to altered feeding behavior, viz. reduced feeding rates, foraging location change, and false prey detection (Cole and Galloway 2015; Desforges et al. 2015). MPs alter prey–predator dynamics. When predators consume prey laden with MPs, it affects their hunting efficiency and disrupts their energy balance. The potential impact of this phenomenon is reduced number of predators, leading to the increase of prey number (Huang et al. 2020).
- **Social Interactions:** MPs cause behavioral changes affecting the social interactions within animal communities. Altered feeding behavior impacts the interactions between parent and offspring. In animals, consumed MPs, low locomotor activity, higher anxiety level, lack of protective social aggregation, and low risk assessment perception in presence of potential predator were observed in the open field tests (da Costa Araújo and Malafaia 2021).

12.3.3.5 Reproductive Effects

- **Hormonal Disruption and Effects on Fertility and Egg Development:** Ingestion of MPs have an impact on an animal's ability to reproduce and the growth of its eggs. Ingested MPs may build up in reproductive organs and might damage the viability of eggs or sperm production. As a result of exposure in male mice, MPs diminish concentration of FSH, LH, and T, while considerably increasing the amount of estradiol (Wei et al. 2022). This disrupts the HPG (hypothalamic-pituitary-gonadal) axis (Wang et al. 2019). Therefore, the delayed gonadal maturation and the changed sex hormone ratio that hampered reproductive development are among the reproductive abnormalities brought on by MPs owing to HPG axis disruption (Wang et al. 2019). According to several research, PS-MPs cause the liver and gonads of male zebrafish to produce more reactive oxygen species (ROS). Their exposure to MPs induces testicular apoptosis, which affects the generation of gametes (Qiang L and Cheng 2021) and interaction with plasma membrane permeability of gametes, preventing gamete binding and offspring growth (Wang et al. 2019).

- **Embryonic Development:** By interacting with the embryonic environment, MPs may have an effect on embryonic development. This could result in aberrant child development or lower infant survival rates. According to several aquatic researches, MPs can be bi-accumulated in female fish eggs and yolk sacs, altering the physiology of progeny in female fish (Pitt et al. 2018a, 2018b). Cross-generational transmission of MPs is made possible by polystyrene MPs' capacity to interfere with plasma proteins linked to oocytes (Pitt et al. 2018b).

12.3.4 Nutrient Cycling and Microplastics

Ecosystems' nutrient cycling is impacted by MPs, which have a detrimental effect on the structural and functional aspects of soil. Microplastics have an indirect impact on soil nutrient cycling because they lower the microbial metabolic status, enzyme activity, coding genes, microbial biomass carbon, and microbial carbon use efficiency (Zhang et al. 2020; Wijesooriya et al. 2023). Changes in the cycling of nutrients impair the growth of plants, primary productivity, and the overall dynamics of ecosystems. Depending on the types and concentration of MP, soil pH is impacted, which is a crucial factor in determining soil nutrient mobility and plant nutrient uptake. Due to their low density, MPs are moved upward by precipitation and irrigation; however, MPs may also move downward because of the ecocorona created by microbial colonization, nutrients, and the absorption of organic pollutants on the MP surface (Li et al. 2021b).

- **Nutrient Release and Sorption:** MPs may break down and release substances into the environment over time. If these substances contain plastic additives or other chemicals that are attached to plastic particles, they may disrupt microbial activity and nutrient availability, which may have an impact on the nutrient cycle. MPs can absorb nutrients and other substances from the water around them due to their large surface area. This adsorption may result in decreased nutrient availability for aquatic species, which may have an impact on their production and development. MPs are easily absorbed hydrophobic contaminants from the aquatic system because of their huge surface-to-volume ratio. Because of its negative impact, particularly on the health and biota of the ocean, microplastic pollution is therefore a growing problem (Chatterjee and Sharma 2019). MNPs emit chemical additives that change soil's physicochemical makeup, nutrient availability, and microbial activity-diversity and elevate GHG emissions (Kumar et al. 2021; Xu et al. 2022).

- **Biofilm Formation:** Biofilms, which are intricate populations of microorganisms, may develop on MPs as substrates. By storing and digesting nutrients differently than natural substrates, these biofilms may have an impact on the processes of the nutrient cycle. The *Bacterium* genera isolated from the surface biofilm was the Enterobacteriaceae family, *Pseudomonas*, *Flavobacterium*, *Caulobacter*, *Corynebacterium*, and others (Stabnikova et al. 1991; Donderski et al. 1999; Fiebig 2018). Gram-negative bacteria make up the majority of the bacteria in the microbial communities that coat the microplastic particles in biofilms.
- **Detrital Pathways:** The organic matter decomposition process is altered by MPs consumed by detritivores (organisms that feed on decaying organic matter), which has an impact on the release of nutrients from detritus. While both fungi and bacteria decompose organic matter, fungi are in charge of the breakdown of refractory organic matter, while bacteria play a significant role in regulating the carbon cycle in soils (Bardgett and Mcalister 1999; Fierer et al. 2007; Rashid et al. 2016; Zhang et al. 2021). Fungi react more strongly to MPs than do bacteria, and the diversity of fungal communities is more susceptible to MP effects than that of bacterial communities. Researchers discovered that MPs had a major impact on fungi similar to plant growth based on the FUNGuid tool (Fan et al. 2022).

12.3.5 Ecosystem Services and Microplastics

Microplastics can disrupt ecosystem services, the benefits to humans and other organisms like water purification, pollination, carbon sequestration, and more. The presence of microplastics in ecosystems can interfere with these services, leading to negative consequences for both the environment and human well-being. The ways in which microplastics can disrupt ecosystem services are discussed next.

- **Water Quality and Purification:** Microplastics contaminate water bodies and affect water quality and reduce the natural water purification capacity of ecosystems. Microplastics adsorb pollutants and transfer those into aquatic environments hindering the ability of ecosystems to filter water and impacting drinking water quality and water for aquatic lives. According to numerous studies (Barboza and Gimenez 2015; Galloway and Lewis 2016; Wright and Kelly 2017; Carbery et al. 2018; Robin et al. 2020), microplastics are common in the marine environment. Although drinking water is a potential source of microplastic consumption for humans, there aren't many thorough reports on the subject (Kosuth et al. 2018; Koelmans et al. 2019).
- **Pollination:** Nowadays, being an integral part of ecosystem, MPs can impact the pollinators, viz. bees and other insects, by affecting their behavior, navigation, and reproductive success. Reduced pollination leads to decreased crop yields and ecosystem disruption. Micro- and nanoplastics impact the health of pollinators (bees and other insects) by disrupting the gut microbiota, and the level of effects depends on the chemical composition of the MPs (Shah et al. 2023). Microplastics' presence in soils can impact soil structure,

microbial communities, and nutrient cycling, potentially affecting carbon storage capacity and the ability of ecosystems to sequester carbon.

- **Aesthetic, Recreational and Cultural Value Disruption:** The presence of microplastics in natural environments diminish their aesthetic and recreational value. Over 260,000 tonnes of plastic debris is floating on the world's ocean surface due to improper waste disposal (Eriksen et al. 2014). Along the shorelines, accumulated plastic debris, due to anthropogenic activities affect the enjoyment of outdoor spaces and impact tourism (Adams 2005; Richmond 2015). Because terrestrial and aquatic ecosystems are intertwined, modifications to one system will inevitably affect the other, similarly to environmental degradation and pollution. Plastic litter is more common in ocean basins than other debris like glass, cloth, paper, food waste, metal, rubber, items related to personal hygiene and medicine, firework items, and wood because of the special qualities of plastics (such as their long shelf life, which makes them easily transported by wind and water currents) (Rosevelt et al. 2013).

12.4 CONCLUSION

The terrestrial effects of microplastic contamination are a growing concern with far-reaching implications for ecosystems, wildlife, and human well-being. The pervasiveness of microplastics in terrestrial environments underscores the urgency of understanding their impacts and implementing effective strategies to mitigate their presence. Here are some key concluding remarks:

- **Ecological Fragility:** Terrestrial ecosystems, from forests and grasslands to urban areas, are vulnerable to microplastic contamination. MPs can disrupt soil health, nutrient cycling, and plant–microbe interactions, which are fundamental to ecosystem functioning.
- **Wildlife Impacts:** MPs can have detrimental effects on soil invertebrates, plants, and animals. MP ingestion leads to physical harm, altered feeding behavior, and potential toxic effects due to associated pollutants.
- **Nutrient Cycling Disruption:** Microplastics can interfere with nutrient cycling processes, affecting primary production, decomposition, and carbon sequestration. These disruptions can ripple through ecosystems, impacting overall ecosystem health.
- **Human Exposure:** MPs' presence in soils and their potential to enter the food chain raise concerns about human exposure to these particles. The possibility of microplastics reaching crops and groundwater sources necessitates thorough research on potential health risks.
- **Bioaccumulation and Trophic Transfer:** MPs can accumulate in organisms across trophic levels, potentially magnifying their effects through the food chain. This poses risks not only to wildlife but also to humans who consume contaminated organisms.
- **Interactions with Pollutants:** MPs can absorb and transport chemical pollutants, compounding the ecological and health risks associated with their presence.

- **Research and Awareness:** To fully comprehend the effects of MP pollution on terrestrial ecosystems, more research is needed. Decisions about policy, management techniques, and public awareness campaigns can all benefit from this knowledge. Overall, mitigating the effects of microplastic contamination in terrestrial ecosystems requires a united effort to curb plastic pollution at its source, raise awareness, conduct research, and implement sustainable solutions. By doing so, we can protect the integrity of our planet's soils, wildlife, and the services they provide for current and future generations.

CONFLICT OF INTEREST

The authors declare that the review work is solely for academic purpose and there is no commercial and any financial conflict of interest.

REFERENCES

Adams, S.M. 2005. Assessing cause and effect of multiple stressors on marine systems. *Marine Pollution Bulletin* 51:8–12. www.sciencedirect.com/science/article/pii/S0025326X04004667

Agamuthu, P. 2018. Marine debris, plastics, microplastics and nano-plastics: What next? *Waste Management & Research* 36(10):869–871. doi:10.1177/0734242X18796770

Alexy, P., Anklam, E., Emans, T., Furfari, A., Galgani, F., Hanke, G., Koelmans, A., Pant, R., Saveyn, H., Sokull, B. 2020. Managing the analytical challenges related to micro- and nanoplastics in the environment and food: Filling the knowledge gaps. *Food Additives & Contaminants: Part A* 37(1):1–10. doi:10.1080/19440049.2019.1673905

Allen, S., Deonie, A., Vernon, R.P., Gaël, L.R., Pilar, D.J., Anaëlle, S., Stéphane, B., Didier, G. 2019. Atmospheric transport and deposition of microplastics in a remote mountain catchment. *Nature Geoscience* 12:339–344. doi:10.1038/s41561-019-0335-5

Awet, T.T., Kohl, Y., Meier, F., Straskraba, S., Grün, A.L., Ruf, T. 2018. Effects of polystyrene nanoparticles on the microbiota and functional diversity of enzymes in soil. *Environmental Sciences Europe* 30:11. doi:10.1186/s12302-018-0140-6

Azeem, I., Adeel, M., Ahmad, M.A., Shakoor, N., Jiangcuo, G.D., Azeem, K., Ishfaq, M., Shakoor, A., Ayaz, M., Xu, M., Rui, Y. 2021. Uptake and accumulation of nano/microplastics in plants: A critical review. *Nanomaterials (Basel).* Nov 2;11(11):2935. doi:10.3390/nano11112935

Baho, D.L., Bundschuh, M., Futter, M.N. 2021. Microplastics in terrestrial ecosystems: Moving beyond the state of the art to minimize the risk of ecological surprise. *Global Change Biology* 27:3969–3986. doi:10.1111/gcb.15724

Banaee, M., Soltanian, S., Sureda, A., Gholamhosseini, A., Haghi, B.N., Akhlaghi, M., Derikvandy, A. 2019. Evaluation of single and combined effects of cadmium and micro-plastic particles on biochemical and immunological parameters of common carp (Cyprinus carpio). *Chemosphere* 236(2):124335. doi:10.1016/j.chemosphere.2019.07.066

Barboza, L.G.A., Gimenez, B.C.G. 2015. Microplastics in the marine environment: Current trends and future perspectives. *Marine Pollution Bulletin* 97:5–12. doi:10.1016/j.marpolbul.2015.06.008

Barboza, L.G.A., Vieira, L.R., Branco, V., Figueiredo, N., Carvalho, F., Carvalho, C., Guilhermino, L. 2018. Microplastics cause neurotoxicity, oxidative damage and energy-related changes and interact with the bioaccumulation of mercury in the European seabass, Dicentrarchus labrax (Linnaeus, 1758). *Aquatic Toxicology* 1:49–57. doi:10.1016/j.aquatox.2017.12.008

Bardgett, R., McAlister, E. 1999. The measurement of soil fungal: Bacterial biomass ratios as an indicator of ecosystem self-regulation in temperate meadow grasslands. *Biology and Fertility of Soils* 29:282–290. doi:10.1007/s003740050554

Biswas, A. 2019. A systematic review on arsenic bio-availability in human and animals: Special focus on the rice-human system. *Reviews of Environmental Contamination and Toxicology.* doi:10.1007/398_2019_28

Bläsing, M., Amelung, W. 2018. Plastics in soil: Analytical methods and possible sources. *Science of the Total Environment* 612:422–435. doi:10.1016/j.scitotenv.2017.08.086

Boots, B., Russell, C.W., Green, D.S. 2019. Effects of microplastics in soil ecosystems: Above and below ground. *Environmental Science & Technology* 53(19):11496–11506. doi:10.1021/acs.est.9b03304

Bosker, T., Bouwman, L.J., Brun, N.R., Behrens, P., Vijver, M.G. 2019. Microplastics accumulate on pores in seed capsule and delay germination and root growth of the terrestrial vascular plant Lepidium sativum. *Chemosphere* 226:774–781. doi:10.1016/j.Chemosphere.2019.03.163

Bouaicha, O., Tiziani, R., Maver, M., Lucini, L., Miras-Moreno, B., Zhang, L., Trevisan, M., Cesco, S., Borruso, L., Mimmo, T. 2022. Plant species-specific impact of polyethylene microspheres on seedling growth and the metabolome. *Science of the Total Environment* 840:156678. doi:10.1016/j.scitotenv.2022.156678

Büks, F., Kaupenjohann, M. 2020. Global concentrations of microplastic in soils—a review. *SOIL* 2020:1–26. doi:10.5194/soil-2020-50

Cao, D., Wang, X., Luo, X., Liu, G., Zheng, H. 2017. Effects of polystyrene microplastics on the fitness of earthworms in an agricultural soil. In *IOP Conference Series: Earth and Environmental Science* (p. 12148). doi:10.1088/1755-1315/61/1/012148

Carbery, M., O'Connor, W., Thavamani, P. 2018. Trophic transfer of microplastics and mixed contaminants in the marine food web and implications for human health. *Environment International* 115:400–409. doi:10.1016/j.envint.2018.03.007

Carpenter, E.J., Smith, K.L. 1972. Plastics on the Sargasso sea surface. *Science* 175(4027):1240. doi:10.1126/scien ce.175.4027.1240

Chatterjee, S., Sharma, S. 2019. Microplastics in our oceans and marine health. *Field Actions Science Reports [Online]* 19:2019. doi: http://journals.openedition.org/facts reports/5257

Cole, M., Galloway, T.S. 2015. Ingestion of nanoplastics and MPs by Pacific Oyster Larvae. *Environmental Science & Technology* 49:14625–14632. doi:10.1021/acs.est.5b04099

Colzi, I., Renna, L., Bianchi, E., Castellani, M.B., Coppi, A., Pignattelli, S., Loppi, S., Gonnelli, C. 2022. Impact of microplastics on growth, photosynthesis and essential elements in Cucurbita pepo L. *Journal of Hazardous Materials* 423:127238. doi:10.1016/j.jhazmat.2021.127238

Conkle, J.L., del Valle, C.D.B., Turner, J.W. 2018. Are we underestimating microplastic contamination in aquatic environments? *Environmental Management* 61:1–8. doi:10.1007/s00267-017-0947-8

da Costa Araújo, A.P., Malafaia, G. 2021. Microplastic ingestion induces behavioral disorders in mice: A preliminary study on the trophic transfer effects via tadpoles and fish. *Journal of Hazardous Materials* 401(2):123263. doi:10.1016/j.jhazmat.2020.123263

de Souza Machado, A.A.S., Lau, C.W., Kloas, W., Bergmann, J., Bachelier, J.B., Faltin, E., Becker, R., Görlich, A.S., Rillig, M.C. 2019. Microplastics can change soil properties and affect plant performance. *Environmental Science & Technology* 53(10):6044–6052. doi:10.1021/acs.est.9b01339

de Souza Machado, A.A., Lau, C.W., Till, J., Kloas, W., Lehmann, A., Becker, R., Rillig, M.C. 2018. Impacts of microplastics on the soil biophysical environment. *Environmental Science & Technology* 52(17):9656–9665. doi:10.1021/acs.est.8b02212

Desforges, J.-P.W., Galbraith, M., Ross, P.S. 2015. Ingestion of MPs by zooplankton in the Northeast Pacific Ocean. *Archives of Environmental Contamination and Toxicology* 69:320–330. doi:10.1007/s00244-015-0172-5

Ding, L., Zhang, S., Wang, X., Yang, X., Zhang, C., Qi, Y., Guo, X. 2020. The occurrence and distribution characteristics of microplastics in the agricultural soils of Shaanxi Province, in north-western China. *Science of the Total Environment* 720:137525. doi:10.1016/j.scitotenv.2020.137525

Donderski, W., Walczak, M., Mudryk, Z., Kobylinski, M. 1999. Neustonic bacteria number, biomass and taxonomy. *Polish Journal of Environmental Studies* 8:137–141.

Edwards, W.M., Shipitalo, M.J., Owens, L.B., Norton, L.D. 1990. Effect of Lumbricus terrestris L. burrows on hydrology of continuous no-till corn fields. *Geoderma* 46:73–84. doi:10.1016/0016-7061(90)90008-W

Eriksen, M., Lebreton, L.C.M., Carson, H.S., Thiel, M., Moore, C.J., Borerro, J.C. 2014. Plastic pollution in the world's oceans: More than 5 trillion plastic pieces weighing over 250,000 tons afloat at sea. *PLoS One* 9(12):e111913. doi:10.1371/journal.pone.0111913

Evangeliou, N., Grythe, H., Klimont, Z., Heyes, C., Eckhardt, S., Lopez Aparicio, S., Stohl, A. 2020. Atmospheric transport is a major pathway of microplastics to remote regions. *Nature Communications* 11(1):3381. doi:10.1038/s41467-020-17201-9

Fan, P., Tan, W., Yu, H. 2022. Effects of different concentrations and types of microplastics on bacteria and fungi in alkaline soil. *Ecotoxicology and Environmental Safety* 229:113045. doi:10.1016/j.ecoenv.2021.113045

Ferrante, M.C., Monnolo, A., Del Piano, F., Raso, G.M., Meli, R. 2022. The pressing issue of micro- and nanoplastic contamination: Profiling the reproductive alterations mediated by oxidative stress. *Antioxidants* 11(2):193. doi:10.3390/antiox11020193

Fiebig, A. 2018. Role of Caulobacter cell surface structures in colonization of the air-liquid interface. *Journal of Bacteriology* 201:e00064–19. doi:10.1128/jb.00064-19

Fierer, N., Bradford, M.A., Jackson, R.B. 2007. Toward an ecological classification of soil bacteria. *Ecology* 88:1354–1364. doi:10.1890/05-1839

Fuller, S., Gautam, A. 2016. A procedure for measuring microplastics using pressurized fluid extraction. *Environmental Science & Technology* 50(11):5774–5780. doi:10.1021/acs.est.6b00816

Galloway, T.S., Lewis, C.N. 2016. Marine microplastics spell big problems for future generations. *Proceedings of National Academy of Sciences USA* 113:2331–2333. doi:10.1073/pnas.1600715113

Giorgetti, L., Spanò, C., Muccifora, S., Bottega, S., Barbieri, F., Bellani, L., Castiglione, M.R. 2020. Exploring the interaction between polystyrene nanoplastics and Allium cepa during germination: Internalization in root cells, induction of toxicity and oxidative stress. *Plant Physiology and Biochemistry* 149:170–177. doi:10.1016/j.plaphy.2020.02.014

Haibo, Z., Qian, Z., Yang, Z., Chen, T., Yongming, L. 2016. Raising concern about microplastic pollution in coastal and marine environment and strengthening scientific researches on pollution prevention and management. *Bulletin of the Chinese Academy of Sciences* 31(10):1182–1189.

Hoffman, M.J., Hittingaler, E. 2017. Inventory and transport of plastic debris in the Laurentian Great Lakes. *Marine Pollution Bulletin* 115(1):273–281. doi:10.1016/j.marpolbul.2016.11.061

Huang, Q., Lin, Y., Zhong, Q. Ma, F, Zhang, Y. 2020. The impact of microplastic particles on population dynamics of predator and prey: Implication of the Lotka-Volterra model. *Scientific Reports* 10:4500. doi:10.1038/s41598-020-61414-3

Huang, Y., He, T., Yan, M., Yang, L., Gong, H., Wang, W., Qing, X., Wang, J. 2021. Atmospheric transport and deposition of microplastics in a subtropical urban environment. *Journal of Hazardous Materials* 416:126168. doi:10.1016/j.jhazmat.2021.126168

Iyare, P.U., Ouki, S.K., Bond, T. 2020. Microplastics removal in wastewater treatment plants: A critical review. *Environmental Science: Water Research & Technology* 6(10):2664–2675.

Jiang, X.J., Chen, H., Liao, Y., Ye, Z., Li, M., Klobučar, G. 2019. Ecotoxicity and genotoxicity of polystyrene microplastics on higher plant Vicia faba. *Environmental Pollution* 250:831–838. doi:10.1016/j.envpol.2019.04.055

Jiang, X.J., Liu, W., Wang, E., Zhou, T., Xin, P. 2017. Residual plastic mulch fragments effects on soil physical properties and water flow behavior in the Minqin Oasis, north western China. *Soil and Tillage Research* 166:100–107. doi:10.1016/j.still.2016.10.011

Kabir, M.S., Wang, H., Luster-Teasley, S., Zhang, L., Zhao, R. 2023. Microplastics in landfill leachate: Sources, detection, occurrence, and removal. *Environmental Science and Ecotechnology* 16:100256. doi:10.1016/j.ese.2023.100256

Karami, A., Romano, N., Galloway, T., Hamzah, H. 2016. Virgin microplastics cause toxicity and modulate the impacts of phenanthrene on biomarker responses in African catfish (Clarias gariepinus). *Environmental Research* 151:58–70. doi:10.1016/j.envres.2016.07.024

Koelmans, A.A., Nor, N.H.M., Hermsen, E., Kooi, M., Mintenig, S.M., France, J.D. 2019. Microplastics in freshwaters and drinking water: Critical review and assessment of data quality. *Water Research* 155: 410–422. doi:10.1016/j.watres.2019.02.054

Kosuth, M., Mason, S.A., Wattenberg, E.V. 2018. Anthropogenic contamination of tap water, beer, and sea salt. *PLoS One* 13:e0194970. doi:10.1371/journal.pone.0194970

Kumar, M., Chen, H., Sarsaiya, S., Qin, S., Liu, H., Awasthi, M.K., Kumar, S., Singh, L., Zhang, Z., Bolan, N.S., Pandey, A., Varjani, S., Taherzadeh, M.J. 2021. Current research trends on micro- and nano-plastics as an emerging threat to global environment: A review. *Journal of Hazardous Materials* 409:124967. doi:10.1016/j.jhazmat.2020.124967

Laist, D.W. 1997. Impacts of marine debris: Entanglement of marine life in marine debris including a comprehensive list of species with entanglement and ingestion records. In *Marine Debris, Sources, Impacts, and Solutions* (Coe, J.M., Rogers, D.B., editors, pp. 99–140). New York, NY: Springer-Verlag.

Larson, R.N., Morin, D.J., Wierzbowska, I.A., Crooks, K.R. 2015. Food habits of coyotes, gray foxes, and bobcats in a coastal southern California urban landscape. *Western North American Naturalist* 75:339–347. doi:10.3398/064.075.0311

Lee, W.S., Cho, H.J., Kim, E., Huh, Y.H., Kim, H.J., Kim, B., Kang, T., Lee, J.S., Jeong, J. 2019. Bioaccumulation of polystyrene nanoplastics and their effect on the toxicity of Au ions in zebrafish embryos. *Nanoscale* 11(7):3173–3185. doi:10.1039/C8NR09321K

Li, Z., Li, Q., Li, R., Zhou, J., Wang, G. 2021a. The distribution and impact of polystyrene nanoplastics on cucumber plants. *Environmental Science and Pollution Research* 28:16042–16053. doi:10.1007/s11356-020-11702-2

Li, S., Wang, T., Guo, J., Dong, Y., Wang, Z., Gong, L., et al. 2021b. Polystyrene microplastics disturb the redox homeostasis, carbohydrate metabolism and phytohormone regulatory network in barley. *Journal of Hazardous Materials* 415:125614. doi:10.1016/j.jhazmat.2021.125614

Li, J., Liu, H., Paul Chen, J. 2018. Microplastics in freshwater systems: A review on occurrence, environmental effects, and methods for microplastics detection. *Water Research* 137:362–374. doi:10.1016/j.watres.2017.12.056

Li, L., Luo, Y., Li, R., Zhou, Q., Peijnenburg, W.J.G.M, Yin, N., Yang, J., Tu, C., Zhang, Y. 2020. Effective uptake of submicrometre plastics by crop plants via a crack-entry mode. *Nature Sustainability* 3:929–937. doi:10.1038/s41893-020-0567-9

Lian, J., Liu, W., Meng, L., Wu, J., Chao, L., Zeb, A., Sun, Y. 2021. Foliar-applied polystyrene nanoplastics (PSNPs) reduce the growth and nutritional quality of lettuce (Lactuca sativa L.). *Environmental Pollution* 280:116978. doi:10.1016/j.envpol.2021.116978

Lian, J., Wu, J., Xiong, H., Zeb, A., Yang, T., Su, X., Su, L., Liu, W. 2020. Impact of polystyrene nanoplastics (PSNPs) on seed germination and seedling growth of wheat (Triticum aestivum L.). *Journal of Hazardous Materials* 385:121620. doi:10.1016/j.jhazmat.2019.121620

Liang, Y., Lehmann, A., Ballhausen, M.B., Muller, L., Rillig, M.C. 2019. Increasing temperature and microplastic fibers jointly influence soil aggregation by saprobic fungi. *Frontiers in Microbiology* 10:2018. doi:10.3389/fmicb.2019.02018

Liu, H.F., Yang, X.M., Liu, G., Liang, C., Xue, S., Chen, H., Ritsema, C.J., Geissen, V. 2017. Response of soil dissolved organic matter to microplastic addition in Chinese loess soil. *Chemosphere* 185:907–917. doi:10.1016/j.chemosphere.2017.07.064

Lu, K., Qiao, R., An, H., Zhang, Y. 2018. Influence of microplastics on the accumulation and chronic toxic effects of cadmium in zebrafish (Danio rerio). *Chemosphere* 202(4):514–520. doi:10.1016/j.chemosphere.2018.03.145

Lwanga, E.H., Gertsen, H., Gooren, H., Peters, P., Salanki, T., van der Ploeg, M., Besseling, E., Koelmans, A.A., Geissen, V. 2017. Incorporation of microplastics from litter into burrows of Lumbricus terrestris. *Environmental Pollution* 220:523–531.

Mahon, A.M., O'Connell, B., Healy, M.G., O'Connor, I., Officer, R., Nash, R., et al. 2017. Microplastics in sewage sludge: Effects of treatment. *Environmental Science & Technology* 51:810–818. doi:10.1021/acs.est.6b04048

Mendenhall, E. 2018. Oceans of plastic: A research agenda to propel policy development. *Marine Policy* 96:291–298. doi:10.1016/j.marpol.2018.05.005

Monkul, M.M., Özhan, H.O. 2021. Microplastic contamination in soils: A review from geotechnical engineering view. *Polymers (Basel)* 13(23):4129. doi:10.3390/polym13234129

Morey, P.S., Gese, E.M., Gehrt, S. 2007. Spatial and temporal variation in the diet of coyotes in the Chicago metropolitan area. *American Midland Naturalist* 158:147–161. doi:10.1674/0003-0031(2007)158[147:SATVIT]2.0.CO;2

Naidu, S.A. 2019. Preliminary study and first evidence of presence of microplastics and colorants in green mussel, Perna viridis (Linnaeus, 1758), from southeast coast of India. *Marine Pollution Bulletin* 140:416–422. doi:10.1016/j.marpolbul.2019.01.024

Neogy, A., Singh, Z., Mushahary, K.K.K., Yadav, S.R. 2021. Dynamic cytokinin signaling and function of auxin in cytokinin responsive domains during rice crown root development. *Plant Cell Reports* 40:1367–1375. doi:10.1007/s00299-020-02618-9

Nizzetto, L., Langaas, S., Futter, M. 2016. Do microplastics spill on to farm soils? *Nature* 537(7621):488. doi:10.1038/537488b

Nolte, T.M., Hartmann, N.B., Kleijn, M., Garnæs, J., van de Meent, D., Hendriks, A.J., Baun, A. 2017. The toxicity of plastic nanoparticles to green algae as influenced by surface modification, medium hardness and cellular adsorption. *Aquatic Toxicology* 183:11–20. doi:10.1016/j.aquatox.2016.12.005

Pflugmacher, S., Sulek, A., Mader, H., Heo, J., Noh, J.H., Penttinen, O.P., Kim, Y., Kim, S., Esterhuizen, M. (2020). The influence of new and artificial aged microplastic and leachates on the germination of Lepidium sativum L. *Plants* 9(3):339.

Piehl, S., Leibner, A., Löder, M.G.J., Dris, R., Bogner, C., Laforsch, C. 2018. Identification and quantification of macro- and microplastics on an agricultural farmland. *Scientific Reports* 8:17950.

Pitt, J.A., Trevisan, R., Massarsky, A., Kozal, J.S., Levin, E.D., Di Giulio, R.T. 2018a. Maternal transfer of nanoplastics to offspring in zebrafish (Danio rerio): A case study with nanopolystyrene. *Science of the Total Environment* 643:324–34. doi:10.1016/j.scitotenv.2018.06.186

Pitt, J.A., Kozal, J.S., Jayasundara, N., Massarsky, A., Trevisan, R., Geitner, N., Wiesner, M., Levin, E.D., Di Giulio, R.T. 2018b. Uptake, tissue distribution, and toxicity of polystyrene nanoparticles in developing Zebrafish (Danio rerio). *Aquatic Toxicology* 194:185–194. doi:10.1016/j.aquatox.2017.11.017

Prata, J.C., da Costa, J.P., Lopes, I., Duarte, A.C., Rocha-Santos, T. 2019. Effects of microplastics on microalgae populations: A critical review. *Science of the Total Environment* 665:400–405. doi:10.1016/j.scitotenv.2019.02.132

Qi, R., Jones, D.L., Li, Z., Liu, Q., Yan, C. 2020. Behavior of microplastics and plastic film residues in the soil environment: A critical review. *Science of the Total Environment* 703:2020. doi:10.1016/j.scitotenv.2019.134722

Qiang, L., Cheng, J. 2021. Exposure to polystyrene microplastics impairs gonads of zebrafish (Danio rerio). *Chemosphere* 263:128161. doi:10.1016/j.chemosphere.2020.128161

Qiang, L., Xudan, X., Wei, H., Xiaoqun, X., Lu, S., Jiangning, Z. 2017. Research advances on the ecological effects of microplastic pollution in the marine environment. *Acta Ecologica Sinica* 37(22):7397–7409.

Rainieri, S., Conlledo, N., Larsen, B.K., Granby, K., Barranco, A. 2018 Combined effects of microplastics and chemical contaminants on the organ toxicity of zebrafish (Danio rerio). *Environmental Research* 162(11):135–143. doi:10.1016/j.envres.2017.12.019

Rashid, M.I., Mujawar, L.H., Shahzad, T., Almeelbi, T., Ismail, I.M.I., Oves, M. 2016. Bacteria and fungi can contribute to nutrients bioavailability and aggregate formation in degraded soils. *Microbiological Research* 183:26–41.

Richmond, R. 2015. Coral reefs: Present problems and future concerns resulting from anthropogenic disturbance. *Integrative and Comparative Biology* 33(6):524–536. https://academic.oup.com/icb/article/33/6/524/2107143

Rillig, M.C., Lehmann, A. 2020. Microplastic in terrestrial ecosystems. *Science* 368(6498):1430–1431. doi:10.1126/science.abb5979

Robin, R.S., Karthik, R., Purvaja, R., Ganguly, D., Anandavelu, I., Mugilarasan, M., Ramesh, R. 2020. Holistic assessment of microplastics in various coastal environmental matrices, southwest coast of India. *Science of the Total Environment* 703:134947. doi:10.1016/j.scitotenv.2019.134947

Rodriguez-Seijo, A., Lourenço, J., Rocha-Santos, T.A.P., da Costa, J., Duarte, A.C., Vala, H. et al. 2017. Histopathological and molecular effects of microplastics in Eisenia Andrei Bouché. *Environmental Pollution* 220:495–503. doi:10.1016/j.envpol.2016.09.092

Rosevelt, C., Los Huertos, M., Garza, C, Nevins, H.M. 2013. Marine debris in central California: Quantifying type and abundance of beach litter in Monterey Bay, CA. *Marine Pollution Bulletin* 71:299–306. doi:10.1016/j.marpolbul.2013.01.015

Roy, T., Dey, T.K., Jamal, M. 2022. Microplastic/nanoplastic toxicity in plants: An imminent concern. *Environmental Monitoring and Assessment* 195(1):27. doi:10.1007/s10661-022-10654-z

Santana, E.M., Armstrong, J.B. 2017. Food habits and anthropogenic supplementation in coyote diets along an urban-rural gradient. *Human-Wildlife Interaction* 11:156–166.

Schwab, F., Rothen-Rutishauser, B., Petri-Fink, A. 2020. When plants and plastic interact. *Nature Nanotechnology* 15:729–730. doi:10.1038/s41565-020-0762-x

Shah, S., Ilyas, M., Li, R., Yang, J., Yang, F.L. 2023. Microplastics and nanoplastics effects on plant–pollinator interaction and pollination biology. *Environmental Science & Technology* 57(16):6415–6424. doi:10.1021/acs.est.2c07733

Sjollema, S.B., Redondo-Hasselerharm, P., Leslie, H.A., Kraak, M.H.S., Vethaak, A.D. 2016. Do plastic particles affect microalgal photosynthesis and growth? *Aquatic Toxicology* 170:259–261. doi:10.1016/j.aquatox.2015.12.002

Smith, M., Love, D.C., Rochman, C.M., Neff, R.A. 2018. Microplastics in seafood and the implications for human health. *Current Environmental Health Reports* 5:375–386. doi:10.1007/s40572-018-0206-z

Stabnikova, E.V., Gordienko, A.S., Ksenofontov, B.S., Poberezhniy, V.Y., Stavskaya, T.Z., Ivanov, V.N. 1991. *Interaction between the Cells and Water–Air Interface* (pp. 1–196). Kiev, Ukraine: Naukova Dumka.

Stabnikova, O., Stabnikov, V., Marinin, A., Klavins, M., Klavins, L., Vaseashta, A. 2021. Microbial life on the surface of microplastics in natural waters. *Applied Sciences* 11(24):11692. doi:10.3390/app112411692

Sun, H., Lei, C., Xu, J., Li, R. 2021. Foliar uptake and leaf-to-root translocation of nanoplastics with different coating charge in maize plants. *Journal of Hazardous Materials* 416:125854. doi:10.1016/j.jhazmat.2021.125854

Sun, X.D., Yuan, X.Z., Jia, Y., Feng, L.J., Zhu, F.P., Dong, S.S., Liu, J., Kong, X., Tian, H., Duan, J.L. 2020. Differentially charged nanoplastics demonstrate distinct accumulation in Arabidopsis thaliana. *Nature Nanotechnology* 15:755–760. doi:10.1038/s41565-020-0707-4

Sun, Y., Duan, C., Cao, N., Li, X., Li, X., Chen, Y. 2022. Effects of microplastics on soil microbiome: The impacts of polymer type, shape, and concentration. *Science of the Total Environment* 806:150516. doi:10.1016/J.SCITOTENV.2021.150516

Thompson, R.C., Olsen, Y., Mitchell, R.P., Davis, A., Rowland, S.J., John, A.W.G., McGonigle, D., Russell, A.E. 2004. Lost at sea: Where is all the plastic? *Science* 304(5672):838. doi:10.1126/science.1094559

Toussaint, B., Raffael, B., Angers-Loustau, A., Gilliland, D., Kestens, V., Petrillo, M., Rio-Echevarria, I. M., Van den Eede, G. 2019. Review of micro- and nanoplastic contamination in the food chain. Food additives & contaminants. *Part A, Chemistry, Analysis, Control, Exposure & Risk Assessment* 36(5):639–673. doi:10.1080/19440049.2019.1583381

UNEP, A., & ASSESSMENT, I. R. R. (2016). The rise of environmental crime. Nairobi: UNEP.

Vaid, M., Mehra, K., Gupta, A. 2021. Microplastics as contaminants in Indian environment: A review. *Environmental Science and Pollution Research* 28:68025–68052. doi:10.1007/s11356-021-16827-6

Verla, A.W., Enyoh, C.E., Verla, E.N., Nwarnorh, K.O. 2019. Microplastic–toxic chemical interaction: A review study on quantified levels, mechanism and implication. *SN Applied Sciences* 1:1400. doi:10.1007/s42452-019-1352-0

Waldman, W., Rillig, M. 2020. Microplastic research should embrace the complexity of secondary particles. *Environmental Science & Technology* 54:7751–7753. doi:10.1021/acs.est.0c02194

Wang, J., Li, Y., Lu, L., Zheng, M., Zhang, X., Tian, H., 2019. Polystyrene microplastics cause tissue damages, sex-specific reproductive disruption and transgenerational effects in marine medaka (Oryzias melastigma). *Environmental Pollution* 254:113024. doi:10.1016/j.envpol.2019.113024

Watt, E., Picard, M., Maldonado, B., Abdelwahab, M.A., Mielewski, D.F., Drzal, L.T., Misra, M., Mohanty, A.K. 2021. Ocean plastics: Environmental implications and potential routes for mitigation—a perspective. *RSC Advances* 11:21447–21462. doi:10.1039/D1RA00353D

Wei, Z., Wang, Y., Wang, S., Xie, J., Han, Q., Chen, M. 2022. Comparing the effects of polystyrene microplastics exposure on reproduction and fertility in Male and female mice. *Toxicology* 465:153059. doi:10.1016/j.tox.2021.153059

Wijesooriya, M., Wijesekara, H., Sewwandi, M., Soysa, S., Rajapaksha, A.U., Vithanage, M., Bolan, N. 2023. Microplastics and soil nutrient cycling. In *Microplastics in the Ecosphere: Air, Water, Soil, and Food* (Vithanage, M., Vara Prasad, M.N., editors). doi:10.1002/9781119879534.ch19

Windsor, F.M., Durance, I., Horton, A.A., Thompson, R.C., Tyler, C.R., Ormerod, S.J. 2019. A catchment-scale perspective of plastic pollution. *Global Change Biology* 25(4):1207–1221. doi:10.1111/gcb.14572

Wright, S.L., Kelly, F.J. 2017. Plastic and human health: A micro issue? *Environmental Science & Technology* 51:6634–6647. doi:10.1021/acs.est.7b00423

Xiaoxia, S. 2016. Progress and prospect on the study of the ecological risk of microplastics in the ocean. *Advances in Earth Science* 31(6):560–566.

Xu, A., Shi, M., Xing, X., Su, Y., Li, X., Liu, W., Mao, Y., Hu, T., Qi, S. 2022. Status and prospects of atmospheric microplastics: A review of methods, occurrence, composition, source and health risks. *Environmental Pollution* 303:119173. doi:10.1016/j.envpol.2022.119173

Yang, X., Bento, C.P.M., Chen, H., Zhang, H., Xue, S., Lwanga, E.H., Zomer, P., Ritsema, C.J., Geissen, V. 2018. Influence of microplastic addition on glyphosate decay and soil microbial activities in Chinese loess soil. *Environmental Pollution* 242(Part A):338–347. doi:10.1016/j.envpol.2018.07.006

Zhang, M., Xu, L. 2020. Transport of micro- and nanoplastics in the environment: Trojan-Horse effect for organic contaminants. *Critical Reviews in Environmental Science and Technology*: 1–37. doi:10.1080/10643389.2020.1845531

Zhang, M., Zhao, Y., Qin, X., Jia, W., Chai, L., Huang, M. 2019. Microplastics from mulching film is a distinct habitat for bacteria in farmland soil. *Science of the Total Environment* 688:470–478. doi:10.1016/j.scitotenv.2019.06.108

Zhang, X.Y., Li, R., Ouyang, D., Lei, J.J., Tan, Q.L., Xie, L.L., Li, Z.Q., Liu, T., Xiao, Y.M., Farooq, T.H., Wu, X.H., Chen, L, Yan, W.D. 2021. Systematical review of interactions between microplastics and microorganisms in the soil environment. *Journal of Hazardous Materials* 418:126288.

Zouxia, L., Xingguang, Y., Xianglong, J., Jianye, R.2017. Progress in marine microplastics pollution research. *Journal of Applied Oceanography* 36(4):586–596.

13 Bioremediation of Microplastic Wastes in Soil

Sakshi Singh and Diksha Singh

13.1 INTRODUCTION

Microplastics are a significant and persistent form of pollution, which has a major effect on the ecology, crop production, and the environment. They may change the physicochemical characteristics of the soil and the mobility of pollutants, which may affect the functionality of the soil ecosystem (Yu et al., 2022). This can lead to changes in essential functions such as litter decomposition, soil aggregates, and nutrient cycles. In addition, microplastics can damage soil biological systems at various terrestrial levels and influence human health by the food web. Although potential negative effects of microplastics are present, there is little groundwork on the reaction of small particles of plastic in the soil. Small particles of plastics are small plastic fragments that are smaller than 5 μm and that are found in the ecosystem due to fragments of plastic pollution (Rillig, 2012). It is found in various products, including cosmetics, microfiber from clothing, plastic bags, tires, and bottles. Many of these products are easily discarded and end up in the environment (Figure 13.1). The hydrogen and carbon particles are linked by the polymer chain to form microplastics. Other components in microplastics involve phthalates, polybrominated diphenyl ethers (PBDEs), and tetraromobisphenol A (TBBPAs), which are released from plastics when entering the environment (Dris et al., 2016).

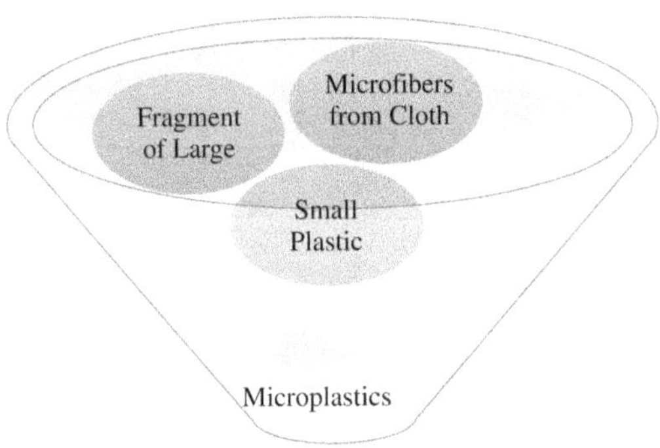

FIGURE 13.1 Different sources of microplastics from the environment.

DOI: 10.1201/9781032684574-13

13.2 MICROPLASTICS SOURCES OF SOIL

Small fragments of plastics are categorized into two types: primary microplastics and secondary microplastics. Primary microplastics show up in the environment directly through several sources, including the use of products such as personal care products that are washed from households into wastewater systems. They are also lost by accident by spills during production or transport or by abrasion during activities such as washing synthetic textile clothing. Secondary microplastics result from the effects of the decomposition of larger plastics, which occurs often when larger polymers are exposed to weather, such as wind, waves, and ultraviolet rays from sunlight (An et al., 2020).

13.2.1 **Agriculture Film:** This is a plastic film used in farming. It is usually made up of polyethylene (PE) and polyethylene chloride (PVC). PE films are lightweight and have good UV transmission, while polyethylene chloride films have better heat retention but have bad light transmission and can release toxic and dangerous compounds when burned. The growing adoption of film-modifying agricultural technology is expected to lead to a 5.7% increase in global film coverage. Due to the low recovery rate and recycling rate of agricultural films, large amounts of waste films will accumulate over a long duration in the ecosystem (Brodhagen et al., 2014) (Figure 13.2).

13.2.2 **Sewage Sludge:** In one study, after treatment of the wastewater, the fragment of microplastics in the wastewater was decreased by 0.25–0.04 ppm/L, and its elimination rate was over 98%. However, the microplastics removed do not suffer significant degradation and remain

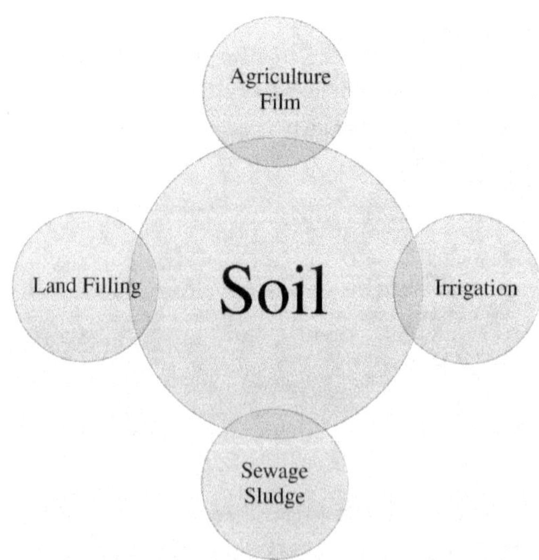

FIGURE 13.2 Different sources of microplastics in the soil.

in sludge. Sludge can be used in composting and applied to agricultural soils because it has a rich nitrogen and phosphorus content, which can improve soil formation and fertility. However, the used of sludge as fertilizer results in microplastic accumulation in the soil. Research shows that sludge contains dangerous persistent bacteria, heavy metals, antibiotic, pathogenic bacteria, and parasite eggs (Lwanga et al., 2022).

13.2.3 **Irrigation:** Microplastics are small particles of plastic that have a harmful effect on the ecosystem and human life. Personal care products and detergents are the main contributors of microplastics in residential waste. Many of the market-available facial cleaning and shower gels contain microplastics, with polyethylene being the most common component. These small particles can also be found in groundwater, with an average content of 0–7 pcs/m3. The main components of groundwater microplastics are polyethylene, polyamide, polyethylene terephthalate, and polyvinyl chloride, with particles of 50–150 m. According to some studies, surface water is also an important source of soil microplastic pollution. Road tire wear can introduce rubber particles into road soil through air deposits or surface flow (Dris et al. (2016).

13.3 MICROPLASTICS MIGRATION IN SOIL

Urbanization and industrialization have exposed the environment to harmful pollutants. Pollution from industries is a main cause of ecosystem pollution. During industrial production, heavy metals of various types and quantities are emitted as byproducts of further manufacturing processes. The characteristics of soil structure, such as macropores (>75m) and land cracks, as well as agricultural practices like haring and harvesting, can influence the fragment of microplastics in soil. The use of fertilizers, insecticides, and herbicides in agriculture results in pollutants such as aluminums, coppers, zincs, nickels, leads, and arsenics being released into the surrounding soil (Rillig, 2012). It has been discovered that soil microarthropods have a greatly enhanced capacity to transport and disseminate microplastics by predator–prey interactions. It can be inferred that the food network in soil ecosystems, which consists of different complex species connections, can increase the movement of microplastics more than the involvement of a single species (Huerta Lwanga et al., 2018).

13.4 IMPACT OF MICROPLASTICS IN SOIL

The characteristic of the soil plays an effective role in the migration of microplastics. The existence of microplastics in soil affects its properties, including its structure, function, and microbial diversity. This can potentially affect flora and fauna growth and raise concerns about food safety and quality. In the end, it could be a threat to human health and safety. It has been noticed that the existence of small plastic films in the soil can decrease the saturated hydraulic conductivity of soil, decrease the activity and abundance of soil microorganisms, and ultimately lead to changes in soil fertility (J. Yu et al., 2022).

13.4.1 **Soil Structure:** Microplastics have various interactions with different soil types, making soil nature the most suitable indicator of threats to terrestrial ecosystems by microplastics. Microplastic particles could bind to soil aggregates and fragment shapes, as well as more solidly in linear forms. There are no discernible trends in the ways that polyethylene, and polyacrylic acid both affect water retention capacity. However, it has been discovered that polyester fibers greatly improve the ability to retain water while simultaneously lowering density and stability in water accumulation. Depending on their composition, microplastics have different effects on the soil (Guo et al., 2020).

13.4.2 **Fertility and Nutrients in the Soil:** Several soil biochemical activities are directly linked to high catalytic soil enzymes. These enzymes are important for monitoring the recovery of elements such as carbon, nitrogen, and phosphorus in the soil and are used to quantify soil fertility. Microplastics have been found in recent research to have a considerable impact on the function of soil enzymes like urease, catalase, and fluorescein diacetate hydrolase (FDAse). These effects may result in short-term alterations in soil quality. Due to their effect on soil density in bulk, plastic particles may lead to an underestimation of soil carbon storage (Sajjad et al., 2022).

13.4.3 **Soil Microbes:** Research has shown that the microbial activity in soil is dependent on soil characteristics and nutrients. Modification to the physical environment of the soil, particularly soil aggregation, has been observed to have a different effect on microbial evolution when compared to non-microfiber-structured soil. When the soil microbe was alive, there was a significant increase in the formation of de-novo-generated aggregates, which was not observed in the sterilized remedy. This positive impact, however, was lost after the microfiber remedy. Microplastic-induced modifications to soil porosity and moisture could affect oxygen flow, altering microbial dispersion (Isari et al., 2021; J. Yu et al., 2022).

13.4.4 **Soil Contamination:** In addition to affecting soil function and health, microplastics further alter the biophysical characteristics of soil, impacting the behavior of other contaminants. Microplastics are not just made up of harmful additives like diethylhexyl phthalate (DEHP) that are commonly used during plastic production. They also absorb other harmful elements such as heavy metals like zinc, copper, and lead, antibiotics, and harmful organic substances including perfluorochemicals (PFOS) and polybrominated diphenyl ether (PBDE). This is because microplastics have a large specific surface area that allows them to adsorb more contaminants (Isari et al., 2021).

13.5 SOIL MICROPLASTICS ARE DISPERSED TO THE OTHER ECOSYSTEM

Microplastics in soil can be transmitted to other via means of either natural or man-made phenomena including wind, particles, sediments, and surface runoff, the environment including the air and water, are affected. Microplastics, especially microfibers, can easily be carried by wind and air movements and remain dissolved in the air for a considerable amount of time (Guo et al., 2020). Furthermore, microplastics in soil can potentially infiltrate subsurface receptors such as aquifers. A recent study has demonstrated that the number of wet-dry cycles is directly proportional to the depth of microplastic penetration. According to the study, higher than 60% of microplastics found in the soil will make their way into water bodies, creating a danger to the water ecosystem. Soil is both an origin and a destination of plastic particles (J. Yu et al., 2022).

13.6 BIOREMEDIATION OF MICROPLASTICS IN SOIL

Microplastics, due to their chemical composition and additives, such as plasticizers, stabilizers, and fire retardants, can potentially have adverse effects on soil quality, soil nature, and biodiversity, and contribute to composite contaminate with other soil contamination. Furthermore, because microplastics travel and alter readily in the environment, they can endanger the lives of animals and people across the food supply chain. Plastics can be recycled, disposed of in landfills, or degraded thermally, mechanically, or biologically. However, because of their small size and broad integration, the existence of microplastics in soil causes considerable issues. It is difficult to cycle (Y. Yu et al., 2022) or enrich microplastics by conventional means. Biodegradation technology is now considered the most suitable option for managing the contamination of microplastics because it is capable of efficiently dismantling microplastics over a wide range of areas without jeopardizing the local ecosystem or biota. Animal degradation by insects, microbes, and enzymes is a popular microplastic biodegradation process (Y. Yu et al., 2022).

13.6.1 **Degradation of Microplastics in Soil by Animals:** The ecosystem's material cycles depend on soil animals because they can actively assimilate elements from their surroundings for their development, growth, and other purposes. In this cycle, soil organisms are critical patrons. Certain insects, including invertebrates and social insect species, have been shown to feed and eat plastic products, using them as their only form of carbon dioxide, decomposing plastic particles into carbon dioxide and water via physical methods such as biting, masticating, and digestion, as well as several biochemical processes (Ayilara & Babalola, 2023). Table 13.1 focuses on several studies of microplastic breakdown by soil animals, and most of the current research focuses on wax worm larvae, wheat pests, and grain beetles. The symbiotic intestines of these insects are mainly responsible for degrading microplastics. For example, *Aspergillus flavus* is the main component of plastic particle degradation by *G. mellonella* due to its ability to produce multicopper oxides such as Lac, and *Lac. Bacillus* and *Enterobacteriaceae* also contribute to the destruction of the PE film by the larvae of the waxworm (*P. interpunctella*) in the intestine (Y. Yu et al., 2022).

13.6.2 **Degradation of Microplastics in Soil by Microbes:** Microorganisms are an efficient and eco-friendly solution for removing microplastics.

TABLE 13.1
Degradation of Microplastics in Soil by Animals (Y. Yu et al., 2022)

Animals	Plastic	Degradation Effect
Tenebrio molitor	Polyethylene	Digested PE was converted to CO_2.
Tenebrio molitor	Polystyrene	The rate of degradation was 50% per day.
Z. atratus	Polystyrene	Degrade 0.58 mg/day/superworm
G. mellonella	Polyethylene, polypropylene	92% of microplastics are degraded, the decreasing rate is 1.84 mg/day/worm.
Achatina fulica	Polystyrene	Polystyrene was ingested by a snail.
Zophobas atratus	Expanded polystyrene, low-density polyethylene	In 33 days, LDPE and PS foams consumption
Plodia interp unctella	Polyethylene	In 60 days, two bacterial strains obtained from worms damaged the PE films.
Achroia	Griselda high-density polyethylene	PE intake rose throughout eight days with ingestion.

They can thrive in almost any environment and can decompose various organic pollutants. Microplastics are used by microorganisms during the decomposition process as substrates for the development of biofilms. The molecular framework of the microplastics dissolves as the biofilm thickness increases, and enzymes released by bacteria and fungi break down the particles of microplastic via specific and nonspecific mechanisms. Microbes in biofilms can absorb microplastics under 600 kDa. After that, enzymes can further break down absorbed fragments into small molecules $(CO_2, N_2, CH_4, H_2O, H_2S)$ and then use microbes as energy sources before being released into the environment (Huerta Lwanga et al., 2018). Table 13.2 summarizes research on the remediation of microplastics, including the type and source of bacteria that digest the plastic and the type of plastic. PE is the primary target material, with studies regarding other polymers being limited. Studies have shown that fungi are more effective in decomposing microplastics than bacteria (Ayilara & Babalola, 2023). Fungi produce more enzymes, and their mycelium can bind more strongly to the furthest layer of microplastics and even penetrate the particles. Furthermore, by promoting the growth of bacteria, fungi can reduce the hydrophilic properties of microplastics. On the other hand, bacteria need less stability in the outer environment than fungi, indicating that they have higher decomposition potential (Ayilara & Babalola, 2023).

13.6.3 **Microplastic Degradation in Soil by Enzymes:** Plastic biodegradation is a complex process based on various microorganisms and enzymes. These enzymes can be found within or outside the cell. Plastic-degrading enzymes can be classified as intracellular or extracellular, depending on how they work. Esters and lipases, for example, are internal enzymes that dissolve ester compounds in polymer esters.

TABLE 13.2
Degradation of Microplastics Through Microbes (Huerta Lwanga et al., 2018; Y. Yu et al., 2022).

Microorganisms	Microbial Sources	Plastic	Degradation Effect
Bacillus bacterial	Mangrove environments	Polyethylene, PET, polystyrene, polypropylene	Decompose the mixture of various plastics.
Ideonella sakaiensis 201-F6	Natural microbial communities exposed to PET	PET	Produces two types of enzymes capable of converting PET into two environmentally beneficial monomers.
Zalerion maritimum	Marine environment	Polyethylene	Reduces the pellets' bulk and size.
Bacillus subtilis	Nutrient medium	Polyethylene	In 30 days, it shed 9.26% of its body weight.
Kocuria palustris M16, Bacillus pumilus M27, Bacillus subtilis H1584	Arabian sea	Polyethylene	After 30 days, lost 1%, 1.5%, and 1.75% weight.
Pestalotiopsis microspora	Ecuadorian rainforest	Polyurethane	Deteriorating the polyester diol component, yielding breakdown byproducts that were employed as a carbon source for polymer mineralization.
Aspergillus terreus, Aspergillus sydowii	Rhizosphere soil of *Avicennia marina*	Polyethylene	The most efficient polythene deterioration was fungal.
C. pseudocladosporioides	Mineral medium	Polyurethane	After 14 days, it deteriorated to 87%.

Esterase, for example, damages PE, PET, and PVC, whereas lipase damages PET and PU. Extracellular enzymes also break down complicated polymers into smaller pieces like oligomers, monomers, and dielectrics. These small units can then be absorbed by microorganisms in the body and further processed by intracellular enzymes. Proteases, lipases, keratoses, insects, manganese peroxidase, lignin peroxidase, and alkane hydroxylase can degrade microplastics (Isari et al., 2021; J. Yu et al., 2022).

13.7 FUTURE RESEARCH AND CONCLUSION

Plastic pollution is a global risk that needs to be addressed with a systematic approach for the future. Current research on the matter is fragmented, which makes it difficult to fully comprehend the impact of microplastics. Through bio-physiochemical processes, these microscopic particles harm animals, plants, and soil microbes, as well as interfering with food supplies. Finally, it has the potential to be hazardous to human health. However, relevant research is still in its early stages, and no unified theoretical or methodological structure has been formed as a result of its late start. In the future, extensive studies will be performed on the following topics:

1. Current techniques for detecting soil microplastics are restricted and insufficient for determining their distribution and origins. As a result, it is critical to research and standardize methods for separating and detecting microplastics in soil. It is critical to build a systematic framework for a consistent methodology for microplastic separation and detection.
2. Because soil animals influence microplastic transport and breakdown, it is critical to examine the infliction of toxicity by microplastics on these species. The majority of current laboratory simulation methodologies rely on variables like production rate, rate of growth, and rate of survival. Future clinical trials are required to investigate the toxicity impacts of plastic particles on soil organisms and the migratory process in-vivo.
3. The impact of plastic on soil ecosystems, especially agricultural ecosystems, is worth investigating. Microplastics can affect soil microbial biomass, functional diversity, and microbial activity, as well as plant growth. This has the potential to cause unpredictable consequences such as the inhibition of seed germination and seed development. Soil, groundwater, and fertility should be investigated as soon as possible, as this may have an unexpected impact on seed germination and growth.
4. The impact of plastic particles on soil organisms is an important field of research. The kind of microplastics presents influences on microorganism adhesion to soil microplastics. It is critical to understand both the mechanisms of microplastic breakdown and their impact on the microbial population.
5. Develop guidelines, standards, and recommendations to address microplastic pollution. Theoretical and empirical support for plastic waste management and treatment and microplastic emission and reporting are needed to raise public awareness of household waste distribution and to strengthen research and the construction on plastic waste.

REFERENCES

An, L., Liu, Q., Deng, Y., Wu, W., Gao, Y., and Ling, W. (2020). Sources of microplastic in the environment. *Emerging Contaminants and Major Challenges*, 143–159. doi: 10.1007/698_2020_449.

Brodhagen, M., Peyron, M., Miles, C., and Inglis, D. (2014). Biodegradable plastic agricultural mulches and key features of microbial degradation. *Applied Microbiology and Biotechnology* 99. doi: 10.1007/s00253-014-6267-5.

Dris, R., Gasperi, J., Saad, M., Mirande, C., and Tassin, B. (2016). Synthetic fibers in atmospheric fallout: A source of microplastics in the environment? *Marine Pollution Bulletin* 104 (1–2), 290–293. doi: 10.1016/j.marpolbul.2016.01.006.

Huerta Lwanga, E., Thapa, B., Yang, X., Gertsen, H., Salánki, T., Geissen, V., and Garbeva, P. (2018). Decay of low-density polyethylene by bacteria extracted from earthworm's guts: A potential for soil restoration. *Science of the Total Environment* 624, 753–757. doi: 10.1016/j.scitotenv.2017.12.144.

Lwanga, E. H., Beriot, N., Corradini, F., Silva, V., Yang, X., Baartman, J., Rezaei, M., van Schaik, L., Riksen, M., and Geissen, V. (2022). Review of microplastic sources, transport pathways and correlations with other soil stressors: A journey from agricultural sites into the environment. *Chemical and Biological Technologies in Agriculture* 9 (1), 1–20. doi: 10.1186/s40538-021-00278-9.

Rillig, M. C. (2012, June 19). Microplastic in terrestrial ecosystems and the soil? *Environmental Science & Technology* 46 (12), 6453–6454. doi: 10.1021/es302011r; Epub: 2012 May 31; PMID: 22676039.

Sajjad, M., Huang, Q., Khan, S., Khan, M. A., Liu, Y., Wang, J., Lian, F., Wang, Q., and Guo, G. (2022). Microplastics in the soil environment: A critical review. *Environmental Technology & Innovation* 27, 102408. doi: 10.1016/j.eti.2022.102408.

Yu, J. R., Adingo, S., Liu, X. L., Li, X. D., Sun, J., and Zhang, X. N. (2022). Micro plastics in soil ecosystem—a review of sources, fate, and ecological impact. *Plant, Soil and Environment* 68, 1–17.

Yu, Y., Luan, Y., and Dai, W. (2022). Biodegradation of microplastics in soil. In *Proceedings of the 3rd International Conference on Green Energy, Environment and Sustainable Development (GEESD2022)* (866–872). IOS Press. doi: 10.3233/ATDE220362.

Yu, Y., Luan, Y., and Dai, W. (2022). Biodegradation of microplastics in soil. doi: 10.3233/ATDE220362.

14 Extraction of Microplastics from Rhizosphere

Ishita Kundu and Avijit Ghosh

14.1 INTRODUCTION

Conventional plastics are widely used due to low cost, versatility, ease of manufacture, and user orientation. The world has become the victim of plastic toxicity for creating pollution in soil and ecosystems. Substances that are used to enhance plastic flexibility, expansibility, and workability mainly release toxins into the soil, and thus soil pollution is created [1]. Microplastic is a term that implies microscaled plastic. Tiny plastic particles with less than 5 mm (0.2 inch) diameter are known as microplastics [2–4]. Microplastics can adhere to soil particles and interact with plant roots, creating complex challenges for their isolation. The devastating side effect of microplastics is that, after harvesting, microplastic is left in the soil [5]. Despite their small size, microplastics can accumulate within the rhizosphere, posing significant threats to the plant-soil ecosystem and influencing humans, plants, and animals. The intricate interplay between microplastics and the rhizosphere introduces risks such as altered nutrient availability, pollutant entrapment, and interactions with root exudates. The rhizosphere is a dynamic realm where plant roots converse with the living and non-living with a 1 mm wide zone of a plant root where biochemical features of soils are determined by the root [6]. The rhizosphere consists of high organic matter with large microbial diversity, which is a nexus where MP profoundly influences the physicochemical properties and soil fertility. This influence extends to microbial diversity, encompassing various types of bacteria, archaea, fungi, and protists and to shaping intricate plant-soil-microbial interactions [7]. Figure 14.1 shows microplastic contaminants in the rhizosphere soil.

14.1.1 Microplastics: Sources, Hazardous Impacts on the Environment

Microplastics are categorized into large microplastic (3–5 mm), medium microplastic (1–3 mm), and small microplastic (<1 mm) [8]. Several ordinary consumer products are the source of it. Sources are categorized as primary and secondary [9, 10].

14.1.1.1 Primary Sources

Commercially used plastic products are generally known as primary sources of microplastic [11, 12].

DOI: 10.1201/9781032684574-14

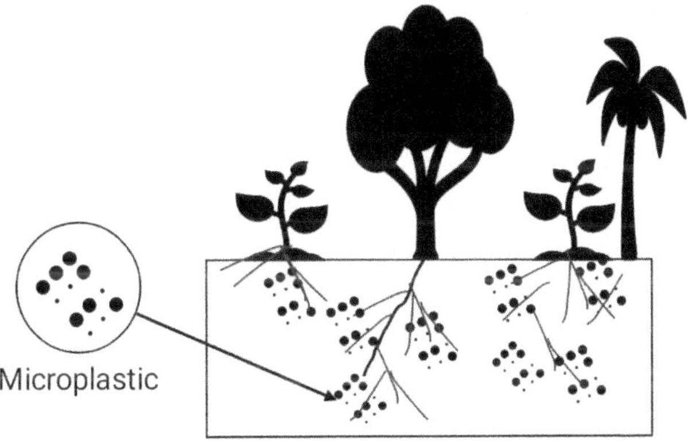

FIGURE 14.1 Microplastic contaminants in the rhizosphere soil.

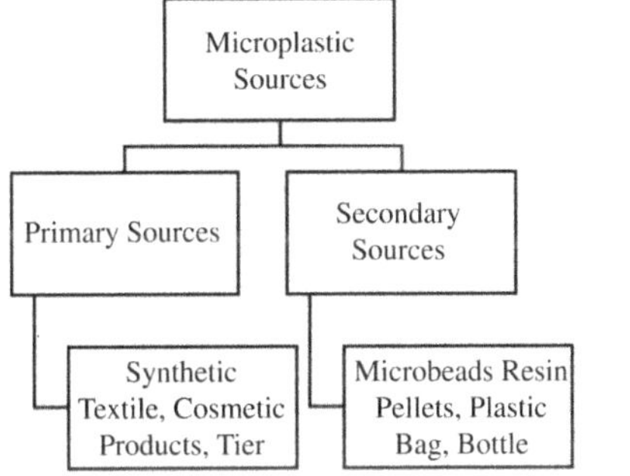

FIGURE 14.2 Sources of microplastics.

- **Synthetic Textile:** Synthetic textiles contain polyester, nylon, and acrylics. Synthetic textiles are deposited in the soil for day-to-day activities. The concentration of deposition is 1586–11,130 days/m³, which is harmful to the environment. Fleeces release 170% more than others, and it easily mixes in plant soil ecosystems, creating hazardous contamination [13, 14].
- **Cosmetic Industries:** Cosmetics industry contains polypropylene, polyethylene, and nylon. Acrylate polymer causes toxic effects in the ecosystem. Face wash, hand soap, facial scrubber, and various kinds of cosmetics are the main sources of this category, which also create pollution in the soil rhizosphere when they mix in the ecosystem [13, 14].
- **Tiers:** Wear and tear from tires contribute most of the pollution in soil 5500–14,000emissions of microplastic, which causes harm to the environment. This content natural rubber, synthetic rubber, and fiber [13, 14]. Figure 14.2 shows all the primary sources.

14.1.1.2 Secondary Sources

The smaller plastic particle that breaks down from more oversized plastic items is called a secondary microplastic source [15, 16].

- **Microbeads Resin Pellet:** This contains nonylphenols types of material. Concentrations of NP are widely distributed in the soil and cause harmful pollution in the environment [13, 14].
- **Plastic Bag:** It is a synthetic or semisynthetic material made of Bakelite, polyvinyl chloride. The plastic deposition rate is about 1.7 mg/kg in the soil ecosystem. The various kinds of plastic bags that are used widely in daily life, when broken down into smaller particles and mixed with nature create hazardous impacts [13, 14].
- **Bottle:** Polypropylene types of materials constitute 14,600 to 455,000 particles per capita per day in 48 regions. These are daily used PET material bottles, silicon water bottles for babies, etc. [13, 14]. Figure 14.2 shows all the secondary sources.

14.1.1.3 Other Sources

Washing synthetic clothes that contaminate the soil ecosystem can cause microplastic pollution. Other polymeric products that are exhibited in the soil also create microplastic pollution in the rhizosphere [17, 18].

In Table 14.1 [21, 22], microplastic polymer names are discussed. Polyethylene can be categorized into HDPE and LDPE [23, 24]. All the other polymers like PP, PVC, PC, PVA, PS, PA, and PET are also discussed with their chemical properties and effects in rhizosphere soil.

14.1.1.4 Specific Effects on Rhizosphere Properties

Microplastics can be potentially impactful for soil contamination. It can affect several important properties of the rhizosphere, including:

- **Water-Holding Capacity:** Microplastics might influence the soil's ability to retain water, affecting the water-holding capacity of the rhizosphere.
- **Hydraulic Conductivity:** The ability of soil to conduct water or hydraulic conductivity can be influenced by microplastics.
- **Soil Aggregation:** Microplastics, potentially altering the structure of the soil, can affect the arrangement of soil particles into aggregates [25, 26].

14.1.2 Rhizosphere: Microplastic's Contamination with Rhizosphere

14.1.2.2 Microplastic's Impact on the Rhizosphere: Disturbance Unveiled

The rhizosphere is a layer of soil with various phytopathogens; microbes perform numerous growing activities for plants. In this layer of the soil, mycoparasitic fungi, nitrogen-fixing bacteria, and rhizobacteria have an important effect on plant health [6]. The effects of MPs on the organisms have three aspects [27]: physical damage, biochemical damage, and biological damage. However, soil organelles,

TABLE 14.1
Different Properties of Polymers and Contamination with Soil

[a]—[19], [b]—[20].

Polymer Name [a]	Monomer	Chemical Formula	Density (g/cm³)	Source [b]	Soil Contamination
I. PE (polyethylene)	Ethylene	$(C_2H_4)n$	0.91–0.97	Facial cleanser, toothpaste,	Alpha diversity of rhizosphere plants
I.I. HDPE (high-density polyethylene)			0.93–0.97	milk juice cans	Terrestrial ecological pollutant
I.II. LDPE (low-density polyethylene)			0.91–0.93	plastic bags, plastic bottles, fishing net	
II. PP (polypropylenes)	Propene/propylene	$(C_3H_6)n$	0.89–0.92	Rope, bottle cap	Prevents seeping of water into the soil
III. PVC (polyvinyl chloride)	Vinyl chloride	$(H_2C\text{–}CHCl)_n$	1.10–1.47	Plastic film, plastic cup	Preserves harmful additives in rhizosphere soil
IV. PC (polycarbonates)	Bisphenol	$C_{15}H_{16}O_2$	1.20–1.22	Food containers, sippy cups	Releases harmful chemicals in soil
V. PVA (polyvinyl alcohol)	Vinyl acetate	$[CH_2CH\ (OH)]_n$	1.19	Dishwasher, laundry pods	Inhibits the activities of microbes
VI. PS (polystyrenes)	Styrene	$C_6H_5CH{=}CH_2$	0.96–1.05	Food containers, plastic utensils	Forms styrofoam, which creates toxicity in the roots of the plant
VII. PA (polyamides)	Benzene-1, 4-dicarboxylic acid	CO-NH	1.15	Fishing net	Consists of amide bond, which is adverse for ecological degradation
VIII. PET (polyethylene terephthalate)	Ethylene terephthalate	$(C_{10}H_8O_4)n$	1.37–1.38	Bottles	Forms harmful chemical compound

bacteria, archaea, fungi, arthropods, nematodes, and earthworm's activities within the rhizosphere are impactful and efficient through the intricate web of relationships [28]. Despite being small, microplastics can easily combine in the rhizosphere, adversely affecting the plant-soil ecosystem and harming humans, plants, and animals. Further, microplastics destroy soil structure, water-holding capacity, and plant nutrient cycling. Enzyme activity is essential for plant growth, and the ecosystem's vitality. Microplastics hamper nutrient enzyme interactions, disrupting soil equilibrium and impeding plant nutrient access. Soil organisms from earthworms to fungi suffer, compromising growth and even causing DNA damage. Therefore, extraction of microplastic is necessary to avoid toxic contamination of soil. The extraction process is a process where obtaining impurities or, more specifically, plastic from a contaminated medium are distinguished [29].

14.1.3 Importance of Extraction

Microplastic is a toxic contaminant that can affect both human and plant health. In Figure 14.3, diseases created by microplastics are shown in the block diagram for better understanding. It is shown that microplastics have a dangerous impact on human health and living species. Therefore, there is a requirement to extract microplastic.

14.2 SEVERAL METHODS OF EXTRACTION

Currently, no published studies focus on microplastic sampling and analysis in the rhizosphere [30, 31]. However, some recent literature has discussed the chemical analysis of microplastic in soil. The rhizosphere contains SOM (solid organic matter) [32] smaller particles due to microbial and plant effects and higher aggregation due to microorganism and exopolysaccharide actions. Adapting bulk soil microplastic analytical methods, for rhizosphere soil is challenging due to soil complexity [33, 34]. Therefore, dynamic sampling of the rhizosphere complicated the study more. Detection of microplastic is generally done by sampling, extraction, cleanup,

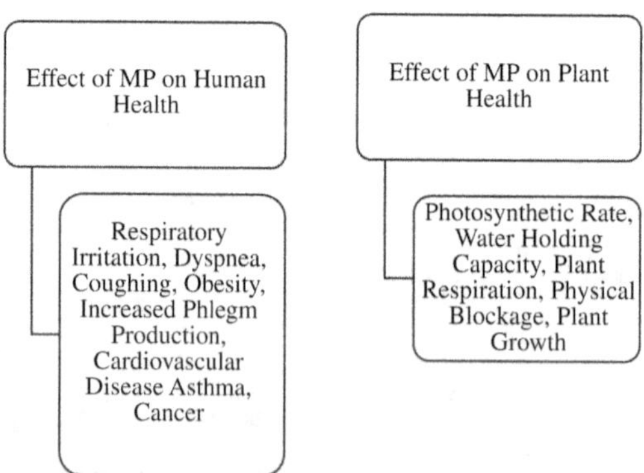

FIGURE 14.3 Effect of microplastics on human and plant health.

identification, and quantification and several processes [35]. Microplastics need to be extracted from the soil to maintain plant physiology, quality, fertility, and other parameters of soil. Several methods are used to extract microplastic.

14.2.1 Separation Process

- **Density Separation:** In aeration, using bubbles from an air pump in extraction solutions creates detachment of initially adhered microplastics from soil or sediment of density range 0.90–1.45 g/cm^3 [36]. The movement of the bubble assists microplastics to transport upward by froth flotation. Remarkably, the recovery rate reached is about 98 – 100% [37].
- **Oil separation**: In the process, water is initially mixed with environmental soil samples in containers like conical flasks. After a settling period, oils like canola oil are added to the mixture [38]. The liquid mixture is then transferred to separatory funnels, along with rinsing the container with distilled water. During the transfer, shaking the funnel helps shift microplastic items into the oil layer based on their lipophilicity. After settling, microplastics accumulate in the upper oil layer while sediment or soil remains in the lower water layer. The water layer is separated, and the microplastics-contained soil layer is isolated for further analysis [39]. Dispensing funnels may be used to prevent microplastic loss or contamination. Residual oil interfering with microplastic identification is removed by rinsing with detergents and reagent alcohol. This method maintains the integrity of the FTIR spectra after treatment.
- **Electrostatic Separation:** The electrostatic separation approach involves mainly the electrical properties of solid matrices like soil and sediment (charged or corona) versus nonconductive microplastics. Electrostatic differences enable efficient separation using an electro-discharge process, aided by a high-voltage generator and corona electrode. A conveyor delivers microplastics-containing samples to a charged area; moisture control is crucial. Electrostatic separation offers automation, convenience, and high accuracy. Simulation tests demonstrate effective separation, removing up to ~90% [54, 55] of the original sample [40, 41].
- **Solvent Extraction Separation:** Extraction methods like Soxhlet, reflux, and reprecipitation are used for plastic analysis [42, 43]. Tetrahydrofuran is employed for residual PS polymer extraction in biodegradation studies [44]. Pressurized fluid extraction (PFE) is efficient for separating water-insoluble organic compounds from soil, sediment, and environmental waste. PFE has been applied for plastic analysis, using hot solvents for emulsification or solubilization and subsequent reprecipitation. PFE successfully quantified microplastics (<30 μm) from waste and soil samples achieving a 100% recovery rate [45].
- **Magnetic Separation:** The magnetic separation of microplastics involves binding magnetic nanoparticles to MPs, enabling their separation in a magnetic field. This approach effectively separated MPs and even microalgae-attached nanoparticles [46]. Hence successful magnetic extraction of MPs

TABLE 14.2
Comparison of Different Types of Separation Methods and Their Properties

Separation Method	Equipment	Density of MP Detected	Recovery Rate	Method Type	Reference
Density separation	Pump	>11 μm	98–100%	Froth flotation	[36, 37]
Oil separation	Funnel Flask FTIR microscope	>200 μm	low	Settling	[38, 39]
Electrostatic separation	High-voltage generator	>63 μm	99%	Electro-discharge process	[40, 41]
Solvent extraction Separation	Soxhlet reflux reprecipitation	0–5000 μm	100%	Traditional extraction	[42–45]
Magnetic separation	Magnetic field	200–1000 μm	49–90%	Magnetic filtration	[46–48]

from suspensions, with recovery rates varying based on MP size and environmental factors. Recovery rates ranged from 49 to 90% for medium-sized MPs (0.2–1 mm) and reached 90% for fine MPs (<20 μm). Factors like MP pollutant attachment, pH, and nanoparticle properties influence separation efficiency [47, 48]. Table 14.2 shows various separation methods along with other properties to understand the effectiveness of the various methods and their recovery rate.

14.2.2 Flotation

The method involves analyzing microplastics (MPs) using sieve analysis, microscopy, and filtration. Light-density polyethylene (LDPE) with 1 g/cm³ density is collected. Different soil types are sieved through a 2 mm mesh. MPs are mixed with soil at varying weight percentages under 20% moisture. The mixture is placed in an aluminum cup and incubated at 4°C. Distilled water (15 ml) is added to a 10 g sample and left overnight. Deposits are filtered through 3 μm filter paper after ultrasonic treatment and dried at 60°C. This method assesses MPs in various soils [49].

14.2.3 Chemical Digestion (Using Chemical Reagents)

The most usable processes using chemical agents are categorized into two processes:

1. Large Particle recovery
2. Small particle recovery

- **Large Particle Recovery:** A microplastic sample (0.08 g) is shaken with log soil in a sealed funnel, and then water is added. After sedimentation, microplastics are vacuum-filtered and treated with ethanol to remove clay.

The residue is digested with hydrogen peroxide, yielding a 90% recovery of large microplastics.

- **Small Particle Recovery:** Microplastic extraction involves incineration at varying temperatures. Polyethylene and polypropylene show > 98% efficiency, while PVC achieves 74%efficiency. This is mainly done by NaCl incineration [50, 51]. Soil is mixed with NaCl, treated with hydrogen peroxide at 50°C for 72 hours, and then filtered. The remaining microplastics are filtered again at 600°C for 30 minutes. This method separates microplastics based on density and thermal stability.

14.2.4 Centrifugation

Microscopic analysis and identification involve several steps:

Soil samples are sieved to 2 mm for analysis. Zn and Cu identification involves centrifuging 5 g soil with water for 15 minutes at 2000 rpm, followed by NaCl and $ZnCl_2$ treatments and filtration.

Density separation under FTIR microscopy is performed using a glass tube with 1 g soil and 10 ml $ZnCl_2$. Ultrasonication, agitation, and shaking are employed before vacuum filtration. Microscopic analysis is conducted for soil sample resolution at 8 cm^{-1}. This method combines centrifugation, density separation, and microscopic analysis to identify and analyze microplastics in soil samples [51, 52].

14.2.5 Visual Identification (Microscopic Analysis)

Microplastic particles mixed with soil are placed on a slide for identification using a microscope. The slide is heated at 130°C for 325 seconds. Impurities are identified, and heating helps distinguish microplastic particles from the soil matrix. This method aids in quantifying and characterizing microplastics in soil samples [53, 54].

This was also examined in UV-Vis, NIR microscope, and NMR spectroscopy [55]. Regrettably, soil organic matter (SOM) shares chemical features with many plastics, hindering molecular spectroscopies like FTIR, NIR, and Raman from reliably identifying microplastics in organic-rich soils. These methods also struggle to detect particles smaller than 1 µm. NIR's detection limit is notably high (15 g MP per kg of soil, 0.5–1.0 mm particle size).

Table 14.3 shows the representation of different methods of microplastic extraction and their comparison.

In Figure 14.4, the percentages and efficiencies of several extraction methods are shown.

14.3 DIFFERENT BENEFITS OF EXTRACTION

14.3.1 Physical Properties

The importance of extracting MP from soil rhizosphere is first to maintain the quality of the soil. Bacteria, archaea, fungi, arthropods, nematodes, and earthworms under the soil can process their activities in the ecosystem.

TABLE 14.3
Different Types of Extraction Process

Extraction Process	Equipment	Process	Reference
Flotation	Aluminum cup, incubator, ultrasonicator	• MPs of 1gm/cm^3 collected • Mixed with soil in water • Filtration • Sonication	[49]
Using chemical reagent	Funnel, filter	• MPs shaken with soil sample in a funnel with water • Microplastics vacuum-filtered and treated with ethanol to remove clay • Residue digested with hydrogen peroxide • Soil mixed with NaCl, treated with hydrogen peroxide at 50°C for 72 hours • MP filtered again at 600°C for 30 minutes	[50, 51]
Centrifugation	Centrifuge, ultrasonicator, agitator, microscope	• Soil samples are sieved to 2 mm • Zn and Cu identification done by centrifuging 5 g soil with water for 15 minutes at 2000 rpm • Density separation under FTIR microscopy performed using a glass tube with 1 g soil and 10 ml $ZnCl_2$	[51, 52]
Microscopic analysis/visual identification	UV-Vis, FTIR, Raman, NIR	• MP and soil mixture taken in a slide • Heated at 130°C for 325 seconds • Examined under a microscope	[53–55]

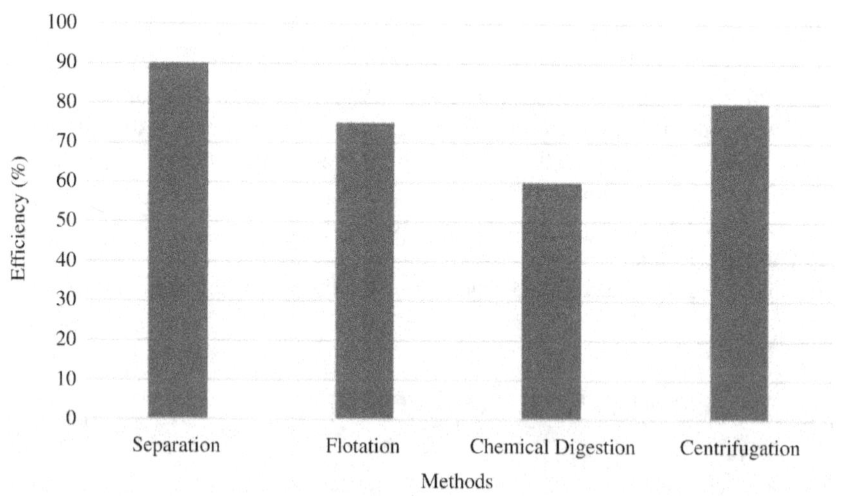

FIGURE 14.4 Graphical representation of the efficiencies of extraction methods.

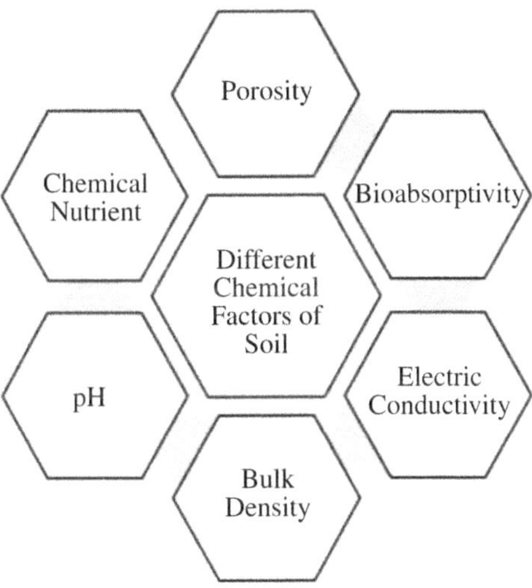

FIGURE 14.5 Various chemical properties of soil.

14.3.2 Different Chemical Factors of Soil for Functional Benefit

Figure 14.5 represents the various chemical properties that are to be maintained for good soil quality. The roles of various properties are discussed here.

- **pH:** The acidic or alkaline nature of the soil, represented by pH to maintain soil quality
- **Bulk Density:** The compactness of soil, known as bulk density—an indispensable parameter for soil
- **Electric Conductivity:** The ability to conduct electrical current. It depends on some factors like soil moisture and ion concentration and is a crucial parameter for assessing soil quality and salinity.
- **Bioabsorbtivity:** The capacity to absorb and hold nutrients, water, and other essentials for plant growth. It is influenced by soil texture, organic matter content, and microbial activity in agriculture and the ecosystem.
- **Porosity:** Measures the amount of open space or voids between soil particles. It influences water retention, drainage, and aeration, which is crucial for plant growth and soil health.
- **Chemical Nutrients:** Essential elements and compounds like nitrogen, phosphorus, and potassium, are vital for plant growth and development. These are the chemical nutrients. Soil nutrient levels affect crop yield, and overall soil fertility depends on it.

All these parameters can be disrupted by the contamination of microplastics in the rhizosphere. If all the extraction methods are done properly, all these parameters of soil are maintained, and the extraction will be beneficial economically and functionally [35].

14.3.3 Several Economical Environmental Benefits

- **Economical:** Agricultural growth is done, and farmers benefit from crop production on fertile soil in bulk quantity.
- **Environmental**: Prevention of soil pollution is done naturally with these various extraction phenomena, and the fertility of soil is maintained.

14.4 TECHNICAL, ECONOMIC CHALLENGES, AND FUTURE PERSPECTIVE

There have been no studies regarding microplastic extraction from the rhizosphere till now, so researchers should focus on this area. The rhizosphere has an important role in nutrients, plant growth, and several factors so it should remain pollution-free. Therefore, the extraction of microplastics has a large perspective for a sustainable future. It should operate largely at effective cost. Still, extraction methods have several technical challenges, so scientists should focus on creating more methods for cost-effective extraction. Education campaigns and regulations are to be done to reduce plastic usage and improve waste management practices that can help prevent further soil contamination in the long term. In summary, with continued research and innovation, continued collaboration among scientists, engineers, and environmentalists is crucial to refine existing methods and develop new, more efficient techniques for extracting microplastics from rhizosphere soil matrices.

14.5 CONCLUSION

It is observed that plastic pollution creates a harmful impact on terrestrial ecosystems, and it is a growing concern due to its threat to agroecosystems, food security, and human health. Therefore, the extraction of microplastics is essential for a better ecosystem. Several extraction methods and their future perspectives are discussed. It can be noted that remedial action shall be taken to restrict microplastic usage and develop biodegradable mulches to ensure the sustainability of agriculture while safeguarding soil health. The following effective methods are summarized for effective separation:

- Standardize separation and detection methods for accurate comparison and risk assessment.
- Investigate microplastics' distribution within soil aggregates to predict transport and behavior.
- Study the sorption of pollutants by microplastics and their effects on soil and organisms.
- Understand plasticizers' impact on soil health and their potential entry into the food chain.
- Examine the long-term fate and legacy of conventional plastics in soil.
- Determine critical limits for microplastic pollution and conduct ecotoxicological studies.
- Establish quality control measures for consistent microplastic experimentation.

- Ensure that studies use representative doses and provide unbiased reporting.
- Conduct longer-term field trials to assess microplastic impacts over time.
- Explore bio- and phytoremediation for cleaning up plastic pollution in soil.
- Compare environmental impacts of macro- and microplastics and use appropriate controls.
- Develop models to understand microplastic movement and persistence at larger scales [8, 20]. The present study shows separation, flotation, microscopic analysis, chemical reagents, and centrifugation for extracting microplastics. It also discussed several important functional economic environmental benefits of extraction.

REFERENCES

1. Chen, Y., Wu, C., Zhang, H., Lin, Q., Hong, Y. and Luo, Y. (2012). Empirical estimation of pollution load and contamination levels of phthalate esters in agricultural soils from plastic film mulching in China. *Environmental Earth Sciences*, 70(1), pp. 239–247. doi: https://doi.org/10.1007/s12665-012-2119-8.
2. Yu, L., Zhang, J., Liu, Y., Chen, L., Tao, S. and Liu, W. (2021). Distribution characteristics of microplastics in agricultural soils from the largest vegetable production base in China. *Science of The Total Environment [Online]*, 756, p. 143860. doi: https://doi.org/10.1016/j.scitotenv.2020.143860.
3. Oliveira, M., Ameixa, O.M.C.C. and Soares, A.M.V.M. (2019). Are ecosystem services provided by insects 'bugged' by micro (nano)plastics? *TrAC Trends in Analytical Chemistry*, 113, pp. 317–320. doi: https://doi.org/10.1016/j.trac.2019.02.018.
4. Collignon, A., Hecq, J.-H., Galgani, F., Collard, F. and Goffart, A. (2014). Annual variation in neustonic micro- and meso-plastic particles and zooplankton in the Bay of Calvi (Mediterranean–Corsica). *Marine Pollution Bulletin*, 79(1–2), pp. 293–298. doi: https://doi.org/10.1016/j.marpolbul.2013.11.023.
5. Brodhagen, M., Goldberger, J.R., Hayes, D.G., Inglis, D.A., Marsh, T.L., Miles, C. (2017). Policy considerations for limiting unintended residual plastic in agricultural soils. *Environmental Science & Policy*, 69, pp. 81–84. doi: https://doi.org/10.1016/j.envsci.2016.12.014.
6. Hinsu, A.T., Panchal, K.J., Pandit, R.J., Koringa, P.G. and Kothari, R.K. (2021). Characterizing rhizosphere microbiota of peanut (Arachis hypogaea L.) from pre-sowing to post-harvest of crop under field conditions. *Scientific Reports*, 11(1). doi: https://doi.org/10.1038/s41598-021-97071-3.
7. de Souza Machado, A.A., Lau, C.W., Kloas, W., Bergmann, J., Bachelier, J.B., Faltin, E., Becker, R., Görlich, A.S. and Rillig, M.C. (2019). Microplastics can change soil properties and affect plant performance. *Environmental Science & Technology [Online]*, 53(10), pp. 6044–6052. doi: https://doi.org/10.1021/acs.est.9b01339.
8. Qi, R., Jones, D.L., Li, Z., Liu, Q. and Yan, C. (2020). Behavior of microplastics and plastic film residues in the soil environment: A critical review. *Science of the Total Environment [Online]*, 703, p. 134722. doi: https://doi.org/10.1016/j.scitotenv.2019.134722.
9. Ghosh, S., Sinha, J.K., Ghosh, S., Vashisth, K., Han, S. and Bhaskar, R. (2023). Microplastics as an emerging threat to the global environment and human health. *Sustainability*, 15(14), pp. 10821–10821. doi: https://doi.org/10.3390/su151410821.
10. Kalogerakis, N., Karkanorachaki, K., Kalogerakis, G.C., Triantafyllidi, E.I., Gotsis, A.D., Partsinevelos, P. and Fava, F. (2017). Microplastics generation: Onset of fragmentation of polyethylene films in marine environment mesocosms. *Frontiers in Marine Science*, 4. doi: https://doi.org/10.3389/fmars.2017.00084.

11. Suardy, N.H., Abu Tahrim, N. and Ramli, S. (2020). Analysis and characterization of microplastic from personal care products and surface water in Bangi, Selangor. *Sains Malaysiana [Online]*, 49(9), pp. 2237–2249. doi: https://doi.org/10.17576/jsm-2020-4909-21.
12. Nava, V. and Leoni, B. (2021). A critical review of interactions between microplastics, microalgae and aquatic ecosystem function. *Water Research*, 188, p. 116476. doi: https://doi.org/10.1016/j.watres.2020.116476.
13. Xu, C., Zhang, B., Gu, C., Shen, C., Yin, S., Aamir, M. and Li, F. (2020). Are we underestimating the sources of microplastic pollution in terrestrial environment? *Journal of Hazardous Materials*, 400, p. 123228. doi: https://doi.org/10.1016/j.jhazmat.2020.123228.
14. Karbalaei, S., Hanachi, P., Walker, T.R. and Cole, M. (2018). Occurrence, sources, human health impacts and mitigation of microplastic pollution. *Environmental Science and Pollution Research*, 25(36), pp. 36046–36063. doi: https://doi.org/10.1007/s11356-018-3508-7.
15. Lambert, S. and Wagner, M. (2016). Characterisation of nanoplastics during the degradation of polystyrene. *Chemosphere [Online]*, 145, pp. 265–268. doi: https://doi.org/10.1016/j.chemosphere.2015.11.078.
16. Dong, H., Chen, Y., Wang, J., Zhang, Y., Zhang, P., Li, X., Zou, J. and Zhou, A. (2021). Interactions of microplastics and antibiotic resistance genes and their effects on the aquaculture environments. *Journal of Hazardous Materials*, 403, p. 123961. doi: https://doi.org/10.1016/j.jhazmat.2020.123961.
17. Kurniawan, S.B., Abdullah, S.R.S., Imron, M.F. and Ismail, N. (2021). Current state of marine plastic pollution and its technology for more eminent evidence: A review. *Journal of Cleaner Production*, 278, p. 123537. doi: https://doi.org/10.1016/j.jclepro.2020.123537.
18. Othman, A.R., Hasan, H.A., Muhamad, M.H., Ismail, N. and Abdullah, S.R.S. (2021). Microbial degradation of microplastics by enzymatic processes: A review. *Environmental Chemistry Letters*, 19(4), pp. 3057–3073. doi: https://doi.org/10.1007/s10311-021-01197-9.
19. Ryedale District Council. (2019). *Different Types of Plastics and Their Classification* [WWW Document]. Environment. www.ryedale.gov.uk/attachments/article/690/Different_plastic_polymer_types.pdf. accessed 4.25.19.
20. Jin, M., Liu, J., Yu, J., Zhou, Q., Wu, W., Fu, L., Yin, C., Fernandez, C. and Karimi-Maleh, H. (2022). Current development and future challenges in microplastic detection techniques: A bibliometrics-based analysis and review. *Science Progress*, 105(4), p. 003685042211321. doi: https://doi.org/10.1177/00368504221132151.
21. Marsh, K. and Bugusu, B. (2007). Food packaging—roles, materials, and environmental issues. *Journal of Food Science [Online]*, 72(3), pp. R39–R55. doi: https://doi.org/10.1111/j.1750-3841.2007.00301.x.
22. Radford, F., Zapata-Restrepo, L.M., Horton, A.A., Hudson, M.D., Shaw, P.J. and Williams, I.D. (2021). Developing a systematic method for extraction of microplastics in soils. *Analytical Methods [Online]*, 13(14), pp. 1695–1705. doi: https://doi.org/10.1039/D0AY02086A.
23. Patel, K., Chikkali, S.H. and Sivaram, S. (2020). Ultrahigh molecular weight polyethylene: Catalysis, structure, properties, processing and applications. *Progress in Polymer Science [Online]*, 109, p. 101290. doi: https://doi.org/10.1016/j.progpolymsci.2020.101290.
24. Restrepo-Flórez, J.-M., Bassi, A. and Thompson, M.R. (2014). Microbial degradation and deterioration of polyethylene—a review. *International Biodeterioration & Biodegradation [Online]*, 88, pp. 83–90. doi: https://doi.org/10.1016/j.ibiod.2013.12.014.
25. Hou, J., Xu, X., Yu, H., Xi, B. and Tan, W. (2021). Comparing the long-term responses of soil microbial structures and diversities to polyethylene microplastics in different

aggregate fractions. *Environment International [Online]*, 149, p. 106398. doi: https://doi.org/10.1016/j.envint.2021.106398.
26. Boots, B., Russell, C.W. and Green, D.S. (2019). Effects of microplastics in soil ecosystems: Above and below ground. *Environmental Science & Technology*, 53(19), pp. 11496–11506. doi: https://doi.org/10.1021/acs.est.9b03304.
27. Hou, J., Xu, X.Y., Cheng, F., Miao, L.Z., Liu, Z.L., 2020. Research progress on interaction and environment behavior between microplastics and organic pollutants (in Chinese). *Journal of Hohai University (Natural Sciences)*, 48(1), 22–28. doi: https://doi.org/10.3876/j.issn.1000-1980.2020.01.004.
28. Ahkami, A.H., Allen White, R., Handakumbura, P.P. and Jansson, C. (2017). Rhizosphere engineering: Enhancing sustainable plant ecosystem productivity. *Rhizosphere*, 3, pp. 233–243. doi: https://doi.org/10.1016/j.rhisph.2017.04.012.
29. Coppock, R.L., Cole, M., Lindeque, P.K., Queirós, A.M. and Galloway, T.S. (2017). A small-scale, portable method for extracting microplastics from marine sediments. *Environmental Pollution [Online]*, 230, pp. 829–837. doi: https://doi.org/10.1016/j.envpol.2017.07.017.
30. He, D., Luo, Y., Lu, S., Liu, M., Song, Y., Lei, L. (2018). Microplastics in soils: Analytical methods, pollution characteristics and ecological risks. *TrAC Trends in Analytical Chemistry*, 109, pp. 163–172. doi: https://doi.org/10.1016/j.trac.2018.10.006.
31. Thomas, D., Schütze, B., Heinze, W.M., Steinmetz, Z. (2020). Sample preparation techniques for the analysis of microplastics in soil—a review. *Sustainability*, 12. doi: https://doi.org/10.3390/su12219074.
32. Bronick, C.J. and Lal, R. (2005). Soil structure and management: A review. *Geoderma [Online]*, 124(1–2), pp. 3–22. doi: https://doi.org/10.1016/j.geoderma.2004.03.005.
33. Möller, J.N., Löder, M.G.J. and Laforsch, C. (2020). Finding microplastics in soils: A review of analytical methods. *Environmental Science & Technology*, 54(4), pp. 2078–2090. doi: https://doi.org/10.1021/acs.est.9b04618.
34. Hinsinger, P., Bengough, A.G., Vetterlein, D. and Young, I.M. (2009). Rhizosphere: Biophysics, biogeochemistry and ecological relevance. *Plant and Soil*, 321(1–2), pp. 117–152. doi: https://doi.org/10.1007/s11104-008-9885-9.
35. Bouaicha, O., Mimmo, T., Tiziani, R., Praeg, N., Polidori, C., Lucini, L., Vigani, G., Terzano, R., Sanchez-Hernandez, J.C., Illmer, P., Cesco, S. and Borruso, L. (2022). Microplastics make their way into the soil and rhizosphere: A review of the ecological consequences. *Rhizosphere [Online]*, 22, p. 100542. doi: https://doi.org/10.1016/j.rhisph.2022.100542.
36. Zobkov, M.B. and Esiukova, E.E. (2017). Evaluation of the Munich plastic sediment separator efficiency in extraction of microplastics from natural marine bottom sediments. *Limnology and Oceanography: Methods*, 15(11), pp. 967–978. doi: https://doi.org/10.1002/lom3.10217.
37. Imhof, H.K., Schmid, J., Niessner, R., Ivleva, N.P. and Laforsch, C. (2012). A novel, highly efficient method for the separation and quantification of plastic particles in sediments of aquatic environments. *Limnology and Oceanography: Methods [Online]*, 10(7), pp. 524–537. doi: https://doi.org/10.4319/lom.2012.10.524.
38. Crichton, E.M., Noël, M., Gies, E.A. and Ross, P.S. (2017). A novel, density-independent and FTIR-compatible approach for the rapid extraction of microplastics from aquatic sediments. *Analytical Methods*, 9(9), pp. 1419–1428. doi: https://doi.org/10.1039/c6ay02733d.
39. Mani, T., Frehland, S., Kalberer, A. and Burkhardt-Holm, P. (2019). Using castor oil to separate microplastics from four different environmental matrices. *Analytical Methods*, 11(13), pp. 1788–1794. doi: https://doi.org/10.1039/c8ay02559b.

40. Felsing, S., Kochleus, C., Buchinger, S., Brennholt, N., Stock, F. and Reifferscheid, G. (2018). A new approach in separating microplastics from environmental samples based on their electrostatic behavior. *Environmental Pollution [Online]*, 234, pp. 20–28. doi: https://doi.org/10.1016/j.envpol.2017.11.013.
41. Silveira, A.V.M., Cella, M., Tanabe, E.H. and Bertuol, D.A. (2018). Application of tribo-electrostatic separation in the recycling of plastic wastes. *Process Safety and Environmental Protection*, 114, pp. 219228. doi: https://doi.org/10.1016/j.psep.2017.12.019.
42. Ceccarini, A., Corti, A., Erba, F., Modugno, F., La Nasa, J., Bianchi, S. and Castelvetro, V. (2018). The hidden microplastics: New insights and figures from the thorough separation and characterization of microplastics and of their degradation byproducts in coastal sediments. *Environmental Science & Technology*, 52(10), pp. 5634–5643. doi: https://doi.org/10.1021/acs.est.8b01487.
43. La Nasa, J., Biale, G., Mattonai, M. and Modugno, F. (2021). Microwave-assisted solvent extraction and double-shot analytical pyrolysis for the quali-quantitation of plasticizers and microplastics in beach sand samples. *Journal of Hazardous Materials*, 401, p. 123287. doi: https://doi.org/10.1016/j.jhazmat.2020.123287.
44. Song, Y., Qiu, R., Hu, J., Li, X., Zhang, X., Chen, Y., Wu, W.-M. and He, D. (2020). Biodegradation and disintegration of expanded polystyrene by land snails Achatina fulica. *Science of The Total Environment [Online]*, 746, p. 141289. doi: https://doi.org/10.1016/j.scitotenv.2020.141289.
45. Fuller, S. and Gautam, A. (2016). A procedure for measuring microplastics using pressurized fluid extraction. *Environmental Science & Technology*, 50(11), pp. 5774–5780. doi: https://doi.org/10.1021/acs.est.6b00816.
46. Xu, L., Guo, C., Wang, F., Zheng, S. and Liu, C.-Z. (2011). A simple and rapid harvesting method for microalgae by in situ magnetic separation. *Bioresource Technology*, 102(21), pp. 10047–10051. doi: https://doi.org/10.1016/j.biortech.2011.08.021.
47. Rhein, F., Scholl, F. and Nirschl, H. (2019). Magnetic seeded filtration for the separation of fine polymer particles from dilute suspensions: Microplastics. *Chemical Engineering Science*, 207, pp. 1278–1287. doi: https://doi.org/10.1016/j.ces.2019.07.052.
48. Grbic, J., Nguyen, B., Guo, E., You, J.B., Sinton, D. and Rochman, C.M. (2019). Magnetic extraction of microplastics from environmental samples. *Environmental Science & Technology Letters*, 6(2), pp. 68–72. doi: https://doi.org/10.1021/acs.estlett.8b00671.
49. Zhang, S., Yang, X., Gertsen, H., Peters, P., Salánki, T. and Geissen, V. (2018). A simple method for the extraction and identification of light density microplastics from soil. *Science of the Total Environment*, 616–617, pp. 1056–1065. doi: https://doi.org/10.1016/j.scitotenv.2017.10.213.
50. Kononov, A., Hishida, M., Suzuki, K. and Harada, N. (2022). Microplastic extraction from agricultural soils using canola oil and unsaturated sodium chloride solution and evaluation by incineration method. *Soil Systems*, 6(2), p. 54. doi: https://doi.org/10.3390/soilsystems6020054.
51. Corradini, F., Casado, F., Leiva, V., Huerta-Lwanga, E. and Geissen, V. (2021). Microplastics occurrence and frequency in soils under different land uses on a regional scale. *Science of the Total Environment*, 752, p. 141917. doi: https://doi.org/10.1016/j.scitotenv.2020.141917.
52. Hidalgo-Ruz, V., Gutow, L., Thompson, R.C. and Thiel, M. (2012). Microplastics in the marine environment: A review of the methods used for identification and quantification. *Environmental Science & Technology*, 46(6), pp. 3060–3075. doi: https://doi.org/10.1021/es2031505.
53. Primpke, S., Wirth, M., Lorenz, C. and Gerdts, G. (2018). Reference database design for the automated analysis of microplastic samples based on Fourier transform infrared (FTIR) spectroscopy. *Analytical and Bioanalytical Chemistry*, 410(21), pp. 5131–5141. doi: https://doi.org/10.1007/s00216-018-1156-x.

54. Li, C., Gan, Y., Zhang, C., He, H., Fang, J., Wang, L., Wang, Y. and Liu, J. (2021). 'Microplastic communities' in different environments: Differences, links, and role of diversity index in source analysis. *Water Research [Online]*, 188, p. 116574. doi: https://doi.org/10.1016/j.watres.2020.116574.
55. Corradini, F., Bartholomeus, H., Huerta Lwanga, E., Gertsen, H. and Geissen, V. (2019). Predicting soil microplastic concentration using vis-NIR spectroscopy. *Science of the Total Environment*, 650, pp. 922–932. doi: https://doi.org/10.1016/j.scitotenv.2018.09.101.

15 Microplastics in Food Products

Anirudh Gururaj Patil, Soumitra Banerjee, Farhan Zameer, A. B. Hemavathi, Sneha Kagale, Pallavi Patil, Gouri Patil, Shalini Billa, and Snehal Yadav

15.1 INTRODUCTION

15.1.1 History and Statistics Related to Microplastics

The plastic manufacturing sector emerged and thrived in the 1940s and 1950s, experiencing exponential growth over the past 60 years, with production surging from about 0.5 million tonnes to over 260 million tonnes today. Plastic's affordability, durability, flexibility, strength, and excellent heat resistance make it suitable for human usage. During the mid-20th century, there was a significant expansion in the variety and forms of plastics being produced in the United States, with a particular focus on industrially manufactured synthetic polymers. These plastics are crafted from polymer chains, and, to achieve the desired shapes, forms, and characteristics, additives like monomers and plasticizers were introduced at various stages of the manufacturing process to tailor the material properties of specific plastics (Geyer et al., 2017). In contrast to earlier materials like celluloid and Bakelite, which had limitations in terms of shape and color, the new plastics of the mid-20th century possessed the ability to become translucent, mimicking the appearance of materials such as glass, and they could be produced in vibrant colors, making them an attractive choice for creating innovative furniture, dinnerware, and children's toys. Novel polymers, known for their durability, strength, and mimicry capabilities, were developed and utilized to manufacture a wide array of products, varying from vinyl seat covers, to records, construction materials, and clothing. While the surge in plastic production is often associated with consumer capitalism, it's worth noting that socialist regimes of the mid-20th century also displayed a keen interest in the potential of this new material.

The year 1907 marked the inception of Bakelite, the very first synthetic plastic ever formulated. While plastic production was reasonably manageable from 1950 to 1970, the quantity of plastic waste had tripled by around 1990, primarily due to the escalation in plastic manufacturing. The increase in plastic waste generated over a decade surpassed the rate of increase between 1950 and 1990 (Mazhandu & Muzenda, 2019). Microplastics were first discovered in the environment in the 1970s. These particles are suspected to enter the food chain and eventually find their way

DOI: 10.1201/9781032684574-15

into the human diet through trophic transfer. Consequently, there has been extensive research into the contamination of fish and aquaculture products by microplastics. Many investigations have examined their presence in various food items, ranging from salt, sugar, drinking water to vegetables. For instance, microplastics from locations like the North Pacific Central Gyre and the Portuguese coast have been discovered to contain polychlorinated biphenyls (PCBs) and polycyclic aromatic hydrocarbons (PAHs) at levels as high as 2856 and 44,800 ng/g, respectively. It's important to note that certain PCBs and PAHs are recognized carcinogens. Environmental plastic materials are typically categorized by their size, including microplastics (>25 mm), mesoplastics (25–5 mm), microplastics (5 mm–0.1 m), and nanoplastics (0.1 m). Nevertheless, the scientific community continues to debate the establishment of a universally accepted definition for these particles, especially microplastics, that would include attributes such as shape, color, solubility in water, and other factors (Alqattaf, 2020). The increased utilization of plastic products in contemporary society has led to the widespread presence of microplastic contamination, which consists of synthetic plastic particles smaller than 5 mm, in nearly all environmental settings. Microplastics have been detected in coastal areas, oceans, soils, sediments, and freshwater systems. Consequently, there is a risk that global microplastic pollution could find its way into our diets through the consumption of a variety of food products (Yu et al., 2022).

Microplastics may potentially endanger public health through their inherent composition, the persistence of manufacturing residues, or the absorption of toxic substances from the environment following human exposure.

15.1.2 Microplastics Intervention in the Food Chain

Microplastics have the potential to enter the human body through various exposure routes, including ingestion (from consuming food), skin contact (dermal contact), and inhalation, similar to other such substances. It is estimated that the annual intake of microplastics through food consumption could be as high as 52,000 MP/year. When inhalation is considered, this figure can increase to a maximum of 120,000 MP/year. Areas where fish and shellfish consumption is prevalent have shown increased oral exposure to microplastics. Notably, microplastics from settling household dust onto meals can significantly contribute to dietary exposure, potentially reaching up to 68,415 MP/year, which is much higher compared to exposure from contaminated mussel tissues, standing at 4,620 MP/year. Given that common items people interact with, like textiles, personal care products (e.g., toothpaste, cosmetics, facial scrubs, and cleansers), may contain microplastics, there is ongoing concern about their assessment as potential sources of microplastics for various exposure routes (H Ya et al. 2021). Furthermore, plastic polymers used in the production of various medical devices and pharmaceutical applications, such as the use of synthetic biodegradable polymers like poly[lactic-co-glycolic] acid for drug delivery, present potential sources of exposure. Consequently, researchers have shown significant interest in assessing the impact of particle exposure from plastic medical devices like prosthetic replacements and implants. The projected annual average intake of microplastics (ANMP) stands at 102,527 MP/year, with the global average range of microplastics ingested (GARMI) per capita falling in the range of 7.7–287 g (or 0.1–5 g/week). Notably, the primary contributor to these exposures was found to be drinking water.

The exact nature and scope of the detrimental health effects resulting from human exposure to microplastics remain a topic of ongoing research and debate. While there is a good understanding of the prevalence of microplastics, their toxicity and how they interact with biological organisms are less clear. Variables such as hydrophilicity, surface chemical composition, charge, and shape can influence the movement and toxicity of absorbed microplastics within the body. Research suggests that, after ingestion, microplastics can interact with biological systems, but the specifics of these interactions are not yet fully understood. Concerns related to food safety and potential toxicity of microplastics are primarily associated with their polymer components, which include monomers, residual impurities, physical degradation, and the adverse effects caused by plastic additives. Moreover, microplastics have the ability to concentrate chemicals like persistent organic pollutants (POPs) and heavy metals while providing a conducive surface for microbial attachment. Both factors may contribute to polymer toxicity. For instance, studies have shown that fish (*Carassius auratus*) exposed to polystyrene nanoplastics (size: 24 nm) through trophic transfer exhibited behavioral changes (e.g., delayed feeding time and motility) and alterations in fat metabolism (changes in cholesterol, weight loss, and redistribution of cholesterol between muscle and liver). These effects became evident around the 22nd day of the fish consuming contaminated zooplankton. Researchers also found a correlation between the presence of bisphenols in wild fish liver and muscle and the amount of microplastics ingested by the fish. However, the estimated daily intake of bisphenols from a diet of fish, for both adults and children, was lower than the European Food Safety Authority's (EFSA) recommended oral reference dose. Nevertheless, the authors suggested that consuming fish contaminated with bisphenol-containing microplastics could lead to elevated exposure to these chemicals compared to uncontaminated fish. According to the United Nations Environment Programme (UNEP), single-use plastics account for 50% of all plastic waste. An additional study reports that a total of 8.3 billion tons of plastic have been produced to date, with half of this production occurring since 2004. UNEP states that 60% of all plastic waste is either in landfills or in the natural environment. Furthermore, the annual production of plastic waste has reached 300 million tonnes, equivalent to the combined weight of the world's population. Predictions indicate a 70% increase in global waste production within the next 30 years. The United States and South Africa are the world's leading countries in terms of plastic consumption, with approximately 0.34 kg and 0.24 kg of plastic per person, respectively. In contrast, China ranks fifth, with approximately 0.12 kg of plastic per person. However, despite these figures, South Africa is ranked 11th globally in terms of plastic pollution, while the United States is not on the list. The top five contributors to plastic pollution in the oceans are China, Indonesia, the Philippines, Vietnam, and Sri Lanka, with South Africa ranking 11th out of 192 coastal nations studied. These findings are based on plastic data from 2010, and the authors also made projections for plastic waste expected in 2025. It's important to note that most of the plastic waste generated worldwide used to be sent to China for processing into valuable products required by various industries until early 2018. In 2016, plastic waste imports into China increased by 8.1 million tons

(12%), in addition to the 67 million tons of locally produced waste. Rivers in Asia were responsible for the largest share of plastic entering the ocean 57.7%, followed by Africa (7.8%), with the Australia Pacific region contributing the least.

While the petrochemical industry plays a pivotal role in plastic production, it is also seen as an indicator of global economic development, with plastic processing and manufacturing being crucial components. Polyolefins like polyethylene, polystyrene, and polypropylene dominate plastic production in terms of tonnage.

In India, leading polyolefin manufacturers include Reliance Industries, Indian Oil Corporation Limited, Haldia Petrochemicals, Bharat Petroleum Corporation Limited, and Gas Authority of India Limited. Exports in 2018 exceeded $8 billion USD, marking a 9.5% increase from previous years, with forecasts suggesting continued growth due to increased domestic production. However, despite the promising growth in plastic production benefiting Indian businesses, there is a lack of an organized system for managing the 15,342 tons of plastic waste generated daily in the country. According to the Central Pollution Control Board (CPCB), plastic waste constitutes 8% of the total solid waste, with Delhi being the largest contributor, followed by Kolkata and Ahmedabad. Furthermore, research shows that only 60% of the total plastic waste is recycled (Rafey & Siddiqui, 2021).

This chapter delves into the origins of microplastics (MP), their characterization, and the hazardous compounds generated by microplastics. It also explores the interaction of microorganisms with microplastics, methods for detecting them, their presence in the food chain, and potential solutions for mitigating the issue.

15.2 ORIGIN OF MICROPLASTICS FROM SYNTHETIC POLYMERS

Synthetic polymers are used for packaging and construction due to their exceptional material characteristics, wide range of applications, lower production and processing costs, and flexibility in formation. Plastic pollution is a significant issue in our contemporary lifestyle, and its pervasive presence has extended to even infiltrating our food, including fruits and vegetables. Because of their widespread use, their quantity is on the rise each year. To identify and evaluate these particles, the European working group on marine litter has introduced a classification system. This system categorizes polymers measuring up to 25 mm as macroplastics, those between 5 and25 mm as mesoplastics, and particles ranging from 5 to 1 mm as microplastics (Fendall & Sewell, 2009). Although plastics have made our lives easier than ever, its huge consumption has resulted in catastrophes our world cannot bear. Most of the plastic produced globally are used for food and beverage packaging. During its course of use, plastic wears out and breaks into small fragments, especially when subjected to heat and environmental conditions, called microplastics. Some of these, such as fillers and stabilizers, are deliberately added by the manufacturers, and some of them accumulate as byproducts. Microplastics have been found in the sea ice of Antarctica, in the digestive tracts of marine organisms residing in the planet's deepest ocean trenches, and within drinking water and food supplies worldwide. One study approximates the presence of around 24.4 trillion microplastic fragments in the Earth's oceans (Isobe et al., 2021). These particles can directly contaminate food crops. A 2020 study

TABLE 15.1
Polymers Used in Food Packaging

Polymers	Application Site	Cause for Degradation
Polyethylene (LDPE/HDPE)	Grocery bags, food wrappers	Poor weather resistance
Polypropylene	Yogurt, cheese/sour cream containers, RTE microwavable packs	UV sensitive, high thermal expansion coefficient
Polystyrene	Bowls, beverage cups, straws	Chemical instability, UV degradation
Polyvinyl chloride	Blister packaging, sandwich, burger covers	Heat instability
Polyethylene terephthalate	Baked goods, soft drink, and sauce bottles	Prone to oxidation

identified the presence of micro- and nano-plastics in fruits and vegetables available in supermarkets in Sicily, Italy. Notably, apples and carrots exhibited the highest levels of microplastic contamination among the tested produce (Conti et al., 2020). Microplastics accumulate in the cells and tissues resulting in chronic ailments such as gastrointestinal disorder, cancer, infertility, and mutation in the genes. Hence it is necessary to ensure foot safety as well as to control plastic usage, abiding by strict rules and regulations set by the governing authorities (Al Mamun et al., 2023). The source of microplastics is mainly from the packaging material used. In earlier days, food was packed in cloths, paper bags, and aluminum foils. Due to less availability and direct metal contact, the packaging materials are now made of polymers/plastic. Polyethylene, polystyrene, polypropylene, polyethylene terephthalate, and polyvinyl chloride are among the widely utilized polymers. Table 15.1 lists the advantages of these polymers and how they are land up as microplastics in the food we consume.

As mentioned in Table 15.1, one must avoid conditions to which a polymer is sensitive to avoid its degradation and formation of microplastics and to keep food free from microplastic contamination. Microplastics are generally categorized into two main groups: primary and secondary microplastics. Primary microplastics encompass microbeads present in personal care items, pellets or nurdles utilized in industrial manufacturing, and microfibers employed in synthetic textiles. These enter the environment directly through various channels such as wastewater drainage from households, industrial drainage, abrasion during textile washing or laundry process. Secondary MPs, such as fragments, ropes, filaments, or films, form due to the breakdown of larger plastics when they undergo weathering through exposure to wind, water waves, and sunlight/UV degradation (Sridharan et al., 2021).

15.3 TOXIC ADDITIVES RELEASED FROM MICROPLASTICS

Microplastics are found throughout the environment, allowing for their easy integration into the food chain. Their entry into the food chain is predominantly attributed to food packaging and the surrounding environment. Microplastics can release harmful additives that pose a danger to human health, thus rendering their presence in food

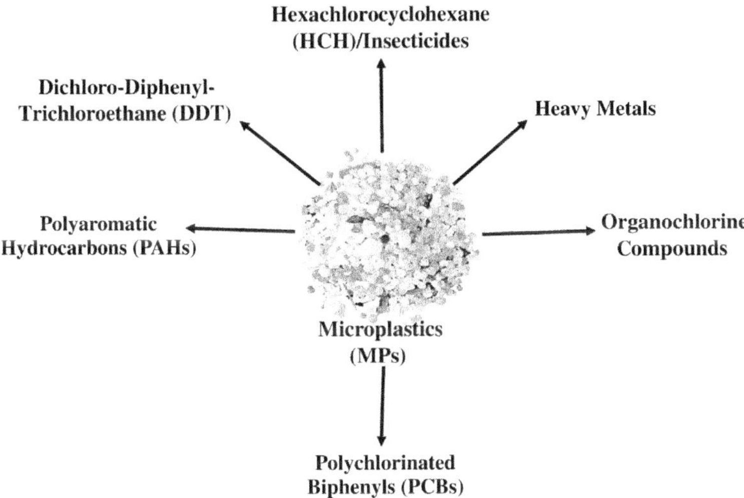

FIGURE 15.1 Toxic substances released from microplastics

a potential hazard (Al Mamun et al., 2023). Once in contact with the environment, plastic materials release harmful additives that subsequently enter the food chain. Prominent instances encompass harmful substances like polychlorinated biphenyls (PCBs), organochlorine compounds, polycyclic aromatic hydrocarbons, DDT, HCH insecticides, and heavy metals (Figure 15.1). Ingestion/inhalation of these additives paves a way for numerous health conditions like digestive toxicity, liver toxicity, reproductive toxicity, risk of cancer, and immune problems, the most common being endocrine disruptors.

15.3.1 Polychlorinated Biphenyls (PCBs)

Polychlorinated biphenyls (PCBs)are manufactured as oily solids or clear to yellow liquids, devoid of taste or odor. These stable mixtures were once widely used in electrical equipment as coolants and lubricants, but their commercial production was banned in 1979 due to associated health risks. PCBs have entered the environment through various means such as spills, improper disposal, leaks, and inadequate storage. Once released, PCBs have the ability to disperse and to bind to soil and sediment, making them persistent environmental contaminants. However, levels of PCBs in both the environment and the food chain have been decreasing since their production ceased. PCBs are pervasive in the environment, leading to potential exposure for everyone. Entry into the body can occur through inhalation of contaminated air, consumption of tainted foods and beverages, or even skin contact. PCBs are readily absorbed by the human body and are stored in fat tissues. Due to poor elimination, they tend to accumulate in the body. People who have consumed contaminated dairy products, fish, and meat typically exhibit higher PCB levels. The uptake of PCBs by plants from the soil is limited, resulting in lower PCB levels in dairy products and grazing animals compared to seafood. PCBs do not readily dissolve in water, so exposure to water containing PCBs is less of a concern. Low levels of PCBs can be found in contaminated dust that can settle on the surfaces of fruits and vegetables. A study on PCBs in food (Saktrakulkla et al., 2020) examined the concentration of 205 PCB congeners in 26 different food items. Salmon exhibited the highest PCB concentration, followed by canned tuna. However, it's worth noting that fish is not the primary source of PCB exposure. Over recent years, a decline in PCB concentrations has been observed across various food types and congeners. This decrease in PCB levels in food suggests that dietary exposure is on par with inhalation exposure to PCBs. The extent of exposure and the quantity of PCBs both factor into the determination of health effects. Conditions such as changes in liver function, chloracne, and impacts on the immune, reproductive, and endocrine systems can result from PCB exposure. PCBs are categorized as potential human carcinogens, and the presence of PCBs can be confirmed through blood tests. Reducing PCB exposure can be achieved by avoiding contact with contaminated soil and by following proper cleaning procedures before cooking, as well as washing fruits and vegetables before consumption.

15.3.2 Organochlorine Compounds

Organochlorine compounds are a group of persistent organic pollutants (POPs). They are capable of bioaccumulation in the food chain, persistent, are transported long distances, and have toxic properties. Organochlorine compounds are allied with suspended water particles that act as contaminants to the sediments. The most examined class of POPs are hexachlorocyclohexanes (HCHs), dichlorodiphenylethanes (DDTs), and PCBs. Organochlorine compounds are found in contaminated sediments, river discharges, coastal sewage disposals, agricultural and industrial processes. They are also present in most of the paper products that are formed during chemical pulp bleaching with chlorine compounds like hypochlorite, chlorine dioxide, and chlorination of fresh water. Organochlorine pesticides are utilized to kill and resist weeds, bacteria, fungi, and bacteria (Asem et al., 2023). Due to the need to protect against pests, low-cost organochlorine insecticides are used, but the overuse or misuse of these compounds causes adverse effects on human health. Organochlorines in the form of organochlorines pesticides (OCP) and agrochemicals used in aquaculture pose potential health threats. OCPs are absorbed by the digestive and respiratory tracts and can penetrate through skin because of high lipophilicity (excluding DDT). The nervous system is the principal target of organochlorine pesticides (OCP), where they inhibit GABAA-gated chloride channels, giving rise to neuronal hyperexcitability. Extended exposure to OCPs can lead to liver and kidney damage, owing to the chlorine atoms within their molecular structure, and may result in irritation of the skin and mucous membranes. To mitigate these health concerns, it is essential to remain mindful of the foods we consume and our exposure to organochlorine compounds.

15.3.3 Polyaromatic Hydrocarbons

Polyaromatic hydrocarbons (PAH) are naturally available chemicals that are found in crude oil, gasoline, and coal. Incomplete combustion of organic matter, such as the burning of wood, tobacco, or garbage, can also produce polyaromatic hydrocarbons, which bind to the particles in air. Cooking at high temperature will form polyaromatic hydrocarbons in the meat and other food. Exposure to PAH is commonly seen due to their ubiquitous occurrence in the environment. Breathing contaminated air, smoke from wood/cigarettes, charred and grilled meats all pave a way for the entry of PAHs. Aquatic ecosystems are contaminated through biologically toxic PAHs (Vijayanand et al., 2023).

Large amounts of PAH in the air can irritate the respiratory tract and eyes. Studies show that workers exposed to PAH by either breathing or skin contact via liquid form potentially developed liver and blood abnormalities. Certain mixtures of PAHs are known to possess carcinogenic properties. Even in the presence of UV-radiation-induced photochemical degradation, the bioavailability of PAHs in water can still be detected under conditions simulating the gastrointestinal tract. This indicates that they maintain their ability to be mobilized and absorbed (Schiferle et al., 2023).

15.3.4 DDT

DDT, also known as dichloro-diphenyl-trichloroethane, was originally created as a synthetic insecticide for the purpose of combating diseases carried by insects. It is generally used as a pesticide because of its effectiveness and low cost to manufacture. However, it was cancelled due to bioaccumulation, carcinogenicity, and health concerns on wildlife. A few insects also show resistance to DDT and have the capability to metabolize it into a lower-toxicity product DDE (dichloro-diphenyl-dichloroethylene). DDT is fat soluble and is rapidly passed through the food chain, leading to biomagnification. Exposure to DDT occurs by consuming DDT-contaminated foods, meats, fish, dairy products and by breathing or touching products that contain DDT. DDT is toxic to humans/mammals when ingested. It impacts the nervous system by disturbing nerve signals. Mammals exposed to moderate concentrations of DDT developed liver tumors. Studies have found how the collective pollution of MPs and toxic organic chemical influence the metabolism and growth by growing *E.coli* in a well-defined complex media treated with dichloro-diphenyl-trichloroethane (DDT) and polystyrene microplastics (PS MPs) at concentrations relevant to human concentrations (Liu et al., 2020). The results exhibited that DDT showed dose dependency and that PS reduced the growth and metabolic interfering effect of DDT on *E.coli*.

15.3.5 HCH Insecticides

Hexachlorocyclohexanes (HCH) is a potent insecticide that is mixture of eight or more stereo isomers. Most of the countries stopped using HCH as they pose serious environmental issues. The humongous dumps of unused HCH residing in the soil cause health hazards. The persistence of HCHs in the environment for such a long term enables them to enter the food chain. HCHs are highly lipophilic compounds with extended biological half-lives. The γ isoform (γ-HCH), commonly known as lindane, is a potent convulsant and neurostimulator (Jackovitz & Hebert, 2015). HCH, being a substituted cyclohexane, induces oxidative stress and causes degenerative changes in the hepatocytes due to its hepatotoxic nature. HCHs present lipophilic properties and persistence in the surroundings; beta-HCH followed by alpha-HCH and to a slighter degree gamma-HCH is biomagnified and bioaccumulated up to the food chain.

Due to its persistence, the contamination of hexachlorohexane (lindane) in water and soil has adverse effects. Research has identified the isolation of a bacterium capable of degrading the γ isomer of HCH, known as *Bacillus cereus* SJPS-2, from the Yamuna River. The growth curve analysis illustrated the efficient degradation of ϒ-HCH by *B. cereus* SJPS-2, with an impressive degradation rate of 80.98% (Jaiswal et al., 2023).

15.3.6 Heavy Metals

Metal pollution in soil can be remediated by understanding the mechanism of the soil-food crop.

Heavy metals like Hg, As, Cd, Cr, Pb, etc. can interrupt human metabolomics, causing diseases and even mortality. Heavy metals enter the food chain via plants that can absorb trace amounts of heavy metals from soil and seafood. Food security has been a problem since rapid urbanization, industrialization, and changes in the use of land. The prime sources of heavy metals in agriculture and soil environment are atmospheric deposition, irrigation with polluted water, livestock manure, metal pesticides or herbicides, sewage sludge usage, and phosphate-based fertilizers. The consumption of crops, fruit, vegetables, and inhalation of soil contaminated by heavy metals causes gastrointestinal cancer. Prolonged exposure to heavy metals through dietary sources can result in both non-cancerous and cancer-related health effects, particularly in infants, given their underdeveloped systems and the ratio of food intake to body weight (Bair, 2022). Remediation methods like vitrification, soil replacement, soil isolation, electrokinetic remediation, phytoextraction, and immobilization can be followed to eliminate contamination of soil by heavy metals.

15.4 INTERACTION OF MICROORGANISMS AND MICROPLASTICS

A type of organism known as a microorganism is small, prolific, has a wide geographic distribution, and is simple to culture. Numerous important functions for microorganisms in their surroundings and host health. For instance, soil microorganisms can reduce the level of heavy metals in soil mostly using their own unique enzymes, organic acids, and other compounds that can complex and disintegrate heavy metal ions. Among the microbes, the microbiota of the intestines can activate the immune system, encourage tissue differentiation in the host, and shield the host from infections. Bacteria from the intestinal tract serve as a vital ‘microbial organ’ of animals, aiding in host health by eliminating disease, stimulating growth and development, and breaking down and absorbing nutrition (Liu et al., 2022). In recent years, there has been a significant increase in research examining the connection between microplastics (MPs) and microorganisms. These studies have unveiled that microorganisms often thrive on the surface of MPs, using them as a habitat. Nonetheless, the exact mechanisms of their influence and the potential associated risks are not yet completely comprehended. Previous research we have conducted has established associations between exposure to polystyrene MPs and health issues such as gut inflammation in zebrafish, disruptions in hepatic lipid metabolism, and disturbances in the gastrointestinal barrier in mice (Jin et al., 2019; Yang et al., 2023). However, there have been assessments of the relationship between microplastics (MPs) and environmental microorganisms. Nevertheless, it is imperative to conduct further studies on the impact of MPs on animal and human health, particularly the interactions between MPs and bacteria as well as gut microbiota. In this comprehensive overview, we primarily consolidate and analyze recent research concerning the interplay between MPs and microorganisms, while also examining the implications for the well-being of animals and humans.

15.4.1 MPs Pollution and Their Biological Effects

Microplastics (MPs) have recently gained recognition as an environmental pollutant, and there has been a growing emphasis in the past year on understanding their effects on living systems. It's evident that MPs have become widespread worldwide, as they are widely distributed in seawater, coastal waters, freshwater ecosystems, soil, and even polar regions (Lu et al., 2019). Furthermore, reports by Nomura et al. in 2016 have confirmed the presence of microplastics (MPs) in bivalves meant for human consumption, drinking water sources, table salt, and bivalves that come into direct contact with people. Concerns about the biological toxicity of MPs, particularly concerning aquatic organisms, have been steadily mounting in recent years. Presently, numerous studies have indicated that fish and various organisms, particularly those in marine environments, may ingest MPs due to their small size and resistance to degradation. These studies predominantly examine the effects of MPs on factors like tissue distribution, growth rates, and oxidative damage. Moreover, certain microorganisms, such as bacteria and fungi, exhibit detrimental interactions with MPs. For instance, polystyrene nanoparticles can have lethal effects on yeast cells. MPs have the potential to hinder the growth of the marine bacterium *Halomonas alkaliphila* and disrupt ecological balances (Dissanayake et al., 2022; Sun et al., 2018; Xia et al., 2022).

15.4.2 Effects of MPs on Marine Organisms

In the context of marine life, microplastics (MPs) can influence plankton, which includes plant-like organisms in the marine environment. For instance, the ocean's phytoplankton may be prevented from absorbing sunlight by the floating MPs, which may impact their capacity to produce food and oxygen for marine life. Additionally, MPs can combine with oceanic microalgae, which causes marine creatures to consume MPs while eating algae. MPs can also engage in interactions with freshwater microalgae, which primarily result in complex development. Intriguingly, microalgae-associated genes that regulate the pathways for sugar production were highly overexpressed. Additionally, MPs have several negative effects on marine creatures, including fish, amphibians, seabirds, and invertebrates like zooplankton and other coastal animals. The main factors that affect zooplankton include digestive system obstruction, decreased appetite, the impact of eating, which can result in malnutrition, delayed growth, or even mortality; because they are so prevalent in deep-sea sediments, marine benthic species like mussel and oysters are particularly affected by MPs (Lagarde et al., 2016; Cunha et al., 2019; Xu et al., 2020).

Through endocytosis, mussels introduce MPs into their digestive tract, which eventually causes inflammation and reduces the stability of lysosomal membranes, showed that MPs can harm the development of progeny and impair reproduction, decrease growth, increase oyster mortality, and disrupt energy absorption. Fish are primarily affected by intestinal blockage, which enters the circulatory system and causes tissue buildup. These further impact in-vivo enzyme activity and metabolism. MPs mainly affect *Xenopus laevis* and other amphibians in terms of reproduction and growth. Seabirds unintentionally consume MPs when they eat marine life, just as MPs infect marine creatures. Studies have revealed that seabirds' MPs come in a variety of sizes, posing a grave threat to their survival (Qiao et al., 2019; Zhang et al., 2022).

15.4.3 Interaction Between MPs and Environmental Microorganisms

Over the past few years, there has been a growing global focus on the connection between microbes and microplastics (MPs). The unique characteristics of MP surfaces, including their hydrophilic nature, durability, and buoyancy, make them conducive to the proliferation of microbial communities and the creation of biofilms in aquatic environments. Furthermore, microplastics (MPs) serve as a distinctive ecological niche for viruses, bacteria, and microorganisms. Notably, Vibrio bacteria are found in high concentrations on the outer surfaces of MPs and are the most prevalent microorganisms on MP surfaces, despite their low abundance in the surrounding saltwater. Additionally, MPs collected by bacteria and viruses may be more biotoxic than regular MPs. It is simple to infect an organism once it has entered it. Furthermore, there is documented evidence of various organic pollutants, including dichlorodiphenyl trichloroethanes (DTTs), polychlorinated biphenyls (PCBs), polycyclic aromatic hydrocarbons (PAHs), and hexachlorocyclohexane (HCHs), interacting with microplastics (MPs). For instance, tetrachlorodibenzofuran, a chronic environmental pollutant, has been shown to influence the morphology, population, and composition of gut microbiota in mice. Given the small size and strong hydrophobic properties of microplastics, which can be influenced by environmental factors like salinity, temperature, and pH, MPs can act as carriers for the transportation of these harmful pollutants. Importantly, multiple species consume both MPs and organic pollutants, potentially heightening toxicity for animals. This raises the possibility that these organic contaminants could introduce microplastics into the food chain (Horton et al., 2017; Gamarro & Costanzo, 2022; Heskett et al., 2012).

Microplastics (MPs) interact with microbes through ingestion, colonization, chemical adsorption, and degradation. Previous studies have revealed that gut bacteria, specifically *Pseudomonas* species, can degrade low-density polyethylene films. Additionally, a bacterial consortium isolated from the intestines of *Lumbricus terrestris*, as demonstrated by Huerta Lwanga et al. in 2018, has shown the potential to significantly reduce the size of low-density polyethylene. These findings provide valuable insights into the role of microorganisms in addressing polymer-based waste in the environment and form the basis for understanding the microbial contributions to microplastic degradation within soil microcosms (Lwanga et al., 2018).

The adsorption interaction between microplastics (MPs) and microorganisms can result in the attachment of potentially harmful bacteria to MP surfaces, which raises concerns for both human and animal health, as highlighted by Wang et al. in 2021. Consequently, this association could give rise to additional ecological implications.

Therefore, it is imperative to conduct comprehensive investigations into the interactions between microorganisms and MPs. To better maintain the equilibrium of microecology, researchers should focus on comprehending these interactions, including the microbial degradation of MPs and the colonization of microorganisms on MP surfaces.

As a type of microbe, microalgae are responsible for preserving the equilibrium of the marine ecosystem. As a result of the MPs' adsorption on the marine alga *Skeletonema costatum*, which has numerous caveolae on its surface, the MPs may become entrenched in the algal cell wall and impact the development of microalgae. According to research findings, the initial exposure to high concentrations of microplastics (MPs) larger than 400 μm did not have an immediate effect on the growth of the freshwater microalgae *Chlamydomonas reinhardtii* within the first day. Furthermore, it did not trigger any changes in the expression of three genes associated with the stress response even after 78 days. However, after 20 days of exposure to polypropylene, there was the formation of hetero-aggregates composed of microalgae, MPs, and extracellular polysaccharides. Additionally, polystyrene MPs were observed to inhibit the microalgae's ability to photosynthesize and halt their growth. The interaction between MPs and algae leads to a situation where an organism consuming microalgae also consumes MPs, resulting in negative consequences for the organism. Hence, there is no doubt that the balance of aquatic ecosystems can be disrupted by the impact of MPs on microalgae, as supported by research (Zhang et al., 2022; Lagarde et al., 2016; Lins et al., 2022).

15.4.4 Effect of MPs on Gut Microbiota of Animals

The term 'gut microbiota' refers to the vast number of microorganisms residing in an animal's digestive tract that rely on the host for their intestinal habitat and play a crucial role in various physiological and biochemical functions. In the human microbiota, bacteria outnumber human cells by 10^{14} times, and about 70% of all identifiable microorganisms in the human body are in the colon. The major components of the human gut microbiota consist of *Firmicutes* and *Bacteroidetes*, with *Actinobacteria*, *Proteobacteria*, *Fusobacteria*, *Verrucomicrobia*, and *Cyanobacteria* making up a smaller portion (Satapathy et al., 2019). Similar patterns are observed in the gut microbiota of zebrafish and mice, where *Proteobacteria* and *Fusobacteria* dominate over *Firmicutes* and *Bacteroidetes*. Increasingly, research has shown that gut microbiota play a pivotal role in maintaining the health of their hosts and actively contribute to regulating various physiological processes. For instance, the ability of many arthropods to metabolize nutrients heavily relies on their gut microbiota. The intestinal microbiota is instrumental in regulating intestinal secretions and motility, breaking down complex polysaccharides in food, aiding in nutrient digestion and absorption, maintaining the structural integrity of the gastrointestinal epithelial barrier, and supporting the normal development and functioning of the immune system (Jin et al., 2018; Wu et al., 2018; Meng et al., 2020).

The composition of the gut microbiota can be influenced by environmental contaminants such as antibiotics, toxic metals, persistent organic pollutants, pesticides, nanomaterials, and food additives. Due to the sensitivity of the gut microbiota, it has emerged as a novel target for toxicological studies regarding some environmental toxins, including fungicides. It may be possible to indirectly impact host health through the gut microbiota. For instance, based on our previous research, alterations in the gut microbiota induced by fungicides could lead to inflammation and metabolic issues in mice, resulting in physiological dysfunction and the development of specific diseases. Changes in the gut microbiota composition induced by epoxiconazole have been linked to liver damage. Consequently, the gut microbiota serves as a vital indicator for toxicological assessments, and toxicological investigations should prioritize this marker as a key factor in examining host health concerns.

Research findings indicate that exposure to microplastics (MPs) resulted in a significant reduction in bacterial diversity within the gut of soil collembolans compared to the surrounding soil. It also brought about alterations in the gut microbiota of these soil-dwelling organisms. There are documented reports concerning the effects of MPs on the gut microbiota of soil-dwelling animals. The primary goal is to increase the overall abundance of Firmicutes while decreasing the relative prevalence of Bacteroides in the intestines of collembolans after their exposure to MPs (Takarina et al., 2022; Zhu et al., 2018).

Microplastics (MPs) have the potential to alter the gut microbiota in aquatic organisms, particularly zebrafish, leading to a variety of potential consequences. Our prior research has revealed that polystyrene MPs can bring about changes in the gut microbial composition in both adult and larval zebrafish, resulting in gut inflammation in adults and metabolic disruptions in the larvae. Larval zebrafish exposed to polystyrene MPs exhibited an imbalance in the ratio of Bacteroidetes to Firmicutes, a pattern associated with obesity. Additionally, polystyrene MPs led to an increase in the abundance of *Flavobacterium* in adult zebrafish, even though *Flavobacterium* is known as a common pathogen that can cause diseases in various fish species. Furthermore, alterations in *Polynucleobacter* were closely linked to a genetic disorder related to glycogenolysis. In earlier research, we investigated the toxic effects of polystyrene MPs of various sizes on mice. The findings indicated that these polystyrene MPs could disrupt the balance of the gut microbiota, primarily characterized by changes in the composition and diversity of gut microbes. This disruption could result in various potential impacts on the host, including abnormal physiological parameters and metabolic disorders (Jin et al., 2018; Wan et al., 2019).

Importantly, significant changes at the genus level in several vital gut bacteria are associated with health and disease. For instance, microbial causes of rheumatoid arthritis are linked to altered *Prevotella*, and modifications in *Anaerostipes* can lead to bacteremia and infections of the bypass graft. The direct entry and accumulation of MPs in the colon are the primary factors that contribute to the impact of MPs on gut flora. Therefore, MPs have the potential to affect animal health by altering gut microbiota, which is a novel area of toxicological research (Deng et al., 2017;

Sangkham et al., 2022). More research is needed to assess the potential health risks to the host and to understand the mechanisms through which MPs influence the gut flora.

15.5 IDENTIFICATION OF MICROPLASTICS IN FOOD

Plastic is not just one molecule but rather a broad class of polymer materials produced for various uses and distinguished by characteristics. As a result, the prevalence of microplastic materials in the environment, like polyethylene, polysterene, and polyethyleneterephthalate, resides on the production volumes, consumption habits, emission sources, and the material characteristics that affect the material's sensitivity to potential fragmentation into microplastics. It is crucial to identify the plastic material of the microplastics in environmental samples for understanding these interactions. Plastic pollution in the environment and the food chain has been on the rise, with its origins dating back to the 20th century (Thompson et al., 2009).

Plastics have become fragmented over time and are not completely degraded and can be classified as microplastics. Microplastics are particles of less than 5 mm in size. Microplastics can be further divided into small microplastics whose particle size is 20 μm–1 mm, large microplastics whose particle size is <5 mm (Hanke et al., 2013). The accumulation of microplastics can be identified using different methods. This section of the chapter focuses on the major and common methods employed for the identification of the microplastics on the basis of physical and chemical characteristics; visual identification, FTIR technique, Raman spectroscopy techniques, IR spectroscopy are discussed. An overview of identification methods based on the characteristics of microplastics is shown in Figure 15.2. The visual identification

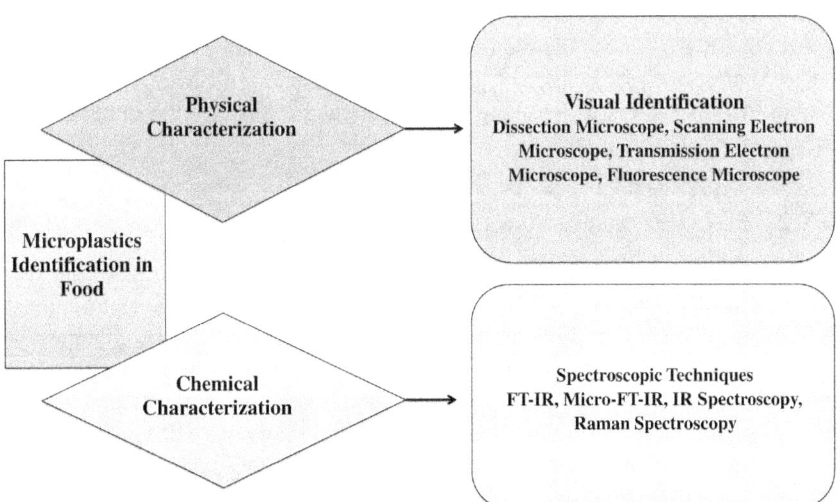

FIGURE 15.2 Overview of identification methods based on the characteristics of microplastics.

techniques that are used to identify the microplastics involve the dissection microscope, fluorescence microscope, atomic force microscope, polarized and scanning electron microscope, which reveals the physical characteristics of microplastics. Techniques for analyzing the chemical properties of microplastics include FT-IR spectroscopy, Raman spectroscopy, energy-dispersive X-ray spectroscopy, and thermal analysis methods such as gas chromatography/mass spectrometry (GC-MS) (Shim et al., 2017). Also, the density-based separation with subsequent analysis of carbon, hydrogen, and nitrogen content, pyrolysis-GC/MS technique can also be employed (Bergmann et al., 2015; Elkhatib & Oyanedel-Craver, 2020).

15.5.1 Visual Identification

A quick and easy identification method for microplastic analysis is visual inspection (Lee et al., 2013). Microplastics with a particle size of 1–5 mm are typically identified using this technique. Microplastics are typically seen with the unaided eye or with the aid of a microscope. According to the color and shape of the microplastics in the environment medium, they are categorized and recognized. However, there are considerable drawbacks to using the visual inspection method, including the inability to visually detect numerous microplastics and the impact of experimentalists' visual preferences on the identification outcomes. Identification of microplastics is done with the naked eyes or utilizing optical microscopy for identification with the objective lens power in the range of 10–50× (McCormick et al., 2016). Visual detections can also be done under dissection microscope with or without staining. The stereomicroscope, along with the photographs, can also be employed for the identification of microplastics equipped with 1–4× objective. The results can be given as the overall number of particles as there was difficulty in characterizing the types of microplastics based on their morphology under the mentioned magnification. An instance for utilizing the stain named Nile red staining dye was photographed and observed (Prata et al., 2020). Since Nile red is a lipophilic dye that can attach to the lipids and to natural and synthetic polymers, false positive results with Nile red staining are possible. An alternative to lowering false positives is to purify the material before further analysis (Zarfl, 2019).

15.5.2 Spectroscopic Techniques

For the investigation of the chemical composition of microplastics, Fourier transform infrared spectroscopy (FTIR) and Raman spectroscopy are the techniques that are frequently utilized. mid-infrared spectroscopy alone and/or in combination with chemometrics is also being employed for the identification of microplastics. In one study, mid-infrared spectroscopy combined with chemometrics was employed for identifying and quantifying the microplastics in the range of 3–100 μm. This study used a rapid analytical tool for identifying and quantifying high levels (1–10%) of microplastic contamination of chicken meat (Huang et al., 2020).

15.5.2.1 FTIR and Micro-FT-IR Techniques

Microplastics are frequently identified using FTIR through attenuated total reflection (ATR), transmission, and reflection. The most reliable surface spectral data is provided by attenuated total reflection (ATR). Microplastics with particle sizes greater than or equal to 300 mm are frequently found using ATR. The analysis can be finished very precisely in 1minute. High-resolution spectra can be obtained through transmission, but doing so necessitates infrared light penetrating the material. For the analysis of opaque, heavier samples, the reflection mode is employed. The advantage of FTIR analysis is that it can quickly and precisely detect the different types of plastic particles and is not impacted by fluorescence interference (Tagg et al., 2015).

One well-known method is micro-FTIR, which combines infrared spectroscopy with an optical microscope approach. It is a reliable, nondestructive method for the particle size less than or equal to 1 μm.In one study to determine the compositional spectra of particles, the micro-FT-IR procedure was used for the surface chemical mapping and the references to an infrared library database. Small microplastics (particles less than or equal to 1 mm) were found in all the samples. Overall abundances ranged from the 2175 to 672 small microplastics kg^{-1} d.w., typically exhibiting higher concentrations in the lagoons. Polyethylene and polypropylene were the most prevalent of the ten types of polymers found, making up more than 82% of all small microplastics. The size range 30–500 μm had the highest frequency (93% of the microplastics that were observed). The metal pollution index and the finer sediment fraction both showed a significant correlation with total small microplastics (Vianello et al., 2013). To image and identify various microplastics types, the micro-FT-IR/Raman imaging technique has been used. The technique can identify small microplastics (those that are less than 5 mm in size) and lowers analysis time; however, microplastics separation protocol is required (Ghosal et al., 2017). Micro-FTIR spectroscopy allows for direct particle visualization in addition to identification. Another value is that the micro-FTIR permits the identification of the sample from the filter membrane. Unfortunately, the instrument is costly, and also the procedure is time-consuming as the particles that are of relevance for material identification must first be removed visually from the sample. Recent advancements in (FPA)focal-plane-array-based micro-FTIR spectroscopy detectors have addressed this issue by enabling the simultaneous generation and recording of numerous chemical spectra within a single measurement, even of an entire filter, as demonstrated, for instance, with samples of marine plankton and sediments (Mintenig et al., 2016; Löder et al., 2015). This high-throughput analysis facilitates the detection of particles as small as 20 μm without bias in visual resolution and cuts the time needed for imaging the sample from several days to less than 9 hours (Mintenig et al., 2016)

15.5.2.2 Raman Spectroscopy and Micro-Raman Spectroscopy

Higher resolution is given by Raman spectroscopy's appealing attribute in comparison to m-FTIR, which is important for recognizing extremely tiny microplastics (20 μm). There is an immediate need for a quick and simple monitoring solution. Raman spectroscopy and imaging are two of the most popular ways to determine the type of polymer and the sizes and distributions of the particles in MP detections. Raman has a high spatial resolution (1 μm) and is unaffected by obtrusive signals from water and CO_2 in the atmosphere. To overcome the fluorescence obstruction brought on by microbiological, organic (humic substance), or inorganic (clay minerals, imperfect crystal structures) components in the samples, Raman spectroscopy must be used. Subsequently, the sample must also undergo a purifying step (H_2O_2

pretreatments to reduce the organic material or density separation to reduce the inorganic content), which is required to prevent unwanted sample alteration before the Raman analysis. Additionally, the issue of an unwelcome strong fluorescent background could be avoided by selecting the proper recognition parameters (Elert et al., 2017). The micro-Raman spectroscopy identified microplastics in the tissues of edible fish (Collard et al., 2017), bottled water, and table salt (Gündoğdu, 2018; Schymanski et al., 2018). The ability of micro-Raman spectroscopy to identify the small microplastics is better than that of the micro-FTIR as micro-Raman can identify particles sizes with <20 μm. In one study, the micro-Raman spectroscopy technique used was the μ-Raman alpha 300R Raman microscope, and the protocol employed was without any preselection, involved automatic scanning of the whole filter area in 10 μm step, and automatic library matching. The results included the detection that the minimum particle size was 5 μm, with the complete analysis time, in per mm^2, was 38h/mm^2(Käppler et al., 2015). Also, Raman mapping provides a way to locate microplastics in the 1–20 μm range. But the economic 'edge' of Raman is anticipated to vanish in upcoming years, as advancements in equipment offers an inexpensive and time-efficient devices for accurate Raman imaging (Araujo et al., 2018).

15.6 OCCURRENCE AND SOURCES OF MICROPLASTICS IN THE HUMAN DIET

Methods for quantitatively determining microplastics in water, air, food, and cosmetics are in their infancy yet, which makes it difficult to determine the risks that microplastics pose to human health. This lack of precise information on exposure doses is a major obstacle. The fact that the available spectroscopic techniques report microplastic fibers, items, or particles in terms of size, number, and shape is a complicating factor because exposure science typically reports doses in mass. Food and water were believed to be significant wellsprings of microplastics openness. A study found that 74,000–121,000 microplastics were consumed annually by individuals through food, water, dust, and air inhalation (Cox et al., 2019). A study estimated that 39,000–52,000 items of microplastics were consumed annually by each person, with 4000 from tap water, 37–1000 coming from sea salt and 11,000 coming from shellfish (Prata, 2018). The sources to which a human can get exposed to microplastics via food include water and beverages, marine creatures, sugars and salt, processed food, and honey, as well as the plastic materials that are in contact with the food substance.

15.6.1 Water

Microplastics have been found in the freshwater resources in bottled as well as in tap water in almost 14countries having the mean average of 5.45 particles/L (Kosuth et al., 2018). According to one study, groundwater contains fewer microplastics than tap or bottled water, indicating the microplastics originate during distribution of water or in the bottling processes (Koelmans et al., 2019). The majority of the microplastics in bottled water(80% of them) were polyethylene teraphthalate and polypropylene with sizes ranging from 5 to 20 μm. Potable water from single-use containers had an average microplastic content ±14 particles/L, respectively. The release of microplastics from the plastic container itself was the cause of the high concentrations of microplastics found in bottled water. However, a study demonstrated that glass-bottled water also contained significant amounts of microplastics with the mean being 50 particles/L and then suggested that microplastics came from bottle caps made of plastic (Schymanski et al., 2018). The sources of microplastics in water may be the pipeline for transportation made of polyvinyl chloride or polyethylene, the storage tank, and packaging material.

15.6.2 Beverages

One study investigated several drinks, including cold tea, energy drinks, and soft drinks (samples 4, 8, and 19, respectively). A total of 4024.53 microplastics particles, on average, were found among the 19 samples of soft drinks that have been analyzed. An average of 145.79 microplastics were discovered among the 8 samples of energy drinks that were evaluated. An average of 115.26 microplastics particles were found in the 4 cool tea samples that were examined. Moreover, only fibers were found in the three examined beverages, with blue-colored fibers predominating, trailed by brown (besides energy drinks) and red-colored fibers. The size range of the particles was 100 to 3000 mm. As can be observed, the three drinks' biggest proportion is made up of microplastics particles smaller than 1000 μm. The lowest number is between 2000 and 3000 μm (Shruti et al., 2020). Milk samples have also included microplastics. A total of 150 microplastics particles were detected in 23 tested milk samples (1 child and 22 adults) from 5 international and 3 brands of Mexico, an average concentration being 6.5 ± 2.3 particles L^{-1}. Water used during preparation of beer, direct contact of beverages with the atmosphere and the packaging materials, and the plastic equipment in contact with beverages can also lead to pollution (Kutralam-Muniasamy et al., 2020).

15.6.3 Marine Creatures like Fish

The sources of dietary consumption of microplastics appear to be marine food products (such as fishes, crustaceans, bivalves), and reports have shown that microplastics are often detected in commercial fish from most parts of the globe. It has been noted that microplastics have been prevalently detected in several types of commercial fish and geographic areas (like Europe and China). The contaminated samples had microplastics particles that ranged in size from 20 μm to 5 mm. Microplastics in bivalves were found to range from 0 to 10.5 items/g. Mussels have more microplastics than other fish, while oysters have the lowest microplastic content per gram. Microplastics are most frequently found in the digestive systems of the bivalves that are typically ingested by humans. Similar to those found in fish, the microplastics detected in bivalves consist of polypropylene, polyethylene, polystyrene, and polyethyleneteraphthalate. Fibers are the most prevalent particle in terms of morphology across all samples, followed by fragments (Jin et al., 2021). In one study, fish purchased from the Beibu Gulf in the South China Sea had 57.7% of the microplastics in the stomach, while the intestine and gills contained 34.6% and 7.7%of the microplastics, respectively (Koongolla et al., 2020). The microbeads in domestic products are the primary source of microplastics in the marine environment (GESAMP Joint Group of Experts on the Scientific Aspects of Marine Environmental Protection, 2016). Industrial abrasive, pellets of the preproduction process in the plastic industry are also an important source. Microfibers, tire dust, and large plastics that eventually degrade are the secondary sources (Duis & Coors, 2016).

15.6.4 Salts

The occurrence of microplastics in salt was investigated from 17 brands from different countries and was found to be in the range of 1–10 microplastics/kg of salt (Karami et al.,2017). Microplastics found were in fragment, fibers, filaments, granules, sheets, pellets, and foams. The common polymers in the studies were polyethylene terephthalate, Polypropylene, polyethylene. The sources include the contaminated water, the evaporation process from air, and the packaging process (Lee et al., 2019).

15.6.5 Processed Foods

Microplastic particles have been identified in a variety of consumables, including beer (Kosuth et al., 2018), honey (Liebezeit & Liebezeit, 2013), milk (Kutralam-Muniasamy et al., 2020), dried products such as traditional medicinal substances (Lu et al., 2020), and even seafood products subjected to processing, such as sprats, seaweed, and teabags (Kwon et al., 2020). Notably, the presence of elevated microplastic levels in traditional Chinese medicinal products has been attributed to the accumulation of these particles in the organisms used in their preparation. This accumulation occurs throughout the processing, packaging, and equipment involved in the production of these consumables, primarily composed of plastic materials, which, in turn, leads to the contamination of these items with microplastics (Kwon et al., 2020).

15.6.6 Packaged Food

Take-out containers are made up of polyethylene theraphthalate, polystyerene, polyethylene, polypropylene, nylon, rayon, acrylic, and polyester. In one study, someone who takes out food four times a week may ingest 12–203 pieces of microplastics from the packaged food. The source is the entire material, which is evidently made up with the plastic package material contaminating the food with microplastics. The concern of the direct interaction and potential release of microplastic particulates into food items has been exacerbated by the extensive adoption of plastic food vessels, packaging materials, plastic water receptacles, disposable drinkware, infant feeding utensils, plastic-coated metallic substrates, and paper-based containers. Human utilization of plastic receptacles encompasses their deployment for containment, conveyance, culinary preparation, and consumption of edibles and beverages. However, a significant segment of the populace may remain oblivious to the inadvertent introduction of microplastic contaminants from these receptacles and packaging materials into their dietary and liquid consumables (Du et al., 2020).

15.7 AVERAGE CONSUMPTION OF FOOD PRODUCTS AND GLOBAL EXPOSURE

Consumption of food is a periodic pattern and is activated at different moments throughout the day by factors like sensory stimulation, social framework, need, etc. Satiation is the inhibitory process that terminates the meal. The amount of consumption depends on the satiety cascade. Humans with eating disorders may consume less or too much; for example, in anorexia nervosa, people constantly think of food and food consumption that ends up them in eating less though there is appetite. The association of food consumption, cultural practices, and diseases is the significant focus of nutritional anthropology. Assessing food consumption inclinations involves a few sources. Food balance sheets (FBS) give information about the average availability/person that is the measure of food reaching the consumer. FBS has food availability data for all the countries. Individual dietary surveys are conducted to monitor nationwide dietary patterns. Projections of daily per capita food consumption are in terms of caloric intake per individual (John Kearney, 2010). Globalization, urbanization, and aging induce challenges to achieving good nutritional balance. We can observe that the MPs are present all over the environment and exposure to them is not an escape. Hence, considering sustainable food consumption patterns is crucial, and that should ensure food dense in micro-/macronutrients, free of MP contamination and derived from healthy agriculture.

15.8 MITIGATING THE EXPOSURE RISK OF MICROPLASTICS

15.8.1 Improving Production Efficiency of Plastic Products

One of the most efficient methodologies for the restoration of marine ecosystems lies in the reduction of plastic debris input through source reduction and enhanced waste management practices. This approach focuses on reducing energy and resource consumption, mitigating harmful emissions, and minimizing the influx of poorly managed plastic waste into marine environments. A holistic waste management strategy emphasizes the hierarchy of the four Rs: reduce, reuse, recycle, and recover (Prata et al., 2019; Tejaswini et al., 2022).

These strategies can be classified into three primary domains: (1) plastics production, involving industrial enhancements that consider life cycle assessments; (2) plastics consumption, which aims to reduce the allure of plastic products, especially through educational initiatives; and (3) waste management, which advocates for the retrieval and recycling of plastic waste (Figure 15.3). The insights garnered from this research have been distilled into recommendations for stakeholders, offering guidance for the development of a circular business model, encompassing the entire lifecycle of plastic products and thus reducing the influx of mismanaged plastic waste into marine ecosystems (Kershaw et al., 2019).

15.8.2 Improving Production Efficiency of Plastic Products

During the manufacturing phase, reducing plastic usage can be achieved through various methods: (1) adopting alternative materials like glass, recycled substances,

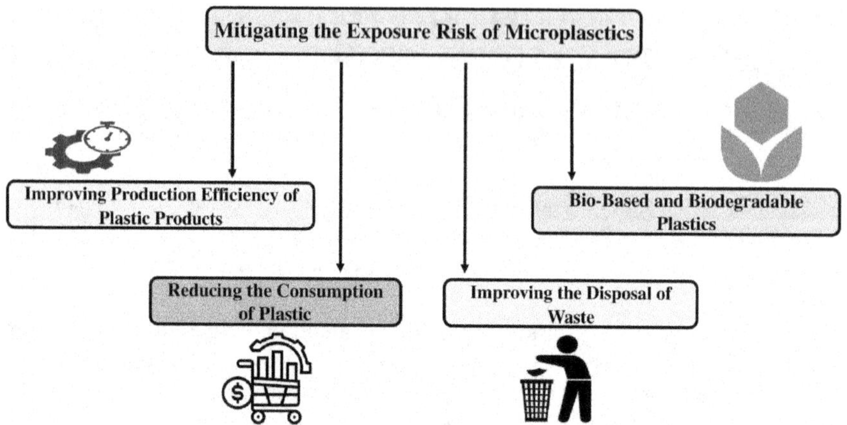

FIGURE 15.3 Mitigating the exposure risk of microplastics.

or biodegradable materials;(2) improving product design to minimize plastic content, extend product lifespan, promote repair and reusability, and enhance material recyclability by controlling polymer and additive diversity;(3) implementing bans on specific categories of single-use plastics.

For example, optimizing the design of plastic bottle caps to make them inseparable for proper disposal is an innovative approach. However, this may unintentionally impede recycling processes due to the presence of different polymer types. Therefore, advancements in product design are needed to simultaneously benefit businesses by reducing their raw material requirements. Furthermore, the promotion of viable alternatives, even if currently limited in availability, is essential. One approach involves incorporating a designated proportion of recycled components, such as 10% by weight, either voluntarily as a marketing strategy or as a mandatory requirement. It is crucial, however, to strike a balance in this percentage, as exceeding it could lead to inefficiencies in each subsequent recycling cycle (Liu et al., 2018; Walker & Fequet, 2023).

15.8.3 Mitigating Plastic Consumption

The reduction of excessive product consumption is a favorable endeavor following streamlined production processes. Nevertheless, implementing such measures may pose challenges considering considerations related to food safety and consumer convenience. However, it remains feasible to circumvent superfluous packaging, including double packaging, or to opt for environmentally sustainable alternatives. To instigate a long-term strategy for curtailing plastic consumption, initiatives can be undertaken to encourage consumers to opt for microbead-free substitutes, a task that can be greatly facilitated through precise product labeling (Beitzen-Heineke et al., 2017).

Raising consumer consciousness regarding the environmental repercussions of their choices can be achieved through formal means, such as educational programs within academic institutions, as well as informal channels, including news broadcasts and community cleanup initiatives. Corporations should be compelled to revise their product offerings in response to the growing demand for plastic-free alternatives, although it is worth noting that these alternatives remain somewhat limited and warrant further developmental strategies (Charlebois et al., 2022).

15.8.4 Enhancing Waste Disposal Practices

The organizational structure that supports the fundamental principles of the four Rs—reduce, reuse, recycle, and recover—forms the foundation of an all-encompassing waste management system. While the primary emphasis is placed on waste reduction and reuse, it is acknowledged that residual waste will persist, warranting judicious management under the aegis of an integrated waste management system (IWMS). The implementation of an IWMS, which necessitates retrieval, sorting, and repackaging, is predominantly confined to high-value commodities such as electronics and automobiles (Soong et al., 2022).

Consequently, waste that eludes recovery mechanisms should be repurposed as a resource or harnessed for energy recovery, with only terminal waste materials, such as ash, being relegated to landfills. Effective solid waste management not only contributes to a decluttered environment with fewer disposable items but also serves to mitigate the proliferation of microplastics. Taiwan has effectively reduced waste disposal rates and observed a significant decline in the prevalence of plastic bags, plastic bottles, and metal beverage cans on its beaches. This achievement is attributed to initiatives such as the Plastic Restriction Policy and the mandatory waste sorting required by the recycling act and compulsory trash-sorting policy (Wu et al., 2017).

Nonetheless, it is imperative to acknowledge that the deployment of integrated waste management systems can be costly and intricate. Developing nations that lack sophisticated waste management infrastructures may necessitate a transitional phase wherein containment of waste is prioritized to avert public health hazards and marine litter generation. Consequently, incineration and landfills may temporarily serve as primary waste management methods, with an evolution toward more environmentally sustainable approaches over time. Given the global nature of marine debris, international collaboration becomes paramount, with the sharing of knowledge and the provision of resources for waste management infrastructure in underdeveloped nations offering potential solutions (Charlebois et al., 2022).

15.8.5 Bio-Based and Biodegradable Plastics: Sustainable Alternatives to Traditional Plastics

The pursuit of environmentally responsible materials has intensified in response to the growing emphasis on sustainability, which seeks to supplant petroleum and natural gas with renewable energy sources and elevate recycling and waste management objectives. In this context, it is crucial to distinguish between the often misused terms ‘bio-based,’ ‘degradable,’ and ‘biodegradable.’ Bio-based polymers, commonly referred to as bioplastics, encompass plastics derived from renewable feedstocks, such as biomass, and are inherently non-biodegradable (e.g., biopolyethylene).

In contrast, biodegradable plastics, including photochemically biodegradable plastics, typically feature additives or polymers that undergo degradation triggered by composting processes or exposure to UV light, with the notable example of polylactic acid (PLA). An alternative to recycling and composting is waste-to-energy, a method that generates renewable energy by incinerating bio-based plastics (Aryan et al., 2023; Wu et al., 2017).

Moreover, conventional plastics, when subjected to plastic-degrading organisms like *Zalerion maritimum*, may experience biodegradation, yielding valuable bioproducts or organic matter Mitrano & Wagner, 2022). This phenomenon presents a sustainable option for managing contaminated or degraded plastics. Biodegradable polymers find relevance in applications characterized by limited shelf life, potential environmental pollution, or a preference for composting, such as packaging, disposable cutlery, and agricultural films. For instance, life cycle assessments (LCAs) of water bottles in Thailand reveal that cassava-based PLA containers generally exert a lower environmental impact than petroleum-based PET, except for potential issues related to eutrophication and acidification stemming from agricultural processes. However, PLA food trays exhibit higher environmental footprints compared to polystyrene (PS). The environmental impact of a material depends on various factors, including production methods, processing, and waste management practices. LCAs can provide valuable insights into these intricate considerations (Soong et al., 2022; Solano et al., 2022).

15.9 CONCLUSION AND FUTURE PROSPECTS

Microplastics (MPs) are becoming the subject of escalating scientific investigation, particularly in the context of their environmental distribution. Concerns have been raised by numerous experts regarding the potential detrimental effects of MPs on various organisms, irrespective of the ecosystem in question, whether it be marine, freshwater, or terrestrial. Several studies have provided empirical evidence of the adverse consequences of MPs, with a notable emphasis on their impact on biota, particularly aquatic species. The toxicological investigations focus mostly on mortality, behavioral effects, oxidative stress, neurological damage, DNA damage, fertile dysfunction, and even liver damage. There are not many reports of MPs' impacts on mammals right now, and it's still not understood how they work. However, MPs have been found in foods and beverages that people consume, such as tap water, beer, sugar, salt, and shellfish. Spatial differences can be seen in the pollutant concentrations on microplastics in maritime and coastal settings. Pollutant concentrations on microplastics from rural regions are lower than those in large cities. HCH and DDT concentrations on microplastics are considerably lower than PAH and PCB concentrations in worldwide ecosystems. Furthermore, it is imperative to acknowledge that the nature and attributes of microplastics, as well as the types and quantities of pollutants present, and the experimental conditions, including factors such as salinity and pH, may significantly influence the sorption behaviors of microplastics. The environmental processes of weathering and aging play a pivotal role in the degradation of microplastics. These processes primarily induce alterations in the color, surface morphology, particle size, crystallinity, and density of microplastics.

A comprehensive evaluation has been conducted to assess the pros and cons of diverse methodologies, such as flotation, membrane separation, chemical treatments, enzymatic processes, and extraction techniques. Various identification techniques, including scanning electron microscopy (SEM), Raman spectroscopy, Fourier transform infrared spectroscopy (FTIR), thermo-analytical methods, and hyperspectral imaging, have received significant scrutiny.

Future research endeavors are poised to explore the inherent properties and biodegradability of microplastics. Despite the initial strides in basic research, further investigations are imperative to comprehensively delineate the diverse polymer types within microplastics and their potential implications for human health within the food chain. This necessitates a concerted effort in two key areas: (1) the standardization of a singular analytical method for assessing microplastics in food samples and (2) the development of a more expeditious and precise methodology for quantifying microplastics per gram of food sample.

In addition to these scientific considerations, other critical concerns such as environmental implications, regulatory frameworks, and risk mitigation strategies require heightened attention. The unregulated and improper use of reusable plastics has led to significant environmental accumulations, exacerbating the worldwide problem of plastic pollution. This poses a substantial risk to ecosystems, biodiversity, and human well-being. To mitigate the uncontrolled spread of plastic materials in the environment at all stages of their lifecycle, this report provides a thorough assessment of the current situation in these areas and presents ten recommendations for stakeholders. A comprehensive waste management framework, guided by the principles of the four Rs (reduce, reuse, recycle, and recover), should be rigorously adhered to. This entails the imperative need to curtail (micro)plastic production and consumption through strategies encompassing improved design, alternative materials adoption, and the creation of long-lasting, reusable items. However, it is acknowledged that waste generation is an intrinsic facet of human societies. In instances where recycling is unfeasible, recovered plastics can be strategically repurposed as chemical feedstock or channeled into waste-to-energy initiatives. Landfilling is a last-resort measure exclusively designated for waste stemming from these operations. The orchestration of governance-driven regulatory measures, economic incentives, voluntary corporate commitments, and shifts in consumer behavior is paramount for the realization of these strategies. Given the transboundary nature of plastic marine debris, the establishment of effective waste management systems is not confined to national borders but necessitates international collaboration. Comprehensive cleanup endeavors are envisioned to facilitate the removal of plastics from the environment, transitioning them into structured waste management processes. These initiatives serve dual purposes, aiding in the restoration of ecosystems adversely affected by plastic pollution and addressing the percentage of polymers present in marine environments.

ACKNOWLEDGMENTS

This endeavor was possible with the constant support from resourceful people from Annasaheb Dange College of Engineering & Technology, Astha–India, Centre for Incubation, Innovation, Research and Consultancy (CIIRC), Jyothy Institute of Technology, Bengaluru–India, Alva's Ayurveda Medical College & ATMA Research Centre, Vidyagiri, Moodubidire–India, department of Polymer Science and Technology, Sri Jayachamarajendra College of Engineering (SJCE), JSSSTU, Mysore–India, and Department of Biotechnology Engineering, Kolhapur Institute of Technology, Kolhapur–India.

REFERENCES

Al Mamun, A., Prasetya, T.A.E., Dewi, I.R. and Ahmad, M., 2023. Microplastics in human food chains: Food becoming a threat to health safety. *Science of the Total Environment*, 858, p. 159834.

Alqattaf, A., 2020, November. Plastic waste management: Global facts, challenges and solutions. In *2020 Second International Sustainability and Resilience Conference: Technology and Innovation in Building Designs (51154)* (pp. 1–7). IEEE.

Araujo, C.F. et al., 2018. Identification of microplastics using Raman spectroscopy: Latest developments and future prospects. *Water Research*, 142, pp. 426–440.

Aryan, Y., Kumar, A. and Samadder, S.R., 2023. Environmental and economic assessment of waste collection and transportation using LCA: A case study. *Environmental Research*, p. 116108.

Asem, D., Kumari, M., Robindro Singh, L. and Bhushan, M., 2023. Pesticides: Pollution, risks, and abatement measures. *Emerging Aquatic Contaminants*, 11, pp. 307–326.

Bair, E. C. 2022. A narrative review of toxic heavy metal content of infant and toddler foods and evaluation of United States Policy. *Clinical Nutrition*, 9.

Beitzen-Heineke, E.F., Balta-Ozkan, N. and Reefke, H., 2017. The prospects of zero-packaging grocery stores to improve the social and environmental impacts of the food supply chain. *Journal of Cleaner Production*, 140, pp. 1528–1541.

Bergmann, M., Gutow, L. and Klages, M., 2015. Marine anthropogenic litter. *Marine Anthropogenic Litter*. https://doi.org/10.1007/978-3-319-16510-3.

Charlebois, S., Walker, T.R. and Music, J., 2022. Comment on the food industry's pandemic packaging dilemma. *Frontiers in Sustainability*, 3, p. 1.

Collard, F. et al., 2017. Microplastics in livers of European anchovies (Engraulis encrasicolus, L.). *Environmental Pollution*, 229, pp. 1000–1005.

Conti, G.O., Ferrante, M., Banni, M., Favara, C., Nicolosi, I., Cristaldi, A., Fiore, M. and Zuccarello, P., 2020. Micro-and nano-plastics in edible fruit and vegetables. The first diet risks assessment for the general population. *Environmental Research*, 187, p. 109677.

Cox, K.D. et al., 2019. Human consumption of microplastics. *Environmental Science and Technology*, 53(12), pp. 7068–7074.

Cunha, C., Faria, M., Nogueira, N., Ferreira, A. and Cordeiro, N., 2019. Marine vs freshwater microalgae exopolymers as biosolutions to microplastics pollution. *Environmental Pollution*, 249, pp. 372–380.

Deng, Y., Zhang, Y., Lemos, B. and Ren, H., 2017. Tissue accumulation of microplastics in mice and biomarker responses suggest widespread health risks of exposure. *Scientific Reports*, 7(1), p. 46687.

Dissanayake, P.D., Kim, S., Sarkar, B., Oleszczuk, P., Sang, M.K., Haque, M.N., Ahn, J.H., Bank, M.S. and Ok, Y.S., 2022. Effects of microplastics on the terrestrial environment: A critical review. *Environmental Research*, 209, p. 112734.

Du, F. et al., 2020. Microplastics in take-out food containers. *Journal of Hazardous Materials*, 399(February), p. 122969.

Duis, K. and Coors, A., 2016. Microplastics in the aquatic and terrestrial environment: Sources (with a specific focus on personal care products), fate and effects. *Environmental Sciences Europe*, 28(1), pp. 1–25.

Elert, A.M. et al., 2017. Comparison of different methods for MP detection: What can we learn from them, and why asking the right question before measurements matters? *Environmental Pollution*, 231, pp. 1256–1264.

Elkhatib, D. and Oyanedel-Craver, V., 2020. A critical review of extraction and identification methods of microplastics in wastewater and drinking water. *Environmental Science & Technology*, 54(12), pp. 7037–7049.

Fendall, L.S. and Sewell, M.A., 2009. Contributing to marine pollution by washing your face: Microplastics in facial cleansers. *Marine Pollution Bulletin*, 58(8), pp. 1225–1228.

Gamarro, E.G. and Costanzo, V., 2022. Dietary exposure to additives and sorbed contaminants from ingested microplastic particles through the consumption of fisheries and aquaculture products. *Microplastic in the Environment: Pattern and Process*, 261.

GESAMP Joint Group of Experts on the Scientific Aspects of Marine Environmental Protection., 2016. Sources, fate and effects of microplastics in the marine environment: Part 2 of a global assessment. (IMO, FAO/UNESCO-IOC/UNIDO/WMO/IAEA/UN/UNEP/UNDP). In: Kershaw, P.J. (Ed.). *Reports and Studies GESAMP* (93), p. 96.

Geyer, R., Jambeck, J.R. and Law, K.L., 2017. Production, use, and fate of all plastics ever made. *Science Advances*, 3(7), p. e1700782.

Ghosal, S. et al., 2017. Molecular identification of polymers and anthropogenic particles extracted from oceanic water and fish stomach a Raman micro- spectroscopy study. *Environmental Pollution*, pp. 1–12.

Gündoğdu, S., 2018. Contamination of table salts from Turkey with microplastics. *Food Additives and Contaminants—Part A Chemistry, Analysis, Control, Exposure and Risk Assessment*, 35(5), pp. 1006–1014.

Hanke, G., Galgani, F., Werner, S., Oosterbaan, L., Nilsson, P., Fleet, D., et al., 2013. MSFD GES technical subgroup on marine litter. Guidance on monitoring of marine litter in European Seas. Luxembourg: Joint Research Centre–Institute for Environment and Sustainability, Publications Office of the European Union.

Heskett, M., Takada, H., Yamashita, R., Yuyama, M., Ito, M., Geok, Y.B., Ogata, Y., Kwan, C., Heckhausen, A., Taylor, H. and Powell, T., 2012. Measurement of persistent organic pollutants (POPs) in plastic resin pellets from remote islands: Toward establishment of background concentrations for International Pellet Watch. *Marine Pollution Bulletin*, 64(2), pp. 445–448.

Horton, A.A., Walton, A., Spurgeon, D.J., Lahive, E. and Svendsen, C., 2017. Microplastics in freshwater and terrestrial environments: Evaluating the current understanding to identify the knowledge gaps and future research priorities. *Science of the Total Environment*, 586, pp. 127–141.

Huang, Y. et al., 2020. Rapid measurement of microplastic contamination in chicken meat by mid infrared spectroscopy and chemometrics: A feasibility study. *Food Control*, 113, p. 107187.

Isobe, A., Azuma, T., Cordova, M.R., Cózar, A., Galgani, F., Hagita, R., Kanhai, L.D., Imai, K., Iwasaki, S., Kako, S.I. and Kozlovskii, N., 2021. A multilevel dataset of microplastic abundance in the world's upper ocean and the Laurentian Great Lakes. *Microplastics and Nanoplastics*, 1, pp. 1–14.

Jackovitz, A.M. and Hebert, R.M. 2015. Wildlife toxicity assessment for hexachlorocyclohexane (HCH). In *Wildlife Toxicity Assessments for Chemicals of Military Concern* (pp. 473–497). Elsevier.

Jaiswal, S., Singh, D.K. and Shukl, P., 2023. Degradation effectiveness of hexachlorohexane (ϒ-HCH) by bacterial isolate Bacillus cereus SJPS-2, its gene annotation for bioremediation and comparison with Pseudomonas putida KT2440. *Environmental Pollution*, 318, p. 120867.

Jin, M. et al., 2021. Microplastics contamination in food and beverages: Direct exposure to humans. *Journal of Food Science*, 86(7), pp. 2816–2837.

Jin, Y., Lu, L., Tu, W., Luo, T. and Fu, Z., 2019. Impacts of polystyrene microplastic on the gut barrier, microbiota and metabolism of mice. *Science of the Total Environment*, 649, pp. 308–317.

Jin, Y., Xia, J., Pan, Z., Yang, J., Wang, W. and Fu, Z., 2018. Polystyrene microplastics induce microbiota dysbiosis and inflammation in the gut of adult zebrafish. *Environmental Pollution*, 235, pp. 322–329.

Käppler, A. et al., 2015. Identification of microplastics by FTIR and Raman microscopy: A novel silicon filter substrate opens the important spectral range below 1300 cm^{-1} for FTIR transmission measurements. *Analytical and Bioanalytical Chemistry*, 407(22).

Karami, A. et al., 2017. The presence of microplastics in commercial salts from different countries. *Scientific Reports*, 7(April), pp. 1–11.

Kearney, J., 2010. Food consumption trends and drivers. *Philosophical Transactions of the Royal Society*, 365, p. 1554.

Kershaw, P.J., Turra, A. and Galgani, F., 2019. Guidelines for the monitoring and assessment of plastic litter in the Ocean-GESAMP reports and studies no. 99. GESAMP Reports and Studies.

Koelmans, A.A. et al., 2019. Microplastics in freshwaters and drinking water: Critical review and assessment of data quality. *Water Research*, 155, pp. 410–422.

Koongolla, J.B. et al., 2020. Occurrence of microplastics in gastrointestinal tracts and gills of fish from Beibu Gulf, South China Sea. *Environmental Pollution*, 258, p. 113734.

Kosuth, M., Mason, S.A. and Wattenberg, E.V., 2018. Anthropogenic contamination of tap water, beer, and sea salt. *PLoS One*, 13(4), pp. 1–18.

Kutralam-Muniasamy, G. et al., 2020. Branded milks—are they immune from microplastics contamination? *Science of the Total Environment*, 714, p. 136823.

Kwon, J.H. et al., 2020. Microplastics in food: A review on analytical methods and challenges. *International Journal of Environmental Research and Public Health*, 17(18), pp. 1–23.

Lagarde, F., Olivier, O., Zanella, M., Daniel, P., Hiard, S. and Caruso, A., 2016. Microplastic interactions with freshwater microalgae: Hetero-aggregation and changes in plastic density appear strongly dependent on polymer type. *Environmental Pollution*, 215, pp. 331–339.

Lee, H., Joon, W. and Kwon, J., 2013. Science of the total environment sorption capacity of plastic debris for hydrophobic organic chemicals. *Science of the Total Environment* [Preprint].

Lee, H. et al., 2019. Microplastic contamination of table salts from Taiwan, including a global review. *Scientific Reports*, 9(1), pp. 1–9.

Liebezeit, G. and Liebezeit, E., 2013. Non-pollen particulates in honey and sugar. *Food Additives and Contaminants—Part A*, 30(12), pp. 2136–2140.

Lins, T.F., O'Brien, A.M., Kose, T., Rochman, C.M. and Sinton, D., 2022. Toxicity of nanoplastics to zooplankton is influenced by temperature, salinity, and natural particulate matter. *Environmental Science: Nano*, 9(8), pp. 2678–2690.

Liu, C., Li, X., Mansoldo, F.R., An, J., Kou, Y., Zhang, X., Wang, J., Zeng, J., Vermelho, A.B. and Yao, M., 2022. Microbial habitat specificity largely affects microbial co-occurrence patterns and functional profiles in wetland soils. *Geoderma*, 418, p. 115866.

Liu, S., Fang, S., Xiang, Z., Chen, X., Song, Y., Chen, C. and Ouyang, G., 2020. Combined effect of microplastics and DDT on microbial growth: A bacteriological and metabolomics investigation in Escherichia coli. *Journal of Hazardous Materials*, 407, p. 124849.

Liu, Z., Adams, M. and Walker, T.R., 2018. Are exports of recyclables from developed to developing countries waste pollution transfer or part of the global circular economy? *Resources, Conservation and Recycling*, 136, pp. 22–23.

Löder, M.G.J., Kuczera, M., Mintenig, S., Lorenz, C. and Gerdts, G., 2015. Focal plane array detector-based micro-Fourier-transform infrared imaging for the analysis of microplastics in environmental samples. *Environmental Chemistry*, *12*(5), pp. 563–581.

Lu, L., Luo, T., Zhao, Y., Cai, C., Fu, Z. and Jin, Y., 2019. Interaction between microplastics and microorganism as well as gut microbiota: A consideration on environmental animal and human health. *Science of the Total Environment*, 667, pp. 94–100.

Lu, S. et al., 2020. Prevalence of microplastics in animal-based traditional medicinal materials: Widespread pollution in terrestrial environments. *Science of the Total Environment*, 709, p. 136214.

Lwanga, E.H., Thapa, B., Yang, X., Gertsen, H., Salánki, T., Geissen, V. and Garbeva, P., 2018. Decay of low-density polyethylene by bacteria extracted from earthworm's guts: A potential for soil restoration. *Science of the Total Environment*, 624, pp. 753–757.

Mazhandu, Z.S. and Muzenda, E., 2019, November. Global plastic waste pollution challenges and management. In *2019 7th International Renewable and Sustainable Energy Conference (IRSEC)* (pp. 1–8). IEEE.

McCormick, A.R. et al., 2016. Microplastic in surface waters of urban rivers: Concentration, sources, and associated bacterial assemblages. *Ecosphere*, 7(11).

Meng, Z., Liu, L., Yan, S., Sun, W., Jia, M., Tian, S., Huang, S., Zhou, Z. and Zhu, W., 2020. Gut microbiota: A key factor in the host health effects induced by pesticide exposure? *Journal of Agricultural and Food Chemistry*, 68(39), pp. 10517–10531.

Mintenig, S.M. et al., 2016. AC SC. *Water Research* [Preprint].

Mitrano, D.M. and Wagner, M., 2022. A sustainable future for plastics considering material safety and preserved value. *Nature Reviews Materials*, 7(2), pp. 71–73.

Nomura, T., Tani, S., Yamamoto, M., Nakagawa, T., Toyoda, S., Fujisawa, E., Yasui, A. and Konishi, Y., 2016. Cytotoxicity and colloidal behavior of polystyrene latex nanoparticles toward filamentous fungi in isotonic solutions. *Chemosphere*, 149, pp. 84–90.

Prata, J.C., 2018. Airborne microplastics: Consequences to human health? *Environmental Pollution*, 234, pp. 115–126.

Prata, J.C., Silva, A.L.P., Da Costa, J.P., Mouneyrac, C., Walker, T.R., Duarte, A.C. and Rocha-Santos, T., 2019. Solutions and integrated strategies for the control and mitigation of plastic and microplastic pollution. *International Journal of Environmental Research and Public Health*, 16(13), p. 2411.

Prata, J.C. et al., 2020. Identification of microplastics in white wines capped with polyethylene stoppers using micro-Raman spectroscopy. *Food Chemistry*, 331(December 2019), p. 127323.

Qiao, R., Sheng, C., Lu, Y., Zhang, Y., Ren, H. and Lemos, B., 2019. Microplastics induce intestinal inflammation, oxidative stress, and disorders of metabolome and microbiome in zebrafish. *Science of the Total Environment*, 662, pp. 246–253.

Rafey, A. and Siddiqui, F.Z., 2021. A review of plastic waste management in India–challenges and opportunities. *International Journal of Environmental Analytical Chemistry*, pp. 1–17.

Saktrakulkla, P., Lan, T., Hua, J., Marek, R.F., Thorne, P.S. and Hornbuckle, K.C., 2020. Polychlorinated biphenyls in food. *Environmental Science and Technology*, 54, p. 18.

Sangkham, S., Faikhaw, O., Munkong, N., Sakunkoo, P., Arunlertaree, C., Chavali, M., Mousazadeh, M. and Tiwari, A., 2022. A review on microplastics and nanoplastics in the environment: Their occurrence, exposure routes, toxic studies, and potential effects on human health. *Marine Pollution Bulletin*, 181, p. 113832.

Satapathy, P., Khan, K., Devi, A.T., Patil, A.G., Govindaraju, A.M., Gopal, S., Prasad, M.N.N., More, V.S., Kakarla, R.R., Raghu, A.V. and Hudeda, S., 2019. Synthetic gutomics: Deciphering the microbial code for futuristic diagnosis and personalized medicine. In *Methods in Microbiology* (Vol. 46, pp. 197–225). Academic Press.

Schiferle, E.B., Ge, W. and Reinhard, B.M., 2023. Nanoplastics weathering and polycyclic aromatic hydrocarbon mobilization. *ACS Nano*, 17, pp. 5773–5784.

Schymanski, D. et al., 2018. Analysis of microplastics in water by micro-Raman spectroscopy: Release of plastic particles from different packaging into mineral water. *Water Research*, 129, pp. 154–162.

Shim, W.J., Hong, S.H. and Eo, S.E., 2017. Identification methods in microplastic analysis: A review. *Analytical Methods*, 9(9), pp. 1384–1391.

Shruti, V.C. et al., 2020. First study of its kind on the microplastic contamination of soft drinks, cold tea and energy drinks—future research and environmental considerations. *Science of the Total Environment*, 726.

Solano, G., Rojas-Gätjens, D., Rojas-Jimenez, K., Chavarría, M. and Romero, R.M., 2022. Biodegradation of plastics at home composting conditions. *Environmental Challenges*, 7, p. 100500.

Soong, Y.H.V., Sobkowicz, M.J. and Xie, D., 2022. Recent advances in biological recycling of polyethylene terephthalate (PET) plastic wastes. *Bioengineering*, 9(3), p. 98.

Sridharan, S., Kumar, M., Singh, L., Bolan, N.S. and Saha, M., 2021. Microplastics as an emerging source of particulate air pollution: A critical review. *Journal of Hazardous Materials*, 418, p. 126245.

Sun, X., Chen, B., Li, Q., Liu, N., Xia, B., Zhu, L. and Qu, K., 2018. Toxicities of polystyrene nano-and microplastics toward marine bacterium Halomonas alkaliphila. *Science of the Total Environment*, 642, pp. 1378–1385.

Tagg, A.S., Sapp, M., Harrison, J. P. and Ojeda, J.J., 2015. Identification and quantification of microplastics in wastewater using focal plane array-based reflectance micro-FT-IR imaging. *Analytical Chemistry*, 87(12), pp. 6032–6040.

Takarina, N.D., Purwiyanto, A.I.S., Rasud, A.A., Arifin, A.A. and Suteja, Y., 2022. Microplastic abundance and distribution in surface water and sediment collected from the coastal area. *Global Journal of Environmental Science and Management*, 8(2), pp. 183–196.

Tejaswini, M.S.S.R., Pathak, P., Ramkrishna, S. and Ganesh, P.S., 2022. A comprehensive review on integrative approach for sustainable management of plastic waste and its associated externalities. *Science of the Total Environment*, 825, p. 153973.

Thompson, R.C. et al., 2009. Plastics, the environment and human health: Current consensus and future trends. *Philosophical Transactions of the Royal Society B: Biological Sciences*, 364(1526), pp. 2153–2166.

Vianello, A. et al., 2013. Estuarine, coastal and shelf science microplastic particles in sediments of Lagoon of Venice, Italy: First observations on occurrence, spatial patterns and identification. *Estuarine, Coastal and Shelf Science*, 130, pp. 54–61.

Vijayanand, M., Ramakrishnan, A., Subramanian, R., Issac, P.K., Nasr, M., Khoo, K.S., Rajagopal, R., Greff, B., Wan Azelee, N.I., Jeon, B.-H., Chang, S.W. and Ravindran, B., 2023. Polyaromatic hydrocarbons (PAHs) in the water environment: A review on toxicity, microbial biodegradation, systematic biological advancements, and environmental fate. *Environmental Research*, 227, p. 115716.

Walker, T.R. and Fequet, L., 2023. Current trends of unsustainable plastic production and micro (nano) plastic pollution. *TrAC Trends in Analytical Chemistry*, p. 116984.

Wan, Z., Wang, C., Zhou, J., Shen, M., Wang, X., Fu, Z. and Jin, Y., 2019. Effects of polystyrene microplastics on the composition of the microbiome and metabolism in larval zebrafish. *Chemosphere*, 217, pp. 646–658.

Wang, J., Lu, J., Zhang, Y., Wu, J. and Luo, Y., 2021. Unique bacterial community of the biofilm on microplastics in coastal water. *Bulletin of Environmental Contamination and Toxicology*, 107, pp. 597–601.

Wu, S., Jin, C., Wang, Y., Fu, Z. and Jin, Y., 2018. Exposure to the fungicide propamocarb causes gut microbiota dysbiosis and metabolic disorder in mice. *Environmental Pollution*, 237, pp. 775–783.

Wu, W.M., Yang, J. and Criddle, C.S., 2017. Microplastics pollution and reduction strategies. *Frontiers of Environmental Science & Engineering*, 11, pp. 1–4.

Xia, B., Sui, Q., Du, Y., Wang, L., Jing, J., Zhu, L., Zhao, X., Sun, X., Booth, A.M., Chen, B. and Qu, K., 2022. Secondary PVC microplastics are more toxic than primary PVC microplastics to Oryzias melastigma embryos. *Journal of Hazardous Materials*, 424, p. 127421.

Xu, S., Ma, J., Ji, R., Pan, K. and Miao, A.J., 2020. Microplastics in aquatic environments: Occurrence, accumulation, and biological effects. *Science of the Total Environment*, 703, p. 134699.

Ya, H., Jiang, B., Xing, Y., Zhang, T., Lv, M. and Wang, X., 2021. Recent advances on ecological effects of microplastics on soil environment. *Science of the Total Environment*, 798, p. 149338.

Yang, Q., Dai, H., Wang, B., Xu, J., Zhang, Y., Chen, Y., et al., 2023. Nanoplastics shape adaptive anticancer immunity in the colon in mice. *Nano Letters*, 23(8), pp. 3516–3523.

Yu, J., Adingo, S., Liu, X., Li, X., Sun, J. and Zhang, X., 2022, January 12. Micro plastics in soil ecosystem–a review of sources, fate, and ecological impact. *Plant, Soil and Environment,* 68(1), pp. 1–7.

Zarfl, C., 2019. Promising techniques and open challenges for microplastic identification and quantification in environmental matrices. *Analytical and Bioanalytical Chemistry*, 411, pp. 3743–3756.

Zhang, X., Jin, Z., Shen, M., Chang, Z., Yu, G., Wang, L. and Xia, X., 2022. Accumulation of polyethylene microplastics induces oxidative stress, microbiome dysbiosis and immunoregulation in crayfish. *Fish & Shellfish Immunology*, 125, pp. 276–284.

Zhu, L., Bai, H., Chen, B., Sun, X., Qu, K. and Xia, B., 2018. Microplastic pollution in North Yellow Sea, China: Observations on occurrence, distribution and identification. *Science of the Total Environment*, 636, pp. 20–29.

16 Current Microplastic Scenario and Its Adverse Effects on the Ecosystem

Nirmala Ganesan, Harithra V., and Christina Prince

16.1 INTRODUCTION

The term 'microplastics' (MPs) was originally introduced by Thompson (2004) to describe minuscule, less than 5 mm sized plastic particles in the oceans. Microplastics (MPs) are made up mostly in tiny diameters for specific uses (microbeads) and secondary MPs that are made from enormous plastic trash that has been broken down and fragmented over time due to environmental factors that are chemical, physical, and biological [1]. Microplastics can adsorb other hazardous substances and have an extended residence period, durability, and high splintering potential. Microplastics are a major source of concern for researchers and ecologists because of their enduring sturdiness owing to their polymeric nature and facile transfer between different ecosystems. In terms of primary polymer compounds, polyesters, polyethylene (PE), polyethylene terephthalate (PET), polyurethane (PU), polystyrene (PS), polyvinyl chloride (PVC), polypropylene (PP), and polyamide (PA, nylon) are the most widely used. This enhances the availability of plastic trash for consumption by an assortment of fauna and focuses attention on the emergence of new threats to the environment [2]. Through research efforts, we have made progress in the last few years in our understanding of the pollution caused by microplastics, the transfer of microplastics from urban areas to water bodies through waterway discharge and ferrying to the sea, as well as the maritime spreading of microplastics across basins of the ocean and into deeper ocean layers. There have been numerous efforts to critically evaluate, gather, and provide a wider perspective on the situation of plastic particles in our surroundings today [3]. Microplastics arrive in a variety of shapes, including spheres, pieces, and fibers. The majority of microbeads—apart from those that are deliberately made—are caused by the breakdown of bigger polymers (macroplastics). Over time, microplastics break down into smaller and smaller pieces, eventually producing nanoplastics. Microplastics are thus predominantly a condition of transition between macro- debris and nanomaterials. According to estimates, spherical microplastics could break up into $>10^{14}$ times more nanoparticles. The very surface of the ocean has been the focus of surveys for plastic waste, but work is now being done in deeper seas, deposits, freshwater sources, soils, airborne particles, and biological communities as well. Recently, microplastics were found in places that were thought to be uninhabited, such as the Antarctic, deep ocean trenches, aloof mountain ranges,

 DOI: 10.1201/9781032684574-16

and Artic Sea ice [4]. Researching extensively and staying current with changes in toxicology, extraction, analysis, and behavior related to microplastic sources are essential. The current situation of microplastics is covered in this chapter, along with information on their existence in the environment, their toxicological effects, the need to understand their extent, their interactions with harmful contaminants, the challenges of defining analytical methods, and prospective treatment options.

16.2 SOURCES OF MICROPLASTICS

Both land-based and marine sources can produce microplastics. Only 20% of the total plastic trash in the marine milieu comes from ocean-based sources like fishing for revenue, ships, and other marine activities. The rest of the 80% comes from terrestrial sources of microplastics. Personal grooming items, the air-blasting process, poorly dumped plastics, and landfill leachates are the main contributors of terrestrial pollution. When terrestrial microplastics are released into natural water systems, rivers carry the majority of them to the oceans while freshwater environments, especially remote mountain lakes, hold the remaining particles.

Primary microplastics tend to be discovered in fabrics, pharmaceuticals, and personal grooming products like face and body washes. They are initially created to have a size less than 5 mm. Rivers, water treatment facilities, wind, and surface runoff can all discharge these main microplastics into freshwater or marine habitats.

Due to processes like photodegradation, as well as chemical, physical, and biological interactions, secondary microplastics are produced from fragmented or large debris of plastic. Fishing nets, the manufacturing sector's resin pellets, domestic goods, and other discarded plastic trash are some of the sources of secondary microplastics. Furthermore, it was shown that secondary microplastics make up the bulk of microplastics and that their abundance in the sea would rise concurrently with an increase in the intake of plastic debris from various sources, causing secondary microplastics to continuously change. Due to the nature of nanoscale materials, there is a greater likelihood that microplastics may break down into more hazardous nanoplastics when they are exposed to the environment [5].

16.3 MICROPLASTICS IN THE AQUATIC ENVIRONMENT

One of the issues harming the marine environment that is most pervasive is plastic pollution. In addition, it contributes to climate change and endangers the health of the ocean, human health, food safety and quality, and coastal tourism [6]. Aquatic species can absorb microplastics via a variety of physiological pathways before translocating them to various structures or body parts, particularly the digestive system, gut, and stomach. It has been shown that physically eaten microplastics made of PS (20 μm) are delivered to the hemolymph and gastrointestinal tract in the clam *Scrobicularia plana*. The mussel *Mytilus edulis*'s ingestion of high-density PE microplastics (less than 80 μm) is responsible for the passage of the particles from the gill surface to the intestinal tract (stomach and intestine) and then to the primary as well as secondary ducts of the digestion tubules. The lysosomal system is where the particles ultimately assemble [7]. The health of the aquatic environment is supported by consistent sustainable productivity, and the structuring of aquatic food chains depends on primary manufacturing. The extensive use of plastic particles as a result of change in the environment brought on by human activity may lower the efficiency of ecosystems by obstructing nutrient synthesis and cycling. These creatures were used as model organisms, and environmentally appropriate levels of microplastics have been identified in the intestinal pellets. In likely future scenarios, up to 72% of intestinal pellets may contain microplastics, which could reduce the efficiency of biological processes in promoting the ocean's retention of human-made carbon dioxide. Additionally, interactions that occur between the microorganisms *Skeletonemacostatum* and microplastics over time, such as adsorption and aggregation, lead to lower algal chlorophyll concentrations and reduced photosynthetic efficiency. The total amount of biomass of this maritime principal producer has been reduced as a result of the discovery that algae development is being hindered [8].

The presence and kind of MPs consumed, as well as their distribution along each trophic stage of the food chain in the marine environment, are determined by the feeding tactics of each creature [9].

16.3.1 Fish

Recent studies focused heavily on the ingesting of microplastics by fish due to their significant role in the diet of humans and the rising popularity of fish from aquaculture. Additionally, important to maritime ecosystems, food chains, fish are also essential to the ocean's biological cycle [10]. Some fish have a highly individualized diet and may only occasionally absorb plastics, but others could be predatory and dine on many different kinds of prey, making them more vulnerable to MP ingestion. Sardines, for example, are unable to pick up the particles they consume, whereas anchovies are. As a result, during the breeding season, sardines may consume small planktonic creatures and floating MPs travelling toward the surface water without discrimination [9]. Seabed fish live on the ocean floor and eat tiny sediment species, whereas pelagic creatures are found in the water layer and devour planktivorous animals. Variations in tiny plastic particles intake, both qualitatively and quantitatively, are anticipated due to variances in feeding behaviors and the dispersion of microplastic from superficial into deeper waters. Plankton feeders *E. encrasicolus* and *S. pilchardus* and coastal feeders *C. lyra* and *M. surmuletus* were identified. For *S. pilchardus* and *E. encrasicolus*, fish eggs and copepods served as their main sources of food. *M. surmuletus*'s food pattern is predominantly made up of suprabenthic amphipods and polychaetes, as opposed to *C. lyra*, which consumed a range of little benthic organisms such ophiurids, pagurids, and polychaetes. Polyethylene (PE) was the plastic that was frequently recognized in the intestinal tracts of all fish, which includes oceanic and demersal. Polypropylene (PP) was the second-highest prevalent polymer made from plastic that has been identified. Low-density plastics are made available to aquatic creatures through polluting processes and have applications in bottle caps and packaging, while polyethylene is frequently employed in packaging. Some 87% of *E. engraulis* and S. *pilchardus*, 79% *of C. lyra*, and 60% of *M. surmuletus* have been determined to have ingested microplastics. The distribution of the 100 microplastics found in the stomachs was as follows: 25 were found in *E. encrasicolus*, 23 in *S. pilchardus*, 38 in *C. lyra*, and 14 in *M. surmuletus*. Microplastics were divided into two distinct groups: pieces (12% of the total) and filaments (88% of the total). Consuming fragments was also observed in *E. encrasicolus* and *S. pilchardus* but at a lower rate (36 and 13% of the absorbed microplastics, respectively). Only filaments were absorbed by *C. lyra* and *M. surmuletus.* Transparent was the most frequent color in maritime species *E. encrasicolus* (56%) and *S. pilchardus* (35%), whereas blue was the most prevalent color in benthic genera *C. lyra* (35%) and *M. surmuletus* (43%). Lesser quantities of other colors, including black, green, red, and white, were also spotted. The typical filament size was 1.11 ± 0.68 mm in *E. encrasicolus*, 1.46 ± 0.83 mm in *S. pilchardus*, 1.78 ± 1.24 mm in *C. lyra*, and 1.20 ± 0. 80 mm in *M. surmuletus.* The average fragment size in *S. pilchardus* was 0.53 ± 0.23 mm and 0.57± 0.29 mm in *E. encrasicolus* [11].

16.3.2 Bivalves

Microplastic ingestion was discovered in mussels (*M. galloprovincialis*) in a recent study. Plastic particles were primarily blue (54.4%) and pink (29.4%) in color, with little amounts of black (4.4%), yellow (4.4%), or translucent (4%), and a trace amount of green (1.4%), in the bivalve samples. There were no appreciable differences across species, while snails taken from the facility contained significantly more blue-colored microplastics than snails taken from Corfu Port. The components used in oyster farm facilities are predicted to have considerably greater amounts of blue microplastics than those found in port oysters. Every step of the long-line mollusk cultivation process involves the use of plastic and rubber, including the chains, lines, suspended buoys, and mesh enclosures. The Greek mollusk industry typically uses floated kegs and blue mesh plastic bags, which suggests that cultured shells ingest a significant amount of blue plastic. In the Adriatic Sea and other studies conducted across European Seas, polyethylene was the most prevalent kind of polymeric in the investigated shellfish (75%). The Mediterranean Sea and the rest of the world both have the highest concentrations of polyethylene waste materials, which largely comes from supermarket bags and beverages and is the most common type of plastic on a global scale [12]. Eastern oysters *Crassostrea virginica* absorbed polystyrene through the gills and accumulated it in the digestive diverticula, with 50 nm particles accumulating more than 3 micrometers particles. The ingestion and deposition rates increased with concentration and exposure length [13]. PET was the most common MP in the scallops studied; however, polyethylene and polypropylene were also found. FTIR microscopy revealed that MPs constituted 16 – 60% of the particles characterized, with PET and PE being the most typically observed. Micro-FT-IR spectroscopy was performed on scallop sample particles chosen at random. In scallops, 372 unknown substances were recovered from tissues, and 101 of these were chemically characterized [14].

16.3.3 Crustaceans

The most commercially important crustacean in Europe is *Nephrops norvegicus* (Linnaeus, 1758), sometimes known as Dublin Bay Prawn or Norway Lobster. This benthic burrowing species lives on the muddy bottoms of the North Atlantic and Mediterranean [15]. The main components of *Nephrops norvegicus*'s dietary habits were decapod crustaceans, which include fish remains (with significant amounts of vertebrae, otoliths, and scales), gastropods, polychaetes, and anonymous items, mostly transparent tender sections that were challenging to identify with certainty but occasionally resembled molluscan tissue [16]. Three morphological distinctions of MP particles were chemically and mechanically defined (shape, major dimension) of *N. norvegicus* (gut, hepatopancreas, and tail). The largest MP particles were discovered in the stomach as fibers, whereas the smallest were found in the tail samples as pieces. There were 388 MP particles in total, 284 of which were fragments and 104 of which were fibers [17]. Plastic films accounted for 72% of all MPs, following pieces and filaments, respectively, by 67.0%and 14%. There are 22 polymers, with polypropylene (PP, 14%) and polyester (PES, 7%) coming in second and third, respectively [18]. The majority of the polyamide 6 (PA-6) was found as pieces. The most prevalent polymerics were PY (75 MPs), PA-6 (57 MPs), polyethylene (PE; 57 MPs), and polyvinyl chloride (PVC; 49 MPs). These polymers were discovered in all of the subjects, and their cumulative amount accounts for approximately 61% of all MPs recognized [17]. Some 10% of the entire amount was made up of other polymers, which consisted of acrylic particles, silicon, polyisoprene, polystyrene, and acrylonitrile-butadiene-styrene (ABS) [18]. The three main components that make up fishing and mussel farm nets, which together make up around 38% of all MPs, are PA-6, PE, and PP (polypropylene). The other 50% of the total plastic waste is made up of food wrapping, plastic containers, containers, and plates or other types of kitchen ceramics. The main components of these plastic-based goods, which make up more than 60% of the MPs identified by *N. norvegicus*, are PVC, PET (polyethylene terephthalate), PE, PY, and PP. The most prevalent type of fiber has been found to be transparent (67.0%), followed by yellowed and brownish filaments (8.9%), black fibers (8.8%), red fibers (7.1%), and blue fibers (5.9%). Only one individual with a large percentage of green fibers $(78-98\%)$ was detected in the Gulf of Cadiz [16]. MP contamination could come from a variety of sources, including aquaculture and fisheries, as well as the management of solid waste and synthetic fabric washing. Under natural conditions, the transfer of MP from the gastrointestinal tract to other body parts, cellular accumulation, and the alimentary chain transmission are all possible. It has long-term consequences, which may raise concerns for seafood security and safety [17].

16.4 MICROPLASTICS IN FRESHWATER

Plastics in freshwater systems degrade physically and environmentally while being subjected to lighter physical stresses than in marine environments (Figure 16.1).

Some environmental factors may have a greater influence on freshwater; for example, plastic fragments may weather rather quickly due to high UV penetration in nutritionally deficient lakes. However, overall microplastic degradation patterns in freshwater were discovered to be comparable to those in the marine environment: cracks, pits, and adhering particles [19]. Rain and storm events have an important role in the mobilization of a variety of indirect contaminants, such as sediment, fertilizers, and pathogens. Similar processes are speculated to be involved in the mobilization and transport of microplastics. During rainfall and storm events, the pathways for microplastic entry into rivers may be altered. Rainfall on hard surfaces may increase microplastics in stormwater systems, which include damaged road paints, tire, and brake pad fragments, particles deposited by air deposition, and diminished

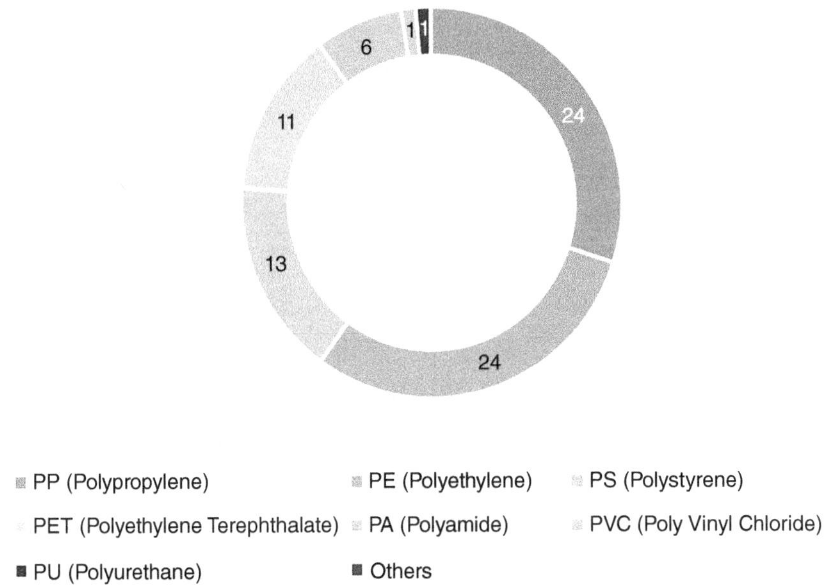

FIGURE 16.1 Microplastics content in freshwater samples.

debris [20]. Around the world, microplastics are becoming more prevalent in sediments and bodies of water, with rivers and lakes having the highest concentrations. Additionally, studies show that, in addition to the United States, microplastics are frequently discovered in the waterways of Australia, England, and Spain. Investigations have been made into the quantity and distribution of microplastics detected in the water from the Pearl River, encompassing the estuary and the metropolitan length of Guangzhou. Microplastic contamination in the Pearl River estuary, especially in its urban section along Guangzhou, to see if urban tributary waters act as accumulation systems that hold more microplastic than estuaries do. In the present investigation, the specimens from Guangzhou had a significantly higher concentration of microplastics than samples from the Pearl River estuary, suggesting that urban tributaries may serve as microplastic accumulation systems and that wastewater discharges from urban cities may be an important contributor to microplastics in the Pearl River. Granules are commonly used in facial cleaners and plastic manufacturing. Numerous plastic granules can be found in cosmetic items such as face scrub and toothpaste. Due to the elevated level of people's behaviors along the Pearl River, these microplastics, which might take the shape of film or granules, are most likely transported by urban wastewater or produced by the deterioration and disintegration of plastic litter. A large portion of the microplastics were below 500 μm in size. The most frequently encountered microplastic shapes were films, fragments, and fibers, which tend to be translucent or blue in hue. Furthermore, polypropylene and nylon were the two most prevalent forms of polymeric substances in this plastic debris [21]. Microplastics are present in wastewater and may become more toxic as a result of adsorption of harmful agents such as medicines and infectious organisms. Despite the fact that wastewater is an important contributor to microplastics, the available

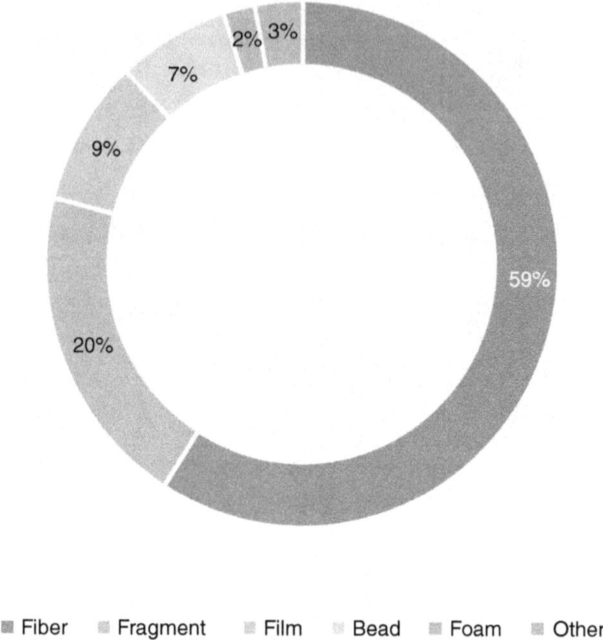

FIGURE 16.2 Microplastics content in freshwater samples.

material is poorly worded and scant. Cosmetics could contain up to 5% primary microplastic beads with a median thickness of 0.25 mm. Glitter, contact lens cleaners, minuscule buttons, and jewelry are examples of consumer products that may discharge microplastics into wastewater systems. The main treatment removes the majority of microplastics $(78-98\%)$, while secondary treatment results in a slight drop in concentration (up to 28%), and the final stage of treatment has barely any effect on the quantity of plastic debris.

Because of their own settling or trapping, microplastics are generally caught during the removal of wastewater solids by skimming and sedimentation procedures. Smaller microplastics are more likely to leave wastewater treatment plants, whereas fibers are more easily trapped. Although cleaned effluents contain only a few microplastics per liter, the large volume of effluent released each day causes significant contamination of aquatic ecosystems [22] (Figure 16.2).

16.5 MICROPLASTICS IN THE TERRESTRIAL ENVIRONMENT

Microplastic contamination of terrestrial habitats comes from a variety of sources and is brought on by specific anthropogenic practices. Pollution can occur in one of the two scenarios, depending on how the sources are differentiated: either through the usage of first-generation microplastics or through secondary microplastics that are produced.

Primary microplastics are generated in large quantities for a variety of uses, including abrasives in industry and cosmetics and aesthetic items. They are specifically made in shorter (5 mm) diameters for a variety of functions. These enter the ground or surroundings via personal gutters, as well as through many additional channels like those employed by traditional drainage networks and wastewater treatment plant effluent. Examples include pellets, home cleansers, beauty and cosmetic goods that contain small beads, and a number of other comparable goods. Secondary microplastics are created during the decomposition of larger-sized plastics (macroplastics), which may have been abandoned in landfills or garbage from the community. Urban solid trash is gathered, processed, and transported; during these procedures, further microplastics are discharged into the natural world. In addition, it can be dispersed by air or runoff, which transports microplastics from dump sites into various ecosystems. Agricultural practices which include using polythene plastic as mulch, creating poly tunnels, using packages, wrapping and fencing supplies, and using fodder bails, etc. can also contaminate soil. Meso- and macroplastic wastes are broken down into pieces by various weathering and deterioration processes, which create secondary microplastics [23]. MPs in terrestrial ecosystems may come from engineering operations, sewage disposal, sludge disposal, airborne dust migration, or sludge disposal. Animals' food supplies were most likely the cause of the MPs contamination, which suggests that the settings in which the animals lived were polluted with various MPs, particularly microfibers. Given the fact that creatures on lower trophic levels consume MPs and are then consumed by predators, it seems likely that MPs were transported via food chains. Important proof that MPs can spread from terrestrial habitats to soil, animals, and possibly people is shown by this study [24] (Figure 16.3).

A single species has been used in the majority of studies looking at the distribution of microplastics by terrestrial living things (86%) and the biological implications of microplastics (50%). A few examples are earthworms (*Lumbricusterrestris*), birds (*Gallus gallusdomesticus, Falco tinnunculus, Buteo buteo,* and *Milvus migranslineatus Zhao*), plants (*corn, Zea mays, soybean,* and *ryegrass*), and snails (*Helix aspersa, H. aperta, and H. pomatia*) [25].

16.5.1 Plants

The primary pathway for microplastics to reach plant communities in terrestrial environments is soil. The entrance of surface runoff and crop irrigation water resources, the lingering breakdown of agricultural mulching film and plastic debris, the land usage of municipal sludge, and accumulating microplastics from the surroundings are the main sources of microplastics observed in plant communities. The airborne MPs among them have the potential to directly harm plant roots and above-ground parts. The presence of plastic film residues can alter the aggregate structure and porosity of the soil, which may therefore have an impact on the activity of enzymes and the microbiological functional diversity of the soil. Water from the surface, groundwater, and purified sewage are the main sources of water for irrigation, and all of these contain microplastics in various shapes

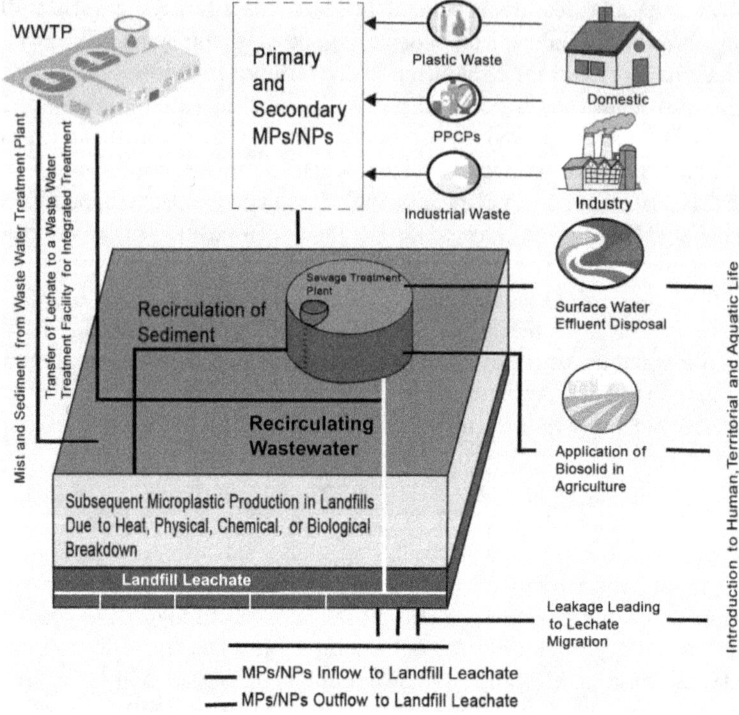

FIGURE 16.3 Microplastics in the terrestrial and marine environments.

(fibers, pieces, and particles), as well as materials (polyethylene terephthalate, polyethylene, polypropylene, and polyvinyl alcohol). Depending on the source of the irrigation water, the amount of microplastic contamination varied from 4.83 × 102 to 1.25 × 105 particles/m^3. Microplastics cannot be removed by the standard sludge extraction treatment methods employed in sewage treatment plants (such as sedimentation, drying, sterilization, composting, etc.). Sludge compost, serving as the main source of microplastics in soil since it is added to farming soil as fertilizer, adds to the total quantity of microplastics that sludge spilled into farming soil [26].

16.5.2 Seed

MPs affect the soil's structure in addition to reducing plant seed germination rates. High-concentration MPs can actually enhance and repair the sprouting rate of wheat seeds. MPs can restrict the sprouting rate of celery seeds primarily because they physically close the holes in the seeds. The adsorption and absorption of MPs by wheat seeds rely on the charge, particle size, and aggregation of MPs, which may be the cause of high concentrations of MPs showing a specific influence on the recovery of wheat seeds. An organic substance known as a plasticizer can make artificial resins and latex more malleable [27].

16.5.3 Rhizome

In the hydroponics experiment, the root length shrank as polystyrene and MP concentrations increased. Root development is inhibited by polystyrene MPs in soil culture conditions. Additionally, the biomass of wheat roots and the root-branch ratio both drastically decrease with the addition of polyethylene MPs. The particular surface area of MPs will grow as the size of the particles decreases. It will thus be simpler to bind to the surfaces of seeds or plant roots, further preventing water absorption. The respiration of seeds is impacted by the shrinkage of MPs, which also impacts the growth and development of roots and buds. Due to their low density, MPs increase soil aeration and reduce the bulk density of the soil. This can further inhibit root penetration into the soil, resulting in a decline in plant quality at the community level with drought. Plant stems and leaves can receive some of the MPs that have accumulated in the roots. Through root pressure and transpiration tension, MPs taken up by the roots may spread to other regions of the plant. Additionally, MPs can enter the wheat through the sprout areas of the lateral roots of the wheat. The transpiration rate encourages the absorption of the plastic particles as they are then transferred from the root to the shoots [27].

16.5.4 Earthworms

Eisenia fetida may consume PP (8–125, 71–383, and 761–1660 μm) and HDPE (28–145, 133–415, and 400–1464 μm) MPs utilized in this investigation. Numerous investigations have shown that earthworms (*Lumbriculus terrestris*) absorb MPs. MPs were castings for the two lesser classes. An anecic species known as *L. terrestrisis* has a fully developed gizzard, frequently resides in persistent vertical dungeons, and consumes a lot of soil. Therefore, *L. terrestris* may unintentionally consume a lot of MPs when feeding and burrowing. *E. fetida*, on the other hand, is an epigeic species that frequently lives in the topsoil without excavating any long-term grubs, uses very little soil, and relies on decomposing decaying matter. Transcriptomic analysis of *E. fetida* exposed to HDPE (28–145 micrometers) and PP (8–125 micrometers) MPs revealed that exposure to MPs significantly disrupted several pathways of metabolism linked to neurodegeneration, cellular damage, and inflammatory reactions. These metabolic pathways included prion-related disorders, complement and coagulation cycles, arachidonic acid metabolism, and tryptophan metabolism. The main mechanisms of industrial-grade HDPE and PP MPs' cytotoxicity in *E. fetida*, according to our findings from pharmacological research, may be neurotoxicity, stress caused by oxidation, and inflammation. These findings increase our understanding of the molecular processes at work in earthworms [28]. The soil ecosystem is not complete without bacteria and fungi. They serve as the foundation of the soil structure, as well as consumers of essential soil elements, and they perform a significant part in the breakdown and mineralization of biological matter [29].

16.5.5 Snails

Three species were represented by the samples: 220 *Helix aspersa*, 185 *Helix aperta*, and 20 *Helix pomatia*. All of the species of *H. aspersa* and *H. pomatia* snails were in the life cycle's active phase; however, all of the *H. aperta* snails were in the dormant phase (operculated). The selection of these three helix-related snail species is based on their physical and behavioral variations as well as the fact that they are the most well-known and extensively grown for use in food and cosmetic manufacturing. *H. aspersa* creates a thin epiphragma while it is resting, making it more resistant than the other two species. Instead, *H. aperta* and *H. pomatia* create a substantial and thick epiphragma during the resting phase, which is referred to as an "operculum" since they are more sensitive to temperature and humidity. In contrast to *H. pomatia* and *H. aspersa*, which occasionally burrow into the ground for a few centimeters, *H. aperta* burrows into the soil for up to tens of centimeters [30].

These findings are important in light of the fate of plastic litter and its biodegradation in soil habitats, as land snails are one of the most well-liked and quickly multiplying terrestrial creatures [31].

16.5.6 Birds

In this study, GI tract tissue included 0.3 (0.1) microplastics on average per gram, whereas 11.9 (2.8) microplastics per species of raptor were present. A total of 1197 pieces of plastic were counted. Microfibers accounted for 86% of all plastics, followed in importance by microfragments (13%), the macroplastics (0.67%), and small beads (0.33%). *Buteo lineatus* and *M. asio* were the two species with the greatest mean microplastic quantity per gram of intestinal tissue for fibers and pieces. The two species with the fewest mean fibers per gram of GI tract tissue were *Strix varia* and *B. jamaicensis*. In both *B. jamaicensis and B. lineatus* one microbead was discovered. Two bands of rubber, an imaginary adhesive hand, a piece of a disposable diaper's covering, and a jagged bit of green glass were all found in one bird, belonging to the only species known to have digested macroplastics, *Coragypsatratus*. The red-shouldered hawk, *Buteo lineatus*, had significantly more plastic particles (both fiber and fragment) per gram of gastrointestinal (GI) tissue than the osprey, *P. haliaetus*. For *B. lineatus*, there was a significant relationship between type and shading. Blue and transparent microfibers dominated the samples, with a mean of 8.0 and 6.0 filaments per bird, respectively. The average numbers of black and red fibers produced by *B. lineatus* were 4.5 and 2.0, respectively. There were significantly fewer and comparable amounts of blue and clear microfragments per individual (1.6 and 1.5, respectively). The microplastic particles identified in the gastrointestinal tissue of *P. haliaetus* displayed similar patterns in the results. Type and color interacted significantly with one another. The studied osprey had few microfragments overall; the only color to have more than one microfragment per bird was clear [32]. There have been numerous accounts of animals, including whales and seabirds, dying because of plastic that was detected in their bodies. Plastic may play a role in an animal's demise, even though this is not yet established as the primary cause. Hundreds of seabirds, 100,000 aquatic mammals, and countless fish perish each year due to plastic waste in the waters. Countless seabirds, 100,000 marine mammals, and untold numbers of fish perish each year as a result of plastic waste in the waters. Fish, turtles, mammals, birds, and even humans can become entangled in floating plastic in the ocean. In Procellariiforms, Brazil, 110 seabirds of 10 distinct genera were gathered, and

64.54% of them contained plastic fragments in their stomachs. In the digestive tract, 890 fragments of plastic were found, the majority of which were granules (34.95%) and users of plastic (62.92%), which comprised monofilament line and fragments of plastic containers. *Puffinus griseus*, *Daption capensis*, *Thalassarchemelanophrys*, *Puffinus giganteus*, *Puffinus gravis*, *Thalassarchemelanophrys*, *Fulmarus glacialoides*, *Pachyptilabelcheri*, and *Procellaria aequinoctialis* were among the species that were identified. Reproductive issues are a result of plastics and microplastics in seabirds [33].

16.6 SIZE CLASSIFICATION OF PLASTIC

16.6.1 Morphological Properties of Microplastics

The morphological properties of microplastics include three primary components, which are outlined below.

1. Compact size (the largest diameter under 0.005 mm)
2. Microplastic particles are classified into five types based on their shape: fiber, pellet, foam, film, and fragment. Every piece of plastic debris should have a consistent and indistinguishable thickness.
3. The particle's color should be quite consistent. Assuming the fragments are clear or translucent, a high-magnification instrument ought to be employed for analyzing them [34].

16.6.1.1 Shapes

The ecosystem contains an array of forms and dimensions of plastic debris. Microplastics typically come in the following morphologies: prills, residue, foamy consistency, fiber, and sheet. The original form of the initial microplastics, the procedures that cause the material's outer layer to deteriorate and break down, and the duration of life the plastic particle spends in the environment all affect these morphologies. Water from river and lake deposits has been reported to include microplastics of diverse shapes, including pieces, foam, fiber, and sheet [34].

16.6.2 Fiber

The fiber is the type of plastic debris that has the greatest prevalence in water body deposits. The most typical shape is fiber [34]. When it comes to abundance in the effluent, fiber often comes in first, followed by fragments, pellets, and beads. The most common kind of microplastic contamination is made up of blue, red, and black fibers. The size range of microplastic particles is between 30 mm < dpl < 100 mm, and the length of fibers is between 100 and 1000 mm. The majority of the initial and intermediate wastewater treatment plant sample MPs primarily consisted of fiber (65.6%), followed by pieces (28.1%) and chunks 95%. Granules, foam, and sheets made up only a small percentage of the MPs found in these samples (0.45%, 0.22%, and 0.20%, respectively). According to research on suspended atmo spheric microplastics (SAMPs), microfibers made up 67% of all SAMPs, prior to fragments and granules, which made up 30% and 3% of the total, respectively. This suggests a strong connection between the polluting of other habitats by MP air fall-out. The significant number of plastic fibers found in effluent demonstrates the significant influence that washing, drying, fabricated textile use, and fabric handling activities have on microplastic emissions. Fiber in coastal areas, however, has various origins. Dubaish and Liebezeit discovered that the fibers may come from rope fiber due to significant ship traffic; as a result, they would find more rope debris in surface waters. According to the findings, synthetic fibers are a problem that requires immediate response [35].

16.6.3 Fragment

MP films and fragments vary in size from 25 to 300 μm. It's possible that bigger plastic objects like packaging, containers, and cleaning supplies exposed to stress, tiredness, or UV light are the source of fragmented microplastics [36]. For instance, the surface indicating fragments and pellets with holes, fissures, attached particulates, and crumbles is a clear indication of the structural degradation and chemical corrosion that produced tiny plastic particles [34].

16.6.4 Foam and Film

The film, which is used in farm cultivation, is mostly made of plastic. The film, which is used in farm cultivation, is mostly made of plastic. Foamed microplastics have an erratic shape and a small size. Foams are widely used in the production, building, wrapping, and thermal insulation industries due to their outstanding heat retention and durability against impact. As a result, these foams might be made by wrapping insulation materials and cushions from different goods [37]. Microplastics in the shape of foam are created when Styrofoam is damaged. Granules that were spilled while processing transportation will probably be entirely pristine, or it could be spherules and microbeads used in air-blasting agents, industrial cleaners, cosmetic goods, and sandblasting medium [34]. Foams and films only make up roughly $5-10\%$ of the data. It is distinct from pellet and fragment forms, which are discovered in wastewater in amounts ranging from 25 to 70% [35].

16.6.5 Microbeads

The majority of microbead particles have an uneven form; however, some are completely spherical. The study showed that no sample's primary bead shape was elliptical but instead were present in only a small overall fraction (1.4%). Microbeads are frequently found at all phases of the process of treating wastewater, often having an exceedingly large share, and in some treatment facilities for wastewater, they are adequate in size to evade filtering. Even though their varied sizes range from 8 to 2 mm, 85–186 mm, and less than 100 mm in cosmetic products, microbeads do in fact have unique properties that create a smooth surface. A significant portion of microbeads—between 6 and 7% of the final product—come from cosmetic and consumer personal care items [35].

16.6.6 Size

Microplastics appear in a wide range of sizes, and, as technology advances, the smallest size of those that have been found gets smaller. In general, water from the atmosphere deposits include more microplastics with an average particle diameter less than 0.001 mm, whereas their abundances tend to decline as particle size increases [34]. The observed microplastics were divided into five groups based on their sizes: 50, 50–500, 500–1000, 1000–1500, and >1500 micrometers. The highest concentration of microplastics is between 0.05 and 0.5 mm, and the lowest percentage is between 1.5 mm and above. These five sorts of tiny plastic particles made up, in order, 20.9%, 1.1%, 63.9%, 4.8%, and 9.3% of the total. This is due to the fact that bigger microplastics can produce numerous smaller fragments during degradation, making small and medium-sized microplastics more prevalent Due to the techniques used, plastic debris that is less than 0.01 mm in diameter were also not found. Because they resemble the size of some low-nutrition creatures, the proliferation of smaller-sized microplastics in stormwater raises serious concerns because they could endanger aquatic life by increasing the likelihood that they will be ingested. Additionally, some contaminants can more easily adhere to the surface of microscopic microplastics, which makes them potential major carriers of pollutants in the environment [37].

16.6.7 Color

White, transparent, black, blue, green, and red microplastics were found.

The three distinct types that made up the ultimate ranking were black, polychromatic, and white (including translucent) due to the small amount of tiny plastic particles that were present in the hues green, blue, transparent, and red. Plastic debris comes in an assortment of hues, which also represents a variety of reasons for pollution. Polychromatic, black, and white (including translucent) microplastics accounted for 44.1%, 14.6%, and 41.3% of the total. The bulk of microplastics are white (including transparent) and multicolored, and polychromatic microplastics make up the majority of the material; white (including translucent) and black plastic particles constitute slightly less than the majority of the total microplastics. On the contrary, the use of transparent plastic goods on agricultural land, such as grocery bags, polyamide webs, and film, would lead to a significant amount of white microplastic production. Contrarily, when polychromatic plastic debris becomes brittle and discolored by heat, sunlight, fluids, and biological reactions in liquid, white microplastics are generated [37].

16.6.8 Polymer

The most prevalent types of polymers, the common ones were PET (8%), PS (9%) PE (26%) and PP (27%): also included were PP, PE, PET, PS polyamide and acrylic. Along with shape and size, polymer density significantly affects the way microplastics are distributed among various matrices in aquatic habitats. As a result, sediments had higher concentrations of high-density polymer types than water samples did, such as acrylic, polyurethane (PU), PVC, polyester, polyamide, The water samples examined contained higher concentrations of PP, PS, and PE. In the debris, higher concentrations of acrylic, polyester, polyamide, PVC, and polyurethane (PU) were detected. The prevalence of these polymer types may be explained by their usage in the production of textile fiber, fabric, textiles, materials for packaging, and reusable goods (PP and PET), as well as by the significant consumption of polypropylene- and PE-based bags of trash and food boxes. In a few studies, markings for roads and styrene butadiene rubber, in addition to these polymer kinds, have been found to contribute significantly to large portions of microplastics. These materials presumably originated from traffic-related highway regions in metropolitan contexts [38]. PE is the primary polymer found in freshwater sediment, followed by PS and PP, and the toxic substance composition varies greatly over the globe. There is currently no conclusive correlation or justification for the variation in polymer types found in freshwater sediment. To figure out the fact that a dominant group of polymers is present in the microplastic pollution of the water from the river and lake debris and the fact that the composition of the aforementioned polymer class varies depending on the location of the sample and the path that the particles travel, more research is necessary [34].

16.7 CONSEQUENCES OF MICROPLASTIC IN THE FRESHWATER ECOSYSTEM

The effects of microplastics in freshwater environments are causing considerable concern. Microplastics' toxicity potential most likely results from one of the following three pathways:

- **Consuming Microplastics:** Freshwater species that consume microplastics run the risk of having their feeding appendages or digestive systems immediately choked or obstructed. While some species can quickly expel or ingest microplastics, others may not be able to do so, sustaining and soliciting the plastics in their bodies. For instance, *Xenopustropicalis* tadpoles had a 95% depuration rate after being introduced to clean water and were able to consume significant volumes of ingested microspheres. Additionally, when species at trophic levels that are higher consuming species from freshwater, there is a chance that microplastics will go up the food chain.
- **Additive Leakage from Plastics:** According to research, plastic waste can stress a fish's intrinsic defenses by altering its defensive reactions, according to a study on the effect of polycarbonate and PS non-materials on the fathead minnow (*Pimephales promelas*), a freshwater fish. PS microparticles have also been demonstrated to increase lipid levels and promote irritation in zebrafish livers. During the production process, coloring agents like polybrominated diphenyl ethers (PBDEs) and harmful compounds like bisphenol-A are added to plastics. Metals are also added to plastics. The sort and number of plasticizers and other additives used during production determine an item's toxicity in the environment, even though it is made of the same polymer as other objects.

- **Concentration and Spread of Organic Contaminants:** Organic contaminants can be concentrated and transferred using microplastic particles as a medium. Polybrominated diphenylethers (PBDEs), polycylic aromatic hydrocarbons (PAHs), polychlorinated biphenyls (PCBs), dichlorodiphenyltrichloroethane (DDT) are a few examples of organic pollutants that are frequently discovered in freshwater environments. PBTs (persistent bioaccumulative and cytotoxic compounds) can bioaccumulate in medaka fish when consumed by species with digestive fluids. Microplastics and pyrene were shown to impair the action of the enzyme acetylcholinesterase (AChE), which is responsible for neural and neuromuscular transmission in fish. By eating freshwater organisms, people may be exposed to microplastics and related pollutants. Because of the proximity of organic pollutants to the sites where they were first used, the interaction between microplastics and them is a problem in areas with freshwater [39].

16.8 CONSEQUENCES OF MICROPLASTIC IN THE MARINE ENVIRONMENT

MPs have an enormous surface area and strong adsorption, which have been shown to be key carriers for chemicals and bacteria that can pose a serious threat to aquatic life and even the ecosystem.

- **MPs as Pollution Carriers:** MPs have a strong hydrophobic nature and are well-known for being pollutant carriers. The two main types of pollutants carried by MPs are the ones found inside the MPs themselves, such as monomers, additives, and other byproducts, and environmental pollutants, such as heavy metals and hydrophobic chemicals. Through additive polymerization or polycondensation reactions, MPs are created from monomers as the starting material. MPs serve as key carriers for toxins and bacteria that can seriously endanger marine organisms as well as the environment because of their vast area and strong adhesion. vinyl monomers, cross-linking monomers, surface active monomers, and acid monomers, aliphatic binary esters, phosphate epoxy compounds, and phthalates are the most widely used plastic additives. The most prevalent of these is phthalate. According to various uses, additional byproducts, such as dyes, pigments, etc., are added. MPs frequently absorb the heavy metals lead, zinc, copper, chromium, and cadmium. Organic contaminants include bisphenol A, petroleum hydrocarbons, polybrominated diphenyl ethers, polycyclic aromatic hydrocarbons, polychlorinated biphenyls, and organ chlorine insecticides. The capacity for adsorption can also be impacted by weathering and other elements, including the impacts of plastics' own aging and weathering, ion strength, and pH found in water bodies, among others [40].
- **MPs Impact Organisms:** MPs have an impact on organisms because they can enter the intestinal tract of aquatic species and circulate there. Particles with diameters of 3.0 or 9.6 μm have reportedly been discovered in the hemolymph of the blue mussel *M. eduli.* MPs are consumed as food by

aquatic creatures like prawns (*Paratya australiensis*), fish (*Dicentrarchus labrax L.*, *Danio rerio*, *Oreochromis niloticus*, *Carassius auratus*, *Girella laevifrons*), and zooplankton (*Daphnia magna*). These aquatic creatures consume MPs of different types and sizes. which are located in specific areas of tissues or the digestive system. Polluted MPs interact with nutrients, causing psychological distress in living things and endangering the stability and composition of ecosystems. Hemolymph, responses from the immune system, endocrine disturbance, and alterations in genetic expression patterns are all common consequences for bivalve species [41].

16.9 CONSEQUENCES OF MICROPLASTIC IN THE TERRESTRIAL ENVIRONMENT

There was no observable effect on earthworms' capacity to adapt when exposed to 0.25–0.5% polystyrene microplastics. However, worms are affected by plastic particles in various ways depending on the types and amounts present, which can slow down growth and have negative effects on their defenses. In *E. fetida*, earthworms exposed to 20% PE and microplastics made of PS for 14 days, the enzyme catalase peroxidase and the per oxidation of lipids levels increased, whereas the amount of superoxide dismutase and the enzyme glutathione S-transferase levels were reduced. Additionally, soil animals' gut microbiota and the relative Rhizobium abundance can both be altered by microplastics. The soil Oligocaeter *Enchytraeuscrypticus*'s microbiome in its intestines included fewer flavobacteria and other the mold after being exposed to a significant amount of nana-PS plastic pellets. The development and reproduction of collembolan (*Folsomia candida*), as well as the variety of microorganisms and microbiota in the stomach of collembolan, were reportedly decreased by exposure to 0.1% PVC microplastics. *Caenorhabditis elegans*'s dimension, mortality rate, reproduction rate, and oxidative stress-related genes were all strongly impacted by polystyrene (PS), and this influence was size-related. When snails ingested PETs, their levels of the TAOC (comprehensive index representing oxidative stress), the activity of the antioxidant enzyme GP*x*, and the amount of malondialdehyde (MDA) increased. These changes would promote lipid peroxidation, which would harm the snails' digestive systems. Snails' gut microbiota may behave differently and grow more slowly if they consume leaves polluted with nanoplastics. This may also cause histological alterations in the digestive organs of the snails. Nanoplastics may prevent the mung bean plant's roots from growing and cause an accumulation in the leaves. More investigation revealed that, depending on the measurement of the plant, microplastics had different effects on it. For instance, 100 mg L^{-1} of 100 nm microplastics reduced the proliferation of *vicia faba*, and they also caused more severe oxidative stress and genetic damage than 5 micrometers sized microplastics. Additionally, by altering rhizosphere structure, vegetative circumstances, and accessibility to nutrients of soils, microplastics can negatively impact soil plants: for instance, the buildup and mobility of microplastics in plant muscles, the effects of microplastics on plants' well-being and toxic effects, and how vegetation responds to pressure. Consuming crops that have collected microplastics will raise human exposure risks to microplastics [42]. People with inflammatory bowel disease (IBS) have an increased uptake of microparticles compared to healthy individuals, and research on IBS patients has shown this to be caused by the poor eating habits of developed countries. After interacting with microfold cells in the gut, nanoplastic copolymer nanoparticles, such as ammonium palmitoyl glycerol chitin and initial prototypes used in the delivery of drugs, recirculate into the bloodstream and biodistribute preceding excretion, to the organs of digestion, the gall urinary tract, and the lymph system [43].

16.10 FUTURE ASPECTS

Working on terrestrial microplastics is encouraged for researchers with diverse backgrounds. In order to comprehend the outcome for soils containing plastic debris, for example, high-throughput sample processing and precise, highly sensitive, cost-effective, and standardized detection technologies are needed. In addition to the systems of agriculture and water from lakes and rivers, which are the subjects of the majority of current studies, there are examinations of the occurrence of microplastics in arid regions, dry regions, forests, grasslands, and the Arctic tundra, as well as their geographic distribution, dimensions, environmental, and monetary consequences. Furthermore, stricter rules might need to be implemented to limit the consumption of synthetic products. Despite the fact that biodegradable plastics are thought to be an option in agriculture, their risk should also be considered given the challenges associated with getting trash made of plastic out of soils. Currently, most studies are conducted in laboratories. Research on microplastics in the natural world, both with and without controlled circumstances, is required [44].

The current evaluation recommends that all nations monitor the amount of microplastics off their own coasts because doing so will help determine and comprehend the worldwide abundance of microplastics. There has to be more research on the effects of microplastics on ecologically rich regions, including mangroves, coral reefs, and kelp beds. Several creatures, from phytoplankton to whales, have been researched; however, no conclusive evidence or findings on potential trophic transfer or biomagnification have been made [45].

The following viewpoints are suggested for further investigation in light of decades of microplastic toxicity investigations and the drawbacks mentioned:

1. Instead of just presenting the normal physiological characteristics of tiny plastic particles and the reactivity of creatures generally, a complete assessment of prospective consequences must remain driven by the distinguishing characteristics of these plastics and their behavioral and bodily properties in animals.
2. To improve our understanding of the overall toxicity pattern of microplastics, methodologies for specific procedures for the evaluation of pollutants with tiny plastic particles and toxicological characteristics must be tightly controlled, and concurrent comparisons across other research ought to be performed in order to increase our comprehension of the overall toxicity pattern of microplastics.
3. A comprehensive record of the lethality of microplastics for various fauna should be cautiously created, and the applicability of current cytotoxic

assessment techniques for plastic debris must be evaluated. The combined toxicity of environmental pollutants and their long-term ecological effects should be investigated further [46].

16.11 CONCLUSION

Both people and animals have been discovered to be adversely affected by microplastic pollutants and additives introduced during the production process in order to add specific qualities to the finished product. These effects are largely connected to cancer-causing potential and reproductive toxicity. Emerging contaminants known as microplastics have been found in a variety of habitats. It is anticipated that the environmental degradation and associated environmental consequences brought on by microplastics will worsen given the prevailing path of plastic consumption and its worldwide manufacturing, it is anticipated that environmental degradation and associated environmental consequences brought on by microplastics will get worse [1]. Regardless of the consequences, most people concur that we still know very little about microplastics and that additional research is necessary to better understand their fate and consequences. As a result, the reporting of the work must be transparent, and discussions of cautions in the explanation of the techniques must be included. Thus, the field can benefit from the findings. Policymakers and the media are now paying close attention to microplastic research on a worldwide scale, but poorer-quality or out-of-context research has the ability to cast doubt on the subject and prevent timely, scientifically informed discussion and mitigation of microplastic contamination of the environment. In the area of microplastics, we must continue to promote and establish strict, examined, verified, and reliable data [47].

REFERENCES

1. Karthikeyan P, Subagunasekar M. Microplastics pollution studies in India: A recent review of sources, abundances and research perspectives. *Regional Studies in Marine Science*. 2023 Feb 9: 102863.
2. Padervand M, Lichtfouse E, Robert D, Wang C. Removal of microplastics from the environment. A review. *Environmental Chemistry Letters*. 2020 May; 18: 807–828.
3. Bank MS. *Microplastic in the Environment: Pattern and Process*. Springer Nature; 2022.
4. Hale RC, Seeley ME, La Guardia MJ, Mai L, Zeng EY. A global perspective on microplastics. *Journal of Geophysical Research: Oceans*. 2020; 125: e2018JC014719.
5. Li J, Liu H, Chen JP. Microplastics in freshwater systems: A review on occurrence, environmental effects, and methods for microplastics detection. *Water Research*. 2018 Jun 15; 137: 362–374.
6. Kumar S, Rajesh M, Rajesh KM, Suyani NK, Rasheeq Ahamed A, Pratiksha KS. Impact of microplastics on aquatic organisms and human health: A review. *International Journal of Environmental Sciences & Natural Resources*. 2020; 26(2): 59–64.
7. Xu S, Ma J, Ji R, Pan K, Miao AJ. Microplastics in aquatic environments: Occurrence, accumulation, and biological effects. *Science of the Total Environment*. 2020 Feb 10; 703: 134699.
8. Ma H, Pu S, Liu S, Bai Y, Mandal S, Xing B. Microplastics in aquatic environments: Toxicity to trigger ecological consequences. *Environmental Pollution*. 2020 Jun 1; 261: 114089.

9. Elizalde-Velázquez GA, Gómez-Oliván LM. Microplastics in aquatic environments: A review on occurrence, distribution, toxic effects, and implications for human health. *Science of the Total Environment*. 2021 Aug 1; 780: 146551.
10. Bajt O. From plastics to microplastics and organisms. *FEBS Open Bio*. 2021 Apr; 11(4): 954–966.
11. Filgueiras AV, Preciado I, Cartón A, Gago J. Microplastic ingestion by pelagic and benthic fish and diet composition: A case study in the NW Iberian shelf. *Marine Pollution Bulletin*. 2020 Nov 1; 160: 111623.
12. Digka N, Tsangaris C, Torre M, Anastasopoulou A, Zeri C. Microplastics in mussels and fish from the Northern Ionian Sea. *Marine Pollution Bulletin*. 2018 Oct 1; 135: 30–40.
13. Song JA, Choi CY, Park HS. Exposure of bay scallop Argopectenirradians to micro-polystyrene: Bioaccumulation and toxicity. *Comparative Biochemistry and Physiology Part C: Toxicology & Pharmacology*. 2020 Oct 1; 236: 108801.
14. Akoueson F, Sheldon LM, Danopoulos E, Morris S, Hotten J, Chapman E, Li J, Rotchell JM. A preliminary analysis of microplastics in edible versus non-edible tissues from seafood samples. *Environmental Pollution*. 2020 Aug 1; 263: 114452.
15. Hara J, Frias J, Nash R. Quantification of microplastic ingestion by the decapod crustacean Nephrops norvegicus from Irish waters. *Marine Pollution Bulletin*. 2020 Mar 1; 152: 110905.
16. Carreras-Colom E, Cartes JE, Constenla M, Welden NA, Soler-Membrives A, Carrassón M. An affordable method for monitoring plastic fibre ingestion in Nephrops norvegicus (Linnaeus, 1758) and implementation on wide temporal and geographical scale comparisons. *Science of the Total Environment*. 2022 Mar 1; 810: 152264.
17. Martinelli M, Gomiero A, Guicciardi S, Frapiccini E, Strafella P, Angelini S, Domenichetti F, Belardinelli A, Colella S. Preliminary results on the occurrence and anatomical distribution of microplastics in wild populations of Nephrops norvegicus from the Adriatic sea. *Environmental Pollution*. 2021 Jun 1; 278: 116872.
18. Cau A, Avio CG, Dessì C, Follesa MC, Moccia D, Regoli F, Pusceddu A. Microplastics in the crustaceans Nephrops norvegicus and Aristeusantennatus: Flagship species for deep-sea environments. *Environmental Pollution*. 2019 Dec 1; 255: 113107.
19. Li C, Busquets R, Campos LC. Assessment of microplastics in freshwater systems: A review. *Science of the Total Environment*. 2020 Mar 10; 707: 135578.
20. Hitchcock JN. Storm events as key moments of microplastic contamination in aquatic ecosystems. *Science of the Total Environment*. 2020 Sep 10; 734: 139436.
21. Yan M, Nie H, Xu K, He Y, Hu Y, Huang Y, Wang J. Microplastic abundance, distribution and composition in the Pearl River along Guangzhou city and Pearl River estuary, China. *Chemosphere*. 2019 Feb 1; 217: 879–886.
22. Prata JC. Microplastics in wastewater: State of the knowledge on sources, fate and solutions. *Marine Pollution Bulletin*. 2018 Apr 1; 129(1): 262–265.
23. Surendran U, Jayakumar M, Raja P, Gopinath G, Chellam PV. Microplastics in terrestrial ecosystem: Sources and migration in soil environment. *Chemosphere*. 2023 Jan 25: 137946.
24. Lu S, Qiu R, Hu J, Li X, Chen Y, Zhang X, Cao C, Shi H, Xie B, Wu WM, He D. Prevalence of microplastics in animal-based traditional medicinal materials: Widespread pollution in terrestrial environments. *Science of the Total Environment*. 2020 Mar 20; 709: 136214.
25. Baho DL, Bundschuh M, Futter MN. Microplastics in terrestrial ecosystems: Moving beyond the state of the art to minimize the risk of ecological surprise. *Global Change Biology*. 2021 Sep; 27(17): 3969–3986.
26. Yu ZF, Song S, Xu XL, Ma Q, Lu Y. Sources, migration, accumulation and influence of microplastics in terrestrial plant communities. *Environmental and Experimental Botany*. 2021 Dec 1; 192: 104635.

27. Ge J, Li H, Liu P, Zhang Z, Ouyang Z, Guo X. Review of the toxic effect of microplastics on terrestrial and aquatic plants. *Science of the Total Environment.* 2021 Oct 15; 791: 148333.
28. Li B, Song W, Cheng Y, Zhang K, Tian H, Du Z, Wang J, Wang J, Zhang W, Zhu L. Ecotoxicological effects of different size ranges of industrial-grade polyethylene and polypropylene microplastics on earthworms Eisenia fetida. *Science of the Total Environment.* 2021 Aug 20; 783: 147007.
29. Jacques O, Prosser RS. A probabilistic risk assessment of microplastics in soil ecosystems. *Science of the Total Environment.* 2021 Feb 25; 757: 143987.
30. Panebianco A, Nalbone L, Giarratana F, Ziino G. First discoveries of microplastics in terrestrial snails. *Food Control.* 2019 Dec 1; 106: 106722.
31. Song Y, Qiu R, Hu J, Li X, Zhang X, Chen Y, Wu WM, He D. Biodegradation and disintegration of expanded polystyrene by land snails Achatina fulica. *Science of the Total Environment.* 2020 Dec 1; 746: 141289.
32. Carlin J, Craig C, Little S, Donnelly M, Fox D, Zhai L, Walters L. Microplastic accumulation in the gastrointestinal tracts in birds of prey in central Florida, USA. *Environmental Pollution.* 2020 Sep 1; 264: 114633.
33. Susanti NK, Mardiastuti A, Wardiatno Y. Microplastics and the impact of plastic on wildlife: A literature review. In *IOP Conference Series: Earth and Environmental Science* (Vol. 528, No. 1, p. 012013). IOP Publishing; 2020 Jul 1.
34. Yang L, Zhang Y, Kang S, Wang Z, Wu C. Microplastics in freshwater sediment: A review on methods, occurrence, and sources. *Science of the Total Environment.* 2021 Feb 1; 754: 141948.
35. Hamidian AH, Ozumchelouei EJ, Feizi F, Wu C, Zhang Y, Yang M. A review on the characteristics of microplastics in wastewater treatment plants: A source for toxic chemicals. *Journal of Cleaner Production.* 2021 May 1; 295: 126480.
36. Chen G, Feng Q, Wang J. Mini-review of microplastics in the atmosphere and their risks to humans. *Science of the Total Environment.* 2020 Feb 10; 703: 135504.
37. Li J, Ouyang Z, Liu P, Zhao X, Wu R, Zhang C, Lin C, Li Y, Guo X. Distribution and characteristics of microplastics in the basin of Chishui River in Renhuai, China. *Science of the Total Environment.* 2021 Jun 15; 773: 145591.
38. Shruti VC, Pérez-Guevara F, Elizalde-Martínez I, Kutralam-Muniasamy G. Current trends and analytical methods for evaluation of microplastics in stormwater. *Trends in Environmental Analytical Chemistry.* 2021 Jun 1; 30: e00123.
39. Wong JK, Lee KK, Tang KH, Yap PS. Microplastics in the freshwater and terrestrial environments: Prevalence, fates, impacts and sustainable solutions. *Science of the Total Environment.* 2020 Jun 1; 719: 137512.
40. Du S, Zhu R, Cai Y, Xu N, Yap PS, Zhang Y, He Y, Zhang Y. Environmental fate and impacts of microplastics in aquatic ecosystems: A review. *RSC Advances.* 2021; 11(26): 15762–15784.
41. Vo HC, Pham MH. Ecotoxicological effects of microplastics on aquatic organisms: A review. *Environmental Science and Pollution Research.* 2021 Sep; 28: 44716–44725.
42. Ya H, Jiang B, Xing Y, Zhang T, Lv M, Wang X. Recent advances on ecological effects of microplastics on soil environment. *Science of the Total Environment.* 2021 Dec 1; 798: 149338.
43. Vazquez OA, Rahman MS. An ecotoxicological approach to microplastics on terrestrial and aquatic organisms: A systematic review in assessment, monitoring and biological impact. *Environmental Toxicology and Pharmacology.* 2021 May 1; 84: 103615.
44. He D, Bristow K, Filipović V, Lv J, He H. Microplastics in terrestrial ecosystems: A scientometric analysis. *Sustainability.* 2020 Oct 21; 12(20): 8739.

45. Ajith N, Arumugam S, Parthasarathy S, Manupoori S, Janakiraman S. Global distribution of microplastics and its impact on marine environment—a review. *Environmental Science and Pollution Research.* 2020 Jul; 27: 25970–25986.
46. Ma H, Pu S, Liu S, Bai Y, Mandal S, Xing B. Microplastics in aquatic environments: Toxicity to trigger ecological consequences. *Environmental Pollution.* 2020 Jun 1; 261: 114089.
47. Provencher JF, Covernton GA, Moore RC, Horn DA, Conkle JL, Lusher AL. Proceed with caution: The need to raise the publication bar for microplastics research. *Science of the Total Environment.* 2020 Dec 15; 748: 141426.

17 Bioremediation of Microplastics by Cyanobacteria

Rwiddhi Sarkhel, Parthapratim Gupta, and Tamal Mandal

17.1 INTRODUCTION

The increase in plastic pollution has been observed worldwide due to the high use of plastic materials. This affects the life of flora and fauna in marine aqua systems as well as posing a threat to human beings and the environment, affecting its sustainability. Recycling of plastics, different methods such as landfills, and incineration are not always the only optimum solutions for reducing plastic pollution (Chia et al., 2020). This study represents the overview of algae or cyanobacteria as a source of microplastics. Cyanobacteria are a group of phototrophic prokaryotes which can produce polyhydroxyalkanoates (PHAs) using sunlight and carbon dioxide (CO_2) as a form of energy. PHAs resemble an alternative to chemical plastics due to their nature, biocompatibility, and biofortification (Sartori et al., 2021). A cyanobacterium is known as the blue-green algae, which can diminish the increased plastic waste globally with different environmental approaches. Cyanobacteria sp. such as *Anabaena* or *Nostoc* can degrade microplastics through enzymes synthesized by them known as cyanotoxins while using the plastic polymers as carbon sources.

Microplastic production has risen to around 359 million metric tons in 2018, from 245 million metric tons in 2008, and it is expected to be tripled by the year 2050, which accounts for five times the global oil consumption (Garside, 2020). Despite huge production and accumulation of plastics since the 1950s, there are effective and inevitable technologies to deal with the disposal issues and effective strategies to reduce the microplastic pollution affecting sustainability in the environment. A tremendous quantity of plastic piles and debris has caused 'white pollution' in the aquatic ecosystems, posing a serious threat to the life of marine animals and coral reefs. However, this debris has caused several problems for marine species, such as impairing their ability to consume food and death (Stephanis et al., 2013). Nowadays microplastics are also affecting the quality of food and air as they have also been seen in food and air samples increasing the toxicity of environment. Therefore, due to the severe consequences faced by microplastics in the environment, an environmentally feasible and cost-effective process, i.e., bioremediation, has been chosen to degrade the microplastics utilizing cyanobacteria, an area that has attracted researchers and

DOI: 10.1201/9781032684574-17

scientists since plastics are needed for everyday purposes (EPA Plastics, 20209). Cyanobacteria sp. that may grow on any waste surfaces can be a good source of bioplastics production with high lipid accumulation as they do not compete with food sources (Khoo et al., 2020; Yew et al., 2020). Cyanobacteria are both beneficial and detrimental to human health and the environment. Microalgal biomass could produce biofilm and can take part in the purpose of carbon and nitrogen fixing due to their rapid growth (Chen et al., 2015; Ho et al., 2020).

Worldwide usage of plastics accounts for about 99% of petroleum-derived sources with a global estimate of 6.3 gigatons (Gt) by 2015 and may exceed 12 Gt by 2050 (CIEL, 2017; Geyer et al., 2017). Thus biodegradable plastics are the rage as an alternative to conventional plastics due to their nature and reduced issues due to the synthetic polymers (Kamravamanesh et al., 2018). These plastics emerged as a substitute over non-biodegradable polymer film composites, hydrogels with advantages of bioremediation, adsorption, and biocompatibility (Luckochen and Pillai, 2011; Ghosh et al., 2013). Biodegradable polymers and their composites are derived from both natural and synthetic sources (Sarkhel et al., 2020). Composites are comprised of two phases where either or both of the filler and matrix consist of biodegradable materials (Sarkhel et al., 2022). However, these polymer composites have lower mechanical characteristics in comparison with the conventional polymer composites using synthetic polymers owing to incompatibility between the hydrophobic and hydrophilic phases (Rogovina, 2015).

Plastic degradation using cyanobacteria is rarely observed in literature since the bioremediation process by Cyanobacterial sp. is a time-consuming process, and there are various steps through which this process takes place, such as biofragmentation, bioassimilation, biomineralization, and biotransformation. Many studies to date have proven the potential of cyanobacteria in many ways, but they are costly and labor intensive. Therefore, biodegradation by cyanobacteria proves a suitable and cost-effective method to minimize marine pollution. This review mainly focuses on the background study of marine plastic pollution, the role of cyanobacteria on the degradation of plastics, and the applications and future prospectives.

17.2 IMPACT OF MARINE PLASTIC POLLUTION ON THE ENVIRONMENT

Plastics, including bioplastics and microplastics, play a ubiquitous role in the environment as they are a part of the fossil record, a source of the Anthropocene. Marine plastic pollution has a severe effect on the environment due to its disposal (Liquete et al., 2013). The disposal of marine litter into the ecosystem directly or indirectly poses a serious threat to human health and aquatic life (Galloway et al., 2017; Naeem et al., 2016). The increasing amount of plastic waste in the marine ecosystem has been called the marine plastic pollution (Geyer et al., 2017). As time increases, biofragmentation of microplastics takes place into 0.1–5 mm.

Some 192 countries border the coast of the Atlantic, Pacific, and Indian oceans, or Mediterranean and Black seas, producing 2.5 billion tons of waste since 2010. The contribution of plastic waste by the UK, India, and China is about 1 million tons,

4.5 million tons, and 16 million tons, respectively (Kumar et al., 2011). Around the world, 1 million plastics are purchased every day and 5 trillion tons of plastic bags are thrown away worldwide. In early 2000, the amount of plastic waste generated rose about four times the size it had in previous years. India generates around 10 thousand tons of plastic waste (Puri et al., 2013). The yearly generation of plastic waste was assessed at 57 million tons in Europe in 2012. Reusing plastics is considered practical and in fact attainable choice to handle plastic waste. It has been estimated that at least 60% of plastic floating in the ocean is exported from coastal to the open-ocean waters.

The study of plastic degradation focuses on using polymers and creating awareness for using plastic wastes on environmental hazards. Generally, plastic bags and plastic polymers are quite extensively used mainly for their cost-effectiveness and for their large-scale production. Microplastic degradation signifies that the response to plastic is greatest in the environment and must be facilitated in large-scale production. Serious environmental problems and health issues have been caused by the plastic bag waste (Adane and Muleta, 2011). Many of the microbes can degrade polymer aerobically or anaerobically (Singh and Mallick, 2017). Biodegradable polymers represent a promising way to reduce the number of plastics being disposed, reducing risk of pollution to enhance sustainable development of the environment (Raaman et al., 2012). Isolation and identification of polymer-degrading cyanobacteria are important as this knowledge provides valuable information for their use in bioaugmentation processes (Kumar et al., 2011. Marine plastic pollution is mainly caused due to the small fragments of plastics in marine water systems, which can cause a hazard to the environment due to the sharp objects present in the water (Figure 17.1). Marine debris is suspended since they float on the surface of water due to the acclimatization of plastics (micro-, macro-, and nanoplastics) on it.

FIGURE 17.1 Marine plastic pollution due to huge accumulation of plastics.

17.2.1 Effect of Microplastics on Cyanobacteria/Microalgae

Smaller sizes of micro-, macro-, and nanoplastics can be derived from abiotic and anaerobic conditions or from toxic products such as drugs or personal body care. Microplastics can be degraded by biofragmentation into smaller units such as macroplastics and nanoplastics, which are of the size <100 mm. Waller et al. (2017) identified and examined that the least polluted populated areas have also been depleted with microplastic pollution, creating a hazard to the sustainability of the environment. Even in coolest areas, i.e., Arctic and Antarctic regions, contamination of micro- and nanoplastics has been compiled. The sizes and nature of microplastics affect the growth of cyanobacteria or microalgae due to their physical properties and the fact that microplastic particles contain toxins which absorb polyaromatic hydrocarbons and organic pollutants such as phenol, cyanide, and heavy metals (Law, 2017). Various literatures stated that microplastics have an adverse effect on the growth rate of cyanobacteria due to their structure and size; therefore, ecotoxicological study is required. But smaller fragments cause more growth inhibitory effect and simulation is observed (Ansari et al., 2021; Gao et al., 2022). Microplastics also have the capability to assimilate carbon into monomers and polymers as they contain PHAs for the production of metabolic and biosynthetic products by different pathways (Agarwal et al., 2022). The impact of microplastics or the fragmentation of plastics on the cyanobacterial bloom has been depicted in Figure 17.2.

17.2.2 Potential of Algae as a Source of Micro- and Bioplastics

The bioremediation process has been triggered by the fact that microplastics get attached to the algal biomass; therefore, algal enzymes interact with the fragments of monomers of plastics, microplastics, nanoplastics, and bioplastics. Different key factors play a significant role in the deterioration of microplastics, such as pH, temperature, intensity of light, aerobic source (O_2), and humidity. Bioplastics are those that are naturally or synthetically made from the microbial biomass or other renewable sources (Mekonnen et al., 2013). Further, cyanobacteria provide promising results by

FIGURE 17.2 Impact of microplastics on cyanobacteria.

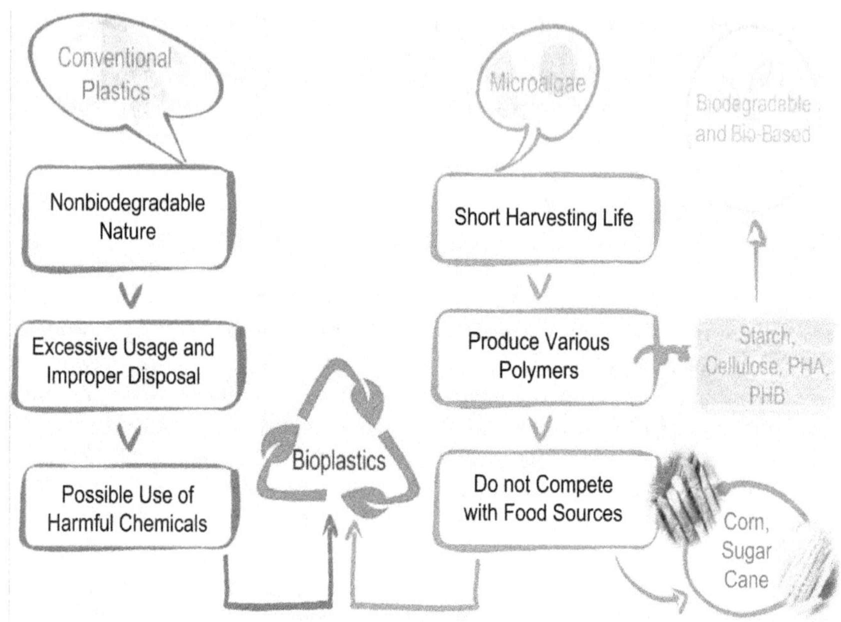

FIGURE 17.3 Cyanobacterial ability as a source of bioplastics.

the degradation of microplastics, but there few challenges for the depletion of natural sources. A cyanobacterium emerges as a potential source for the generation of micro- and bioplastics so as to create awareness and sustainability in the environment (Figure 17.3). The merits of cyanobacteria as a source of microplastics are as follows:

1. Cyanobacteria are readily available throughout the year worldwide.
2. They can withstand high temperature, light intensity, pH, and moisture content.
3. A cyanobacterium has a high resistance and needs much less time for harvesting.
4. Being a photosynthetic prokaryote, it does not need an additional micronutrient or nitrogen source (Chia et al., 2020; Cinar et al., 2020).

There are different types of bioplastics based on their nature and sources. Basically bioplastics are derived from biomass and agricultural waste sources, such as corn, starch, milk proteins, and crop-based sources.

They can be classified into degradable PHA-based plastics, non-biodegradable plastics, and fossil-based petroleum source. The production of algae-based plastic does not compete with food sources and can remediate wastewater utilizing carbon dioxide as a nutrient source. The encapsulation of these types of plastic can offer carbon capture sequestration and blending with other sources, thus developing a sustainability in the environment and eliminating toxicity (Zhang et al., 2019). The production of biofilm from algae is shown in Figure 17.4.

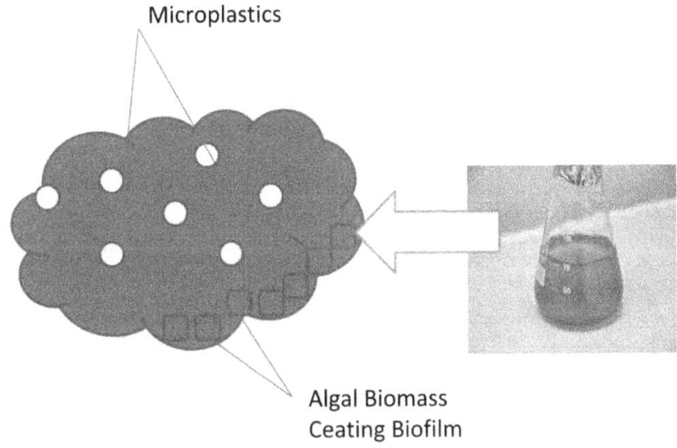

FIGURE 17.4 Biofilm generation by cyanobacteria.

17.2.3 Factors Influencing Degradation of Biodegradable Plastics from Cyanobacteria

The degradation of biodegradable plastics depends on a number of key variables, such as pH, temperature, time, inoculum dosage, concentration, presence of micro- and macronutrients, nature and structure of the microplastic, etc. This also results in weight change and also slight changes in crystalline structure, deformations. Factors that are mainly responsible are due to the characteristics of microplastics, geographical behavior, and the environmental conditions. The impact of environmental issues poses a serious threat and hazard to human health and aquatic life (Adrah et al., 2020). Non-degradable and indispensable plastics are a concern as they are an addition to toxicological nature. Random disposal and recycling microplastics inefficiently have led to the environmental pollution through various components in the past.

Several factors control the growth conditions of cyanobacterium, and changing these growth conditions affects the production of different polyaromatic and polycyclic compounds (Derner et al., 2006; Soni et al., 2017). Microplastics and biobased plastics are an alternative source of petroleum-based synthetic polymers. As surveyed through different literatures, different nutrient supplies, such as nitrogen, potassium, phosphorus, have an effect on the growth of cyanobacteria and the production of biopolymers (Zhang et al., 2019). Due to the acclimatization of phosphate with intracellular polyhydroxyalkanoates, cyanobacteria break the phosphate chains due to nutrient-deficient medium for its growth. Due to the absence of nitrogen present in the cells, a protein is deficient; therefore, these compounds are saved internally as polyhydroxybutanoates (Balaji et al., 2013). The strain of *Synechocystis sp.* is mutant and able to accumulate PHB (Schlebusch and Forchhammer, 2010). The production of adequate carbon sources and selection of material properties with highly active synergistic strains of cyanobacteria also plays a key role responsible for the degradation of microplastics (Koller et al., 2017). According to literatures by Bhati and Mallick, biopolymer degradation

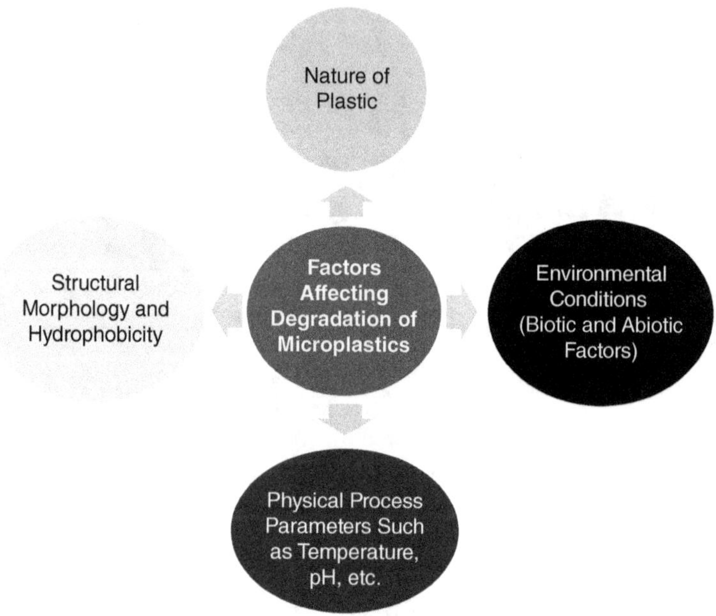

FIGURE 17.5 Factors affecting biodegradation of microplastics.

yields about 0.098 and 0.101 g/l/d biomass under deficient-nutrient conditions (Bhati et al., 2019). Factors affecting degradation of microplastics are shown in the schematic representation in Figure 15.5.

17.3 COMMERCIAL APPLICATIONS AND COST FEASIBILITY

Many advantages come with the production of micro- and bioplastics from cyanobacteria as they are economically as well as environmentally feasible. Cyanobacteria, being photosynthetic prokaryotes, have the ability to possess secondary metabolites that can be further used for the medicinal and biotechnological aspects (Zahra et al., 2022). Biodegradable microplastics, also known as bioplastics, have many commercial applications, such as the biomedical surgery, agricultural and forestry purposes, pharmaceuticals, and food industries. They can also be used as surgical material, heart stents, and bone plates for fracture purposes. Bioplastics can be utilized in the drug delivery process and genetic engineering in pharmaceutical industries and for packaging purposes in the food industries (Markl et al., 2018; Vermaas, 2019; Meixner, 2017). The global market production for polyhydroaromatics has reached to $ 57 million USD, and by 2024 it will reach $ 96 million USD with an increase of 11.2% annual growth rate (AGR) (Markets and Markets, 2019). There is an increase of 33% in global potential substitution of polymeric materials in the market (Singh and Kumar, 2018). Different factors are involved in the production of bioplastics by new technologies so as not to hamper the economic aspects (Singh and Mallick, 2017). Cyanobacterial applications and applications of microplastics in the commercial aspect are shown in Figures 17.6 and 17.7.

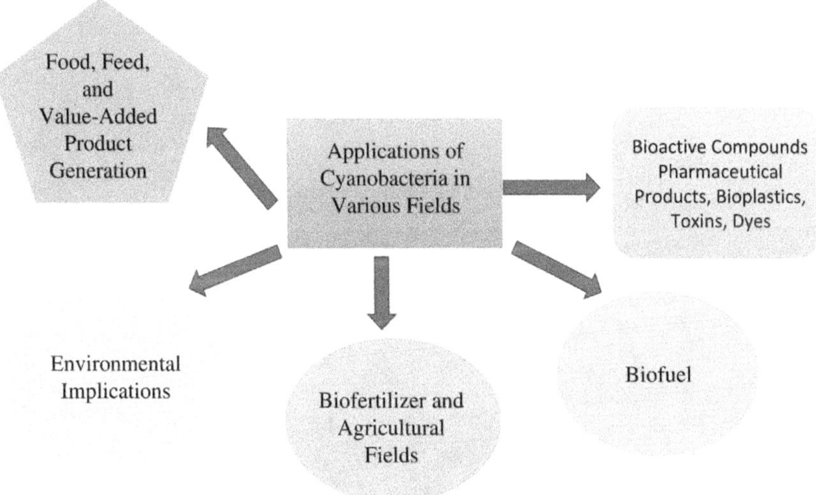

FIGURE 17.6 The diverse applications of cyanobacteria for generation of value added products.

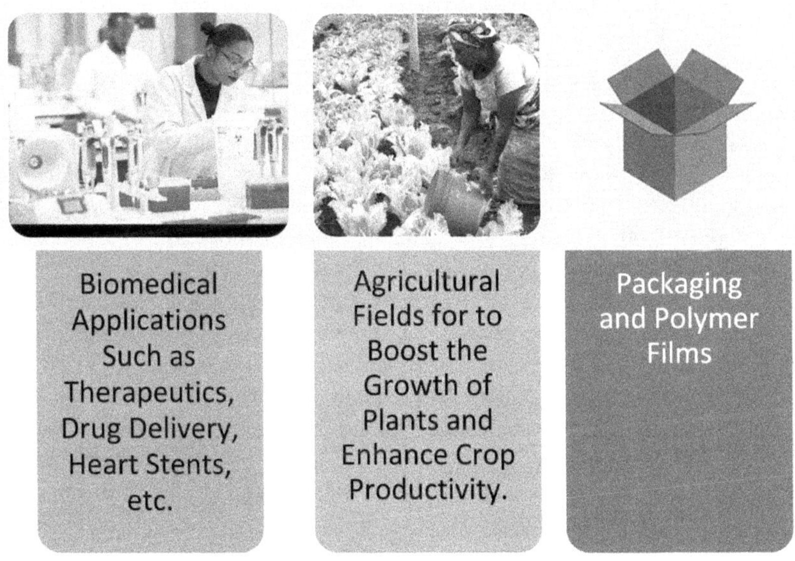

FIGURE 17.7 Cyanobacteria: as a source of biomedical, environmental and packaging purposes.

17.4 FUTURE PROSPECTS AND CONSIDERATIONS

The increasing global demand is for plastic production with enforced green technologies and lower environmental impact. The market exposure is enhanced for the biodegradable polymeric raw materials with the implementation of regulatory

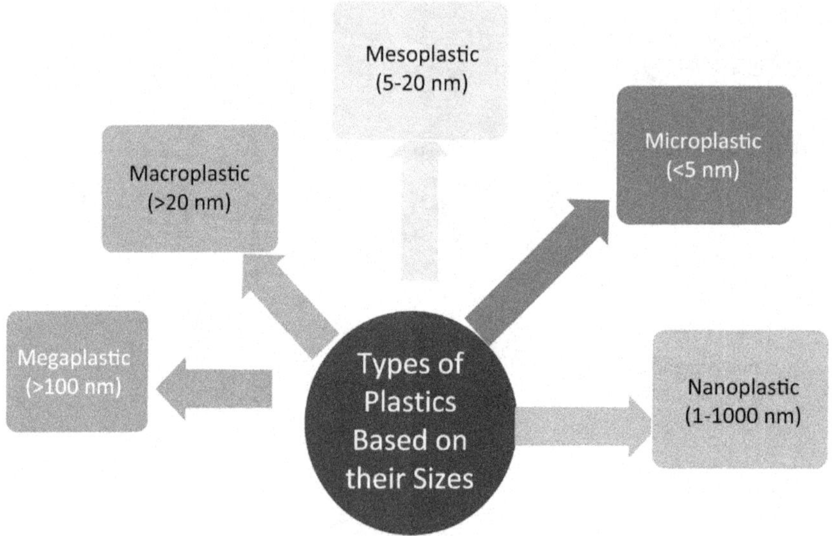

FIGURE 17.8 Different sizes of microplastics for their environmental impact.

policies. Cyanobacteria have the ability to produce bioplastics that vary with their growth parameters. Polyhydroaromatics generated from hydrocarbons is a source of reducing environmental hazards caused by fossil fuels and other synthetic polymers worldwide. Biodegradation of microplastics weaken the enzymatic bonds attached with the polymeric chains. Utilization of cyanobacteria as a source to convert microplastics fragmenting to secondary metabolites such as carbon dioxide, water, and algal biomass has been of keen interest to researchers and scientists. Among the phyla, cyanobacteria possess an environmentally feasible and sustainable option to produce biodiesel and other chemicals. An extensive research study on biodegradation of microplastics by cyanobacteria is yet to be recovered for the benefit of environment and society. Microalgae contains hydrocarbon, lipid, other value-added products are basically used for food products, so they must be devoid of organic pollutants and other pollutants, including microplastics (Khoo et al., 2020). Cyanobacteria can degrade microplastics through enzymes or toxin systems since they have the ability to synthesize themselves by producing cyanotoxins. Cyanobacteria or microalgae (blue-green algae) are the group of phyla belonging to the group of *Proteobacteria* or *Actinobacteria* sp. The economic aspect for the degradation of microplastics has been evaluated by Beckstrom et al. (2020), who intimated that the global emission of greenhouse gas would be reduced to 67–116% as this process is a very cost-effective process. Also, degradation depends up the size of plastics (Figure 17.8). Synthesis of industrial-grade polymers derived from the algal biomass is environmentally benign, and they mitigate and minimize waste disposal and management problems.

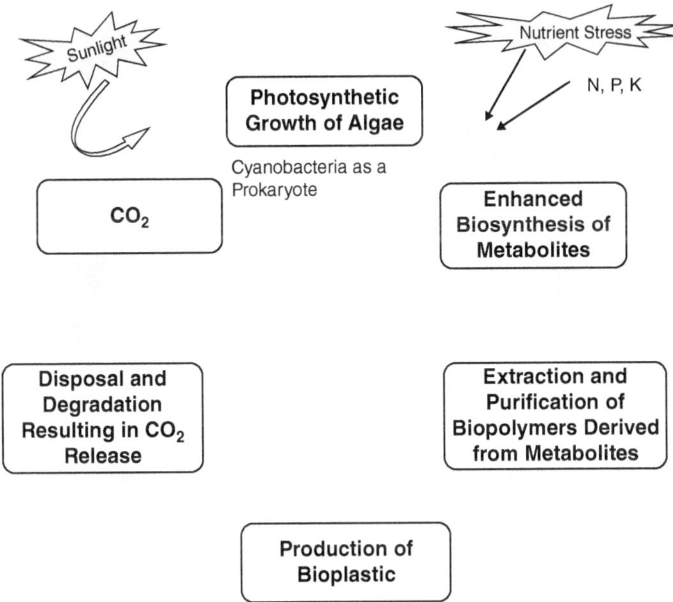

FIGURE 17.9 Role of microplastics in green technology.

17.4.1 Challenges Faced due to Utilization of Algal Micro- and Bioplastics

Several aquatic plants, animals, and many microorganisms act as sources for the production of algal micro- and bioplastics. However, certain drawbacks exist, such as limited supply of nutrients (the nitrogen, potassium, phosphorus mainly required and other macronutrients) for cyanobacterial growth, and, second, cultivation of a huge biomass. Seaweeds and marine plants produce various kinds of polysaccharides (Chea et al., 2023). Despite results obtained from the biodegradation of microplastics by cyanobacteria in the laboratory or commercial scale, there are several challenges faced due to application at industrial scale. The identification of the type of cyanobacterial species and nature of microplastics is keenly promising (presented in Figure 17.9).

The composition of algal biomass varies with the geographical indications. Table 17.1 represents the merits and demerits of algal microplastics utilization.

17.5 CONCLUSION

The decomposition of plastics nowadays has been a serious concern for the solid waste management sector. There are many chemical and physical methods for polymer degradation. But degradation by biological treatment has been an appealing option to the researchers and scientists. Cyanobacteria, also known as blue-green algae (microalgae), as a photosynthetic prokaryote and nitrogen fixer, is also becoming a

TABLE 17.1
Merits and Demerits for the Utilization of Algal Micro- and Bioplastics

Merits	Demerits
1. Energy efficient and cost-effective	1. High amount of energy required
2. Flexible to adjust	2. Brittle
3. No labor cost	3. Labor intensive

TABLE 17.2
Colonization of Cyanobacteria on Microplastics

Sl No:	Names of Cyanobacterial sp.	Nature of Microplastics	Type of Water Bodies	References
1.	*Scenedesmus dimorphus* (green alga), *Anabaena spiroides* (blue-green alga), and *Navicula pupula* (diatom)	Waste plastic bags	Lake near Chennai	Kumar et al., 2017
2.	*Oscillatoria princeps*, *O. acuminate*, *O. subbrevis*, *O. willei*, *O. amoena*, *O. splendida*, *O. vizagapatensis*, *O. limnetica*, *O. earlei*	Microplastics inside water	Waterbody near Silchar, Assam	Sarmah and Rout, 2017
3.	*O. limnetica*, *P. lucidum*, *Phormidium calcicola*, *O. earlei*, *Lyngbya cinerascens*, *Nostoc carneum*, *Nostoc linckia*, *Spirulina major*, *Hydrocoleum* sp., *Chlorella* sp., *Pithophora* sp., *Scenedesmus quadricauda*, *Calothrix fusca*, *Stigeoclonium tenue*	Polythene carry bags	Waste litters	Sarmah and Rout, 2018a
4.	*Phormidium lucidum*, *Oscillatoria subbrevis*, *Lyngbya diguetii*, *Nostoc carneum*, and *Cylindrospermum muscicola*	Waste polythenes	Waste sites near Sikhar, Assam	Sarmah and Rout, 2018b

(Continued)

TABLE 17.2 (*Continued*)
Colonization of Cyanobacteria on Microplastics

Sl No:	Names of Cyanobacterial sp.	Nature of Microplastics	Type of Water Bodies	References
5.	*Phormidium tenue*, *Oscillatoria tenuis*, *Monoraphidium contortum*, *Microcystis aeruginosa*, *Closterium constatum*, *Chlorella vulgaris*, and *Amphora ovalis*	Waste polythene and plastic bags	Ponds and lakes near Rajasthan	Sharma et al., 2014
6.	*Coleochaete scutata*, *Coleochaete soluta*, *Chaetophora*, *Aphanochaete*, *Gloeotaenium*, *Oedogonium*, *Oocystis*, *Oscillatoria*	PVC materials	Water bodies in Uttar Pradesh	Suseela and Toppo, 2007

huge source for the production of PHA and PHB (polyhydroxyalkanoates and polyhydroxybutanoates). Several key factors are responsible for the bioremediation of micro- and bioplastics by the utilization of cyanobacterial species, such as temperature, pH, time of incubation and inoculation, and light intensity, which plays a key role in the bioremediation process as discussed in this chapter. Cyanobacteria also has the ability to produce secondary metabolites. The merits and demerits of microplastics, as well as the utilization of cyanobacteria for the purpose of biodegradation, have also been reviewed.

ACKNOWLEDGMENT

All authors are thankful for the financial support by MHRD scholarship in NIT Durgapur and also the facilities and instruments provided for the research study.

CONFLICT OF INTEREST

None.

DECLARATION OF COMPETING INTERESTS

None.

REFERENCES

Adane L., Muleta D. (2011). Survey on usage of plastic bags, their disposal and adverse effects on the environment: A case study in Ethiopia. *J Toxi Health Ser.* 3(8) (234–238).

Adrah K., Ananey-Obiri D., Tahergorabi R. (2020). Development of bio-based and biodegradable plastics: novelty, advent, and alternative technology, in: O.V. Kharissova, L.M.T. Martínez, B.I. Kharisov (Eds.), *Handbook of nanomaterials and nanocomposites for energy and environmental applications*, Springer International Publishing, Cham (1–25).

Agarwal P., Soni R., Kaur P., Madan M., Mishra R., Pandey J., Singh S. (2022). Cyanobacteria as a promising alternative for sustainable environment: Synthesis of biofuels and biodegradable plastics. *Fron Microbio.* 13 (939347).

Ansari F.A., Ratha S.K., Renuka N., Ramanna L., Gupta S.K., Rawat I., Bux F. (2021). Effect of microplastics on growth and biochemical composition of microalga *Acutodesmus obliquus. Algal Res.* 56 (102296).

Balaji S., Gopi K., Muthuvelan B. (2013). A review on production of poly βhydroxybutyrates from cyanobacteria for the production of bio plastics. *Algal Res.* 2 (278–285).

Beckstrom B.D., Wilson M.H., Crocker M., Quinn J.C. (2020). Bio plastic feedstock production from microalgae with fuel co-products: A techno-economic and life cycle impact assessment. *Algal Res.* 4 (6101769).

Bhati R. (2019). Biodegradable plastics production by cyanobacteria, in: M. Khoobchandani, A. Saxena (Eds.), *Biotechnology products in everyday life, eco production*, Springer, Cham (131–143).

Chea J.D., Yenkie K.M., Stanzione III J.F., Mercado G.J.R. (2023). A generic scenario analysis of end-of-life plastic management: Chemical additives. *J Hazard Mat.* 441 (129902).

Chen W.H., Lin B.J., Huang M.Y, Chang J.S. (2015). Thermo chemical conversion of micro algal biomass into biofuels: A review. *Bioresour Technol.* 184 (314–327).

Chia W.Y., Tang D.Y., Khoo K.S., Lup A.N.K., Chew K.W. (2020). Nature's fight against plastic pollution: Algae for plastic degradation and bioplastics production. *J Env Sci Tech.* 4 (100065).

CIEL (Center for International Environmental Law). (2017). Campaign for stopping plastic pollution. Fossils, plastics, & petrochemical feedstocks, in: *Fueling plastics, center for international environment law* (1–5). https://www.ciel.org/wp-content/uploads/2017/09/Fueling-Plastics-Fossils-Plastics-Petrochemical-Feedstocks.pdf

Cinar S.O., Chong Z.K., Kucuker M.A., Wieczorek N., Cengiz U., Kuchta K. (2020) Bioplastic production from microalgae: A review. *Int J Environ Res Public Health.* 17(11) (3842).

Derner R.B, Ohse S., Villela M., Carvalho S.M.F.R. (2006). Microalgae, products e applicators. *Cienc Rural.* 36 (1959–1967).

EPA. (2020). Best practices for solid waste management: Addressing plastic waste. EPA-530-R-23-11.

Galloway T.S., Cole M., Lewis C. (2017). Interactions of micro plastic debris throughout the marine ecosystem. *Nat Ecol Evol.* 1 (1–8).

Gao H., Liu Q., Yan C., Mancl K., Gong D., He J., Mei X. (2022). Macro/micro plastics as an emerging threat effect crop growth and soil health. *Resour Conserv Recycl.* 186 (106549).

Garside, M. (2020). Production of plastics worldwide from 1950 to 2019. Statista.

Geyer R., Jambeck J.R., Law K.L. (2017). Production, use, and fate of all plastics ever made. *Sci Adv.* 3 (e1700782).

Ghosh S.K., Pal S., Ray S. (2013). Study of microbes having potentiability for biodegradation of plastics. *Int J Environ Sci Poll.* 20 (4339–4355).

Ho S.H., Zhang C., Tao F., Zhang C., Chen W.H. (2020). Micro algal torrefaction for solid biofuels production. *Trends Biotechnol.* 38(9) (1023–1033).

Kamravamanesh D., Lackner M., Herwig C. (2018). Bioprocess engineering aspects of sustainable polyhydroxyalkanoates production in cyanobacter. *Bioengineering.* 5 (111).

Khoo K.S., Chew K.W., Yew G.Y., Leong W.H., Chai Y.H., Show P.L., Chen W.H. (2020). Recent advances in downstream processing of microalgae lipid recovery for biofuels production. *Bioresour Technol.* 304 (122996).

Koller M., Maršálek L., Sousa Dias M.M., Braunegg G. (2017). Producing microbial polyhydroxyalkanoates (PHA) bio polyesters in a sustainable manner. *New Biotechnol.* 37.

Kumar A.A, Karthick K., Arumugam K.P. (2011). Biodegradable polymers and its applications. *Int J Biosci Biochem Bioinform.* 1(3) (173–176).

Kumar R.V., Kanna G., Elumalai S. (2017). Biodegradation of polyethylene by green photosynthetic microalgae. *J Biorem Biodegrad.* 8(2).

Law K.L. (2017). Plastics in the marine environment. *Ann Rev Mar Sci.* 9 (205–229).

Liquete C., Piroddi C., Drakou E.G., Gurney L., Katsanevakis S., Charef A., Egoh B. (2013). Current status and future prospects for the assessment of marine and coastal ecosystem services: A systematic review. *PLoS One.* 8 (e67737).

Luckochen G.E., Pillai S.K. (2011). Biodegradable polymers-A review on recent trends and emerging perspectives. *J Polym Env.* 19 (637–676).

Markets & Markets. (2019). Polyhydroxyalkanoates (PHA) market. Attractive opportunities in PHA market. CH1610. PHA market by type, production methods, application and region - global forecast to 2028.

Markl E., Grünbichler H., Lackner M. (2018). Cyanobacteria for PHB bioplastics production: A review. In Y. Keung Wong (Ed.), *Algae*. IntechOpen. https://doi.org/10.5772/intechopen.81536

Meixner K. (2017). Cyanobacteria for bioplastics production. *Conference: Erasmus International Week at Hochschule Darmstadt. At: Hochschule Darmstadt (GER) Affiliation: University of Natural Resources and Life Sciences, Vienna; Institute of Environmental Biotechnology.*

Mekonnen T., Mussone P., Khalil H., Bressler S. (2013). Progress in bio-based plastics and plasticizing modifications. *J Mater Chem A.* 1 (13379–13398).

Naeem S., Chazdon S., Duffy J.E., Prager C., Worm B. (2016). Biodiversity and human well-being: An essential link for sustainable development. *Proc Royal Soc B.* 283 (20162091).

Puri N, Kumar B, Tyagi H (2013). Utilization of recycled wastes as ingredients in concrete mix. *Int J Innov Technol Explor Eng.* 2(2) (74–78).

Raaman N., Rajitha N., Jayashree A., Jegadeesh R. (2012). Biodegradation of plastic by *Aspergillus sp.* Isolated from polythene polluted sites around Chennai. *J Acad Indus Res.* 1(6).

Rogovina S.Z. (2015). Biodegradable polymer composites based on synthetic and natural polymers of various classes. *J Polym Sci Ser C.* 58 (62–73).

Sarkhel R., Ganguly P., Das P., Bhowal A., Sengupta S. (2022). Synthesis of biodegradable PVA/C polymer composites and their application in dye removal. *Environ Qual Manag.* 32 (1–11).

Sarkhel R., Sengupta S., Das P., Bhowal A. (2020). Comparative biodegradation study of polymer from plastic bottle waste using novel isolated bacteria and fungi from marine source. *J Poly Res.* 27(16).

Sarmah P., Rout J. (2017). Colonisation of oscillatoria on submerged polythenes in domestic sewage water of Silchar town, Assam (India). *J Algal Biomass Utln.* 8 (135–144).

Sarmah P., Rout J. (2018a). Algal colonization on polythene carry bags in a domestic solid waste dumping site of Silchar town in Assam. *Phykos.* 48 (67–77).

Sarmah P., Rout J. (2018b). Biochemical profile of five species of cyanobacteria isolated from polythene surface in domestic sewage water of Silchar town, Assam (India). *Curr Trends Biotechnol Pharm.* 12.

Sartori R.B., Severo I.A., Santos A.M., Zepka L.Q., Lopes E.J. (2021). Biodegradable plastics from Cyanobacteria. *Mat Res Forum.* 99 (269–289).

Schlebusch M., Forchhammer K. (2010). Requirement of the nitrogen starvation-induced protein s110783 for polyhydroxybutyrate accumulation in *Synechocystis* sp. strain PCC 6803. *Appl Environ Microbial.* 76 (6101–6107).

Sharma M., Dubey A., Pareek A. (2014). Algal flora on degrading polythene waste. *CIBTech J Microbial.* 3 (43–47).

Singh A.K., Mallick M. (2017). Advances in cyanobacterial polyhydroxyalkanoates production. *FEMS Microbiol Lett.* 364 (1–13).

Singh P., Kumar R. (2018). Radiation physics and chemistry of polymeric materials, in: V. Kumar, B. Chaudhary, V. Sharma, K. Verma (Eds.), *Radiation effects in polymeric materials*. Springer International Publishing, pp. 35–68. https://doi.org/10.1007/978-3-030-05770-1_2

Soni R.A., Sudhakar K., Rana R.S. (2017). Spirulina—from growth to nutritional product: A review. *Trends Food Sci Technol.* 69 (157–171).

Stephanis R., Gimenez J., Carpinelli E., Exposito C.G., Canadas A. (2013). As main meal for sperm whales: Plastic debris. *Mar Pollut Bull.* 69(1–2) (206–214).

Suseela M., Toppo K. (2007). Algal biofilms on polythene and its possible degradation. *Curr Sci.* 92 (285–287).

Vermaas W. (2019). Production of bioplastics and other biomaterials from the cyanobacterium *Synechocystis*. Research Output: *Patent.*

Waller C.L., Griffiths H.J., Waluda C.M., Thorpe S.E., Loaiza I., Moreno B., Pacherres C.O., Hughes K.A. (2017). Microplastics in the Antarctic marine system: An emerging area of research. *Sci Total Environ.* 598 (220–227).

Yew G.Y., Khoo K.S., Chia W.Y., Ho Y.C., Law C.L., Leong H.Y., Show P.L. (2020). A novel lipids recovery strategy for biofuels generation on microalgae Chlorella cultivation with waste molasses. *J Water Process Eng.* 38 (101665).

Zahra Z., Habib Z., Hyun S., Sajid M. (2022). Nanowaste: Another future waste, its sources, release mechanism, and removal strategies in the environment. *Sustainability.* 14(4).

Zhang C., Show P.L., Ho S.H. (2019). Progress and perspective on algal plastics–a critical review. *Bioresour Technol.* 289 (121700).

18 Critical Evaluation of Remediation Methods for Microplastic Pollution in Sewage Water

Sumanta Bhattacharya

18.1 INTRODUCTION

The accumulation of microplastics in soil and water through sewage water resources diminishes agricultural productivity and lowers soil quality in rural areas. The growth of soil microorganisms, which are crucial to soil fertility, is stifled by the accumulation of microplastics in the soil. This causes a disturbance in the nitrogen and carbon cycles of the atmosphere. Microplastics from sewage irrigation could contaminate the soil over time. If these minuscule plastic particles become embedded in the soil matrix or remain on the soil surface, it could have a negative effect on soil health and the well-being of plants and soil-dwelling creatures. Runoff water from irrigated crops can introduce microplastics into local water sources such as rivers, lakes, and seas. This discharge has the potential to transport microplastics prevalent in aquatic environments to new locations. It has been established that plants can and do absorb microplastics through their roots and then transport them throughout the plant. This raises concerns that humans and other animals may be ingesting microplastics through the plants they consume. Microplastics found in streams can be consumed by fish and other marine organisms. The introduction of microplastics into the food chain can have detrimental effects on the health of these organisms. Microplastics are so tough that they can persist in the environment for decades. Sewage irrigation can release these pollutants into the environment, where they will remain for years and pose a persistent danger to wildlife and ecosystems. Researchers are examining possible solutions to the growing problem of microplastic pollution caused by the sewage irrigation. Methods such as soil supplements, filtration systems, and state-of-the-art wastewater treatment technologies may help reduce microplastic concentrations and the damage they cause to the ecosystem. The problem of microplastic pollution through sewage irrigation requires a systemic response. The concentration of microplastics in sewage can be reduced by increasing public awareness of the problem and implementing efficient wastewater treatment. Research into the long-term effects of microplastics on soil health, plant growth, and wildlife is necessary if this emerging environmental concern is to be addressed effectively.

DOI: 10.1201/9781032684574-18

Disposing of microplastics reduces soil aeration, which in turn reduces the speed at which seeds germinate and hinders the growth of roots and rhizomes [1]. The rural population, especially the poor and those with fewer resources, may not be able to make it through if agricultural output continues to fall. The presence of microplastics in sewage water complicates the process of purifying that water. This reduces the availability of high-quality water for enterprises, which in turn hinders their growth. As they are flushed down the toilet, microplastics pollute both freshwater and saltwater habitats [2]. Another harmful effect of carrying contaminated water is the production of microplastics in plant cells, which leads to a decrease in chlorophyll concentration. In these plant species, photosynthesis is inhibited. Poor water quality is a major danger to marine life. Another factor in its spread to terrestrial and wetland ecosystems is its ability to bioaccumulate and biomagnify through the transfer of mass and energy up and down food chains. Newborn cells contain microplastics, according to recent studies, which raises serious concerns about genetic alteration. [3] Because microplastics can permeate the soil, they can pollute groundwater. The harmful consequences of microplastic pollution are felt most acutely in coastal and wetland ecosystems. The invasion of microplastics into the environment and biodiversity of delta regions is a serious problem. The declines in marine water quality and in wetland agriculture have repercussions for coastal and delta economies. [4] In addition to slowing the water's overall velocity, the accumulation of microplastics inhibits the free flow of water at the surface. Pollutants in the water sample associated with microplastics cannot be eliminated by conventional filtration methods. Microplastic pollution of water sources drives up the cost of remediation. Careful scientific planning can prevent the presence of microplastics in wastewater. Switching to polymers made from biological resources or other biodegradable materials can help reduce microplastic contamination. The potential for diluted sewage water to enter natural water supplies can be mitigated through the creation of an efficient water transportation system that permits the reuse and scientific treatment of sewage water. Microplastics in wastewater can only be removed with state-of-the-art scientific and technological facilities. Microplastic formation can be hindered by switching to jute bags and other textiles made from plastic. Recycled sewage water can be used for irrigation once it has been purified to get rid of microplastics. However, if rainwater collection equipment is established, less farmland would be irrigated with potentially contaminated sewage water and river water. In agriculture, biotic and abiotic stresses can be mitigated by the use of phytoremediating plant species and genetically modified stress-tolerant plant species. The employment of AI and ML allows for the development, evaluation, simulation, monitoring, and optimization of a number of water treatment and management systems. Management of watershed ecosystem security, prevention of water pollution, and improvement of water quality are all covered. These cutting-edge tools can be utilized in the removal of microplastics from wastewater. Despite advances in technology, cleaning water samples of microplastics remains challenging. This underscores the need to fund long-term strategies to reduce microplastic manufacturing. Policies that encourage the use of bioplastics and other forms of biodegradable packaging are needed. Processing and redistributing diverse packaging goods made from vegetable, agricultural, food, and textile wastes will be beneficial to the rural economy and agricultural revenue. The village's industrial sector as a whole will also gain. By spreading information about plastic pollution and its repercussions, stakeholders like NGOs, self-help organizations, and other social organizations can help improve public awareness of the problem. Cross-sector cooperation in the development of sustainable bioplastics and alternative packaging materials can help reduce society's dependency on plastic products. Expanding state support for the establishment of agricultural and vegetable waste recycling facilities and jute processing firms will increase the economy's availability of alternative packaging materials. Additionally, mechanisms for the disposal and collection of plastic waste need to be improved to safeguard sewage water and other bodies of water from plastic pollution. Robots, drones, AI, and other state-of-the-art technologies can be used to increase the effectiveness of plastic trash collection from different bodies of water.

18.2 IMPACT OF MICROPLASTIC CONTAMINATED SEWAGE WATER

Sewage water contamination is largely attributable to people's and businesses' use of plastics. This sewage pollution affects both land and water. It has a negative impact on both terrestrial and aquatic biota. Reduced agricultural output is a direct result of microplastics' negative effect on plant growth. Through bioaccumulation and biomagnification, pollution in the terrestrial and aquatic biomes eventually reaches humans, where it causes cancer and other catastrophic genetic disorders.

As more people use products made from plastic, more trash is generated from these items. It is quite difficult to track down the microplastics that are produced from these plastic discards. Plastic's pervasive presence in modern urban life makes it an essential material. Increased plastic consumption raises the prospect of microplastic formation. Most plastic trash comes from the textile and packaging industries. One of the main sources of microplastics is the littering and attrition of textile items. Through wind flow or atmospheric deposition, these contaminants contaminate urban stormwater runoff. Microplastics end up polluting urban water supplies because rain and snow collect them. Microplastics can also come from other parts of vehicles and the road, as well as from the tires themselves. Plastics are also an indispensable part of daily life. Sewage water is contaminated by plastic waste from domestic use, such as plastic bags and other broken plastic items. Now the aquatic ecosystem in the urban catchment areas is gradually invaded by urban stormwater runoffs such as household wastewater, industrial wastewater, and atmospheric deposition on the water bodies, and then it flows to the river and urban water bodies [5]. Figure 18.1 shows the impact of microplastics on the environment and human body.

Increased plastic use in rural households and small businesses is polluting soil and water resources. The water bodies that are used as the source of irrigation in agricultural fields are contaminated by sewage water that contains microplastic particles. Furthermore, soil contamination results from the disposal of plastic trash in the soil. Due to microplastics' ability to enter soil pores, aeration is reduced. The carbon, phosphorus, and sulfur cycles are disturbed, and this has an impact on the biodiversity of soil microorganisms. Soil fertility declines when the potential for storing carbon, nitrogen, minerals, and other critical nutrients decreases. It slows down the priming of seeds and prevents new crops from growing normally. As the concentration of nutrients in the soil drops due to contamination with microplastics, the

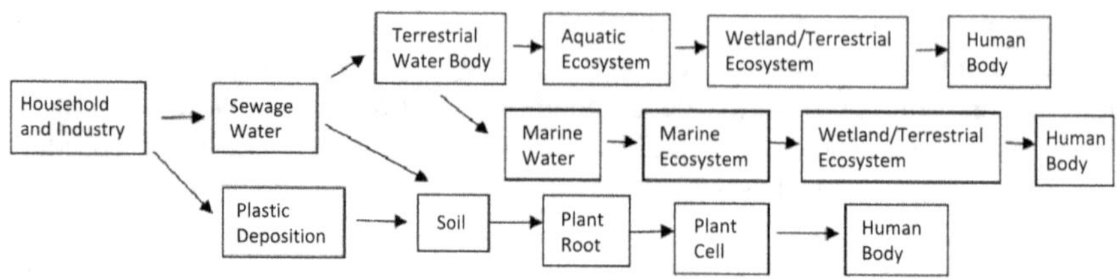

FIGURE 18.1 Contamination of microplastics in the environment and human body.

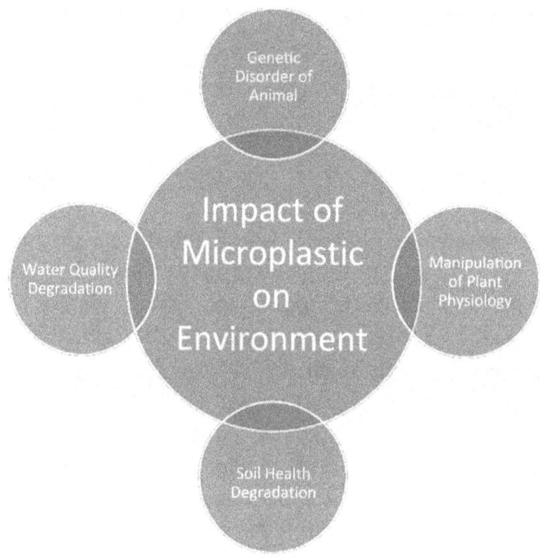

FIGURE 18.2 Impact of microplastics on environment.

bioavailability of the nutrients for the existing plants is also reduced. When the number of rhizospheric bacteria declines, less phytohormone is produced. Microplastics are ingested by plants and eventually build up in plant tissues like roots, stems, and leaves. The rate of photosynthesis is slowed when microplastics contaminate the leaves, decreasing the concentration of chlorophyll pigments. The aquatic ecology is harmed by the introduction of plastics into water sources. When it comes to rural development, the blue economy is crucial. The expansion of aquatic livelihoods, essential components of the blue economy, is stunted by microplastic contamination. Microplastic contamination also has an impact on agriculture-related operations such as pisciculture, sericulture, forest produce gathering, aquafarming, etc. [6]

Food processing, textiles, packaging, and other businesses rely heavily on agricultural produce and aquatic animals. The quality of our raw material supplies is being compromised by the presence of microplastics. The pollution of water sources contributes to the freshwater shortage that threatens a variety of sectors. Figure 18.2 depicts the impact of microplastics on the environment.

Greenhouse gases are released via anaerobic decomposition, which is slowed when microplastics accumulate in the pores of the soil. Microplastic pollution slows plant development, another factor in lowering photosynthesis rates. As a result, it alters the carbon cycle and raises atmospheric concentrations of greenhouse gases like carbon dioxide. The presence of greenhouse gases like nitrogen dioxide in the atmosphere is influenced by the rate of growth of nitrifying bacteria in the soil. Microplastics pollute soil and water because they include plasticizers, dyes, pigments, and other additives. The pollution of water supplies with microplastics reduces the amount of dissolved oxygen, making it harder for aquatic organisms to survive. [7]

Microplastics get into people's bodies via the water supply and the food chain. Microplastics have been found in the placenta and bodies of newborn babies, according to recent studies. Through biomagnification and bioaccumulation, microplastics move from polluted plant and aquatic food items to the human body. Due to their high surface charge density and absorption capacity in cells and tissues, microplastics will have more pronounced impacts on the human body. These have the ability to control cell division. The accumulation of microplastics can also reduce the blood's ability to transport oxygen. Plasticizers, dyes, flame retardants, and so on all contain substances that have been linked to major health problems, including cancer and acute genetic illnesses. Reactive oxygen species are formed when microplastics enter cells, leading to oxidative stress. It can lead to heart disease, breathing problems, and even cancer of the lungs. [8]

18.3 TECHNOLOGICAL CONSIDERATION TO PROTECT SEWAGE WATER FROM MICROPLASTICS CONTAMINATION

Removing microplastics from wastewater treatment plants relies heavily on the utilization of scientific and technological methods. The incidence of contamination is on the rise, and current technologies need to keep up with this trend as the usage of plastics in every industry becomes more commonplace. Improvements are needed in the physical, chemical, and biological methods used to filter microplastics out of wastewater. Figure 18.3 depicts the impacts on technology for the protection of sewage water from microplastic contamination.

Biocatalysts integrated into membrane filtration technology boost microplastic removal efficiency. It has been demonstrated that membrane bioreactor technology is superior to conventional filtration methods. Granular sand filtration is another viable option for cleaning wastewater with microplastic fibers. The adsorption technique is useful for cleaning up nanoplastic contamination. Applying nanotechnology to adsorbents can enhance their pore structure. Microplastics can be removed using two recently developed sorption materials: bentonite (Branany) and zeolite (clinoptilolite). [9]

Filtration through coprecipitation is much more effective when double-layered sorption materials are used. Chitin and graphene oxide have created a compressive sponge that can filter out 90% of polystyrene microplastics in water. Graphene oxide nanosheets are superior to other methods for absorbing microplastics. Aerogel, biochar, nanobiochar, and magnetic biochar are examples of bio-based materials that have been shown to be an efficient and environmentally benign technique for

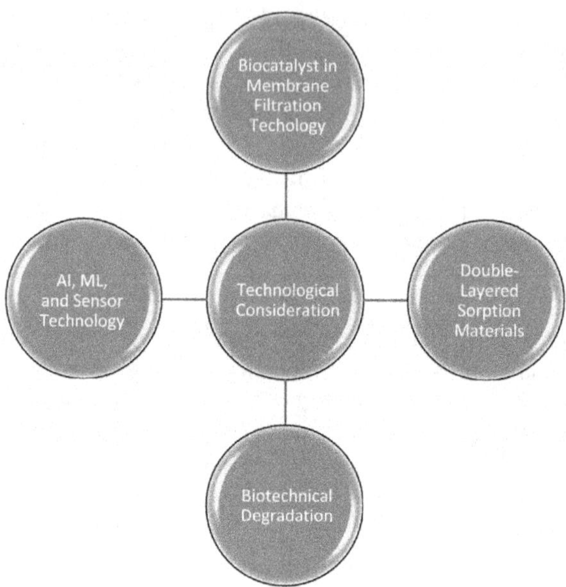

FIGURE 18.3 Technological consideration to protect sewage water from microplastic contamination.

microplastics removal. Because of their hydrophobic nature, magnetic nanoparticles can be used to effectively remove microplastics. More than 90% of microplastics in wastewater may be removed using a magnetic polyoxometalate-supported ionic liquid phase and magnetic biochar. Eco-friendly removal of microplastics from water samples is achieved by using biodegradable flocculants. It's an alternative to harmful coagulants. Recently, a flocculant with increased microplastic remediation capacity, the lysozyme amyloid fibril, has been created. The microplastics are broken down into smaller molecules via the advanced oxidation approach, which includes Fenton-based and photocatalysis technologies. [10]

Microplastic cleanup from wastewater should make use of environmentally friendly technology to cut down on pollution. Starch and biochar, two biological materials low on the environmental toxicity scale, can be employed as coagulants and adsorbents. Microplastics can be effectively removed using a green technique that combines photocatalysis based on nanomaterials with biodegradation based on bacteria. Microplastic cleanup is made more effective by the use of cutting-edge technologies like nanotechnology and biotechnology.

Artificial intelligence, machine learning, sensor technology, the Internet of Things, etc. allow for continuous and accurate water quality monitoring. Nanobiosensors and bioelectrochemical sensors can be used to better monitor water quality. Effective water quality monitoring necessitates technological advancements in the areas of pH sensors, conductivity sensors, residual chlorine sensors, turbidity sensors, dissolved oxygen sensors, ORP sensors, COD sensors, and ammonia nitrogen ion sensors. It is recommended that people utilize green sensor technology such as chlorophyll sensors and blue-green algae sensors.

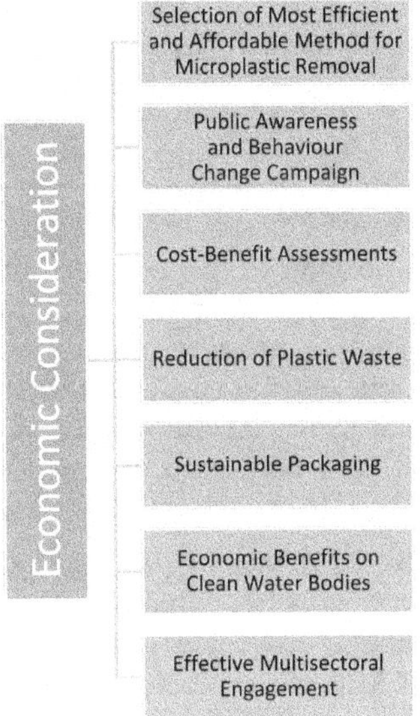

FIGURE 18.4 Economic consideration for reducing microplastic contamination in sewage water.

18.4 ECONOMIC CONSIDERATIONS FOR MICROPLASTIC ISSUES IN SEWAGE WATER

Dealing with the problem of microplastics in sewage water is complicated by financial constraints. Tiny pieces of plastic called microplastics have been shown to disrupt aquatic ecosystems. The expense of developing efficient wastewater treatment processes to eliminate microplastics is the key economic concern. It may be prohibitively expensive to upgrade or construct new facilities at existing wastewater treatment plants in order to remove these microscopic particles. There needs to be a balance struck between the monetary and ecological costs and rewards. Figure 18.4 depicts the economic consideration for reducing microplastic contamination in sewage water.

- **Technology Choice:** This is the process of choosing the most efficient (and affordable) method for microplastic removal. Filtration, sedimentation, and high-tech oxidation are only some of the options. To make rational financial choices, it is necessary to assess their usefulness and compare their prices.
- **Allocating Resources:** This is the next step toward reducing microplastic contamination, which is the responsibility of local governments and wastewater treatment bodies. Funding for study, observation, and corrective action all fall under this category. It is crucial to prioritize measures according to their positive economic and ecological effects. [11] There may be monetary costs associated with complying with environmental rules. Treatment facilities may incur greater expenses to comply with stricter rules for microplastic removal. Because of this, it's possible that user fees or taxes will increase.

- **Public Awareness and Behavior Change Campaigns:** Reducing the amount of microplastics entering wastewater systems may lower treatment costs; therefore, encouraging people to use less plastic and dispose of it correctly is important.
- **Cost-Benefit Assessment:** To assess the financial viability of various microplastic removal strategies, it is necessary to conduct cost-benefit assessments. The long-term economic and environmental benefits, in addition to the immediate expenses, should also be factored into this study.
- **Research and Innovation:** Spending money on research and development can yield more efficient and cost-effective strategies for microplastic cleanup. Such technologies can be developed and promoted through cooperation among governments, businesses, and academic organizations.
- **Waste Reduction-Removal:** Removing microplastics from sewage water is an expensive proposition compared to preventing their introduction in the first place. It can be financially beneficial to encourage companies to reduce plastic output and promote sustainable packaging.[12]
- **Economic Benefits of Mitigation:** Protecting fisheries and tourism sectors that rely on clean water bodies is only one example of the possible economic benefits of reducing microplastic contamination. These monetary gains may help mitigate some of the expenses. Since microplastics can traverse international borders via waterways, it may be vital for countries to work together to find a solution. The economic burden of nations can be lessened through cooperation with neighboring countries to share expenses and best practices.

In conclusion, there needs to be a middle ground approach that takes into account the costs and advantages of dealing with microplastic concerns in sewage water. Environmental and public health preservation needs costly mitigation measures but are worth it in the long run. The only way to effectively deal with this environmental crisis is to evaluate cost-effective solutions, launch public awareness campaigns, and work with partners throughout the world.

18.5 ENVIRONMENTAL CONSIDERATIONS

When designing long-term solutions to microplastic pollution, it's important to keep the environment in mind for the sake of both short- and long-term sustainability. Ecosystems must be protected during the planning, construction, and operation of any sustainable infrastructure. This includes avoiding building in wetland areas, on wetlands, or in locations with a high concentration of species. Planning should be informed by environmental impact assessments performed beforehand. Figure 18.5 depicts the environmental considerations of the development of sustainable infrastructure for the prevention of microplastic contamination.

18.5.1 Stormwater Management

Microplastic runoff can be prevented from entering water bodies by installing and maintaining efficient stormwater management systems. Green infrastructure

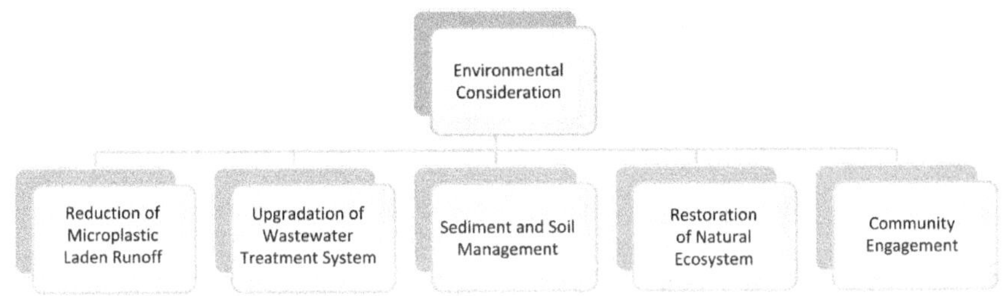

FIGURE 18.5 Environmental considerations of the development of sustainable infrastructure of prevention of microplastic contamination.

(such as permeable pavements and rain gardens) should be incorporated into sustainable infrastructure to slow down and filter stormwater, reducing the amount of microplastic-laden runoff that reaches rivers and the ocean. [13] To stop microplastics from polluting waterways, it's important to upgrade and improve wastewater treatment systems. The removal of microplastics from sewage water before releasing it should be a top priority for every sustainable infrastructure project. Plastic waste reduction strategies to decrease plastic waste at the source should be included in the building of sustainable infrastructure. The use of sustainable and biodegradable building materials is also encouraged, along with the promotion of recycling and the reduction of single-use plastics.

Initiatives to evaluate the efficacy of microplastic prevention methods should be part of a sustainable infrastructure's continuing monitoring and research initiatives. By keeping an eye on the environment on a regular basis, potential problems can be spotted early and solutions implemented. Restoring natural ecosystems is an important step to take once infrastructure construction has disturbed them. Sustainable projects may incorporate habitat restoration or augmentation to offset this effect on the local environment.

18.5.2 Sediment and Soil Management

Microplastics that have been contained in soils and sediments may be released during construction. The best way to stop this leak and lessen the amount of contamination is to use good sediment and soil management practices.[14]

The local community and other stakeholders must be involved in the design and construction of sustainable infrastructure through community engagement. Their insights can be used to spot environmental hazards and guarantee that projects respect local norms and standards. Energy-efficient designs and operations are a top priority for a sustainable infrastructure. Greenhouse gas emissions and the environmental effect of energy production can both be reduced by cutting energy use. To ensure that measures taken to avoid microplastics have lasting effects over time, infrastructure should be constructed with long-term maintenance in mind. The deterioration of infrastructure due to neglect can lead to a net negative impact on the environment.

18.5.3 Adaptation to Climate Change

Resilient, sustainable infrastructure must be built from the ground up to make adaptations to changing precipitation patterns and sea levels brought on by climate change. This can aid in keeping microplastics from entering the environment during times of high winds and rising water levels.

18.5.4 Regulatory Compliance

All environmental laws must be followed and any necessary permissions obtained. In order to protect local ecosystems, sustainable infrastructure projects must follow all applicable environmental regulations and secure all appropriate permissions. [15] Preventing microplastic contamination through the development of sustainable infrastructure calls for an integrated strategy that takes into account environmental impact at every stage of the process, from initial concept to long-term maintenance. Sustainable infrastructure may successfully reduce microplastic pollution and boost ecological health and resilience by taking into account certain environmental factors.

18.6 INDUSTRIAL IMPACT OF THE USE OF MICROPLASTICS IN THE SEWAGE SYSTEM

Microplastics' use in the wastewater treatment system can have far-reaching consequences for industry, both positive and negative. Figure 18.6 depicts the industrial impact of the use of microplastic in the sewage system.

18.6.1 Positive Impact

To begin, the use of microplastics in wastewater treatment procedures can increase the efficiency with which contaminants, such as suspended particles and organic matter, are eliminated. As flocculants or coagulants, they help particles clump together and settle, which boosts therapy effectiveness. Microplastics can remove and concentrate precious resources like metals and organic compounds from wastewater, which can then be recovered and reused. This can pave the way for new recycling and resource-recovery initiatives, which can cut down on the consumption of raw materials and even bring in money. [16] Microplastics can help sewage treatment plants run more efficiently, which in turn reduces the amount of money spent on utilities, repairs, and new facilities. The environmental impact of microplastics can be mitigated by incorporating them into sewage treatment, where they can be captured and removed before being released into natural water bodies.

18.6.2 Negatives Impact

Microplastics used in sewage treatment might cause the release of even more microplastics into the environment, leading to the first problem. These may come from the treatment process itself, in the form of microplastic-laden sludge, or from the breakdown of bigger plastic materials utilized in the treatment systems. Even the release of small amounts of microplastics into the environment can have detrimental

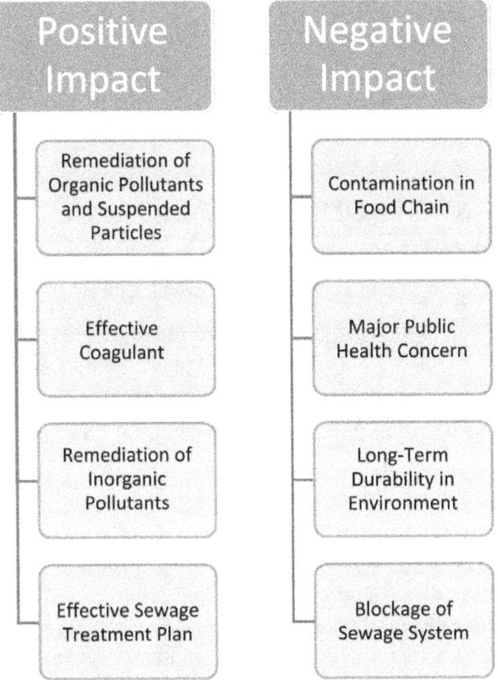

FIGURE 18.6 Industrial impact of the use of microplastic in the sewage system.

ecological effects. Aquatic species may absorb microplastics, introducing them to the food chain and potentially harming ecosystems. [17]

18.6.2.1 Public Health Concerns

Such concerns have been raised with respect to the possibility of microplastics becoming concentrated in human drinking water supplies. While researchers continue to investigate the effects of microplastic exposure on human health, caution and additional study are warranted.

There may be issues with regulation and public perception if microplastics are used in sewage treatment. Industries that employ plastics in environmental operations may feel the effects of stricter laws and public reaction.

18.6.2.2 Waste Management

It can be difficult to dispose of sludge containing microplastics that is produced during the treatment of sewage. To stop microplastics from polluting the environment, proper disposal practices must be used.

Long-term durability is an issue because, although microplastics can increase treatment effectiveness in the short term, they may degrade into even smaller particles that are more difficult to remove from water systems in the long run. There are pros and cons to using microplastics in the industrial sector, making this an extremely nuanced subject. Their potential to worsen microplastic contamination and increase environmental and health problems must be carefully examined, notwithstanding their usefulness in improving wastewater treatment and resource recovery. [18] In addition to maintaining research into alternate materials and methods for wastewater treatment, developing and implementing stringent rules, monitoring standards, and safe disposal methods are crucial for mitigating harmful effects.

18.7 ROLE OF STAKEHOLDERS AND NGOs IN MICROPLASTIC MANAGEMENT AND ITS REDUCTION IN THE SEWAGE SYSTEM

Microplastics management and attempts to minimize microplastics in the sewage system rely heavily on the participation of stakeholders and non-governmental organizations (NGOs). Figure 18.7 depicts the role of different stakeholders in microplastic management and its reduction. Their participation can fuel education, activism, investigation, and implementation. Regulation and enforcement are the responsibility of government authorities, which establish and uphold policies for microplastic management and wastewater treatment. They check for problems and make sure rules are followed. [19] Governments can allocate cash to study microplastics and their effects in order to create efficient treatment procedures. Government agencies can conduct public education program to raise awareness of the dangers of microplastics and the importance of reducing their usage and disposal. Wastewater treatment

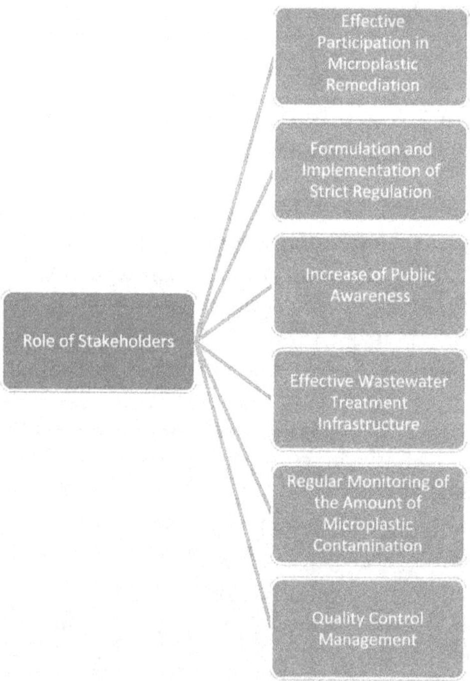

FIGURE 18.7 Role of stakeholders in microplastics management and its reduction.

plants through technology adoption for treatment plants need to invest in and adopt modern technology for microplastic removal, guaranteeing that sewage treatment procedures effectively reduce microplastics. Treatment facilities can better detect trends and enhance their removal operations if they routinely monitor effluent and influent for levels of microplastics. Businesses can play a role in reducing microplastic pollution by implementing material innovation strategies that seek to replace the use of plastics in product and packaging with more sustainable options. Industries can help reduce microplastic pollution by enacting waste reduction initiatives, recycling program, and appropriate disposal practices.

Institutions dedicated to scientific inquiry perform studies on microplastics to learn more about their origins, pathways, and effects on ecosystems and human health. Investigations like this help shape strategies and answers. They can create and improve methods of treating microplastic contamination. Campaigns, educational program, and public outreach initiatives led by NGOs are essential to increasing people's understanding of microplastic contamination and its effects [20]. They support stricter municipal, national, and international rules and restrictions on microplastics. Research and monitoring are conducted by NGOs to evaluate the concentrations of microplastics in water bodies and sediments, which aids in locating the origins and most problematic areas of pollution. To investigate the ecological and health effects of microplastics, NGOs may team together with scientists and academic institutions. Participation of the community NGOs engage local communities in cleanup operations, monitoring program, and citizen science activities, giving people a voice in microplastic management. They host workshops and events to inform the public about the importance of cutting down on plastic consumption and recycling properly. [21]

NGOs frequently work with government agencies, corporations, and other stakeholders to create and implement efficient microplastic management methods. They could work with businesses to spread awareness of sustainable methods and microplastic alternatives. Policy Advocacy—NGOs push for legislation to limit the amount of microplastics allowed in consumer goods, promote safe production and disposal methods, and fund studies of the problem. They can also support innovation by providing money and aid to startups and innovators developing innovative technology for the collection and disposal of microplastics and the minimization of waste. The efforts of stakeholders and NGOs are complementary in reducing microplastic pollution and their accumulation in wastewater treatment facilities. Microplastics have negative effects on the environment and human health, but these groups are working together to raise awareness, push for policy changes, and find effective solutions. [22]

18.7.1 Challenges

Microplastics can be found in a variety of places, including consumer goods, textiles, manufacturing processes, and even decomposing plastic trash. It's challenging to pinpoint and control all of these many origins. Microplastics typically have a diameter of less than 5 mm, making them the second smallest type of litter in the world. Due to their little size, they pose a significant challenge in removing them from wastewater.

It is possible for certain microplastics to enter the environment even after thorough sewage treatment. This can take place when wastewater is released after treatment or when sewage sludge is used as a fertilizer.

Bigger plastic products that get flushed down the toilet can eventually break down into even smaller pieces of plastic. Preventing the production of secondary microplastics requires addressing the source of these bigger plastics. Disposal and use laws for microplastics are sometimes incomplete or nonexistent, which is problem. It can be difficult to create and enforce such laws on a global scale in addition to regional and national ones. There is the issue of economic cost associated with retrofitting or upgrading sewage treatment plants to efficiently remove microplastics. Municipalities and treatment facilities face a difficult trade-off when weighing the costs of action against the benefits to the environment. The current technologies for treating wastewater may not be optimal for removing microplastics. Technological limitations creating practical, efficient, and extensible technological solutions is a never ending task. To accurately measure microplastic concentrations in sewage water, specialized equipment and procedures are required, making monitoring and measurement difficult. The development of standardized monitoring procedures continues. Despite increased public awareness of the issue, it is still difficult to achieve widespread behavioral adjustments to reduce microplastic pollution at the source. Due to the ease with which microplastics can be transported through water, international cooperation is of the utmost importance in the fight against this global problem. International and regional cooperation is difficult but essential. [23] Microplastics have been shown to have negative effects on human health and the environment, although the full scope of these effects is still being studied. It is essential for policymaking and public participation that these effects be understood and quantified. In getting rid of sewage sludge that could be full of microplastics, it's critical to find ways to dispose of waste that don't harm the environment.

Reducing microplastic pollution significantly will require political will and dedication. Policymakers need to make addressing this problem a top priority and invest sufficient resources.

18.7.1.1 Alternative Materials

It can be difficult to find viable replacements to microplastics for use in a wide range of products and industrial processes, as these alternatives must meet certain functional and safety requirements. [24] It will need a concerted effort by governments, businesses, academics, NGOs, and the general public to overcome these obstacles. To make real headway in cleaning up the sewage system of microplastics and protecting the environment and human health, we need to work together, think creatively, and come up with sustainable solutions.

18.7.2 Future Prospective

Effectively tackling this environmental issue requires a long-term perspective on the role of governance and policy in reducing microplastics in the sewage system. The future direction required for the protection of sewage water from microplastic contamination has been depicted in Figure 18.8. Here are some major factors to think about, along with probable future directions:

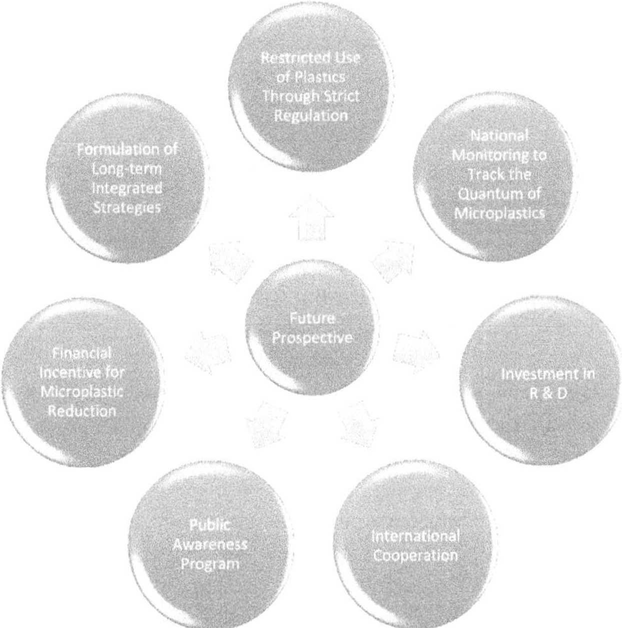

FIGURE 18.8 Future prospective.

- **Government Mandates:** Governments can impose tougher restrictions on the manufacturing and distribution of items that include microplastics. The use of microplastics in cosmetics, household cleaners, and other items could be prohibited or subject to rigorous regulations. Governments and businesses should collaborate on setting guidelines to limit the spread of microplastics during production and promote the adoption of more sustainable materials. To encourage sewage treatment facilities to invest in efficient removal methods, it is important to establish and enforce wastewater discharge criteria that are tailored to microplastics. By mandating that businesses and wastewater treatment facilities record the amount of microplastics they release, governments may ensure that their actions are transparent and accountable. Microplastic concentrations in water bodies are an important indicator of the environmental health of a region; hence it is important to establish national monitoring program to track these levels.
- **Government Funding:** Governments can provide financing for research into the origins, destinations, and effects of microplastics, as well as the creation of cutting-edge cleanup methods. Grants and incentives for new ideas can hasten the introduction of eco-friendly materials and techniques that use fewer plastic microbeads in their production.
- **International Cooperation:** Since microplastics can flow across national borders, international cooperation is crucial. Regulations and best practices can be harmonized through international and regional agreements between governments. To better coordinate a worldwide response to the problem of

microplastic pollution, it is important to support program that encourage the sharing of research findings and data in this area.

- **Informing and Educating the General Public:** Public awareness program are needed to inform the public about microplastics and the ways in which they can help in the fight against this problem. Education about environmental issues, such as microplastic pollution, should be a part of the school curriculum from a young age.
- **Rewards and Sanctions:** Financial incentives, such as tax credits or subsidies, can be offered by governments to businesses that implement microplastic reduction strategies. Establishing sanctions for non-compliance with legislation regarding microplastics can dissuade both businesses and individuals from contributing to pollution. Initiatives for a more circular economy by limiting the use of single-use plastics, increasing recycling rates, and funding research and development into more sustainable product options. Governments can allocate cash to improve wastewater treatment facilities' capacity to remove microplastics by upgrading and retrofitting them. Governments can design long-term, integrated strategies for microplastic management that take into account both sewage systems and other sources of microplastics, like runoff and industrial discharges.
- **Working Together with Non-Profits and Other Stakeholders:** To effectively reduce microplastics, governments can work with NGOs, companies, researchers, and other stakeholders through engagement. Governance and regulatory initiatives that put environmental sustainability and public health first are crucial to the future of microplastic reduction in sewage systems. Microplastic pollution in sewage systems has negative effects on ecosystems and human health, but these can be lessened with the correct regulations and concerted efforts.

18.8 CONCLUSION

The microplastic contamination in water becomes an important challenge for public health, environmental protection, biodiversity maintenance, and the ecosystem. It hampers the terrestrial as well as the aquatic ecosystems. The bioaccumulation and biomagnification of the contaminants cause severe health issues like cancer and genetic disorders in the human body. The research in the field of wastewater mechanisms and the development of eco-friendly alternatives of plastic is still in initial stage. The developing nations are still not able to finance these infrastructures properly. The training of the workforce in collection, segregation, and processing of waste products is necessary to reduce the contamination of plastics with water and soil. Moreover, further research is required to tackle these issues effectively due to the increase in contamination day by day. The research in the field of sustainable polymers will help to innovate new eco-friendly alternatives of the plastics. The technological modification of the wastewater treatment mechanism is also necessary for this purpose. The application of cutting edge technologies like nanotechnology, biotechnology, genetic engineering, analytical chemistry, etc. is necessary for improving the wastewater mechanism.

REFERENCES

1. Issac, M. N.; Kandasubramanian, B. (2021); Effect of microplastics in water and aquatic systems; Springer Link, *Environmental Science and Pollution Research* 28, 19544–19562; DOI: https://doi.org/10.1007/s11356-021-13184-2
2. United Nations Environment Programme (n.d.); Plastic planet: How tiny plastic particles are polluting our soil; www.unep.org/news-and-stories/story/plastic-planet-how-tiny-plastic-particles-are-polluting-our-soil#:~:text=Microplastics%20can%20also%20interact%20with,Science%20Daily%20about%20the%20rcsearch
3. The Gurdian (December 22, 2020); Microplastics revealed in the placentas of unborn babies; www.theguardian.com/environment/2020/dec/22/microplastics-revealed-in-placentas-unborn-babies
4. Kumar, R.; Sinha, R.; Rakib, Md. R. J.; Padha, M.; Ivy, N.; Bhattacharya, S.; Dhar, A.; Sharma, P. (2022); Microplastics pollution load in Sundarban delta of Bay of Bengal; Elsevier, *Journal of Hazardous Materials Advances* 7, 100099; DOI: https://doi.org/10.1016/j.hazadv.2022.100099
5. Osterlund, H.; Blecken, G.; Lange, K.; Marsalek, J.; Gopinath, K.; Viklander, M. (2023); Microplastics in urban catchments: Review of sources, pathways, and entry into stormwater; Elsevier, *Science of the Total Environment* 858 (1); DOI: https://doi.org/10.1016/j.scitotenv.2022.159781
6. Okeke, E. S.; Okoye, C. O.; Atakpa, E. O.; Ita, R. E.; Nyaruaba, R.; Mgbechidinma, C. L.; Akan, O. D. (2022); Microplastics in agroecosystems-impacts on ecosystem functions and food chain; Elsevier, *Resources, Conservation and Recycling* 177, 10591; DOI: https://doi.org/10.1016/j.resconrec.2021.105961
7. Corcoran, P. L. (2020); *Degradation of Microplastics in the Environment; Handbook of Microplastics in the Environment*, Springerlink, pp. 1–12.
8. Thakur, S.; Mathur, S.; Patel, S.; Paital, B. (2022); Microplastic accumulation and degradation in environment via biotechnological approaches; MDPI, *Water* 14 (24), 4053; DOI: https://doi.org/10.3390/w14244053
9. Spacilova, M.; Dytrych, P.; Lexa, M.; Wimmerova, L.; Masin, P.; Kvacek, P.; Solcova, O. (2023); An Innovative sorption technology for removing microplastics from wastewater; MDPI, *Water* 15 (5), 892; DOI: https://doi.org/10.3390/w15050892
10. Gao, W.; Zhang, Y.; Mo, A.; Jiang, J.; Liang, Y.; Cao, X.; He, D. (2022); Removal of microplastics in water: Technology progress and green strategies; Elsevier, *Green Analytical Chemistry* 3, 100042; DOI: https://doi.org/10.1016/j.greeac.2022.100042
11. Fältström, E.; Anderberg, S. (2020); Towards control strategies for microplastics in urban water, *Environmental Science and Pollution Research* 27, 40421–40433; DOI: https://doi.org/10.1007/s11356-020-10064-z
12. Kumar, R.; Verma, A.; Shome, A.; Sinha, R.; Sinha, S.; Jha, P. K.; Kumar, R.; Kumar, P.; Shubham; Das, S.; et al. (2021); Impacts of plastic pollution on ecosystem services, sustainable development goals, and need to focus on circular economy and policy interventions, *Sustainability* 13, 9963; DOI: https://doi.org/10.3390/su13179963
13. Müller, A.; Österlund, H.; Marsalek, J.; Viklander, M. (2020); The pollution conveyed by urban runoff: A review of sources, *Science of the Total Environment*, 709; DOI: https://doi.org/10.1016/j.scitotenv.2019.136125; ISSN 0048–9697
14. UNEP (2021, April); A beginner's guide to ecosystem restoration; www.unep.org/news-and-stories/story/beginners-guide-ecosystem-restoration
15. Griggs, G.; Reguero, B. G. (2021); Coastal adaptation to climate change and sea-level rise, *Water* 13, 2151; DOI: https://doi.org/10.3390/w13162151
16. Kwon, H. J.; Hidayaturrahman, H.; Peera, S. G.; Lee, T. G. (2022); Elimination of microplastics at different stages in wastewater treatment plants, *Water* 14, 2404; DOI: https://doi.org/10.3390/w14152404

17. UNEP (2018, August); Wastewater treatment plants—a surprising source of microplastic pollution; www.unep.org/news-and-stories/story/wastewater-treatment-plants-surprising-source-microplastic-pollution
18. Park, H.; Park, B. (2021); Review of microplastic distribution, toxicity, analysis methods, and removal technologies, *Water* 13, 2736; DOI: https://doi.org/10.3390/w13192736
19. Onyena, A. P.; Aniche, D. C.; Ogbolu, B. O.; Rakib, M. R. J.; Uddin, J.; Walker, T. R. (2022); Governance strategies for mitigating microplastic pollution in the marine environment: A review, *Microplastics* 1, 15–46; DOI: https://doi.org/10.3390/microplastics 1010003
20. Roy, P.; Mohanty, K. A.; Misra, M. (2022); Microplastics in ecosystems: Their implications and mitigation pathways, *Environmental Sciences: Advances*; DOI: https://doi.org/10.1039/D1VA00012H
21. Kumar, R.; Verma, A.; Shome, A.; Sinha, R.; Sinha, S.; Jha, P.K.; Kumar, R.; Kumar, P.; Shubham; Das, S.; et al. (2021); Impacts of plastic pollution on ecosystem services, sustainable development goals, and need to focus on circular economy and policy interventions, *Sustainability* 13, 996; DOI: https://doi.org/10.3390/su13179963
22. Mitrano, D. M.; Wohlleben, W. (2020); Microplastic regulation should be more precise to incentivize both innovation and environmental safety, *Nature Communications* 11, 5324; DOI: https://doi.org/10.1038/s41467-020-19069-1
23. UNEP (2023, April); Microplastics: The long legacy left behind by plastic pollution; www.unep.org/news-and-stories/story/microplastics-long-legacy-left-behind-plastic-pollution
24. Prata, J. C.; Silva, A. L. P.; da Costa, J. P.; Mouneyrac, C.; Walker, T. R.; Duarte, A. C.; Rocha-Santos, T. (2019); Solutions and integrated strategies for the control and mitigation of plastic and microplastic pollution. *International Journal of Environmental Research and Public Health* 16, 2411; DOI: https://doi.org/10.3390/ijerph16132411

19 Investigation of Prevailing Directives Regarding Microplastic Pollution

Research Gaps and Recent Advancements

Arkajyoti Mukherjee, Ved Prakash Ranjan, Gourav Dhar Bhowmick, and Papita Das

19.1 INTRODUCTION

Microplastics (MPs), an emerging environmental pollutant, are composed of synthetic polymers with dimensions of all particles ≤5 mm, and they can be produced both purposely and by dilapidation of bulky plastics (Lechthaler *et al.*, 2021). Globally, the annual discharge of MPs is assessed to be higher than 3 million tonnes, and by 2030, the quantity of MPs is expected to be doubled (Hale *et al.*, 2020). Microplastics can be found in all environmental compartments, and their persistent nature, easy transportability, and potency to cause negative ecological impacts entice the attention of the global scientific community (Usman *et al.*, 2022). Different types of MPs based on their shape, origin, and composition are represented in Figure 19.1.

OECD (Organisation for Economic Co-Operation and Development) has classified MPs into four extensive classes: (1) plastic pellets, flakes, and powders (MPs used for manufacturing plastic products), (2) deliberately added primary MPs (like microbeads in cosmetic products, polymer-coated agricultural products), (3) use-phase secondary MPs (like polymer from degraded tires, microfiber from synthetic textiles, degraded paints, residues from fisheries and aquaculture), and (4) degradation-based secondary MPs (like weathering of large and lost plastics mainly owing to generation by solar UV radiation, which is usually unintentional in the recycling sector) (OECD, 2021). Moreover, Usman *et al.* (2022) reported that secondary microplastics, specifically degraded ones, are the primary reason behind marine plastic pollution. MPs can also be categorized by their form and chemical composition (Picó and Barceló, 2019). MPs can commonly be found as fibers, fragments, and spherical beads. Both in water medium and biota, fibers are the most prevalent types of MPs (Akdogan and Guven, 2019).

DOI: 10.1201/9781032684574-19

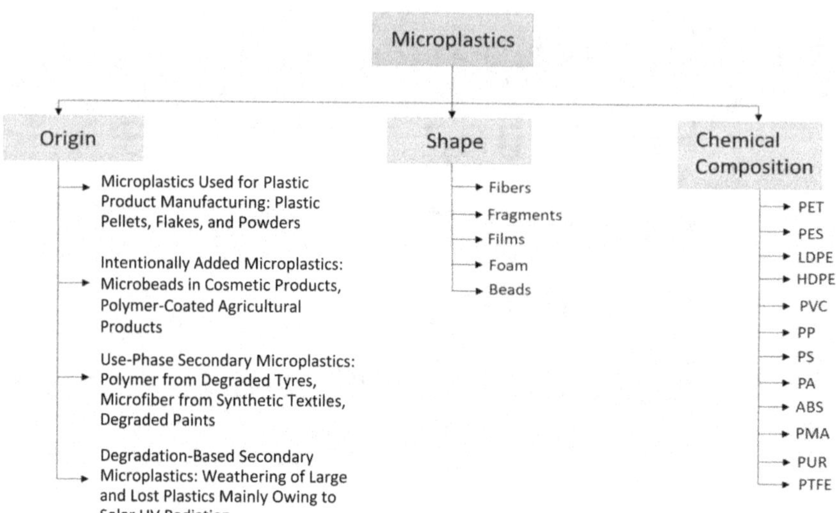

FIGURE 19.1 Types of microplastics.

Legend: PET: polyethylene terephthalate, PES: polyester, LDPE: low-density polyethylene, HDPE: high-density polyethylene, PVC: polyvinylchloride, PP: polypropylene, PS: polystyrene, PA: polyamide, ABS: acrylonitrile-butadiene-styrene, PMA: polymethyl acrylate, PUR: polyurethane, PTFE: polytetrafluoroethylene.

Polyethylene terephthalate (PET), polyester (PES), low-density polyethylene (LDPE), high-density polyethylene (HDPE), polyvinylchloride (PVC), polypropylene (PP), polyamide (PA), polystyrene (PS), acrylonitrile-butadiene-styrene (ABS), polytetrafluoroethylene (PTFE), polymethyl acrylate (PMA), and polyurethane (PUR) are the common polymers of MPs (Silva *et al.*, 2018). These polymers have dissimilar relative densities and life spans. United Nations Environment Programme (UNEP) (2018) assessed the major sources of MPs and their report determined that automotive tire debris, city dust, paints, and synthetic textiles are the foremost origin of MPs in the milieu. MPs from tires can be produced throughout their life cycle from manufacturing and use (tire and road wear particles) to the end-life stage.

These minuscule particles have inordinate impacts on the ecological integrity and functioning of both aquatic and terrestrial ecosystems. However, its impacts are more prominent in aquatic counterparts (Osman *et al.*, 2023). MPs, with great specific surface area, can absorb a wide array of contaminants like POPs, heavy metals, and antibiotics (Zhou *et al.*, 2021). This property of these pollutants aggravates their negative impacts on aquatic ecosystems as they serve as vectors for contaminant transmission to living organisms. Biota associated with aquatic ecosystems can be exposed to MPs through five major pathways: ingestion, inhalation, absorption, physical contact, and trophic transfer (Nordic Council of Ministers, 2022). The presence of MPs in diverse body regions, including the lungs, bloodstream, and heart, indicates the extensive exposure they have undergone (Central Pollution Control Board, 2023). Moreover, mostly owing to its persistent nature, MPs can bioaccumulate themselves in different trophic levels of aquatic food chains (Osman *et al.*, 2023).

Accumulation of these contaminants by aquatic faunae can affect the digestive and immune systems of these animals and eventually cause their death (Andrady, 2011). Additionally, human health can also be directly affected by consuming contaminated fish or other aquatic organisms. Humans can accumulate MPs not only by consuming food but also by drinking water and by inhalation from the air (Yong et al., 2020). ICMR (Indian Council of Medical Research) reported the presence of MP from samples of human blood, lungs, placenta, stool, saliva, and brain (Central Pollution Control Board, 2023). ICMR also attested to the genotoxic and cytotoxic effects of MP in humans and reported that MP can be a source of oxidative stress in humans. Topical research by Yang *et al.* (2023) fascinatingly reported traces of MPs from the heart tissues of humans. Thus, studies on the ecotoxicology of MPs are getting attention from global scientific communities.

India is one of the chief manufacturers of plastic surplus, and thus Indians are more prone to the negative impacts of MP pollution (Vaid *et al.*, 2021). In India, tire abrasion and city dust are the foremost sources of MPs in the atmosphere and hydrosphere, followed by textile washing and road markings (United Nations Environment Programme, 2018). In a painstaking review, Vaid *et al.* (2021) conveyed the presence of MPs from a wide array of consumables, from seafood to tap water.

Nevertheless, the effects of MPs on terrestrial and aquatic food chains are not well understood. Many countries, including India, have adopted many policies to curtail MP concentrations in the environment. Existing policies primarily focus on the enclosure of both lawfully obligatory and voluntary methods and justifiable usage and manufacture of plastics (Nordic Council of Ministers, 2022). However, owing to growing demand and uses, more stringent policies are needed to curtail MP contamination in different mediums. Studies on different restraining policies and future directives are sparse. To bridge this knowledge gap, this chapter brings together disseminated information on the prevailing policies for curtailing MP pollution around the globe. This chapter also recommends the probable sustainable future policies for maintaining ecological balance.

Further, this chapter specifically aims to (1) identify the possible sources of MP in terrestrial and aquatic ecosystems; (2) reveal the threats on plants and animals including humans and assess the ecological implications of MPs in different ecosystems; (3) study the prevailing remediation strategies and current policies (both global and national) for MP elimination; (4) study the research gaps and recent advancements in remediation strategies; (5) recommend future directives in policymaking for MP removal. It is hypothesized that this chapter can shed light on the sources and toxic effects of MPs and will help ecologists, policymakers, and conservationists in framing policies to control and remove MPs from ecosystems and will also enable the researchers to envisage its effects on community well-being and the milieu.

19.2 SOURCES OF MP

An *et al.* (2020) stated that it was very tough to recognize the precise sources of MPs in the environment. However, the sources of MPs can be classified into primary and secondary sources (An *et al.*, 2020) and land- and ocean-based sources (Osman *et al.*, 2023). The sources of MPs are depicted in Figure 19.2.

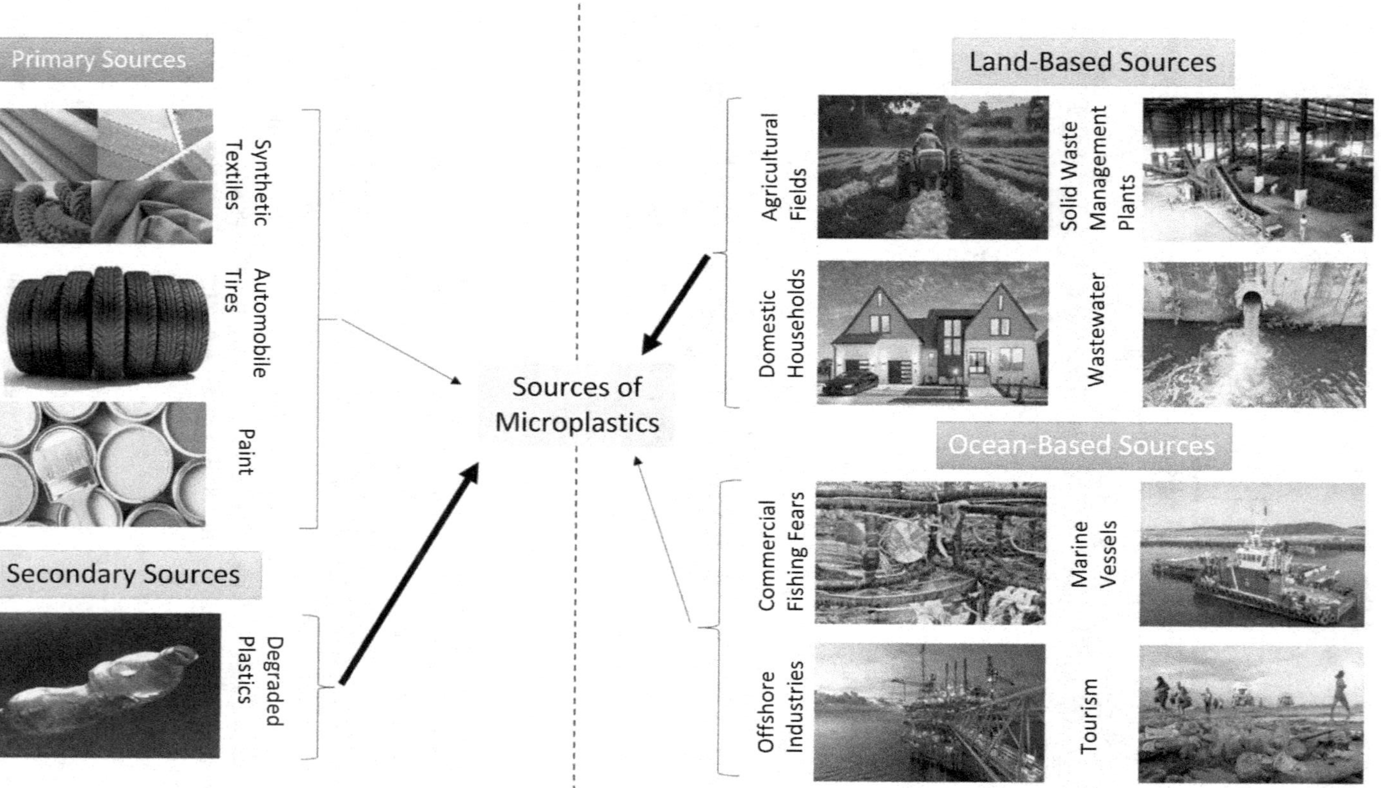

FIGURE 19.2 Sources of microplastics.

Primary sources are unswervingly released into the setting as tiny particles. Cleaning synthetic textiles and abrasion of automobile tires are two primary sources. However, these are estimated to represent only 15–31% of total marine microplastics (Roy *et al.*, 2022). On the contrary, secondary sources are major contributors of marine microplastics, and these sources are comprised of degraded larger plastic objects. Land-based sources include plastic utensils, construction materials, synthetic textiles, and personal care goods (Yang *et al.*, 2021), and these sources are primarily accountable for the MPs in aquatic ecosystems. Agricultural fields, domestic households, wastewater, and solid waste management plants are also possible land-based sources of MPs (Nordic Council of Ministers, 2022). Conversely, ocean-based sources include marine vessels, tourism, commercial fishing equipment, and seaward industries (Karbalaei *et al.*, 2019). Lost or degraded commercial fishing equipment (nylon nets and monofilament lines) are the primary ocean-based sources of MPs, and this also causes the death of many marine animals.

19.3 EFFECTS AND ECOLOGICAL IMPLICATIONS OF MPs POLLUTION

MPs can move between five earth systems by natural processes like wind or water movement and/or by trophic transfers by biota, and they exert a negative impact on the environment and biotic communities due to their heterogeneous and persistent nature. The impact of MPs in a specific ecosystem depends on several factors like chemical composition, concentration, age of MPs, ability to transfer within mediums, and receiver ecosystems (Nordic Council of Ministers, 2022). Organisms are frequently exposed to MPs through ingestion, inhalation, physical contact, absorption, and trophic transfer, and among these pathways, ingestion is the major pathway of MP exposure (Osman *et al.*, 2023). Consequences of MP exposure in biota and ecosystems are clearly depicted in Figure 19.3.

Ma *et al.* (2020) classified MP toxicity in organisms into three major categories: physical impairment owing to accretion within the digestive tract, disturbance in energy budget, and exposure of interior tissues due to translocation within the body. These toxic effects can ultimately lead to neurotoxicity, cytotoxicity, and immunotoxicity. Mohammad and Qusay (2023) classified the toxic effects of MPs into three distinct categories, namely structure-based toxicity, physicochemical toxicity, and microorganism toxicity. Moreover, MPs can act as a carrier of other contaminants due to the aquaphobic exterior and large surface-area-to-volume ratio (Nabi *et al.*, 2022). Combined with other xenobiotics, MPs can show higher levels of toxicity in organisms (Singh *et al.*, 2017).

Humans, being in the top tier of trophic levels, are more prone to MP accumulation and toxicity and both polluted water and food sources can act as potent sources of MP contamination (Osman *et al.*, 2023). Additionally, humans can acquire MPs from the air by inhalation. However, in humans, contaminated seafood is the major contributing agent of MP contamination (Toussaint *et al.*, 2019). Blackburn and Green (2022) recorded cytotoxic and carcinogenic effects of MPs in humans, and they can also cause mutation in deoxyribonucleic acid (DNA) inducing cancers. Osman *et al.* (2023) stated that in humans, MPs at 10 μg/mL and 20 μg/mL concentrations could

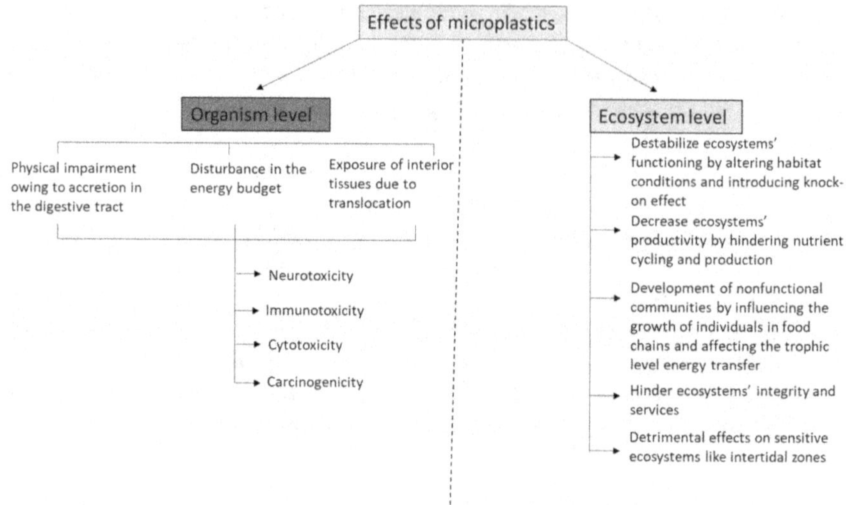

FIGURE 19.3 Effects of microplastics on organization and ecosystem level.

induce cytotoxic and immunological responses, respectively. MPs can also induce redundant changes in gut microbial flora and that might cause negative impacts by changing gut permeability and metabolism (Salim *et al.*, 2013). Pulmonary cytotoxicity and inflammation can also be caused by inhalation of MPs, and this cytotoxicity results from the production of ROS (Halimu *et al.*, 2022). Moreover, MPs can also induce COPD (chronic obstructive pulmonary disease) by impairing pulmonary barriers. Inhalation of MPs in humans can cause a wide array of diseases like cardiovascular diseases, asthma, hypersensitivity, and autoimmune ailments (Campanale *et al.*, 2020).

MPs have the potential to cause direct or indirect negative effects on not only organisms but also ecosystems. MPs in high concentration can potentially destabilize ecosystems' functioning and integrity, and these perturbations are more prominent in aquatic ecosystems (Ma *et al.*, 2020). MPs can induce changes in ecosystem functioning by altering habitat conditions and introducing knock-on effects. Recently, owing to anthropogenic activities, MPs have increased at exponential rates leading to a significant decrease in bionetwork efficiency by hindering nutrient production and cycling. This impacts the food web structures of aquatic ecosystems (Wright *et al.*, 2013). MP influences the growth of individuals in food chains and affects the trophic-level energy transfer, leading to non-functional communities. Owing to bioaccumulation and biomagnification properties, MPs have more impact on the organisms at higher trophic levels. Green (2016) assessed that MPs decreased the abundance of European flat oysters nearly 1.5 times. Effects of MPs are more protruding in sensitive ecosystems like intertidal zones owing to their proximity to MP sources (Ma *et al.*, 2020). Similarly, MPs hinder the ecosystem services rendered by the mangrove ecosystems.

19.4 REMEDIATION STRATEGIES OF MPs

Potential remediation strategies usually focus on the aquatic ecosystems as maximum MPs can be found in these ecosystems and the removal of those MPs is the key challenge to mitigate environmental pollution. Remediation strategies can be generally subdivided into three broad categories: physical, chemical, and biological processes (Tursi *et al.*, 2022). Details of some regularly used methods are described next.

1. **Filtration Methods:** These methods are widely used methods for MP separation globally, and some frequently used methods are rapid sand filtration (RSF), disc filtration (DF), and membrane-based methods (e.g., reverse osmosis and membrane bioreactors) (Poerio *et al.*, 2019). Hidayaturrahman and Lee (2019) stated that RSF is a feasible method for MP removal from water as it has a high flow rate and less sensitivity to water quality.

   The fundamental concept of the membrane filtration method is to use the selective permeability of membrane pores for segregation (Lu *et al.*, 2023). The variance of pressure among the two sides of the membrane facilitates the separation and solvents, mineral ions can pass through the pores while MPs cannot pass. As MPs are hydrophobic and negatively charged, attractive polar forces are neutralized by repulsive electrostatic forces from membrane surface charge of MPs (Liu *et al.*, 2021). Moreover, Enfrin *et al.* (2020) classified this method into three distinct categories based on the particle sizes of MPs in aquatic mediums, namely nanofiltration, ultrafiltration, and microfiltration. Membrane materials, extent of membrane pore, source of contaminated water, flux, and transmembrane pressure are significant factors influencing filtration performance (Poerio *et al.*, 2019).
2. **Adsorption Method:** Adsorption is also a frequently used method for MP removal, and this technique has superior efficiency in MP elimination from wastewater. Being a cost-effective and energy-saving method, adsorption has great potential for separating MPs from contaminated mediums (Osman *et al.*, 2023). This technique can be classified into physical and chemical adsorption (Lu *et al.*, 2023). Physical adsorption works through the interaction of intermolecular forces and shows fast adsorption rates. However, the non-selective features of this procedure are the main downsides of the procedure.

   On the contrary, chemical adsorption is more selective, and it involves the formation and breaking of chemical bonds (Wang and Guo, 2020). Adsorbent capacity (quantity of absorbate removed per unit mass) and adsorbent efficiency (percentage of adsorbate removed) are the imperative parameters to assess the adsorption performance (Lu *et al.*, 2023). Sponges, made of chitin and graphene oxide, effectively remove MPs from wastewater with an efficiency of 70–90%, and this adsorption process is optimal at a slightly acidic range (Badola *et al.*, 2022). The advantage of these sponges is their biodegradable, reusable, and biocompatible nature. Another improved adsorbent is magnetic carbon nanotubes (M-CNTs), which efficiently remove

PET, PE, and PA from wastewater (Chellasamy *et al.*, 2022). The effective ability of reprocessing is the major desirable point of this adsorbent.

Biochar filters utilizes both adsorption and filtration and separate MPs by absorbing and tangling them on their exterior (Siipola *et al.*, 2020). Biochar filters show the effectiveness of the elimination of polystyrene microbeads at a superior level than sand filters.

3. **Coagulation-Flocculation-Sedimentation:** These methods are the most widely used removal method of MPs from contaminated sources. Production of flocs by isolating colloidal particles from the solution by counteracting their charges and removing them by sedimentation or filtration is the central course of this method (Zhou *et al.*, 2021). Diverse sorts or mixture of chemicals are frequently used to augment the competence of coagulation and flocculation, and these coagulants can subvert the colloidal state of MPs by neutralizing their surface charges (Badola *et al.*, 2022; Lu *et al.*, 2023). Rajala *et al.* (2020) stated that the addition of a mixture of ferric chloride, polyaluminum chloride, and cationic polyacrylamide has increased the removal efficiency to >99%. The morphology of MPs also plays crucial part in determining the elimination efficiency, and the elongated and rough MPs are separated earlier than the spherical and smooth ones. Apart from this, the type of coagulant, quantity of the coagulant, and residence time of the coagulant also are the three foremost factors influencing the efficiency of the process. However, the enormous sludge produced by the coagulation process may be comprised of highly toxic particles and treatment of this sludge usually increases the overall cost of this procedure (Osman *et al.*, 2023). Additionally, pH induced sol-gel process can also endorse flocculation and these flocs can be simply parted (Chellasamy *et al.*, 2022).

   Metallic electrodes are used for producing flocs from the cations in the electrocoagulation method, and these eventually form micro-coagulants, leading to a loss of suspended particle stability (Osman *et al.*, 2023). The chief advantages are the generation of a low amount of sludge and the production of water with a lower amount of TDS.

4. **Advanced Oxidative Process (AOPs):** These techniques are innovative methods of MP removal, and these techniques show high efficiency in MP removal (Osman *et al.*, 2023). Highly oxidizing hydroxyl radical is generated during AOPs, which aids in the degradation of MPs, and these AOPs can be classified into several categories based on the free radical production process: photocatalytic degradation, electrochemical oxidation, Fenton oxidation, and so on (Lu *et al.*, 2023). These AOPs can entirely degrade microplastics to CO_2 and H_2O. However, their wide-scale application is costly and tough. The fundamental concept of photocatalytic degradation is the absorbance of visible or ultraviolet light by a semiconductor material, which leads to the generation of free radicals. Such free radicals like superoxide, singlet oxygen, and hydroxyl groups break down the MPs (Zhu *et al.*, 2019). However, the management of sludge is a major problem of this procedure.

   On the other hand, electrochemical oxidation involves two oxidations: anodic oxidation and indirect cathode oxidation. Moreover, this method is cost-effective, sustainable, and does not produce sludge. Thus this method is gathering attention from the global scientific diaspora (Osman *et al.*, 2023). Kiendrebeogo *et al.* (2021) did an excellent study on the efficiency of the method and specified the surface area and materials of the anode, current intensity, and duration of the reaction. Oxidation mediators with redox potential >1.7 V are usually required for the electrochemical oxidation of MPs, and the boron-doped diamond electrode (BDD) is the choicest electrode owing to its degradation and detection efficiency (Martic *et al.*, 2022).

5. **Bioremediation:** Cost-effectiveness, lower energy requirement, and environmental friendliness are the prime positives for this method, and this method can be categorized into two major categories: biodegradation of MPs and adsorption of MPs (Lu *et al.*, 2023). Bacteria and fungi immensely aid bioremediation MPs (Masiá *et al.*, 2020). Bacteria, fungi, and algae cleave the polymer chains of MPs with extracellular enzymes, and degraded polymer chains are gradually metabolized to produce nontoxic materials (Ebrahimbabaie *et al.*, 2022). Lu *et al.* (2023) did a prolific review and noted that a wide variety of bacteria (*Bacillus* sp., *Rhodococcus* sp., *Streptomyces* sp., *Pseudomonas* sp.) and fungi (*Aspergillus* sp., *Zalerion* sp.) have been used for the degradation of MPs. They also concluded that both biological factors (types of microbes, hydrophobicity of cells) and environmental factors (temperature and pH of the medium, humidity, salinity, and UV radiation) influence the degradation procedure.

   Adsorption of MPs by algae and other organisms mainly depends on the charge interactions between MPs and organisms (Cheng *et al.*, 2019). The surface of algae is negatively charged owing to the presence of a carboxyl group, and this helps in strong binding between neutral or positively charged MPs. An exciting study by Sundbæk *et al.* (2018) noted that an edible algae, Bladderwrack (*Fucus vesiculosus*) efficiently absorbed polystyrene particles, and >94 % of MPs can be subsequently eliminated by washing. Additionally, marine microalgae produce an extracellular polymer made of carbohydrates, and these polymers can be used as a gelling agent to potentially remove MPs (Lu *et al.*, 2023).

   Bio-flocculants are excellent alternatives to chemical coagulants as they are safe, biodegradable, and produce sludge without toxic pollutants (Cunha *et al.*, 2019). A study by Cunha *et al.* (2019) attested that extracellular polymers produced by *Cyanothece* sp. showed excellent bio-flocculant properties even at low concentrations.

These methods are often used for MP removal from contaminated waters. Filtration, adsorption, and coagulation methods show high efficiency in MP removal. However, these methods have limitations in treating large quantities of wastewater. On the contrary, bioremediation can decontaminate great extents of water with low efficiency of MP removal. Future research may focus on high-efficacy MP removal from large amounts of wastewater. Moreover, the application of two or more methods simultaneously can increase the efficiency of MP removal and future studies can focus on the sustainable and cost-effective combination of multiple methods (Lu *et al.*, 2023).

Besides the remediation strategies, MP control strategies must be followed to prevent MP contamination in possible mediums. Control strategies focus on prevention followed by the principle of 7 Rs (reducing, reusing, recycling, refusing, rethinking, regifting, recovering) and appropriate disposal (Osman *et al.*, 2023). Moreover, the use of biodegradable plastics can offer a suitable alternative to reduce conventional plastic use. Conventional BPA-based plastics can be easily degraded by enzymatic actions by microorganisms (Sarma *et al.*, 2022). Changes in public acuities and daily life performance can lessen the MPs from water resources. Findings from the research of De Falco *et al.* (2019) attested that usage of natural clothes instead of synthetic clothes can also reduce the MPs from water mediums significantly.

19.5 RECENT DEVELOPMENTS AND CHALLENGES IN MICRO- AND NANO-PLASTIC IDENTIFICATION METHODS

Analysis of microplastics primarily depends on two key characteristics: physical and chemical features. Any method or analytical tool that can effectively measure both is suitable for MP analysis. As it is impossible to obtain both features of MPs using one analytical tool, a combination of multiple tools has to be used to obtain both features. The size of microplastics is a major deciding factor when choosing the identification methods. Microscopy can measure only the physical characteristics of large microplastic particles (of size >0.2 μ) and does not give any information about chemical characteristics, but the combination of spectroscopic (Raman, FTIR) or chromatographic tools (Pyro- GC/MS, LC/MS) with microscopic tools can effectively measure both the features (physical and chemical) of even smaller size (<1 mm) microplastics at the same time. However, as the size of microplastics goes down to nano sizes, their identification becomes more time-consuming and hence not possible for frequent monitoring studies at present.

For assessing the biological risks, there is a growing need to analyze plastic particles at the submicron level. Presently, the analysis of submicron plastic particles in environmental or biological samples is laborious, time-consuming, and requires great instrumental effort. For frequent monitoring of microplastic pollution in the environmental samples, there is a need for improvement in the existing method, which can reduce both effort and identification time. In addition, it is also important to develop reliable and practical identification methods for quantifying and detecting nano-plastics from environmental samples.

One of the major challenges in identifying and characterizing nano-plastics from environmental matrices is proper sampling and processing, which should be performed in the most environmentally friendly way without causing contamination. Proper consideration needs to be paid to the particles that look similar to microplastics in order to avoid confusion.

19.6 INITIATIVES TO REDUCE MPs AND FUTURE DIRECTIONS

In 1972, the London Convention was adopted to avert maritime contamination by dumping waste and to limit all the possible sources of maritime pollution. Later on, in 2012, during the Earth Summit, the Global Partnership on Marine Litter was also formed to curtail ocean plastic pollution loads. In 2022, UNEA Resolution on 'End Plastic Pollution: Towards a Legally Binding Instrument,' established an Intergovernmental Negotiating Committee (INC) that will develop the precise content of the new plastic pollution treaty, and this resolution will be implemented by 2024.[1] India has tabled the resolution, and as of now over 175 countries have already ratified the treaty. This resolution focuses on national and international cooperation and the development of national action plans for the prevention, reduction, and elimination of plastics (Central Pollution Control Board, 2023).

In 2015, the USA adopted the Microbeads-Free Waters Act to ban the production and distribution of over-the-counter drugs that are also cosmetics and that contain plastic microbeads to exfoliate or cleanse human body parts (Central Pollution Control Board, 2023). Moreover, the USA also adopted the Save Our Seas 2.0 Act in 2021 to prevent the production of microfibers (Nordic Council of Ministers, 2022). Other countries also adopted regulations to ban the production of microbeads (Argentina in 2020, Australia in 2021, Bangladesh in 2019, Canada in 2015, China in 2020, Germany in 2012 and 2018, Ireland in 2019, Italy in 2018, New Zealand in 2017, South Korea in 2017 and 2021, United Kingdom in 2017 and 2018) (Badola *et al.*, 2022). In 2021, the European Union espoused the Zero Pollution Action Plan (ZPAP), and this plan endorsed reducing plastic litter pollution by 50% and MPs by 30% by 2030 (Nordic Council of Ministers, 2022). The Association of Southeast Asian Nations adopted the Bangkok Declaration on Combating Marine Debris and the ASEAN Regional Action Plan on Combating Marine Debris to advance regional collaboration on marine debris issues and combat plastic pollution.

The Bureau of Indian Standards (BIS) banned the use of microbeads in cosmetics in October 2017 and implemented the ban in 2020 (Central Pollution Control Board, 2023). India also amended (second amendment) the Plastic Waste Management Act of 2016 in 2022 and banned all single-use plastics with low utility and high littering potentials from July 1, 2022. Moreover, this amendment act also mandates an increase in the thickness of plastic carry bags to over 120 microns from 31st December 2022. Earlier in 2021, the thickness of plastic carry bags increased from 50 to 75 μ. This act also released Extended Producer Responsibility (EPR), which mainly emphasizes two requirements: reduction of virgin plastics materials in packaging and reusing and recycling of plastic packaging materials (Central Pollution Control Board, 2023). The concept of EPR was introduced by the 1st Plastic Waste Management (Amendment) Rules of 2018. The main goals of the EPR guidelines are the sustainable management of plastic waste, consolidation of the circular economy, and promotion of biodegradable plastic packaging. Moreover, it also commands the producer, importer, and brand owner to take sustainable measures in the collection and supervision of non-single-use plastics. Moreover, the United Nations Environment Programme (UNEP), WWF-India, and Confederation of Indian Industry (CII) joined hands to launch the Un-Plastic Collective with the objectives of eliminating unnecessary plastics and using sustainable alternative materials.

Future strategies can be focused on the combination of strategies, cooperation at the national and international levels, lessening plastic consumption, and emphasizing recycling methods (Lu *et al.*, 2023). A snapshot of future directions is shown in Figure 19.4.

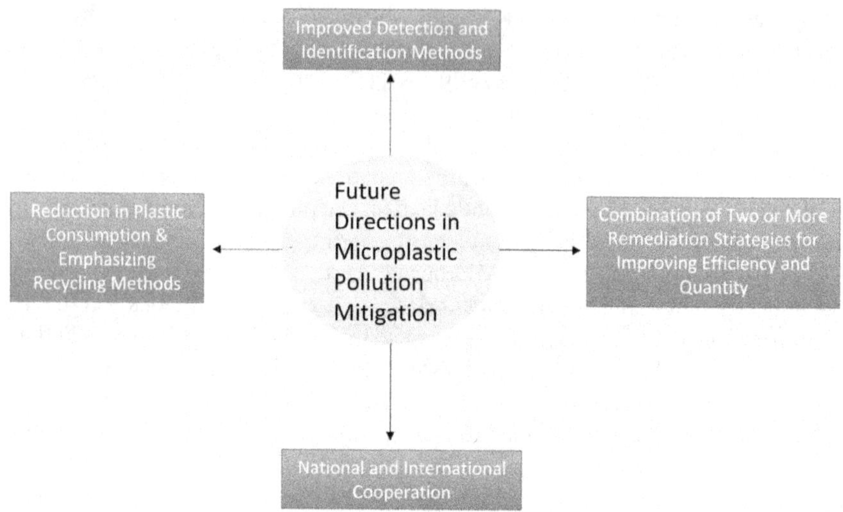

FIGURE 19.4 Future directions in microplastics pollution mitigation.

Chaukura *et al.* (2021) proposed an amalgamation of photodegradation and bioremediation for efficient and large-quantity water treatment. Effective national action plans on MPs and international cooperation are effective steps for the reduction of MPs. The conception of public awareness and collaborative actions by governmental and nongovernmental agencies can ensure less plastic consumption. Promotion of alternative materials and recycling of plastics can also aid in controlling MPs in the environment (Osman *et al.*, 2023).

19.7 CONCLUSIONS

With the growing demand and use of plastic goods, MPs are evolving as a potent and threatening pollutant for any ecosystem. MPs can move easily among earth systems by natural processes like wind or water movement and/or by trophic transfers by biota, and they exert a negative impact on the environment and biotic communities due to their heterogeneous and persistent nature. Recently, researchers have identified the ecological implications of MPs, and their toxic effects can ultimately lead to neurotoxicity, cytotoxicity, and immunotoxicity in organisms. This review provides insights into the sources, ecological implications, remediation measures, and future directions in research of MPs. Remediation strategies of MPs can be generally subdivided into three broad categories: physical, chemical, and biological processes. Conventional methods like filtration, adsorption, and coagulation show inordinate efficiency in MP removal. However, these methods have limitations in treating large quantities of wastewater. On the contrary, bioremediation approaches can decontaminate bulk quantity of waters with low competence. Future research may focus on high-efficacy MP removal from large amounts of wastewater by simultaneously applying two or more methods. Future strategies of MP regulations can be focused on the cost-effective and sustainable combination of strategies, cooperation at national and international levels, lessening plastic consumption, and emphasizing recycling methods.

ACKNOWLEDGMENTS

Authors are grateful to Dr. Subhra Kumar Mukhopadhyay, UGC Emeritus fellow; Prof. Sanjoy Chakraborty, Principal, GCELT, Kolkata; Dr. Utpal Singh Roy, Assistant Professor, P. R. Thakur Govt. College; Dr. Asitava Chatterjee, DFO, Kangsabati South Division, Purulia for their cooperation and continuous encouragement. The authors gratefully acknowledge the Science and Engineering Research Board (SERB), Department of Science and Technology, and Government of India (SRG/2022/002212) for their financial assistance. The first author is thankful to ANRF (formerly SERB) for the National Post-Doctoral Fellowship (PDF/2023/000416).

NOTE

1 Fifth session (INC-5) is scheduled for November 25 to December 1, 2024 in Busan, Republic of Korea.

REFERENCES

Akdogan, Z. and Guven, B. (2019) 'Microplastics in the environment: A critical review of current understanding and identification of future research needs', *Environmental Pollution*, 254, p. 113011. Available at: https://doi.org/10.1016/j.envpol.2019.113011.

An, L. *et al.* (2020) 'Sources of microplastic in the environment', in *Microplastics in Terrestrial Environments: Emerging Contaminants and Major Challenges*, pp. 143–159. Available at: https://doi.org/10.1007/698_2020_449.

Andrady, A.L. (2011) 'Microplastics in the marine environment', *Marine Pollution Bulletin*, 62(8), pp. 1596–1605. Available at: https://doi.org/10.1016/j.marpolbul.2011.05.030.

Badola, N. *et al.* (2022) 'Microplastics removal strategies: A step toward finding the solution', *Frontiers of Environmental Science & Engineering*, 16(1), p. 7. Available at: https://doi.org/10.1007/s11783-021-1441-3.

Blackburn, K. and Green, D. (2022) 'The potential effects of microplastics on human health: What is known and what is unknown', *Ambio*, 51(3), pp. 518–530. Available at: https://doi.org/10.1007/s13280-021-01589-9.

Campanale, C. *et al.* (2020) 'A detailed review study on potential effects of microplastics and additives of concern on human health', *International Journal of Environmental Research and Public Health*, 17(4), p. 1212. Available at: https://doi.org/10.3390/ijerph17041212.

Central Pollution Control Board (2023) 'Report in the matter of Tribunal on its own motion SuoMotu based on the news item published in The Hindu titled "Detecting Microplastics in human blood"', dated March 29, 2022 (O.A. No. 251/2022).

Chaukura, N. *et al.* (2021) 'Microplastics in the aquatic environment—the occurrence, sources, ecological impacts, fate, and remediation challenges', *Pollutants*, 1(2), pp. 95–118. Available at: https://doi.org/10.3390/pollutants1020009.

Chellasamy, G. *et al.* (2022) 'Remediation of microplastics using bionanomaterials: A review', *Environmental Research*, 208(January), p. 112724. Available at: https://doi.org/10.1016/j.envres.2022.112724.

Cheng, S.Y. *et al.* (2019) 'New prospects for modified algae in heavy metal adsorption', *Trends in Biotechnology*, 37(11), pp. 1255–1268. Available at: https://doi.org/10.1016/j.tibtech.2019.04.007.

Cunha, C. *et al.* (2019) 'Marine vs freshwater microalgae exopolymers as biosolutions to microplastics pollution', *Environmental Pollution*, 249, pp. 372–380. Available at: https://doi.org/10.1016/j.envpol.2019.03.046.

De Falco, F. *et al.* (2019) 'The contribution of washing processes of synthetic clothes to microplastic pollution', *Scientific Reports*, (January), pp. 1–11. Available at: https://doi.org/10.1038/s41598-019-43023-x.

Ebrahimbabaie, P., Yousefi, K. and Pichtel, J. (2022) 'Photocatalytic and biological technologies for elimination of microplastics in water: Current status', *Science of the Total Environment*, 806, p. 150603. Available at: https://doi.org/10.1016/j.scitotenv.2021.150603.

Enfrin, M. *et al.* (2020) 'Kinetic and mechanistic aspects of ultrafiltration membrane fouling by nano- and microplastics', *Journal of Membrane Science*, 601(January). Available at: https://doi.org/10.1016/j.memsci.2020.117890.

Green, D.S. (2016) 'Effects of microplastics on European flat oysters, Ostrea edulis and their associated benthic communities', *Environmental Pollution*, 216, pp. 95–103. Available at: https://doi.org/10.1016/j.envpol.2016.05.043.

Hale, R.C. *et al.* (2020) 'A global perspective on microplastics', *Journal of Geophysical Research: Oceans*, 125(1). Available at: https://doi.org/10.1029/2018JC014719.

Halimu, G., Zhang, Q., Liu, L., Zhang, Z., Wang, X., Gu, W., Zhang, B., Dai, Y., Zhang, H., Zhang, C. and Xu, M. (2022) 'Toxic effects of nanoplastics with different sizes and surface charges on epithelial-to-mesenchymal transition in A549 cells and the potential toxicological mechanism', *Journal of Hazardous Materials*, 430. Available at: https://doi.org/10.1016/j.jhazmat.2022.128485.

Hidayaturrahman, H. and Lee, T.-G. (2019) 'A study on characteristics of microplastic in wastewater of South Korea: Identification, quantification, and fate of microplastics during treatment process', *Marine Pollution Bulletin*, 146, pp. 696–702. Available at: https://doi.org/10.1016/j.marpolbul.2019.06.071.

Karbalaei, S. *et al.* (2019) 'Abundance and characteristics of microplastics in commercial marine fish from Malaysia', *Marine Pollution Bulletin*, 148, pp. 5–15. Available at: https://doi.org/10.1016/j.marpolbul.2019.07.072.

Kiendrebeogo, M., Estahbanati, M.K., Mostafazadeh, A.K., Drogui, P. and Tyagi, R.D. (2021) 'Treatment of microplastics in water by anodic oxidation: A case study for polystyrene', *Environmental Pollution*, 269. Available at: https://doi.org/10.1016/j.envpol.2020.116168.

Lechthaler, S. *et al.* (2021) 'Baseline study on microplastics in Indian rivers under different anthropogenic influences', *Water*, 13(12), p. 1648. Available at: https://doi.org/10.3390/w13121648.

Liu, W. *et al.* (2021) 'Review article A review of the removal of microplastics in global wastewater treatment plants: Characteristics and mechanisms', *Environment International*, 146, p. 106277. Available at: https://doi.org/10.1016/j.envint.2020.106277.

Lu, Y. *et al.* (2023) 'Microplastic remediation technologies in water and wastewater treatment processes: Current status and future perspectives', *Science of the Total Environment*, 868(January), p. 161618. Available at: https://doi.org/10.1016/j.scitotenv.2023.161618.

Ma, H. *et al.* (2020) 'Microplastics in aquatic environments: Toxicity to trigger ecological consequences', *Environmental Pollution*, 261, p. 114089. Available at: https://doi.org/10.1016/j.envpol.2020.114089.

Martic, S. *et al.* (2022) 'Emerging electrochemical tools for microplastics remediation and sensing', *Frontiers in Sensors*, 3(August), pp. 1–8. Available at: https://doi.org/10.3389/fsens.2022.958633.

Masiá, P. *et al.* (2020) 'Bioremediation as a promising strategy for microplastics removal in wastewater treatment plants', *Marine Pollution Bulletin*, 156, p. 111252. Available at: https://doi.org/10.1016/j.marpolbul.2020.111252.

Mohammad, R.A. and Qusay, A.-A. (2023) 'Eco-friendly microplastic removal through physical and chemical techniques: A review', *Annals of Advances in Chemistry*, 7(1). Available at: https://doi.org/10.29328/journal.aac.1001038.

Nabi, G. *et al.* (2022) 'The adverse health effects of increasing microplastic pollution on aquatic mammals', *Journal of King Saud University—Science*, 34(4), p. 102006. Available at: https://doi.org/10.1016/j.jksus.2022.102006.

Nordic Council of Ministers (2022) *Addressing Microplastics in a Global Agreement on Plastic Pollution*. Available at: https://pub.norden.org/temanord2022-566/#121825.

OECD (2021) *Policies to Reduce Microplastics Pollution in Water: Focus on Textiles and Tyres*, OECD Publishing, Paris. Available at: https://doi.org/10.1787/7ec7e5ef-en.

Osman, A.I. *et al.* (2023) *Microplastic Sources, Formation, Toxicity and Remediation: A Review*, *Environmental Chemistry Letters*. Springer International Publishing. Available at: https://doi.org/10.1007/s10311-023-01593-3.

Picó, Y. and Barceló, D. (2019) 'Analysis and prevention of microplastics pollution in water: Current perspectives and future directions', *ACS Omega*, 4(4), pp. 6709–6719. Available at: https://doi.org/10.1021/acsomega.9b00222.

Poerio, T., Piacentini, E. and Mazzei, R. (2019). 'Membrane processes for microplastic removal', *Molecules*, 24(22), p. 4148. Available at: https://doi.org/10.3390/molecules24224148.

Rajala, K. *et al.* (2020) 'Removal of microplastics from secondary wastewater treatment plant effluent by coagulation/flocculation with iron, aluminum and polyamine-based chemicals', *Water Research*, 183, p. 116045. Available at: https://doi.org/10.1016/j.watres.2020.116045.

Roy, P., Mohanty, A.K. and Misra, M. (2022) 'Microplastics in ecosystems: Their implications and mitigation pathways', *Environmental Science: Advances*, 1(1), pp. 9–29. Available at: https://doi.org/10.1039/d1va00012h.

Salim, S.Y., Kaplan, G.G. and Madsen, K.L. (2013) 'Air pollution effects on the gut microbiota: A link between exposure and inflammatory disease', *Gut Microbes*, 5(2), pp. 215–219. Available at: https://doi.org/10.4161/gmic.27251.

Sarma, H. *et al.* (2022) 'Microplastics in marine and aquatic habitats: Sources, impact, and sustainable remediation approaches', *Environmental Sustainability*, 5(1), pp. 39–49. Available at: https://doi.org/10.1007/s42398-022-00219-8.

Siipola, V. *et al.* (2020) 'applied sciences Low-Cost Biochar Adsorbents for Water Purification Including Microplastics Removal'.

Silva, A.B. *et al.* (2018) 'Analytica Chimica Acta Microplastics in the environment : Challenges in analytical chemistry—A review', *Analytica Chimica Acta*, 1017. Available at: https://doi.org/10.1016/j.aca.2018.02.043.

Singh, N. *et al.* (2017) 'Synergistic effects of heavy metals and pesticides in living systems', *Frontiers in Chemistry*, 5. Available at: https://doi.org/10.3389/fchem.2017.00070.

Sundbæk, K.B. *et al.* (2018) 'Sorption of fluorescent polystyrene microplastic particles to edible seaweed Fucus vesiculosus', *Journal of Applied Phycology*, 30(5), pp. 2923–2927. Available at: https://doi.org/10.1007/s10811-018-1472-8.

Toussaint, B. *et al.* (2019) 'Review of micro- and nanoplastic contamination in the food chain', *Food Additives and Contaminants—Part A Chemistry, Analysis, Control, Exposure and Risk Assessment*, 36(5), pp. 639–673. Available at: https://doi.org/10.1080/19440049.2019.1583381.

Tursi, A. *et al.* (2022) 'Microplastics in aquatic systems, a comprehensive review: Origination, accumulation, impact, and removal technologies', *RSC Advances*, 12(44), pp. 28318–28340. Available at: https://doi.org/10.1039/d2ra04713f.

United Nations Environment Programme United Nations Environment Programme and Technical University of Denmark (DTU) (2018) *No TitleMapping of Global Plastics Value Chain and Plastics Losses to the Environment: With a Particular Focus on Marine Environment.* Available at: https://wedocs.unep.org/20.500.11822/26745.

Usman, S. *et al.* (2022) 'The Burden of microplastics pollution and contending policies and regulations', *International Journal of Environmental Research and Public Health*, 19(11), p. 6773. Available at: https://doi.org/10.3390/ijerph19116773.

Vaid, M., Mehra, K. and Gupta, A. (2021) 'Microplastics as contaminants in Indian environment: A review', *Environmental Science and Pollution Research*, 28(48), pp. 68025–68052. Available at: https://doi.org/10.1007/s11356-021-16827-6.

Wang, J. and Guo, X. (2020) 'Adsorption isotherm models: Classification, physical meaning, application and solving method', *ECSN*, p. 127279. Available at: https://doi.org/10.1016/j.chemosphere.2020.127279.

Wright, S.L. *et al.* (2013) 'Microplastic ingestion decreases energy reserves in marine worms', *Current Biology*, 23(23), pp. R1031–R1033. Available at: https://doi.org/10.1016/j.cub.2013.10.068.

Yang, Y., Xie, E., Du, Z., Peng, Z., Han, Z., Li, L., Zhao, R., Qin, Y., Xue, M., Li, F., Hua, K. and Yang, X. (2023) 'Detection of various microplastics in patients undergoing cardiac surgery', *Environmental Science & Technology*, 57(30), pp. 10911–10918. Available at: https://doi.org/10.1021/acs.est.2c07179. Epub 2023 Jul 13. PMID: 37440474.

Yang, Z. *et al.* (2021) 'Is incineration the terminator of plastics and microplastics?', *Journal of Hazardous Materials*, 401, p. 123429. Available at: https://doi.org/10.1016/j.jhazmat.2020.123429.

Yong, C.Q.Y., Valiyaveettil, S. and Tang, B.L. (2020) 'Toxicity of microplastics and nanoplastics in Mammalian systems', *International Journal of Environmental Research and Public Health*, 17(5). Available at: https://doi.org/10.3390/ijerph17051509.

Zhou, G. *et al.* (2021) 'Removal of polystyrene and polyethylene microplastics using PAC and FeCl3 coagulation: Performance and mechanism', *Science of the Total Environment*, 752, p. 141837. Available at: https://doi.org/10.1016/j.scitotenv.2020.141837.

Zhu, K. *et al.* (2019) 'Formation of environmentally persistent free radicals on microplastics under light irradiation', *Environmental Science & Technology*, 53(14), pp. 8177–8186. Available at: https://doi.org/10.1021/acs.est.9b01474.

Index

For Product Safety Concerns and Information please contact our EU
representative GPSR@taylorandfrancis.com
Taylor & Francis Verlag GmbH, Kaufingerstraße 24, 80331 München, Germany

www.ingramcontent.com/pod-product-compliance
Lightning Source LLC
LaVergne TN
LVHW020602110826
845149LV00002B/357

* 9 7 8 1 0 3 2 6 8 4 5 6 7 *